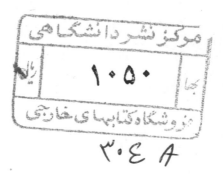

مرکز نشر دانشگاهی

بها ۱۰۵۰ ریال

فروشگاه کتابهای خارجی

۳۰٤٨

—KHASHAYAR—

THE OXFORD
DICTIONARY FOR
WRITERS AND
EDITORS

THE OXFORD
DICTIONARY FOR
WRITERS AND
EDITORS

COMPILED BY

THE OXFORD ENGLISH DICTIONARY DEPARTMENT

CLARENDON PRESS · OXFORD

1981

Oxford University Press, Walton Street, Oxford OX2 6DP

London Glasgow New York Toronto
Delhi Bombay Calcutta Madras Karachi
Kuala Lumpur Singapore Hong Kong Tokyo
Nairobi Dar es Salaam Cape Town
Melbourne Wellington

and associate companies in
Beirut Berlin Ibadan Mexico City

Published in the United States by
Oxford University Press, New York

© Oxford University Press 1981

(Reprinted with corrections, 1982, 1983)
British Library Cataloguing in Publication Data

The Oxford dictionary for writers and editors.
1. English language – Dictionaries
I. Collins, Frederick Howard. Authors' and
printers' dictionary
423   PE1625   80-40506
ISBN 0-19-212970-8

Printed in Great Britain by
William Clowes (Beccles) Limited,
Beccles and London.

PUBLISHER'S NOTE

The present work is the successor to eleven editions of the *Authors' and Printers' Dictionary*, first published under the editorship of F. Howard Collins in 1905. The eleventh edition, published in 1973, has been thoroughly revised and extensively rewritten, and a great deal of new material—everyday words, proper names, and abbreviations—has been incorporated to bring the book up to date in matters of vocabulary and usage and to give it an orientation and purpose that are reflected in its new name. The editorial work has been undertaken by the Oxford English Dictionary Department, and in particular by R. E. Allen, D. J. Edmonds, and J. B. Sykes. The book may now be regarded as belonging to the family of Oxford dictionaries, on whose resources it has extensively drawn.

The policy of the book continues to be a presentation of the house style of the Oxford University Press, and the reader is therefore strongly recommended to use it in conjunction with *Hart's Rules for Compositors and Readers at the University Press Oxford* (38th ed., 1978), which further explains and illustrates that style. 'Writers and editors' are the main concern of the dictionary, and it therefore no longer caters directly or extensively for the needs of printers except in so far as those needs are shared by writers and editors in their dealings with publishers and printers.

Because this is primarily a dictionary of style for written English, guidance on pronunciation has been given only in cases where the reader might otherwise experience difficulty or confusion, and where clarification could not be expected from a normal desk or household dictionary such as *The Concise Oxford Dictionary* (6th ed., 1976), to which the reader is referred for systematic treatment of pronunciation in individual cases.

ACKNOWLEDGEMENTS

The compilers wish to acknowledge assistance of various kinds afforded during the preparation of this dictionary by the British Standards Institution, the Committee of Vice-Chancellors and Principals of the Universities of the United Kingdom, Her Majesty's Stationery Office, and the Law Society; by the county archivists of Cheshire and Lancashire County Councils; and by many District and Borough Councils, and many town clerks, in Wales. The compilers are also indebted to Mr P. M. W. Duffy of the Printing Division of the Oxford University Press and to Mr J. S. G. Simmons of All Souls College, Oxford.

CONVENTIONS
USED IN THE DICTIONARY

The reader will appreciate that in a work of this size a number of special conventions are adopted in the interests of saving space. The most important of these are listed below.

(i) *bold face.* This is used for headwords and for all approved forms and spellings; a rejected form or spelling is printed in light face except when it stands as a headword with cross-reference to the preferred form. A bold italic face is used for words normally to be printed in italics.

(ii) *oblique stroke.* An oblique stroke is used to mark the end of that part of the headword to which bold--face elements introduced with a dash (iii) or a hyphen (iv) later in the same entry should be attached. Thus the entry **collegi/um,** . . . *pl.* **-a** shows that the plural form should be *collegia*, and the inflexions given as **gallop/,** . . . **-ed, -er, -ing** are to be understood as *galloped, galloper,* and *galloping.* Only **bold** elements follow this rule; it should be noted for example in the entry **metal/, -led, -ling, -lize,** *not* -ise, that **-led, -ling,** and **-lize** join the headword, forming *metalled, metalling,* and *metallize,* but the rejected termination -ise is independently given.

(iii) *dash.* A dash is used to stand for the repetition of a single word already given, and the oblique stroke (ii) again marks the end of a common element. Thus in the entry **glove/ box, — compartment, — puppet** (two words), the three combinations are *glove box, glove compartment,* and *glove puppet.* Use of more than one dash indicates the repetition of a corresponding number of words: in the list at **Doctor** beginning **Doctor of/ Canon Law** *or* **Civil Law, DCL; — — Dental Surgery, DDS; — — Divinity, DD,** the two dashes stand for the two words 'Doctor of', and at **Garcilaso de la Vega** all four words are repeated as dashes in the second part of the entry: — — — — (**'the Inca'**), 1540–1616, Spanish historian.

(iv) *hyphen.* In headwords the hyphen serves two purposes: to indicate that a compound is normally hyphenated whenever used, and to introduce a further second element to form a compound with a first element marked off earlier in the entry by means of an oblique stroke (ii). In the second case the intended form of that compound will be clear from scrutiny of the first compound given and by recourse to the comment given in parentheses at the end of the sequence: (hyphen), (hyphens), (one word). The entry **bird/cage, -seed** (one word) therefore indicates that *birdcage* and *birdseed* are each to be written as one word, and the entry **birth/-control, -rate** (hyphens) indicates that *birth-control* and *birth-rate* are both to be written with hyphens. The dash (iii) and the designation (two words) indicate pairs of words each of which is to be written as two words without hyphen: the entry **fellow/ citizen, — men** (two words) therefore indicates the forms *fellow citizen* and *fellow men.*

In this book, a hyphen that falls at the end of a line is repeated at the beginning of the next line to distinguish a compound that is always hyphenated from a word which happens to be hyphenated because it is split at the end of a line.

(v) *parentheses.* In addition to their normal function, parentheses are used to formulate a concise definition that will serve two or more parts of speech, the part in parentheses being included or omitted as appropriate. Thus the definition at **oxytone** (Gr. gram.), (word) with acute accent on last syllable, covers both the noun ('word with acute accent on last syllable') and the adjective ('with acute accent on last syllable').

(vi) *full point.* The full point is used in the normal way to mark an abbreviated form (when appropriate) or the end of a sentence. It is omitted as an end-of--sentence mark from the last sentence or section of *complete entries*; where it occurs in this position it serves some other purpose, usually as part of an abbreviation or contraction in which the point is retained. This device enables the reader to determine at a glance whether an abbreviation which occurs as the last word in an entry (as is frequently the case) has a full point or not:

> **cosine** (math., no hyphen), abbr. **cos**
>
> **cosmogony, cosmography,** abbr. **cosmog.**

(vii) Being a dictionary and not a descriptive handbook, this book is designed to provide information whenever possible at the point in the alphabetical sequence to which the reader may be expected to refer with a particular word in mind. The compilers have thought it useful however to include general entries on broader aspects of usage, especially where this has enabled them to group together representative examples of words and uses of which total coverage in their respective places in the dictionary is precluded for reasons of space. Appended below is a list of longer entries which deal with the broader aspects of usage:

abbreviations and contrac-
    tions
-able
accents
Acts of Parliament
Arabic
Assemblies
astronomy
authorities

book sizes
botany
BSI (list of relevant stand-
    ards)

capitalization
Chinese
Creole

date
decimal currency
decimal fractions
diphthongs
division of words
Dutch

edition
-ei-
elision
ellipsis

footnotes
fractions
French

German
Greek

headlines or running titles
Hebrew

-ie-
impression
imprint
Italian
italic
-ize and -ise

Japanese

Latin
ligature
-like
logic symbols

Mac
manuscript
mathematics

names of persons and places
neo-
non-
numerals

pagination
poetry
Polish

possessive case
preliminary matter
proof
proof-correction marks
punctuation

quotation marks
quotations

re-
reference marks
rule (typ.)
Russian

Saint
SI
Spanish
Swedish
Syriac

titles (cited)

US spellings

well-
Welsh

-yse

zoology

All dates given in the dictionary are after Christ unless otherwise specified.

ABBREVIATIONS
USED IN THE DICTIONARY

Abbreviations used regularly in the rulings and explanations are listed below; others are given in the dictionary itself. Abbreviations of subjects (anthropology, botany, etc.) are also valid for their corresponding adjectival forms (anthropological, botanical, etc.).

abbr.	abbreviation	Can.	Canada, -ian
absol.	absolute, (used) absolutely	cap.	capital
		cc.	centuries
adj.	adjective	Ch.	Church
adv.	adverb	chem.	chemistry, chemical
Afr.	Africa(n)		
Afrik.	Afrikaans	Chin.	China, Chinese
anat.	anatomy	class.	classical
anc.	ancient	collect.	collective(ly)
anthrop.	anthropology	colloq.	colloquial(ly)
apos.	apostrophe	comb.	combination, combining
Arab.	Arabic		
arch.	archaic	comp.	computer type-setting
archaeol.	archaeology		
archit.	architecture	cook.	cookery
astr.	astronomy		
attrib.	attributive, (used) attributively	d.	died
		Dan.	Danish
		dau.	daughter
		dép.	*département* (Fr.)
b.	born		
bacteriol.	bacteriology	derog.	derogatory
bibliog.	bibliography	dress.	dressmaking
bind.	binding	Du.	Dutch
bot.	botany	dyn.	dynamics
Brit.	British		
		eccl.	ecclesiastical
c.	century	elec.	electrical
c.	*circa*	esp.	especially
campan.	campanology	exclam.	exclamation

f., fem.	feminine	mod.	modern
fam.	family	mt.	mountain
fig.	figurative(ly)	mus.	music
Fl.	Flemish	myth.	mythology
foot.	football		
Fr.	French		
		n., neut.	neuter
		naut.	nautical
gen.	general		
geog.	geography		
geol.	geology	obs.	obsolete
geom.	geometry	off.	official
Ger.	German	orig.	original(ly)
Gr.	Greek	ornith.	ornithology
gram.	grammar		
		Parl.	Parliament(ary)
her.	heraldry	partic.	participle, par-
hist.	history		ticipial
hort.	horticulture	path.	pathology
hunt.	hunting	philos.	philosophy
		photog.	photography
		phys.	physics, physi-
Ind.	India(n)		cal
indep.	independent	physiol.	physiology
Ins.	Insurance	pl.	plural
Ir.	Irish	poet.	poetical
It.	Italian	polit.	political
ital.	italic	Port.	Portuguese
		predic.	predicative,
joc.	jocular(ly)		(used) predi-
			catively
Lat.	Latin	prep.	preposition
lit.	literary	pron.	pronounced
liturg.	liturgical, lit-	propr.	proprietary
	urgy	pseud.	pseudonym
		psych.	psychology
m.	masculine		
mag.	magnetism	RC(C)	Roman Catholic
math.	mathematics		(Church)
mech.	mechanics	relig.	religion
med.	medicine	rhet.	rhetoric(al)
meteor.	meteorology	Rom.	Roman
mfr.	manufacture	Russ.	Russian
mil.	military		

s.	small	Swed.	Swedish
S. Afr.	South Africa(n)	techn.	technical
S. Amer.	South America(n)	theat.	theatre
		theol.	theology
Sc.	Scottish	trop.	tropical
Scan.	Scandinavian	Turk.	Turkish
sci.	science	typ.	typography
Scrip.	Scripture		
sig.	signature	US	United States, America(n)
sing.	singular		
Skt.	Sanskrit	usu.	usually
Sp.	Spanish		
spec.	specifically	zool.	zoology

NOTE ON PROPRIETARY TERMS

This dictionary includes some words which are or are asserted to be proprietary names or trade marks. Their inclusion does not imply that they have acquired for legal purposes a non-proprietary or general significance, nor is any other judgement implied concerning their legal status. In cases where the compilers have some evidence that a word is used as a proprietary name or trade mark this is indicated by the words 'propr. term', but no judgement concerning the legal status of such words is made or implied thereby.

A

A, Advanced (Level examination), ampere, *avancer* (on time-piece regulator), the first in a series; film-censorship classification

A., Academician, Academy, Acting (rank), alto, amateur, anna, answer, artillery, Associate

a, are (unit of area) (no point in scientific and technical work), atto-, year

a., accepted (Fr. *accepté*, Ger. *acceptiert*), on bills of exchange; acre, active (in grammar), adjective, anna, area, arrive, (Lat.) *ante* (before)

a, *not* **an,** before all words beginning with a consonant (except silent *h*: e.g. a horse, a hotel; but an heir, an hour) or with a vowel that is pronounced with the sound of *w* or *y*: e.g. a one, a European, a union. Similarly with abbreviations: e.g. an FA Cup match, a Unesco post

Ä, ä, in German, Swedish, etc., may *not* be replaced by *Ae, ae* (except in some proper names), *A, a,* or *Æ, æ*. The first only of two vowels takes the umlaut sign, as *äu*

Å, ångström

å, the 'Swedish *a*', used in Danish, Norwegian, and Swedish

@, at, the 'commercial *a*' used in calculating from prices

A₀ ('A nought' or 'A zero'), paper size. See **DIN**

A1, first class; *see* **Lloyd's**

AA, Advertising Association, Alcoholics Anonymous, anti-aircraft, Architectural Association, Associate in Arts, Automobile Association; film-censorship classification

AAA, Amateur Athletic Association, American Automobile Association

AAC, *anno ante Christum* (in the year before Christ) (s. caps.)

Aachen, W. Germany; cf. **Aix-la-Chapelle**

AAF, Auxiliary Air Force (till 1957), (US) Army Air Force

AAG, Assistant Adjutant-General

AAIA, Associate of the Association of International Accountants

AAMI, Association of Assistant Mistresses, Incorporated (usu. called **AAM**). See also **IAAM**

A. & M., (Hymns) Ancient and Modern

A. & N., Army and Navy (Stores, Club)

a.a.O. (Ger.), *am angeführten Orte* (at the place quoted)

AAR, against all risks (Ins.)

aard/vark, -wolf, different animals (one word)

Aarhus, Denmark

AAS, *Academiae Americanae Socius* (Fellow of the American Academy of Arts and Sciences), *Acta Apostolicae Sedis* (Acts of the Apostolic See), Agnostics' Adoption Society

AAU (US), Amateur Athletic Union

AB, *Artium Baccalaureus* (US Bachelor of Arts); able-bodied (seaman)

ab (Lat.), from

ABA, Amateur Boxing Association, American Bankers' Association, American Bar Association, Antiquarian Booksellers' Association, Association of British Archaeologists

ABAA, Associate of British Association of Accountants and Auditors

abac/us, *pl.* **-uses,** counting-frame, (archit.) plate at top of column

abalone, shellfish

à bas (Fr.), down with

abattoir, slaughter-house

Abb. (Ger.), *Abbildung*

abb., abbess, abbey, abbot

Abbasid, Baghdad caliphs, *not* Abbasid

abbé (Fr. m.), eccl. title (not ital.)

Abbildung| (Ger. typ. f.), an illustration (cap.); *pl. -en*; abbr. *Abb.*

Abbotsinch Airport, Glasgow

abbr., abbreviat/ed, -ion

abbreviat/e, -or

abbreviations and contractions. See the many instances in this book for rulings in individual cases. Subject to these, the following leading types may be noted for·guidance:

(*a*) Abbreviations with several even full or even small capitals, and acronyms in initial capitals and lower case, take no points (BBC, TUC, MA, QC, NW, OHG, *OED*, BC, AD, Aslef, Sogat); similarly the numerical abbreviations 8vo, 4to, etc., 1st, 2nd, etc., but *not* initials of personal names (H. G. Wells, G. B. S.).

(*b*) Abbreviations having a single capital letter take full points (J. Austen, five miles S.); *but* C, F (of temperature) and chemical symbols do *not*.

(*c*) Abbreviations and contractions in a mixture of upper and lower case take full points (A. & M., M.o.D., B.Litt., Kt., Sun., Jan., Yorks.); *but* Dr, Revd, Mr, Mrs, Ms, Mme, Mlle, Hants, Northants do *not*.

(*d*) Single-letter lower-case abbreviations take the full point (l. 5, p. 7, don't be a b. fool); but p (new pence) does *not* (price 50p).

(*e*) Measures of length, weight, time, etc. take the full point *except* when used in specialist scientific or technical writing (cm., min., lb.)

ABC, the alphabet, Argentina, Brazil, and Chile, Associated British Cinemas, Audit Bureau of Circulations, Australian Broadcasting Commission

ABCA, Army Bureau of Current Affairs, *now* **BCA,** Bureau of Current Affairs

ABC Guide

abdicat/e, -ion, -or

abecedarian (adj.), arranged alphabetically; (noun), person learning alphabet

à Becket (Thomas), ?1118–70, Abp. (one *t*)

Abendlied (Ger.), evening song (cap.)

Aberdeen Angus (two caps.)

Aberdeen terrier (one cap.)

Aberdonian, (native) of Aberdeen

Abergavenny, Gwent

Abernethy biscuit (one cap.)

Aberystwyth, Dyfed, *not* -ith

abest (Lat.), he, she, *or* it, is absent; *pl.* **absunt**

abett/er, in law **-or**

ab extra (Lat.), from outside

abgk., abk., abgekürzt (Ger.) =abbreviated

Abhandlungen (Ger. f. pl.), Transactions (of a Society); abbr. *Abh.*

Abidjan, Ivory Coast

abigail, lady's maid

Abilene, Kan., Tex., US

ab incunabulis (Lat.), from the cradle

Abingdon, Oxon. and Va., US

Abington, Cambs., Northampton, Strathclyde, Limerick, and Mass., US

ab | *initio* (Lat.), from the beginning, abbr. *ab init.*; *— intra,* from within

Abl. (Sp.), *abril* (April)

abl., ablative

ablaut, variation in root vowel of a word (e.g. sing, sang, sung)

-able (the suffix). Words ending in silent *-e* tend to drop the *e* before *-able*, as conceivable, debatable. It is retained when it

serves to prevent the modification of the preceding consonant, as changeable, peaceable.
In practice the following words also often retain the *e*: blameable, giveable, hireable, likeable, liveable, nameable, rateable, saleable, sizeable, tameable, timeable, tuneable, unshakeable. Many verbs of more than two syllables ending in *-ate* drop this ending before *-able*, as alienable, calculable, tolerable. See the list in *Hart's Rules*, p. 83

able-bodied (hyphen)

ABM, anti-ballistic-missile missile

Åbo, Swed. for **Turku**

A-bomb, atomic bomb (no point, hyphen)

aboriginal (adj.), indigenous (with init. cap., in techn. sense in Australia); **aborigin/es** (pl. noun, for sing. *use* **-al**) also with init. cap. in Australia

ab origine (Lat.), from the beginning

aboul/ia, loss of will-power, *not* abulia; adj. **-ic**

above/-board, -mentioned, -named (hyphens)

ab ovo (Lat.), from the beginning

Abp., Archbishop

abr., abridged, abridgement

abrégé (Fr. m.), abridgement

abréviation (Fr. f.), abbreviation

abridgement, abbr. **abr.**

abs., absolute(ly), abstract

abscess

Abschnitt/ (Ger. typ. m.), section, part, chapter, or division (cap.); *pl.* **-e**

absciss/a (math.), *pl.* **-ae**

absente reo (Lat.), the defendant being absent; abbr. *abs. re.*

absenter, *not* -or

absent-minded/ (hyphen), **-ly, -ness**

absinth/, the plant; **-e,** the liqueur

absit/ (Lat.), let him, her, *or* it be

absent; — *omen,* let there be no (ill) omen

absolute(ly), abbr. **abs.**

absorption, *not* -btion

abs. re., absente reo, q.v.

abstract, abbr. **abs.**

absunt, see **abest**

Abt. (Ger.), *Abteilung* (division)

abt., about

ABTA, Association of British Travel Agents

Abu Dhabi, United Arab Emirates

abulia, *use* **aboulia**

ab/ uno disce omnes (Lat.), from one (sample) judge the rest; — *urbe condita,* AUC (but A.U.C. in printing classical Latin texts), from the foundation of Rome, 753 BC

Abu Simbel, site of Egyptian rock-temples

abut/, -ment, -ted, -ter (law), **-ting**

abys/s, deep cavity, **-mal**

abyssal, deep in sea; (geol.) plutonic

Abyssinia, *now* **Ethiopia**

AC, Alpine Club, (*or* **A/C**) Aircraftman, Assistant Commissioner, Athletic Club, Companion of the Order of Australia

A/C, (current) account; in printing use **acct.**

AC, ante Christum (before Christ) (s. caps.)

Ac, actinium (no point)

a.c., alternating current (elec.), author's correction

ACA, Associate of the Institute of Chartered Accountants in England and Wales; (*or* Ireland)

Academician, abbr. **A.**

Académie française (one cap.)

Academy, a learned body; abbr. **A.** *or* **Acad.** (cap.)

Academy (the), the Platonic school of philosophy

Acadian, of Nova Scotia (Fr. **Acadie**)

a cappella (It.) (mus.), unaccompanied, *not* alla cappella

ACAS, Advisory, Conciliation, and Arbitration Service

ACC, Army Catering Corps

acc., acceptance (bill), accusative

Acca, Israel, *use* **Acre**

Accademia (It.), Academy

Accadian, *use* **Akkadian**

acced/e, -er

accedence, a giving consent

accelerando/ (mus.), accelerating; as noun, *pl.* **-s;** abbr. **accel.**

accelerat/e, -or

accents and special sorts:
acute (´), grave (`), circumflex (ˆ); vowels: long (ˉ), short (˘), doubtful (˟); diaeresis, umlaut (¨); cedilla (ˌ); tilde (˜); Gr. breathings: asper, rough (ʽ), lenis, smooth (ʼ); Arabic ʽ*ain*, Hebrew ʽ*ayin* (ʽ), Arabic *hamza*, Hebrew *aleph* (ʼ), ancient Egyptian special sign ʽ.

Anglo-Saxon (Old English) Þ, ð (cap. Ð), æ, œ, ā, ē, ī, ō, ū, ǣ, œ̄.

Czech á, č, dʼ (cap. Ď), é, ě, í, ň, ó, ř, š, tʼ (cap. Ť), ú, ů, ý, ž.

Danish æ, ø, å.

Finnish ä, ö, å.

French œ, à, â, ç, é, è, ê, ë, î, ï, ô, ù, û, ü.

Gaelic (Irish) á, é, í, ó, ú; (Scots) à, è, é, í, ò, ó, ú.

German ß (cap. SS), ä, ö, ü.

Hungarian á, é, í, ó, ö, ő, ú, ü, ű.

Icelandic þ (cap. Þ), ð (cap. Ð), æ, œ, á, é, í, ó, ö, ú, ý.

Norwegian æ, ø, å.

Polish ą, ć, ę, ł, ń, ó, ś, ź, ż.

Portuguese à, á, â, ã, ç, è, é, ê, ì, í, ò, ó, ô, õ, ú, ù.

Romanian à, â, ă, è, ì, î, ş, ţ, ù.

Spanish á, é, í, ñ, ó, ú, ü.

Swedish å, ä, ö.

Turkish â, ç, ğ, İ, ı, î, ö, ş, ü, û.

Welsh â, á, ê, ë, î, ï, ô, ö, û, ŵ, ŷ.

African languages may be set in the Latin alphabet using extra phonetic symbols as required: the most common are ɛ, ɔ, ŋ, ɓ, ɗ, ɖ, ọ, ɤ

accepᵒⁿ, acceptance (bill)

acceptance, abbr. **acc.**

accept/er, in law and science **-or**

accessary, use **accessory**

accessible

access/it (Lat.), he, she, *or* it, came near; *pl.* **-erunt**

accessory (noun and adj.), *use now* in all senses, *not* -ary

access time (computing)

acciaccatura (mus.), a grace-note

accidence, inflexions of words

accidental (mus.)

accidentally, *not* -tly

accidie, acedia, listlessness

acclimatize, *not* -ise

ACCM, Advisory Council for the Church's Ministry (*formerly* CACTM)

Acco, Israel, use **Acre**

accolade

accommodat/e, -ion (two *c*s, two *m*s)

accompanist, *not* -yist

accouche/ment, -ur, -use (not ital.)

account, abbr. **A/C;** in printing use **acct.**

Accountant-General, abbr. **A.-G.**

Accra, Ghana, *not* Akkra

acct., account, *or* account current

accumulat/e (two *c*s), **-or**

accusative, abbrs. **acc., accus.**

ACE, after Christian (*or* Common) era (non-Christian usage in place of AD)

ACF, Army Cadet Force

ACGB, Arts Council of Great Britain

ACGBI, Automobile Club of Great Britain and Ireland

ACGI, Associate of the City and Guilds of London Institute

Achaemenid, *not* Achai-

Acheson (Dean Gooderham), 1893–1971, US diplomat

à cheval (Fr.), on horseback

Achilles/ʼ heel (apos.); — **tendon** (no apos.)

Achin, Sumatra, Indonesia, *not* Atchin

4

Achnashellach, Highland
achy (adj.), *not* achey
ACI, Army Council Instruction
ACIA, Associate of the Corporation of Insurance Agents
ACIB, Associate of the Corporation of Insurance Brokers
ACII, Associate of the Chartered
Insurance Institute
ACIS, Associate of the Institute
of Chartered Secretaries and
Administrators
acknowledgement, *not*
-ledgment
ACMA, Associate of the Institute
of Cost and Management Accountants
acolyte, *not* -ite
acoustics, (noun sing.) the
science of sound, (noun pl.)
acoustic properties (of room,
etc.)
ACP, Associate of the College of
Preceptors; Association of Clin-
ical Pathologists; Association of
Correctors of the Press, now part
of **NGA,** q.v.
acquit/, -tal (verdict), **-tance**
(release from debt)
Acre, Israel, *not* Acca, Acco
acre, abbr. **a.**
acriflavine, an antiseptic, *not* -n
Acrilan, propr. term (cap.)
acronym, word formed from
initials, as Anzac, q.v.
acrylic (chem.)
ACS, Association of Commonwealth Students
ACSA, Associate of the Institute
of Chartered Secretaries and
Administrators
ACT, Australian Capital Territory
act., active
Actaeon, a mythical hunter
ACTH, adrenocorticotrophic
hormone
actini/a (zool.), *pl.* **-ae**
actinium, symbol Ac
actinomy/ces (bot.), *pl.* **-cetes**
actionnaire (Fr. m.), a shareholder
active, abbrs. **a., act.**

actor (person), *but* **one-acter**
(play)
acts of a play (typ.), cap. A
only when number follows, as
Hamlet, Act I, sc. ii. *See also*
authorities
Acts of Parliament (cap. A),
cited thus: Factory and Workshop Act, 1891; use arabic numerals for chapter numbers in
Public (General) and Private
Acts (e.g. 3 & 4 Geo. V, c. 12, ss.
18, 19) and lower-case roman
numerals in Public (Local) Acts
(e.g. 3 & 4 Geo. V, c. xii, ss. 18,
19). See also *Hart's Rules*,
pp. 52–3
Acts of Sederunt (Scots
law)
Acts of the Apostles, abbr.
Acts
ACTT, Association of Cine-
matograph, Television and
Allied Technicians
actualité (Fr. f.), present state
actuel/ (Fr.), *fem.* **-le,** present
acute accent (´), *see* **accents**
ACW, Aircraftwoman
AD (*anno Domini*) should always be
placed *before* the numerals; **BC**
after (s. caps.)
ad, advertisement (no point)
ad., adapted, adapter
a.d., after date
a.d., *ante diem* (before the day)
adagio/ (mus.), slow; as noun,
pl. **-s**
Adams (**John**), 1735–1826, US
statesman, President 1797–
1801; — (**John Quincy**), his
son, 1767–1848, President
1825–9
Adam's Peak, Sri Lanka
adapt/able, -er (person), **-or**
(device)
a dato (Lat.), from date
ADC, aide-de-camp
ad captandum vulgus (Lat.),
to catch the rabble, claptrap
Addams (**Charles Samuel**),
b. 1912, US cartoonist; —
(**Jane**), 1860–1935, US social
worker and writer

addend/um, something to be added; *pl.* **-a** (not ital.)

addio (It.), goodbye

Addis Ababa, capital of Ethiopia

additive, (something) in the nature of an addition

addorsed (her.), *not* adorsed, adossed

Addressograph, propr. term for addressing-machine (cap.)

adducible, *not* -eable

Adélie/ Land, Antarctica; — **penguin**

Adenauer (Konrad), 1876–1967, German statesman, Chancellor of W. Germany 1949–63

ad eundem (*gradum*) (Lat.), to the same degree at another university

à deux/ (Fr.), of (or between) two; — — *temps* (*see valse*)

ad/ extra (Lat.), in an outward direction; — *extremum,* to the last; — *finem,* near the end, abbr. *ad fin.* ; — *hanc vocem,* at this word; — *hoc,* for this (purpose); — *hominem,* to the (interests of the) man, personal; — *hunc locum,* on this passage, abbr. *a.h.l.;* — *idem,* to the same (point)

adieu/, *pl.* **-s** (not ital.)

ad/ infinitum (Lat.), to infinity, — *interim,* meanwhile, abbr. *ad int.*

adiós (Sp.), goodbye

Adirondack Mountains (US), *not* -dac

Adj., Adjutant

Adj.-Gen., Adjutant-General

adjectiv/e, abbr. **a., adj.; -al, -ally**

adjudicator, *not* -er

adjutage, a nozzle, *not* aj-

Adjutant/, abbr. **Adj.** (rank, cap.) ; —-**General,** abbr. **A.-G.** *or* **Adj.-Gen.**

adjuvant

Adler (Alfred), 1870–1937, Austrian psychologist

ad lib (adv., noun, and verb)

ad/ libitum (Lat.), at pleasure; — *litem,* appointed for a lawsuit

ad locum (Lat.), at the place, abbr. **ad loc.** (not ital.)

Adm., Admiral, Admiralty

adman (one word)

admin, short for administration (no point)

administrable, *not* -atable

administrat/or, abbr. **admor.; -rix,** abbr. **admix.**

Admiral, in Fr. *amir/al,* pl. *-aux*; abbr. **Adm.**

Admiralty, the, abbr. **Adm.;** *see* **M.o.D.**

admiration (note of), exclamation mark, ! *See* **punctuation** VII

ad misericordiam (Lat.), appealing to pity

admiss/ible, *not* -able, *but* **admittable**

admix., administratrix

ad modum (Lat.), after the manner of

admonitor/, *not* -er; **-y**

admor., administrator

ad nauseam (Lat.), to a sickening degree

ado, work, trouble (one word)

adobe (Sp.), sun-dried brick

Adonai (Heb.), the Lord

Adonais, Keats, in Shelley's elegy, 1821

Adonis (myth.), beloved of Venus

adopter, *not* -or

adorable, *not* -eable

adorsed, adossed, *use* **addorsed**

ADP, automatic data processing

ad/ personam (Lat.), personal; — *referendum,* for further consideration; — *rem,* to the point

adrenalin, *not* -ine

adresse (Fr. f.), address

Adrianople, *now* **Edirne**

à droite (Fr.), to the right

adscriptus glebae (Lat.), bound to the soil (of serfs)

adsorb, adsorption, condensing, condensation, of gases etc. on surface of solid; **adsorbent**

adsum (Lat.), I am present

adulator, *not* -er

ad/ *unguem* (Lat.), perfectly;
— *unum omnes,* all, to a man;
— *usum,* according to custom, abbr. **ad us.**

adv., adverb, -ially, advocate

adv., *adversus* (against)

ad valorem (Lat.), according to value; abbr. **ad val.**

adverb/, **-ially,** abbr. **adv.**

ad verbum (Lat.), to a word, verbatim

adversus (Lat.), against; abbr. **adv.**

advertise, *not* -ize

advertisement, abbr. **ad,** *or* **advt.,** *pl.* **ads,** *or* **advts.**

advis/e, *not* -ize; **-able,** *not* -eable

advis/er, *not* -or; **-ory**

ad/ *vitam aut culpam* (Lat.), for lifetime or until fault; — *vivum,* lifelike

advocaat, liqueur

advocate, abbr. **adv.**

Advocates (Faculty of), the Bar of Scotland

advocatus diaboli (Lat.), devil's advocate (q.v.) (not ital.)

advt/., **-s.,** advertisement, -s

A.E. *or* **AE** (originally Æ), pseud. of George Russell (1867–1935)

Æ, in numismatics = copper (Lat. *aes*); *see* **Lloyd's**

æ (ligature), *see* **diphthongs**

AEA, Atomic Energy Authority

AEC, Army Educational Corps (*now* **RAEC**), (US) Atomic Energy Commission (since 1975 **ERDA**)

aedile, *not* e-

AEF, Amalgamated Union of Engineering and Foundry Workers

Aegean Sea, the Mediterranean between Greece and Turkey

aegis, a shield, protection, *not* e-

aegrot/*at* (Lat.), he, she, is ill (certificate that exam. candidate is ill, or degree thus awarded); *pl.* **-ant**

AEI, Associated Electrical Industries

Ælfric, d. 1020, English writer

Aemilius

Aeneas; *Aeneid*

Aeolian, Aeolic (caps.), *not* E- **aeon,** *not* eon

aepyornis (zool.), *not* epi-, epy-

aequales (Lat.) equal (pl. adj.), (pl. noun) equals (in age or performance); abbr. **aeq.**

AERA, Associate Engraver, Royal Academy

aer/ate etc., **-ial, -ify, -obic, -onaut,** *not* aë-

AERE, Atomic Energy Research Establishment

aerie, *use* **eyrie**

aerodrome (US and increasingly Brit., **airfield**)

aerodynamics (one word)

aero/-elastic, -engine (hyphens)

aerofoil (one word)

aeronaut/, -ical

aeroplane, tending to be superseded by **aircraft** in some contexts

aerosol

aerospace (one word)

Aeschylus, 525–456 BC, Greek playwright

Aesculapi/us, Latin form of the Greek god Asclepius (q.v.); **-an,** *not* E-

Aesop, 6th c. BC, Greek fable-writer

aesthete etc., *not* e-

aestiv/al, -ation, *not* e-

aet. or *aetat.,* *anno aetatis suae* (of age, aged)

Aetheling, *use* **Atheling**

aether, in all meanings *use* **e-**

aetiology, *not* e-

Aetna, *use* **Etna**

AEU, Amalgamated Engineering Union

AF, audio frequency

AFA, Associate of the Faculty of Actuaries

Afars and Issas (French Territory of the), *formerly* Fr. Somaliland, *now* **Jibuti**

AFAS, Associate of the Faculty of Architects and Surveyors

AFC, Air Force Cross, Association Football Club

aff., affirmative, affirming

affaire/ *d'amour* (Fr. f.), love--affair; — (*de cœur*), affair of the heart; — *d'honneur*, duel

affairé (Fr.), busy (accent)

affenpinscher, dog (not ital.)

affettuoso (mus.), with feeling

affidavit, abbr. **afft.**

affranchise, *not* -ize

afft., affidavit

Afghan (hound); **afghan** (blanket or shawl)

Afghanistan, abbr. **Afghan.**

aficionado/ (Sp.), enthusiast for sport or hobby; *pl.* *-s*

afield (one word)

AFM, Air Force Medal

à fond/ (Fr.), thoroughly; — *de train*, at full speed

aforethought (one word)

a fortiori (Lat.), with stronger reason, *not* à — (ital.)

Afr., Africa, -n

A.F.R.Ae.S., Associate Fellow of the Royal Aeronautical Society

Africander, *use* **Afrikander** *or* **Afrikaner**

Afridi, tribe of Cent. Asia

Afrikaans, S. African Dutch language, abbr. **Afrik.**

Afrikander, S. African breed of cattle or sheep

Afrikaner, Afrikaans-speaking white S. African; *formerly* Africander

afterbirth (one word)

after/-care, -effect (hyphens)

afternoon, abbr. **aft.** *or* **p.m.**

aftershave (one word)

afterthought (one word)

AFV, armoured fighting vehicle

AG, Air Gunner, (Ger.) *Aktiengesellschaft* (joint-stock company)

A.-G., Accountant-General, Adjutant-General, Agent-General (of Colonies), Attorney-General

Ag, *argentum* (silver) (no point)

Agadah, *use* **Haggadah**

Aga Khan, leader of Ismaili Muslims

agape, love-feast, divine love

à gauche (Fr.), to the left

age-group (hyphen)

ageing, *not* agi-

agend/a, 'things to be done', freq. list of items for consideration: in this sense *pl.* **-as**

agent, abbr. **agt.**

Agent-General, abbr. **A.-G.**

agent provocateur (Fr.) (ital.); *pl.* *-ts -rs*

ages (typ.), to be printed in numerals; but 'he died in his fortieth year', and, in literary contexts, 'a man of forty'

aggiornamento (It.), bringing up to date

aggrandize, *not* -ise

agitato (mus.), hurried

agitat/or, *not* -er

AGM, Annual General Meeting

agnostic/, -ism (not cap.)

agonize, *not* -ise

agouti, S. American rodent, *not* -y, aguti

AGR, advanced gas-cooled reactor

agreeable, in Fr. *agréable*

agric., agricultur/e, -al, -ist

agriculturist, *not* -alist

agrimony (bot.)

agronomy, rural economy

AGSM, Associate of the Guildhall School of Music

agt., agent

Ag^{to} (Sp.), *agosto* (August)

Agulhas, Cape (S. Africa), *not* L'A-

aguti, *use* **agouti**

AH, *anno Hegirae*, the Muslim era (s. caps.). *See also* **hegira**

ah, when it stands alone, takes an exclamation mark (!). When it forms part of a sentence it is usually followed by a comma, the ! being placed at the end of the sentence: as 'Ah, no, it cannot be!'

AHA, Area Health Authority

aha, exclam. of surprise. *See also* **ha ha**

ahimsa (Skt.), non-violence
a.h.l., *ad hunc locum* (on this passage)
Ahmad/abad, -nagar, India, *not* Ahmed-, Amed-
Ahmadu Bello University, Nigeria
AHMI, Association of Headmistresses, Incorporated. *See also* **IAHM**
Ahriman, Zoroastrian spirit of evil
à huis clos (Fr.), in private
Ahura-Mazda, *use* **Ormuzd**
a.h.v., ad hanc vocem (at this word)
AI, American Institute, Anthropological Institute, artificial insemination
AIA, Associate of the Institute of Actuaries, American Institute of Architects
AIAA, Architect Member of the Incorporated Association of Architects and Surveyors
AIAC, Associate of the Institute of Company Accountants
AIAS, Surveyor Member of the Incorporated Association of Architects and Surveyors
AIB, Associate of the Institute of Bankers
AICC, All India Congress Committee
AICS, Associate of the Institute of Chartered Shipbrokers
AID, Army Intelligence Department; artificial insemination by donor
Aida, opera by Verdi, 1871 (no diaeresis)
aide-de-camp, abbr. **ADC;** *pl.* **aides-——-**
aide-mémoire (Fr. m.), aid to memory; *pl.* **aides-——**
aigrette, a spray, *not* ei-, -et (*see also* **egret**)
AIH, artificial insemination by husband
Ailesbury (Marquess of) (*see also* **Ayl-**)
A.I.Loco.E., Associate of the Institute of Locomotive Engineers

A.I.Mech.E., Associate of the Institution of Mechanical Engineers
A.I.Min.E., Associate of the Institution of Mining Engineers
AIMM, Associate of the Institution of Mining and Metallurgy
AIMTA, Associate of the Institute of Municipal Treasurers and Accountants
Ain, dép. France
'**ain** (Arab.), *see* **accents**
aîn/é (Fr.), *fem.* *-ée,* elder, senior; opposed to *puîné,* *fem.* *puînée,* or *cadet/, -te,* younger
A.Inst.P., Associate of the Institute of Physics
AIQS, Associate of the Institute of Quantity Surveyors
air/-bed, -blast (hyphens)
airborne (one word)
airbus (one word)
Air Commodore, abbr. **Air Cdre.** *See also* **Commodore**
air-conditioning (hyphen)
aircraft/, *sing.* and *pl.* (one word); **-carrier** (hyphen)
aircraft/man, -woman, *not* aircrafts- (one word)
Airedale/, W. Yorks., and terrier; — **(Baron)**
air/field, -flow (one word)
Air Force, *see* **Navy**
airgun (one word)
Air Gunner, abbr. **AG**
airlift, airline, airmail (no hyphens)
Air Ministry, *see* **M.o.D.**
Air (Point of), Clwyd. *See also* **Ayre**
airport (one word)
air raid (two words)
airship, airsick, airspace, airtight (one word)
air-to-air (adj., hyphens)
airworthy (adj.), of aircraft (one word)
AISA, Associate of the Incorporated Secretaries Association
Aisne, dép. France
ait, islet, *not* eyot (cap. when part of name)

aitchbone (one word), *not* H-, edge-

Aix-la-Chapelle (hyphens), *properly* **Aachen**

Aix-les-Bains (hyphens)

Ajaccio, Corsica

Ajax, *pl.* **Ajaxes**

ajutage, *use* **adj-**

AK, Alaska (off. postal abbr.), Knight of the Order of Australia

AKC, Associate of King's College (London)

akimbo (one word)

Akkadian, *not* Acc-

Akkra, Ghana, *use* **Accra**

Aktiengesellschaft (Ger. f.), joint-stock company; abbr. **AG**

AL, Alabama (off. postal abbr.), autograph letter

Al, aluminium (no point)

ALA, Associate of the Library Association

Ala., Alabama (off. abbr.)

à la carte, (a meal) that must be ordered from a wide range of available dishes (not ital.). *See also* **table d'hôte**

Aladdin, *not* Alladin

à la mode, in fashion (not ital.)

Alanbrooke (Viscount) (one word)

Åland Islands, Baltic Sea

à la page (Fr.), up to date

alarm/, *not* alarum except in alarums and excursions; — **clock** (two words)

à la russe, in the Russian style (not cap.)

Alas., Alaska

alas, when it stands alone, takes an exclamation mark (!). When it forms part of a sentence it is usually followed by a comma, the ! being placed at the end of the sentence: as 'Alas, it is true!'

Alaska (baked), sponge confection (cap.)

Alb., Albanian

Alba., Alberta, *use* **Alta.**

Alban., formerly signature of Bp. of St. Albans (full point); *now* **St. Albans:** (colon)

albedo/, fraction of radiation reflected; *pl.* **-s**

Albee (Edward), b. 1928, US playwright

Albigens/es, Manichaean sect; adj. **-ian**

albin/o, *pl.* **-os,** *fem.* **-ess; -ism,** *not* -oism; **-otic** (adj.)

Albrighton, Shropshire

album/en, natural white of egg; **-in,** its chief constituent; adj. **-inous**

Albuquerque, New Mexico, US

Albury, Herts., Oxon., Surrey

Alcaeus of Mytilene (b. *c.*620 BC), Greek lyric poet

alcaic, metre

alcalde (Sp.), magistrate, mayor

Alcatraz, US prison, San Francisco Bay, Calif.

alcayde (Sp. *alcaide*), governor, gaoler

alcázar (Sp.), palace, fortress, bazaar

Alceste, opera by Gluck, 1767

Alcestis, Alcibiades, *not* Alk-

ALCM, Associate of the London College of Music

Alcoran, *use* **Koran**

Alcyone (myth.), *not* Hal-

Aldborough, Norfolk, N. Yorks.

Aldbrough, Humberside

Aldbury, Herts.

Aldeburgh, Suffolk

al dente (It.), cooked but not soft

Alderbury, Wilts.

Alderman, abbr. **Ald.**

Aldermaston, Berks.

Alderney, Channel Islands; in Fr. **Aurigny**

Aldine (adj.), printed by **Manutius,** q.v., who introduced italic type

Aldis lamp, a hand lamp for signalling

Aldsworth, Gloucester, W. Sussex

Aldus Manutius, printer, *see* **Manutius**

Aldworth, Berks.

Alecto (myth.), one of the Furies

alehouse (one word)

Alençon lace (ç)

aleph (Heb.), *see* **accents**

Aleutian Islands, Bering Sea

A level, examination (no hyphen)

alexandrine, metre

ALF, Automatic Letter Facer

Alfa-Romeo, Italian make of car

Alford, Grampian, Lincs., Som.

Alfred, abbr. **A.** *or* **Alf.**

al fresco (Eng. adj. and adv.), two words predic.; one word (no hyphen) attrib.

Alfreton, Derby.

Alg., Algernon, Algiers

alg/a (bot.), *pl.* **-ae** (not ital.)

Algarve (**the**), Portugal

algebra, abbr. **alg.** *See also* **mathematics**

ALGOL (comp.), Algorithmic Language

algology (bot.)

Algonqui/an, of a large group of N. Amer. Indian tribes; **-n,** of one people in this group; *not* -nki-

alguazil (Sp.), a constable

Alhambra, Moorish palace at Granada; adj. **alhambresque**

alia (Lat. pl.), other things

alias/ (noun), *pl.* **-es**; also adv.

alibi/ (noun), *pl.* **-s**

alienator, *not* -er

Aligarh, Uttar Pradesh, India

Alighieri, family name of **Dante**

align/, -ment, *not* aline

alii (Lat. pl.), other people

alimentative/, -ness, *not* alimentive, -ness

aline, *use* **align**

alinéa (Fr. m.), paragraph

Alipore, India

Alipur, India and Pakistan

Alitalia, Italian airline

ali/us (Lat. m.), another person; *pl.* **-i**

Aliwal, E. Punjab, India, *also* S. Africa

alkali/, *pl.* **-s**

alkalize, *not* -ise

Alkestis, Alkibiades, *use* Alc-

Alkoran, *use* **Koran**

alla breve (mus.), 2 or 4 minims in bar

alla cappella (mus.), *use* **a cappella**

Alladin, *use* **Aladdin**

Allahabad, Uttar Pradesh, India

Allah il Allah, corruption of Arab. *la ilaha illa'llah,* There is no god but God (Muslim prayer and war-cry)

Allan-a-Dale, minstrel hero

allargando (mus.), broad, spread out

allée (Fr. f.), alley, avenue

Allegheny, mountains and river, US; *but* **Allegany,** Pittsburgh, Pa.

allegretto/ (mus.), fairly brisk; as noun, pl. **-s**

allegro/ (mus.), brisk, merry; as noun, *pl.* **-s**

allele, alternative genetic character, *not* -l

alleluia, Lat. form (also liturgical) for Heb. (and biblical) **hallelujah**

allemande, Ger. dance

Allendale, Northumb.

Allen (**George**) **& Unwin, Ltd.,** publishers

Allen (**W. H.**) **& Co., Ltd.,** publishers

allerg/y, -ic, sensitive(ness) to certain foods, pollens, etc.

alleviator, *not* -er

Alleynian, member of Dulwich College

All Fools' Day (caps., no hyphen)

allg., allgm., allgemein (Ger.), general (adj.)

All-Hallows, All Saints' Day, 1 Nov. (caps., hyphen)

Allhallows, Kent (one word)

allineation, *not* alin-

Allingham (**Margery**), 1904–66, English novelist

allodium, estate held in absolute ownership, *not* alo-

allons! (Fr.), let us go!, come!

allot/, -ment, -table, -ted, -ting

all' ottava (mus.), an octave higher than written; abbr. **all' ott.**

all right, *not* alright, all-right

all round ('all round the Wrekin'), prep.

all-round ('an all-round man'), adj.

All/ Saints' Day, 1 Nov.; — **Souls College,** Oxford (no apos.) — **Souls' Day,** 2 Nov.; — **Souls' Eve,** 1 Nov. (caps., no hyphens)

allspice (one word)

all together (in a body); *but* **altogether** (entirely)

alluvi/um, *pl.* **-a**

Alma-Ata, cap. of Kazakh Rep., USSR

Alma Mater, fostering mother, one's school or university

almanac, *but* 'Oxford', *also* 'Whitaker's', **Almanack**

Alma-Tadema (Sir Lawrence), 1836–1912, Dutch-born English painter

Almighty (the) (cap.)

Almondbury, W. Yorks.

Almondsbury, Avon

Alnmouth, Northumb.

alodium, *use* **allodium**

A.L.O.E., A Lady of England (Charlotte M. Tucker, 1821–93)

à l'outrance should be **à outrance,** to the bitter end

ALP, Australian Labor Party

Alpes-de-Haute-Provence, dép. France (hyphens)

Alpes-Maritimes, dép. France; abbr. **A.-M.**

Alphaeus, Apostle James's father

alphanumeric (comp.), of instructions containing alphabetic, numerical, and other characters (one word)

alpha/ particle, — ray (two words)

alright, *use* **all right**

ALS, Associate of the Linnean Society; autograph letter signed

Alsace-Lorraine

Alsirat (Arab.), bridge to paradise

alt (mus.), high note

alt., alternative, altitude

Alta., Alberta, Canada, *not* Alba.

alter/ ego (Lat.), one's second self, *pl.* **— egos** ; **— idem,** another self

Altesse (Fr. f.), *Altezza* (It.), Highness

Althing, the Icelandic parliament

Althorpe, Humberside

Althorp Library and **Park** (Northants)

altitude, abbr. **alt.**

alto/ (mus.), *pl.* **-s**; abbr. **A.**

altogether, *see* **all together**

alto-relievo/, high relief; *pl.* **-s**; Anglicized version of It. *alto rilievo*

aluminium, symbol **Al** (no point); US **aluminum**

alumn/us, *pl.* **-i,** *fem.* **-a,** *pl.* **-ae**; abbr. **alum.**

AM, Air Ministry, Albert Medal, amplitude modulation, *Artium Magister* (US Master of Arts); *Ave Maria* (Hail, Mary!) (caps.); Member of the Order of Australia

A.-M., dép. France, Alpes--Maritimes

AM, *anno mundi* (in the year of the world) (s. caps.)

Am, americium (no point)

Am., American

a.m. (Lat.), *ante meridiem* (before noon) (lower case, points)

AMA, Assistant Masters' Association, officially **IAAM;** American Medical Association

amah, child's nurse (in Far East)

Amalek, *not* -ech, -eck

amanuens/is, *pl.* **-es**

amateur, abbr. **A.**

Ambala, India, *not* Umbal(l)a

ambassad/or, *fem.* **-ress**

ambassy, *use* **embassy**

ambergris, waxy substance used in perfumery

ambiance, French spelling of **ambience** (q.v.), used in English esp. of accessory details in a work of art

ambidextrous, *not* -erous

ambienc/e, the surroundings (*not* -y). *See also* **ambiance**

ambivalen/t, -ce

amblyop/ia, impaired vision; adj. **-ic**

Amboina, Indon., *properly* **Ambon,** *not* Amboyna

amboyna (wood)

ambry, *use* **aumbry**

AMDG, *ad majorem Dei gloriam* (for the greater glory of God)

âme damnée (Fr.), a 'cat's paw', devoted adherent

Amedabad, Amednagar, India, *use* **Ahmad-**

Ameer, *use* **Amir** for Arabic, **Emir** for Turkish and Indian titles

amend, to correct an error, make minor improvements. *See also* **emend**

amende honorable (Fr. f.), honourable reparation

a mensa et toro (Lat.), from bed and board, a legal separation, *not* — — — *thoro*

Amer., American

Americanize, *not* -ise

American spellings, *see* **US spellings**

America's Cup (the), yachting trophy

americium, symbol **Am**

Amerindian, American Indian

à merveille (Fr.), perfectly, wonderfully

AMG(OT), Allied Military Government (of Occupied Territory) (Second World War)

Amharic, official lang. of Ethiopia

amicus curiae (Lat.), a friend of the Court, a disinterested adviser; *pl. amici* —

amidships, *not* -ship (one word)

amino acid (chem., two words)

Amir, Arab. title, *not* Ameer. *See also* **Emir**

Amis (Kingsley), b. 1922, English writer

Ammal (Ind.), suffix used to indicate that a name belongs to a woman, e.g. Dr E. K. Janaki Ammal

Amman, Jordan

Ammergau, Bavaria

ammeter, for measuring electric current

amoeb/a, *pl.* **-as** (not ital.)

à moitié (Fr.), half

amok, *not* amuck

Amoor, *use* **Amur**

amortize, *not* -ise

amour propre (Fr.), self--respect, proper pride

amp, ampere

Ampère (André Marie), 1775–1836, French physicist (accent)

ampere, elec. unit of current (no accent), abbr. **A** (sing. and pl.) *or* **amp** (no point)

ampersand = &, may be used in names of firms as Smith & Co., and in **Acts of Parliament,** q.v. In general *use* **and.** For &c. *use* **etc.**

amphetamine (med.)

amphibian, (zool.) member of the Amphibia; vehicle adapted for land and water

amphor/a, a jar; *pl.* **-ae**

ampoule, glass container for hypodermic dose

Amritsar, India, *not* Umritsur

amt., amount

amuck, *use* **amok**

Amundsen (Roald), 1872–1928, Norwegian explorer; S. Pole, 1911

Amur, Siberia, *not* -oor

amygdalin/, *not* -e

AN, Anglo-Norman, autograph note

an, see **a,** *not* **an**

an (arch. *or* dial.), if (so spelt)

ana/, sayings, collective pl.; or sing. with pl. **-s**

-ana, suffix denoting anecdotes

-ana (*cont.*)
  or other publications concerning, as Shakespeariana, Victoriana

anacoluth/on (gram.), *pl.* **-a**, *not* -outhon, -kolouthon, -koluthon

Anacreon/, 6th c. BC, Greek poet; abbr. **Anacr.**

anacreontic, metre

anaemi/a, -c

anaerob/e, -ic

anaesthetize, *not* -ise

anal., analog/y, -ous, analys/e, -er, -is, analytic, -al

analogous, abbr. **anal.**

analys/e, -er, -is, *pl.* **-es**; abbr. **anal.**

analyst

analytic/, -al; abbr. **anal.**

Anam, *use* **Annam**

anamorphos/is, *pl.* **-es**

anapaest, foot of three syllables ($\cup\cup-$)

anastase, *use* **anatase**

anastomos/is, communication by cross-connections; *pl.* **-es**; **-ist**

anat., anatom/y, -ical

anatase, a mineral, *not* anastase

anathema/, a curse, *pl.* **-s**; **-tize**

anatomize, dissect, *not* -ise

anatom/y, -ical, -ist; abbr. **anat.**

anatta (bot.), *use* **annatto**

anc., ancient

ancest/or, *fem.* **-ress**, *not* -rix

anchor/ite, hermit, *not* -et; adj. **-etic**

anchylosis, *use* **ank-**

ancien régime (Fr. m.), the old order of things

ancient, abbr. **anc.**

Ancient Mariner (*Rime of the*), by S. T. Coleridge, 1798

Ancient Order of Druids, abbr. **AOD**; — — — **Foresters, AOF**; — — — **Hibernians, AOH**

and (gram.). Where *and* joins two or more subjects in the singular number the verb must be in the pl., e.g. Jack and Jill *are* going. Where *and* joins two single words

the comma is generally omitted. Where *and* joins the last two words of a list a comma should precede it; e.g. black, white, and green. *See also* **ampersand**

Andalusia, Spain; in Sp. **Andalucía**

andante (mus.), moving easily, steadily

andantino/ (mus.), rather quicker (*formerly* slower) than andante; as noun, *pl.* **-s**

Andersen (**Hans Christian**), 1805–75, Danish story-teller

Andhra Pradesh, Indian state

and/or, signifying either or both of the stated alternatives

Andrea del Sarto, 1486–1531, Italian painter

Androcles, *not* -kles

aneurysm, *not* -ism

Angeles (**Victoria de los**), b. 1923, Spanish soprano

Angelico (**Fra**), 1387–1455, Italian painter

anglais/ (Fr.), *fem.* **-e**, English (not cap.); *but* **Anglais/, -e**, Englishman, Englishwoman

angle, the sign $\angle$; angle between the two lines $\wedge$; right angle $\llcorner$; two right angles $\perp$

angle brackets < > (wide), $\langle \rangle$ (narrow)

Anglesey, Wales, *not* -ea; in Welsh **Ynys Môn**

Anglesey (**Marquess of**)

anglice (Lat.), in English (no accent); abbr. *angl.*

Anglicize, *not* -ise (cap.)

Anglo-/French, abbr. **AF**; —**-Norman**, abbr. **AN**; —**-Saxon**, abbr. **AS**; *see* **accents**

Angora, Turkey, *now* **Ankara**

Angst (Ger. f.), fear, anxiety

Ångstrom (**A. J.**), 1814–74, Swedish physicist

ångström, unit of length; abbr. **Å**; in SI units, 10^{-10} m

aniline, source of dyes. Also adj.

animalcule/, *pl.* **-s**, *not* -ae

animalcul/um, *pl.* **-a**

animé, W. Ind. resin (not ital.)

animé (Fr., mus.), animated

anion (elec.), ion carrying negative charge of electricity

Ankara, Turkey, formerly **Angora**

ankylos/is, fusion of bones, *not* anch-; *pl.* **-es**

Anmerkung/ (Ger. f.), a note (cap.); pl. *-en*; abbr. *Anm.*

ann., annals, *anni* (years), *anno*, annual

anna/, *pl.* **-s,** formerly 16 to rupee, now replaced by decimal coinage (Ind. and Pak.); abbr. **a.**

annales (Fr. pl. f.), annals; abbr. *ann.*

annals, abbr. **ann.**

Annam, *not* Anam, formerly kingdom within Fr. Indo-China, now merged in Vietnam

Annan, Dumfries & Galloway

Ann Arbor, Mich., US

annatto (bot.), orange-red dye, *not* an-, -a

Anne of Cleves, 1515–57, fourth wife of Henry VIII of England

Anne of Geierstein, by Sir W. Scott, 1829

Anne (Queen), 1702–14, b. 1665

Anne (Saint)

annex, verb; **annexe,** noun

anno/ (Lat.), in the year, abbr. **ann.**; — *aetatis suae,* aged, abbr. *aet., or aetat.*; — *Domini* (cap. *D*), abbr. **AD** (s. caps.); — *Hegirae* (Muslim era), abbr. **AH** (s. caps.); — *mundi,* in the year of the world, abbr. **AM** (s. caps.); all three abbrs. to be placed *before* the numerals

annonce (Fr. f.), advertisement

annotat/ed, -or, *not* -er; abbr. **annot.**

annual, abbr. **ann.**

annul/ar, ringlike; **-ate, -ated, -et, -oid**

ann/us (Lat.), year; *pl.* **-i;** *-us mirabilis,* remarkable year

anonymous, abbr. **anon.**

Anouilh (Jean), b. 1910, French playwright

anschluss, union (cap. A in Ger.)

answer, abbr. **A.** *or* **ans.**

Ant., Anthony, Antigua

ant., antonym

Antaeus

antagonize, *not* -ise

Antananarivo, *use* **Tananarive**

Antarctic/, -a

ante (not ital.), stake in card-game

ante/ (Lat.), before; — *bellum,* before the war

antechamber, *not* anti-

ante diem (Lat.), before the day; abbr. *a.d.*

antediluvian (one word)

antemeridian (one word)

ante meridiem (Lat.), before noon; abbr. **a.m.**

ante-mortem (adj., hyphen)

antenatal (one word)

antenn/a (zool.), *pl.* **-ae;** (US) radio, *pl.* **-as**

ante-post, of racing bets (hyphen)

ante-room (hyphen)

Anthony, Anglicized from Lat. Antonius (NB *Antony and Cleopatra*); abbr. **Ant.**

anthropolog/y, -ical; abbr. **anthrop.**

anthropomorphize, *not* -ise

anthropophag/us (noun), cannibal; *pl.* **-i;** adj. **-ous**

antichamber, *use* ante-

Antichrist (cap.), *but* **antichristian**

anticline (geol.)

anticlockwise (one word)

antifreeze (one word)

Antigua, abbr. **Ant.**

anti-hero, principal character lacking traditional heroic qualities

antilogarithm (one word), abbr. **antilog**

antimatter (one word)

antimony, symbol **Sb**

antinomy, conflict of authority

antiparticle (one word)

antipathize, *not* -ise

antiq., antiquar/y, -ian

antiqs., antiquities

Antiqua (Ger. typ. f.), roman type (cap.)

antique paper, a moderately bulky, opaque book paper with roughish surface

antisabbatarian

anti-Semit/e (hyphen, one cap.); adj. **-ic**; noun **-ism**

antistrophe, stanza corresponding to the strophe in Greek dramatic chorus

antistrophon (rhetoric), a retort

antitetanus (adj., one word)

antithes/is, *pl.* **-es**

antithesize, *not* -ise

antitoxin (one word)

antitype, *not* ante-; that which corresponds to the type

Antony, *see* **Anthony**

antonym, a word of opposite meaning; abbr. **ant.**

antrycide (against tsetse fly), *not* -tri-

Anvers, Fr. for **Antwerp,** Belgium, in Fl. **Antwerpen**

Anwick, Lincs.

anybody, any person

any body, any group of persons

any/how, -one, -thing, -where (one word)

any one (person, thing), when each word to retain its meaning (see *Hart's Rules,* p. 77)

anyway, one word in conventional sense, *but note* you can do it any way you like (two words)

Anzac, Australian and New Zealand Army Corps (First World War)

Anzeige/ (Ger. f.), notice, advertisement; *pl.* **-n**

AO, Accountant Officer, Army Order, Officer of the Order of Australia

A.O.C.-in-C., Air Officer Commanding-in-Chief

AOD, Army Ordnance Department, Ancient Order of Druids

AOF, Ancient Order of Foresters

AOH, Ancient Order of Hibernians

aorist, abbr. **aor.**

aort/a (anat.), *pl.* **-ae**

août (Fr. m.), August (not cap.)

à outrance (Fr.), to the bitter end, not *à l'outrance*

AP, Associated Press

Ap., Apostle

a.p., above proof, author's proof

Apache, (US) Indian tribe; **apache,** Paris hooligan

apanage, *not* app-

à part (Fr.), apart

apartheid (Afrik.), segregation

ap/e, -ed, -ing, -ish

Apelles, 4th c. BC, Greek painter

Apennines, Italy (one *p,* three *n*s), in It. **Appennini**

aperçu (Fr. m.), outline (*ç*)

aperitif/, an alcoholic appetizer, *pl.* **-s** (not ital., no accent)

APEX, Association of Professional, Executive, Clerical, and Computer Staff

apex/, *pl.* **-es;** adj. **apical**

apfelstrudel, baked apples in pastry (not ital., not cap.)

aphaere/sis, removal of a sound from the beginning of a word; adj. **-tic**

apheli/on (astr.), *pl.* **-a**

aphid/, *pl.* **-s**

aphi/s, *pl.* **-des**

apical, *see* **apex**

ap. J.-C. (Fr.), AD

à plaisir (Fr.), at pleasure

Apocalypse, abbr. **Apoc.**

Apocrypha (cap. A); abbr. **Apocr.** (for abbrs. of books see under names); adj. **apocryphal**

apodictic, clearly established; *not* -deictic

apodos/is (gram.), *pl.* **-es**

apog/ee (astr.), abbr. **apog.;** adj. **-ean**

Apollinaire (Guillaume), 1880–1918, French poet

Apollinaris (water), mineral water from Ahr Valley, Germany

Apollin/arius (*sometimes* **-aris**), Christian heretic, *c.*310–*c.*390; **-arianism**

Apollo/, Greek god; — **Belvedere**, statue in Vatican; adj. **-nian**

Apollonius Rhodius, Greek poet, *c.*250 BC

Apollos, Acts 18:24, follower of St. Paul

Apollyon, the Devil

apologize, *not* -ise

apophthegm, pithy maxim

a-port (hyphen)

apostasy, *not* -cy

apostatize, *not* -ise

a posteriori, (reasoning) from effects to causes

Apostle, abbr. **Ap.**, *pl.* **App.**

Apostles' Creed (caps.)

Apostroph/ (Ger. m.), apostrophe; *pl.* **-e** (cap.)

apostrophe, *see* possessive case, punctuation VIII, quotations

apostrophize, *not* -ise

apothecaries' weight, signs: ℳ minim; ℈ scruple; ℨ drachm; ℥ ounce. Print quantities in lower-case letters close up behind symbol: final *i* becomes *j*, as vij = 7, j = 1

apothegm, *use* **apophthegm**

apotheos/is, *pl.* **-es**

apotropaic, averting ill luck

App., Apostles

app., appendix

appal/, **-led**, **-ling**

Appalachian Mts., E. North America

appanage, *use* apa-

apparatus/, *pl.* **-es**, *not* -ati (*but* use an alternative, e.g. appliances, when possible)

apparatus criticus, variant readings; abbr. **app. crit.**

apparel/, **-led**, **-ling**

apparitor, officer of ecclesiastical court, *not* -er

app. crit., apparatus criticus

appeal/, **-ed**, **-ing**

appeasement, *not* -sment

appell/ant, **-ate**, **-ation**, **-ative**

append/ix, abbr. **app.**; *pl.* **-ices**, zool. and general, abbr. **apps.**

appetiz/e, *not* -ise; **-er**, **-ing**

appliqué, Eng. noun, adj., and verb; past partic. **appliquéd**

appoggiatura (mus.), type of grace-note, leaning note

appreciat/or, *not* -er; adjs. **-ive**, **-ory**

apprentice, abbr. **appr.**

apprise (to inform)

apprize (to value)

appro, (on) approval (no point)

appro., approbation

approver, one who turns Queen's evidence

approximat/e, **-ely**, **-ion**; abbr. **approx.**

appurts., appurtenances

Apr., April

APRC, *anno post Romam conditam* (in the year after the building of Rome in 753 BC) (s. caps.); *but* use **AUC**, q.v.

après/ (Fr.), after; — **coup**, after the event; — **-midi**, afternoon (hyphen); — **moi** (*or* **nous**) **le déluge**, after me (*or* us) the deluge; — **-ski**, (of) time after day's skiing

April, abbr. **Apr.**; in Fr. **avril** (not cap.)

a/ primo (Lat.), from the first; — **principio**, from the beginning

a priori, (reasoning) from causes to effects (not ital.)

apriorism, doctrine of a priori ideas (one word, not ital.)

apropos (of); **à propos de bottes** (Fr.), beside the mark

APS, Associate of the Pharmaceutical Society of Great Britain

apse/, semicircular recess, esp. in church; *pl.* **-s**

apsi/s, in an orbit, point of greatest or least distance from central body; *pl.* **-des**

apud (Lat.), according to, in the work, or works, of

Apuleius (Lucius), 2nd c. AD, Roman satirist

Apulia, Italy; in It. **Puglia**

aqua/ fortis, — regia (chem.),
— **vitae** (two words, not ital.)

aquarium/, *pl.* **-s**

à quatre mains (Fr. mus.), for
two performers

Aquinas (St. Thomas), 1225–
74, Italian theologian

Æ, in numismatics = silver (Lat.
argentum) (no point)

AR, Arkansas (off. postal
abbr.)

AR, *anno regni* (in the year of the
reign) (s. caps.)

Ar, argon (no point)

a/r, all risks

ARA, Associate of the Royal
Academy, London

Arabi/a, -an, -c; abbr. **Arab.**

*Arabian Nights' Entertain-
ments (The)*

Arabic (typ.), 28 letters (all
consonants, each with several
forms according to position) and
many accents (representing
vowel sounds and other phonetic
features). It is read from right to
left; hence if any passage is
divided, the right-hand words
must go in the first line, and the
left-hand words in the second
line. *See also* **BSI**

arabic numerals, numerals
used in ordinary computation,
as, 1, 2, 3 (not cap.). *See also*
numerals

arach., arachnology

arachnid, (zool.) member of the
Arachnida

arachnoid, (bot.) having long
hairs, (anat.) one of meninges

ARAD, Associate of the Royal
Academy of Dancing

Aragon, *not* Arr-; in Sp. **Aragón**

Araldite, propr. term (cap.)

ARAM, Associate of the Royal
Academy of Music

Aramaic, Semitic language;
abbr. **Aram.**

Aran, Island of, Donegal,
Ireland. *See also* **Arran**

Aran Islands, Galway Bay,
Ireland. *See also* **Arran**

ARAS, Associate of Royal Astro
nomical Society

ARB, Air Registration Board,
Air Research Bureau

ARBA, Associate of the Royal
Society of British Artists

arbit/er, -rary, -ration, abbr.
arb.

arbiter elegantiae (Lat.), a
judge of taste

arbitr/ament, *not* -ement

arbitrator, legal or official word
for **arbiter**

arbor, spindle, axis

arboret/um (bot.), tree-
-garden; *pl.* **-a**

arboriculture, abbr. **arbor.**

arbor vitae, evergreen

arbour, bower

ARBS, Associate of the Royal
Society of British Sculptors

ARC, Aeronautical Research
Council, Agricultural Research
Council, Architects' Registra-
tion Council

arc, sign ∩

ARCA, Associate of the Royal
College of Art

Arcades ambo (Lat.), two with
like tastes

arcan/um, *pl.* **-a**

Arc de Triomphe, Paris

arced (elec.), *not* arck-

arc-en-ciel (Fr. m.), rainbow;
pl. **arcs- — - —**

arch., archa/ic, -ism, archery,
archipelago, architect, -ural,
-ure

archaeolog/y, -ical; abbr.
archaeol.

archangel (one word)

Archangel, USSR; in Russ.
Arkhangel′sk

Archbishop, abbr. **Abp.**

Archd., Archdeacon, Archduke

archetype, *not* archi-

archidiaconal, *not* archide-

archiepiscopal

Archimed/ean, *not* -ian

archipelago/, *pl.* **-s;** abbr.
arch.

architect/, -ural, -ure; abbr.
arch., archit., archt.

architype, *use* **archetype**

arcing (elec.), *not* arck-

ARCM, Associate of the Royal College of Music

ARCO, Associate of the Royal College of Organists

ARCS, Associate of the Royal College of Science

arctic (not cap.), cold

Arctic Circle (two caps.); **Arctic regions** (one cap.)

Ardleigh, Essex

Ardley, Oxon.

ARE, Associate of the Royal Society of Painter-Etchers and Engravers, Arab Republic of Egypt

are, unit of square measure, 100 square metres

aren't, to be printed close up

areol/a, a small area; *pl.* -ae (not ital.)

Arequipa, Peru

arête, mountain ridge

argent (her.), silver, abbr. **arg.**

Argentina, the country, *or* **Argentine Republic**

Argentine (adj.); *also* (noun) the inhabitant. *See also* **Argentinian**

Argentinian (adj.); *also* (noun) the inhabitant; tending to replace **Argentine**; *not* -ean

argentum, silver, symbol (chem.) **Ag,** (numismatics) Æ

argon, symbol **Ar**

argot, slang (not ital.)

arguable, *not* -eable

argumentum ad/ crumenam (Lat.), argument to the purse; —— *hoc,* — for this (purpose); —— *hominem,* — to the man's interests; —— *ignorantiam,* — based on the adversary's ignorance; —— *invidiam,* — to men's hatreds or prejudices; —— *rem,* — to the purpose; —— *verecundiam,* appeal to modesty; *argumentum baculinum,* or — *ad baculum,* — of the stick, club-law; *argumentum e silentio,* — from silence

argy-bargy, dispute, wrangle, *not* argle-bargle (hyphen)

Argyle, Minnesota, US

Argyll/ and Bute District; — and Sutherland Highlanders; — and the Isles (Bp. of); — (Duke of)

Argyllshire, (former county of) Scotland

Arian (theol.), a follower of Arius; (one) born under Aries

ARIBA, Associate of the Royal Institute of British Architects. *See also* **RIBA**

ARICS, Professional Associate of the Royal Institution of Chartered Surveyors

Ariège, dép. France

Aristotel/ian, *not* -ean

Arius, *c.*250–*c.*336, Christian heretic

Ariz., Arizona (off. abbr.)

Ark., Arkansas (off. abbr.)

Arlay, dép. Jura, France

Arle, Glos.

Arles, dép. Bouches-du-Rhône, France

Arm., Armenian, Armoric

armadillo/, *pl.* -s

armchair (one word)

Armenia, ancient country, now in part a Soviet Socialist Republic; adj. **Armenian** (geog., ethnic, or eccl.). *See also* **USSR**

armes blanches (Fr.), side--arms (bayonet, sabre, sword)

armful/, *pl.* -s

armhole (one word)

Arminians, followers of Arminius

Arminius (Jacobus), Latinized form of **Jacob Harmen,** 1560–1609, Dutch theologian

armory, heraldry

armoury, collection of arms

armpit (one word)

ARMS, Associate of the Royal Society of Miniature Painters

arm's length (two words)

Army, *see* Navy

Army/ Dental Corps, *now* **RADC; — Nursing Service, ANS; — Order, AO; — Ord-**

Army (*cont.*)
nance Corps, *now* **RAOC**; — **Service Corps**, *later* **RASC**, *now* **RCT**; — **Veterinary Department**, **AVD**, *now* **RAVC**

Arnold (Edward) (Publishers), Ltd.

Arnold (E. J.) & Son, Ltd., publishers

arnotto, *use* **annatto**

Arola, Piedmont, Italy

Arolla, Switzerland

Arolo, Lombardy, Italy

Aroostook War, 1842, settled border of Maine and New Brunswick

ARP, air-raid precautions

arpeggio/ (mus.), striking of notes of chord in (usu. upward) succession; *pl.* **-s**

ARPS, Associate of the Royal Photographic Society

arquebus, *use* **harq-**

ARR, *anno regni Regis* or *Reginae* (in the year of the King's or Queen's reign) (s. caps.)

arr., arranged, arriv/e, -ed, -es, -als

Arragon, *use* **Aragon**

Arran (Earl of)

Arran (Isle of), Scotland. *See also* **Aran**

arrêt (Fr. m.), decree

Arrhenius (Svante August), 1859–1927, Swedish scientist

arrière/-garde (Fr. f.), rearguard; — **-pensée/**, a mental reservation, *pl.* **-s**

arrivis/me (Fr. m.), ambitious behaviour; **-te**, one who behaves thus

arrondissement (Fr. m.), division of department

arrowhead (one word)

Arrows of the Chace, by Ruskin, *not* Chase

ARSA, Associate of the Royal Scottish Academy, ditto Royal Society of Arts

ARSCM, Associate of the Royal School of Church Music

arsenic, symbol **As**

ars est celare artem (Lat.), the art is to conceal art

ARSL, Associate of the Royal Society of Literature

ARSM, Associate of the Royal School of Mines

ARSS, *Antiquariorum Regiae Societatis Socius*, Fellow of the Royal Society of Antiquaries

ARSW, Associate of the Royal Scottish Society of Painting in Water Colours

art., article, artillery, artist

art deco, decorative art-style of 1920s

artefact, *not* arti-

arteriosclerosis, hardening of the arteries (one word)

arthropod/, member of Arthropoda, animals with jointed body and limbs; *pl.* **-s**

article, abbr. **art.**

articles of roup (Sc. law), conditions of sale (by auction)

articles (titles of), (typ.), when cited, to be roman quoted

artifact, *use* **artefact**

artillery, abbr. **A.** *or* **art.**

artisan, *not* -zan

artiste (either sex), professional singer, dancer, or other performer

artizan, *use* **artisan**

art nouveau (Fr.), 19th-c. ornamental art-style

art paper, a high-quality coated paper used for illustrations

artwork (one word); abbr. **a/w**

Arundel, W. Sussex

Arundell of Wardour (Baron), title now extinct

Arva, Cavan, *not* Arvagh

ARWS, Associate of the Royal Society of Painters in Water Colours

Aryan, Indo-European, *not* -ian

AS, Anglo-Saxon

As, arsenic (no point)

As., Asia, -n, -tic

ASA, Amateur Swimming Association; Associate Member, Society of Actuaries

asafoetida (med.), ill-smelling gum resin; *not* the many variants

Asante, mod. form of **Ashanti,** q.v.

a.s.a.p., as soon as possible

Asbjörnsen (Peter Christian), 1812–85, Norwegian writer

ascendan/ce, *not* -ence; **-cy, -t**

ascender (typ.), top part of letters such as b, d, f, h, k, l

Ascension/ Day (caps., two words); — **Island,** S. Atlantic

ascetic, austere

Ascham (Roger), 1515–68, English writer

ascites (*sing.* and *pl.*), abdominal dropsy

Asclepiad, metre, *not* Ask-

Asclepius, the Greek god, *not* Asklepios, except in Gr. contexts. *See also* **Aesculapius**

a/s de (Fr.), *aux soins de*, c/o

ASDIC, Anti-Submarine Detection Investigation Committee; used for a form of hydrophone (**Asdic,** now officially **Sonar,** q.v.)

ASE, Amalgamated Society of Engineers

ASEAN, Association of South East Asian Nations

as follows: (*use* colon only, *not* : —)

Asgard, the heaven of Norse mythology

ash, Old-English letter æ, also used in phonetic script, Danish, and Norwegian

Ashant/i, Ghana; *not* -ee; *also* **Asante**

Ashby de la Zouch, Leics. (no hyphens)

Ashkenazim, *pl.*, East-European Jews, as distinct from Sephardim

Ashkhabad, cap. of Turkmenistan, USSR

ashlar (archit.), *not* -er

Ashmolean Museum, Oxford

Ashtar/oth, -eth, Bib. and Sem., *otherwise use* **Astarte**

Ashton/-in-Makerfield, —-under-Lyne, Gr. Manchester (hyphens), *not* -Lyme; **Ashton upon Mersey,** Gr. Manchester

ashtray (one word)

Ash Wednesday, first day of Lent (two words)

Asian (adj. and noun), (native) of Asia; *not* Asiatic, which is now freq. pejorative

asinin/e, like an ass, *not* ass-; noun **-ity**

Asir ('the inaccessible'), *see* **Saudi Arabia**

Asklepiad, *use* **Asclepiad; Asklepios,** *see* **Asclepius**

Aslef, Associated Society of Locomotive Engineers and Firemen

Aslib, Association of Special Libraries and Information Bureaux

ASM, air-to-surface missile

Asnières, Paris suburb

Asola, Lombardy, Italy

Asolo, Venetia, Italy

asper, Greek rough breathing (')

asperges (noun sing.), sprinkling with blessed water

asphalt, *not* ash-, -e

asphodel, immortal flower in Elysium

asphyxia, interruption of breathing

aspic, a poisonous serpent, a savoury jelly

ASR (comp.), answer send and receive

ass., assistant

assafoetida, *use* **asa-**

assagai, *use* **assegai**

assai (mus.), very

assailant

assassin

assegai, spear, *not* assa-

Assemblies, National and Federal. In addition to titles in English (Parliament, Assembly, Senate, Diet, etc.) in English-speaking countries and translation of foreign titles (National

Assemblies (*cont.*)
Assembly for Assemblée
Nationale, etc.), the following
titles are correct for particular
countries:
Bulgaria—Subranie.
Denmark—Folketing.
Finland—Eduskunta.
Germany, Federal Republic
of—Bundesrat (Upper House)
and Bundestag (Lower House).
Germany, Democratic Re-
public of—Volkskammer
(not recognized by Western
countries).
Iceland—Althing.
India—Rajya Sabha (Council
of States) and Lok Sabha
(House of the People).
Iran—Senate and Majlis.
Ireland, Republic of—Seanad
Éireann (Senate) and Dáil
Éireann (House of Represen-
tatives)
Israel—Knesset.
Netherlands—Staten Generaal,
comprising Eerste Kamer
(First Chamber) and Tweede
Kamer (Second Chamber).
Norway—Storting, comprising
Lagting (Upper Council) and
Odelsting (Lower Council).
Poland—Sejm.
Spain—Cortes.
Sweden—Riksdag (or Diet).
Switzerland—Nationalrat/
Conseil National and Stän-
derat/Conseil des États
Assembly (Church of Scotland)
(cap. A), *properly* **General As-
sembly** (also in other Presby-
terian Churches). *See also*
Church Assembly
assent/er, one who assents; **-or,**
one who subscribes to a nomin-
ation paper
assert/er, one who asserts; **-or,**
an advocate
assess/able, -or
ASSET, Association of Supervisory Staffs, Executives, and
Technicians
assign/ee, one to whom a right

or property is assigned; **-or,** one
who assigns
Assiniboine, Canada, *not*
Assinn-
Assisi, Italy
assistant, abbr. **ass.** *or* **asst.**
assizer, officer with oversight of
weights and measures; *not* -ser,
-sor, -zor
assoc., associat/e, -ion
ASSR, Autonomous Soviet
Socialist Republic
asst., assistant
Assuan, *use* **Aswan**
assurance, insurance of *life* (the
technical term). *See also* **life
insurance**
assymmetry, *use* **asymmetry**
Assyr., Assyrian
a-starboard (hyphen)
Astarte, Syrophoenician goddess; *not* Ashtar/oth, -eth
astatine, symbol **At**
Asterabad, Iran
asterisk, *see* **reference marks**
Asti, Italian white wine
ASTMS, Association of Scientific, Technical, and Managerial
Staffs
astr., astronom/y, -er
astrakhan, lambskin from
Astrakhan, USSR
astrol., astrolog/y, -er
Astronomer Royal (caps., no
hyphen)
astronomy, abbr. **astr.; —,
planetary signs:** Sun, ☉;
Moon, new, ●; Moon, first
quarter, ☽; Moon, full, ○;
Moon, last quarter, ☾; Mercury,
☿; Venus, ♀; Earth, ⊕; Mars,
♂; Jupiter, ♃; Saturn, ♄; Uranus, ♅; Neptune, ♆; Pluto, ♇;
asteroids in order of discovery,
①, ②, ③, etc.; fixed star, ✶ *or*
✳; conjunction, ☌; opposition,
☍; ascending node, ☊; descending node, ☋. **—, zodiacal
signs:** Aries, ♈; Taurus, ♉;
Gemini, ♊; Cancer, ♋; Leo, ♌;
Virgo, ♍; Libra, ♎; Scorpio,
♏; Sagittarius, ♐; Capricornus,
♑; Aquarius, ♒; Pisces, ♓

astrophysic/s (one word); **-al, -ist**

Asturias, Spain, *not* The —

Asunción, Paraguay

ASVA, Associate of the Incorporated Society of Valuers and Auctioneers

Aswan, Egypt; *not* the many variants

asymmetry, *not* ass-

asymptote, line approaching but not meeting a curve

asyndeton, omission of conjunction

Asyut, Egypt

AT (Ger.), *Altes Testament* (the Old —)

At, astatine (no point)

at., atomic

Atahualpa, the last Inca

Atalanta (Gr. myth.)

atar, *use* **attar**

Atatürk (Kemal), 1881–1938, Turkish statesman

ATC, Air Traffic Control, Air Training Corps

Atchin, Sumatra, Indonesia, *use* **Ach-**

ATCL, Associate of Trinity College of Music, London

atelier, studio (not ital.)

a tempo (mus.), in time; indicates that, after a change in time, the previous time must be resumed

Athabasca, Canada, *not* -ka

Athanasi/us (St.), d. 373, Greek Father; adj. **-an**

atheling, *not* aeth-

Athenaeum, London club

Athol, Canada, New Zealand

Atholl (Duke of)

Atl., Atlantic

Atlanta, Ga., US

atmosphere, gaseous envelope around the earth, the pressure exerted by it; abbr. **atm.**

atoll, ring-shaped coral reef surrounding lagoon

atomize, *not* -ise

atonable, *not* -eable

atri/um, hall of Roman house, cavity in heart; *pl.* **-a** *or* **-ums**

Atropos (Gr. myth.), the Fate that cuts the thread of life

ATS, Auxiliary Territorial Service (*superseded by* **WRAC**)

attaché/, *pl.* **-s** (not ital.)

attar (as of roses), *not* atar, otto, ottar

Att.-Gen., Attorney-General

Attic salt, delicate wit (cap. A)

Attila, *c.* 406–43, king of the Huns

attitudinize, *not* -ise

Attlee (Clement Richard, Earl), 1883–1967, British politician, Prime Minister 1945–51

attn., for the attention of

atto-, (as prefix) 10^{-18}; abbr. **a**

attorn (law), to transfer

Attorney-General, abbr. **A.-G.** *or* **Att.-Gen.**

attractor, *not* -er

ATV, Associated Television

Atwood's machine (physics), *not* Att-

at. wt., atomic weight

AU, ångström unit, astronomical unit

Au, *aurum* (gold) (no point)

aubade (Fr. f.), dawn-piece

auberge (Fr. f.), an inn; **aubergiste** (m. or f.), innkeeper

aubergine, fruit of egg-plant

Aubigné, dép. Deux-Sèvres, France

Aubigny, dép. Nord, France

aubr/ietia, dwarf perennial, *not* -etia

AUC, *anno urbis conditae* (in the year from the building of the city of Rome in 753 BC) (s. caps.; no points *except* in Latin composition); also *ab urbe condita*, from the foundation of the city

Auchinleck (Field-Marshal Sir Claude John Eyre), 1884–1981, British soldier

Auchnashellach, Highland, *use* **Ach-**

au/ contraire (Fr.), on the contrary; **— courant de,** fully acquainted with

Audenarde, *see* Oudenarde

Audi, German make of car

audio/ frequency, — typist (two words), **-visual** adj. (hyphen)

Auditor-General, abbr. **Aud.-Gen.**

auditori/um, *pl.* **-ums**

Audubon (John James), 1785–1851, US ornithologist

AUEW/, Amalgamated Union of Engineering Workers, — **(TASS),** ditto (Technical and Supervisory Section)

au fait (Fr.), thoroughly conversant

Aufklärung (Ger. f.), enlightenment, esp. the eighteenth-century intellectual movement

Auflage/ (Ger. f.), edition, . impression; *pl.* **-n** (cap.); abbr. *Aufl.*; *unveränderte* —, reprint; *verbesserte und vermehrte* —, revised and enlarged edition

au fond (Fr.), at the bottom

auf Wiedersehen (Ger.), till we meet again

Aug., August

aug., augmentative

Augean (stables), filthy (cap.)

auger, tool for boring

aught, anything. *See also* **nau-au/ grand sérieux** (Fr.), in all seriousness; — *gratin,* browned with breadcrumbs or cheese

augur, Roman soothsayer; (verb) foresee, portend

August, abbr. **Aug.**

Augustan, of Augustus or an era compared to his

Augustine, Augustinian monk; **St.** —, 354–430, Bp. of Hippo; **St.** —, d. 604, first Abp. of Canterbury

Augustinian, monk of order following St. Augustine of Hippo

au jus (Fr.), in gravy

auk, diving bird, *not* awk

'Auld Lang Syne' (not ital.)

Auld Reekie, Old Smoky, that is, Edinburgh

Aumale (duc d')

aumbry, closed recess in wall of church, *not* ambry

Aumerle, Duke of (in Shakespeare's *Richard II*)

au/ mieux (Fr.), very intimate; — *naturel,* in its natural state, plainly cooked

aunty, *not* -ie

au/ pair (Fr.), at par, on mutual terms; **au pair girl** (not ital.); — *pied de la lettre,* literally

Aurangzeb, 1618–1707, Mogul emperor, *not* the many variants

aurar, *pl.* of **eyrir,** q.v.

aurea mediocritas (Lat.), the golden mean

aureola, a saint's halo

au/ reste (Fr.), besides; — *revoir,* till we meet again

Aurigny, Fr. for **Alderney**

auror/a austral/is, *pl.* **-ae -es;** — **boreal/is,** *pl.* **-ae -es**

aurum, gold, symbol (chem.)

Au, (numismatics) *N* (no point)

Aus., Austria, -n

Auschwitz, German concentration camp in Poland in Second World War

au sérieux (Fr.), seriously

Ausgabe (Ger. f.), edition; abbr. **Ausg.** (cap.). *See also compounds* **Buch-, Pracht-, Volks-**

Austen (Jane), 1775–1817, English writer

Austin, English form of the name Augustine; — **(Alfred),** 1835–1913, Poet Laureate 1896–1913; — **(John Langshaw),** 1911–60, English philosopher; —, Tex., US

Austral/ia, -ian, abbr. **Austral.; -asia, -asian**

Australian Capital Territory, abbr. **ACT**

Austrasia, Frankish Kingdom of 6th–8th cc.

Austria, abbr. **Aus.**

Austria-Hungary (*but* **Austro-Hungarian**); in Ger. **Österreich-Ungarn**

autarchy, absolute sovereignty

autarky, self-sufficiency

Auteuil, Paris suburb
auteur (Fr. m.), author
auth., authentic, author, -ess,
-ity, -ized
authorities at the end of
quotations, or in notes:
I. In scientific and technical
work use the **Harvard system,** giving:
  1. Name of author in roman.
  2. Date of publication.
  3. *a, b, c,* etc. close up after
date to distinguish between
works of same author and date.
  4. A full alphabetical list in
the style of II below at the end
of the work, or of each chapter.
II. In general setting give:
  1. Name of author in roman.
  2. Name of book in italic, and,
if necessary, the series title in
roman.
  3. Title of article roman in
quotation marks, name of journal in italic.
For detailed examples see *Hart's
Rules,* pp. 50–4.

Act, scene, and line   III. iii. 45
Book   iii
Book and line (long poems)
                                             iii. 25
Canto and line   xvi. 25
Chapter and page   xiv. 25
Chapter and verse (biblical)
                                             2: 34
  (formerly   ii. 34)
Chapter, section, and paragraph (some scientific and
  philosophical works) 22.2.25
Paragraph   ¶68
Section   §5
Volume, chapter, and page
                                    IV. vi. 97
Quote the minimum number
of divisions consistent with clarity; the smallest in arabic numerals, the next in lower case
roman, and so on
authorize, *not* -ise
Authorized Version (of Bible)
(caps.); abbr. **AV**
autis/m, morbid absorption in
fantasy, or mental condition

obstructing response to environment; adj. **-tic**
autobahn/, German arterial
road (not cap., not ital. in
Anglicized form); *pl.* **-s**
autochthon/, original inhabitant; *pl.* **-s**; adj. **-ous**
autocracy, absolute government
auto/-da-fé (Port.), *pl. autos-
-da-fé*; (Sp.) — *de fe*; 'act of
the faith', burning of heretic by
Inquisition
autogiro, *not* -gyro
autolithography (typ.), printing by lithography from stones
or plates prepared by the artist
personally
automaton/, *pl.* **-s** (but **auto-
mata** when used collectively)
autonomy, self-government
autonym, a book published
under author's real name
autore (It.), author; abbr. *aut.*
auto/pista, Spanish arterial
road; *-route,* French ditto
autostrad/a, Italian arterial
road; *pl.* **-e**
autres temps, autres mœurs
(Fr.), other times, other manners
autumn (not cap.)
Auvergne (Fr.), *not* The —
Auvers, dép. Seine-et-Oise,
France
Auverse, dép. Maine-et-Loire,
France
auxiliary, abbr. **aux.** *or* **auxil.**
Auxiliary Forces (caps.)
N, in numismatics = gold (Lat.
aurum) (no point)
AV, Authorized Version (of
Bible)
a/v, ad valorem
av., average
avant/-courier, in Fr. m. *-cour-
rier or -cour/eur, fem. -euse,*
a forerunner; —*-propos* (m.),
preface, *pl.* same (hyphens)
avant-garde, the advanced
guard, the innovators
av. C. (It.), *avanti Cristo* (BC)
avdp., avoirdupois
Ave., Avenue

ave atque vale (Lat.), hail and farewell

Ave Maria, (Hail, Mary!); abbr. **AM**

Avenue, abbr. **Ave.**

average, abbr. **av.**

Averroës, 1126–98, Muslim philosopher, *not* -oes

avertible, *not* -able

avertissement (Fr. m.), notice, warning

aviculture, rearing of birds

Avignon, dép. Vaucluse, France

a vinculo matrimonii (Lat.), full (divorce)

avis au lecteur (Fr.), notice to the reader

avizandum (Sc. law), judge's consideration of case in private

av. J.-C. (Fr.), BC

AVM, Air Vice-Marshal

avocad/o, *pl.* **-os**

avocet, a bird, *not* -set

Avogadro (Amedeo), 1776–1856, Italian scientist, *not* Avro-

avoirdupois, abbr. **avdp.**

à volonté (Fr.), at pleasure, at will; — **votre santé!** (here's) to your health!

Avon/, county of England, and name of several English rivers; — **(Earl of)**

avril (Fr. m., not cap.), April

a/w, (typ.) artwork

aweing, *not* awi-

awesome, *not* aws-

awhile (adv.), *but* for a while

awk, *use* **auk**

AWOL, absent without official leave

AWRE, Atomic Weapons Research Establishment

ax., axiom

axe, *not* ax

axel, jumping movement in skating

ax/is, *pl.* **-es;** (cap., hist.) the German alliance formed in 1939

ay/, yes: 'Ay, ay, sir'; *pl.* **-es:** 'The ayes have it'

ayah, Indian nursemaid

ayatollah, Iranian religious leader (cap. when used as title)

Aycliffe, Co. Durham, 'new town', 1947

aye, ever: 'For ever and for aye'

'ayin (Heb.), *see* **accents**

Aylesbury, Bucks. *See also* **Ail-**

Ayr, Strathclyde

Ayre (Point of), Isle of Man

AZ, Arizona (off. postal abbr.)

az., azure

Azerbaijan/, a Soviet Socialist Republic, *see* **USSR,** and province of NW Iran; **-ian,** the language

Azores Islands; in Port. **Açores**

Azov (Sea of), *not* -off, -of

Azrael (Muslim), the angel of death

Aztec, one of a people dominant in Mexico before the Spanish conquest; adj. **Aztec/, -ian**

B

B, bel, boron, (chess) bishop, (pencils) black (*see also* **BB, BBB**), (mus.) B flat in German system, the second in a series

B., Bachelor, Baron, bass, basso, Bay, Blessed (*Beat/us, -a*), British

b (phys.), barn; (compass), by

b., base, billion, book, born, (cricket) bowled, (cricket) bye, (meteor.) blue sky

BA, Bachelor of Arts, Booksellers Association, British Academy, — Airways (combining BEA and BOAC, 1973), — America, — Association

Ba, barium (no point)

BAA, British Airports Authority, — Astronomical Association

Baalbek, Syria

baas (Afrik.), master, boss

Ba'ath, Iraqi and Syrian political party

Babbitt, by Sinclair Lewis, 1922

Babcock & Wilcox, engineers

babiroussa, a wild hog, *not* baby-, -russa

babu, Indian title, *not* baboo

BAC, British Aerospace Corporation

baccalaureate, university degree of bachelor

baccarat, a card-game; in Fr. m. *baccara*

Bacchanalia/, festival of Bacchus, strictly pl. but used in Eng. as sing.; **-n** (not cap.), riotous or drunken

Bacchant/, male or female follower of Bacchus; **-e,** female only

Bach/, German musical family, more than fifty in number in seven generations, of whom the greatest was — (**Johann Sebastian**), 1685–1750 (of the fifth generation), composer of orchestral, instrumental, and choral music; also important

were his father's cousin, — (**Johann Christoph**), 1642–1703, composer; and of J. S. B.'s twenty children, — (**Wilhelm Friedemann**), 1710–84; — (**Carl Philipp Emanuel,** usu. referred to as **C. P. E.**), 1714–88, developer of sonata and symphony form; — (**Johann Christoph Friedrich**), 1732–95; — (**Johann Christian**), 1735–82, 'the English Bach'

Bacharach, small Rhineland town (wine-producing); — (**Burt**), b. 1929, US composer

Bachelier ès/ lettres (Fr.), Bachelor of Letters, abbr. *B. ès L.*; — — *sciences,* ditto Science, *B. ès S.* (no hyphens)

Bachelor/, abbr. **B.;** — **of Agriculture, B.Agr.** (US); — **Architecture, B.Arch.;** — — **Arts, BA;** — — **Canon and Civil Law, BUJ;** — — **Civil Engineering, BCE;** — **Civil Law, BCL;** — — **Commerce, B.Com.;** — — **Dental Surgery, BDS** *or* **B.Ch.D.;** — — **Divinity, BD;** — — **Education, B.Ed.;** — **Engineering, BE** *or* **B.Eng.** *or* **B.A.I. Dub.** (Dublin); — — **Law, BL:** — — **Laws, B LL;** — — **Letters, B.Litt.** *or* **Litt.B.,** in Fr. *B. ès L.;* — — **Medicine, BM** *or* **MB;** — — **Metallurgy, B.Met.;** — — **Mining Engineering, BME;** — — **Music, B.Mus.** *or* **Mus. B.;** — — **Obstetrics, BAO;** — — **Philosophy, B.Phil.;** — — **Science, B.Sc.,** *or* **BS** (US), in Fr. *B. ès S.;* — — **Surgery, BC, B.Ch., BS,** *or* **Ch.B.;** — — **Technical Science, B.Sc. Tech.;** — — **Theology, B.Th.**

bacill/us, *pl.* **-i** (not ital.)

back (typ.), *see* **margins**

27

back-bench(er) (Parl.) (hyphen)

Backhuysen (Ludolf), 1631–1708, Dutch painter, *not* Bakhuisen

backing up (typ.), printing on the second side

backwoodsman (one word)

Bacon (Sir Francis) (often incorrectly Lord), 1561–1626, Baron Verulam and Viscount St. Albans

Baconian, pertaining to Roger Bacon, died *c.* 1292; or to Sir Francis Bacon; or to the theory that Sir Francis Bacon wrote Shakespeare's plays

bacshish, *use* **baksheesh**

bacteri/um, microscopic organism; *pl.* **-a; -cide**

bade, *see* **bid**

Baden-Baden, W. Germany (hyphen)

Baden-Powell (Robert Stephenson Smyth, Baron), 1857–1941, founder of Boy Scouts, 1908 (hyphen)

badinage (Fr. m.), humorous ridicule

Baedeker (Karl), 1801–59, guidebook publisher

Baffin Bay, NE America, *not* -ns, -n's

BAFU, Bakers', Food, and Allied Workers' Union

bagarre (Fr. m.), brawl

Bagehot (Walter), 1826–77, English writer

Baghdad, Iraq, *not* Bagd-

Bagnères/ de Bigorre, dép. Hautes-Pyrénées; — **de Luchon,** dép. Haute-Garonne, France

bagnio/, bathing-house, oriental prison, brothel; *pl.* **-s**

Bagnoles, dép. Orne, France

Bagnols, dép. Gard, France

bagpip/e(s), -er (one word)

B.Agr. (US), Bachelor of Agriculture

Baha'/i, religion initiated by Bahá'u'lláh (1817–92); **-ism**

Baham/as, W. Indies, indep. 1973; adj. **-ian**

Bahrain, islands, Persian Gulf, *not* -ein

Baiae, Naples

B.A.I. Dub., Bachelor of Engineering, Univ. of Dublin

baignoire (Fr. f.), theatre box at stalls level

Baikal (Lake), USSR

bail, (noun) security for appearance of prisoner, cross-piece over stumps in cricket, (hist.) outer line of fortifications; (verb) to deliver goods in trust, (naut.) to scoop water out of; to secure release on bail. *See also* **bale**

bailee, one to whom goods are entrusted for a purpose

bailer (naut.), one who bails water out, scoop used for bailing. *See also* **bailor, bale**

bailey, outer wall of castle; **Old Bailey,** London's Central Criminal Court

bailie (Sc.), an alderman; **bailiery,** jurisdiction of a bailie, *not* -iary

bailiwick, office or jurisdiction of a bailiff

Baillie (Joanna), 1762–1851, Scottish poet

Baillière, Tindall, Ltd., publishers

bailor (law), one who entrusts goods to a bailee. *See also* **bailer**

Baily's/ Magazine; — Directory

bain-marie, double saucepan; *pl.* **bains-**

Baireuth, Bavaria, *use* **Bayreuth**

Bairut, Lebanon, *use* **Beirut**

baklava, Turkish pastry with nuts and honey

baksheesh (Arab., Turk.), a gratuity, *not* the many variants

bal., balance

Balaclava, *not* -klava

balalaika, Russian stringed instrument

balanceable, *not* -cable

Balbriggan, Co. Dublin; also knitted cotton goods

bale/, -s, abbr. **bl.** or **bls.**

bale, (noun) a package, (arch.) destruction, woe; (verb) to make into a package; — **out,** to escape from an aircraft by parachute. See also **bail**

Bâle, Fr. for **Basle**

Balfour/ (Baron, of Burleigh); — (Baron, of Inchrye); — (Earl of). (Three distinct titles)

Bali/, Indonesia; adj. **-nese**

Baliol, an Anglo-Norman family, not the Oxford College, **Balliol**

balk (verb), not baulk

Balkan Mts., Bulgaria

Balkhan Mts., Transcaspia

Ballaarat, Victoria, Australia, use **-larat**

ballade, medieval French poem, also its imitation

Ballantine (James), 1808–77, artist and poet; — **(William),** 1812–86, serjeant-at-law

Ballantyne (James), 1772–1833, and — **(John),** 1774–1821, Sir W. Scott's printers and publishers; — **Press,** founded by them in 1796; — **(Robert Michael),** 1825–94, Scottish writer for boys

Ballarat, Victoria, Australia, not -aarat

ballet-dancer (hyphen)

ball game (two words)

Balliol College, Oxford. See also **Baliol**

ballistic/ (adj.); **-s,** pl., the science (two ls)

ballon d'essai (Fr. m.), a 'feeler' of any kind

ballot/, -ed, -ing

ballroom (one word)

Ballsbridge, Co. Dublin (one word)

ballyhoo, vulgar publicity

baloney, use **boloney**

BALPA, British Air Line Pilots' Association

Baluchi, a native of Baluchistan, not Be-, Bi-

baluster, upright supporting a rail, commonly of stone and outdoors, the whole structure being a **balustrade.** See **banister**

Balzac (Honoré de), 1799–1850, French writer; — **(Jean Louis Guez de),** 1596–1654, French writer

Bamberg, W. Germany, not -burg

Bamburgh, Northumberland

banal, commonplace

Banaras, Hindi for **Benares**

Band (Ger. m.), a volume, pl. **Bände;** abbr. **Bd.,** pl. **Bde.**

bandanna, a handkerchief, not -ana

b. & b., bed and breakfast

B. & F.B.S., British and Foreign Bible Society

bandit/, pl. **-s**

bandoleer, a belt for cartridges, not -olier, -alier

bandolero, Spanish brigand

Bandung, Indonesia, not Bandoeng

bang, Indian hemp, use **bh-**

Bangalore, Mysore, India

Bangkok, Thailand, not Bankok

Bangladesh, 'Land of the Bengalis', formerly E. Pakistan; indep. 1972

banister, upright supporting a rail; commonly of wood, indoors, on staircase, the whole structure being **banisters** (pl.). See **baluster**

banjo/, pl. **-s**

bank, abbr. **bk.**

banking, abbr. **bkg.**

banknote (one word)

Bankok, use **Bangkok**

bank paper, a thin, strong paper

banlieue (Fr. f.), precinct, suburb

banneret (law), a knight made on the field of battle

bannerette, small banner

banns, not bans

banquet/, -ed, -ing (one t)

banquette, a raised way behind firing-rampart; seat

Bantu, designating a group of Central, E., and S. African native peoples and languages

BAO, Bachelor of Art of Obstetrics

BAOR, British Army of the Rhine

bap., baptized

Baptist, abbr. **Bapt.**

baptistery, *not* -try

baptize, *not* -ise

Bar, called to the (cap.); **bar,** unit of pressure, 1/10 megapascal

bar., barley-corn, baromet/er, -ric, barrel

Barbad/os, *not* -oes, indep. 1966, abbr. **Barb.,** adj. **-ian**

barbarize, *not* -ise

Barbary, N. Africa

barbecue, open-air roasting (party) (*not* -que)

barberry, shrub of genus *Berberis, not* ber-, -ery

barbet, tropical bird

barbette, gun platform

Barbirolli (Sir John), 1899–1970, English conductor

barcarole, Venetian gondolier's song; in Fr. *barcarolle*

B.Arch., Bachelor of Architecture

Barclays Bank, Ltd. (no apos.)

Bareilly, Uttar Pradesh, India, *not* -eli

Barenboim (Daniel), b. 1942, Argentinian-born Israeli musician

Barents Sea, north of Norway and USSR, *not* -'s, -'z

bargain/er, a haggler; **-or** (law), the seller

Bar Harbor, Maine, US

Baring-Gould (Revd Sabine), 1834–1924, English author (hyphen)

baritone (mus.), *not* bary-

barium, symbol **Ba**

bark, vessel (arch. and poet.); in techn. sense *use* **barque**

Barkston, Lincs., N. Yorks.

barleycorn (one word), abbr. **bar.**

bar/maid, -man (one word)

Barmecide, one who offers illusory benefits, *not* -acide

bar mitzvah, Jewish initiation rite

Barmston, Humberside

barn (phys.), unit of area, 10^{-28} m², abbr. **b**

Barnaby Bright, St. Barnabas' Day, 11 June; longest day, Old Style

Barnard (Dr Christiaan), b. 1922, S. African surgeon

Barnardo (Dr Thomas John), 1845–1905, British philanthropist

baro/graph, a recording barometer; **-gram,** the record; **-logy,** science of weight; **-meter, -metric,** abbr. **bar.**

baron, member of lowest order of British nobility; cap. when with name (*but* **Lord** more common); l.c. with foreign name (but in Ger. cap.); abbr. **B.**

baron/ (Fr.), Baron; *fem.* **-ne** (not cap.)

baron and feme (law), husband and wife regarded as one

baronet/, member of lowest hereditary titled British order; cap. when with name; abbr. **Bt.; baronetage,** baronets collectively, book about them; **baronetcy,** patent or rank of baronet

Barons Court, London (no apos.)

baroque, exuberant archit. style of seventeenth and eighteenth centuries, applied also to music and other arts (not ital., not cap.)

barouche, four-wheeled carriage, *not* baru-

barque, vessel with aftermost mast fore-and-aft rigged and remaining masts square-rigged, *not* bark

barquentine, three-masted vessel with foremast only square-rigged, *not* -antine, barke-

barr., barrels, barrister

barrac/uda, large W. Indian sea-fish; **-outa** only with reference to *Thyrsites atun* (Austral. and NZ)

barrator (law), malicious litigant, *not* -ater, -etor

barratry (law), vexatious litigation; (marine law) master's or crew's fraud or negligence

Barrault (**Jean-Louis**), b. 1910, French actor, director, producer

barrel/, -s, abbr. **bar., barr., bl., bls.**

barrel/led, -ling

barrico/, a small cask; *pl.* **-es**

Barrie (**Sir James Matthew**), 1860–1937, Scottish novelist and playwright

Barrie & Jenkins, Ltd., publishers

barrister, abbr. **barr.**

Barrow-in-Furness, Cumbria (hyphens)

Bart., Baronet, *use* **Bt.**

bartender (one word)

Barth (**Heinrich**), 1821–65, German explorer; — (**John**), b. 1930, US novelist; — (**Karl**), 1886–1968, Swiss theologian

Bartholomew Day, 24 Aug. (caps., two words, for London local terminology but **St. Bartholomew's Day** for the massacre, 1572)

Bartók (**Béla**), 1881–1945, Hungarian composer

Bartolommeo (**Fra**), 1475–1517, Italian painter

Bartolozzi (**Francesco**), 1727–1813, Italian engraver

Bart's, St. Bartholomew's Hospital, London

Baruch (Apocr.), not to be abbr.

baruche, *use* **barouche**

barytone, *use* **bari-** except for the Greek accent

bas bleu (Fr. m.), a 'bluestocking'

base, abbr. **b.**

baseball, US national game played by two teams of nine; ball used in it

Basel, Ger. for **Basle**

bashaw, *use* **pasha**

bashi-bazouk (hyphen), Turkish mercenary

Bashkirtseff (**Marie K.**), *properly* **Marya Bashkirtseva,** 1860–84, Russian diarist

BASIC (comp.), Beginners' All-Purpose Symbolic Instruction Code

Basic English (two caps.), 'debabelized' language proposed by C. K. Ogden, 1929

Basildon, Essex, 'new town', 1949

basinet, light steel headpiece, *not* bass-

bas/is, *pl.* **-es**

Baskerville (**John**), 1706–75, typefounder and printer; designer of Baskerville type

Basle, Switz., in Fr. **Bâle**; in Ger. **Basel**

Basra, Iraq, *not* -ah

bas-relief, *not* bass-

Bas-Rhin, dép. E. France (hyphen)

bass, a fish, *not* -e

bassinet, cradle, perambulator, *not* basi-, -ette, berceaunette

basso/ (mus.), abbr. **B.;** — **profundo** (It. mus.), lowest male voice

basso/-riliev/o (It.), bas-relief, *pl.* **-i -i,** Anglicized to **basso-relievo,** *pl.* **-os**

bastaard (S. Afr.), person of mixed white and coloured parentage, whether legitimate or not

bastardize, *not* -ise

bastard/ size (typ.), a non-standard size; — **title,** *see* **half-title**

Bastille (Fr. f.), Paris prison

bastille, a fortified tower, *not* -ile

bastinado/, *not* basto-; *pl.* **-s**

Basuto/, -land, S. Africa. *See* **Lesotho**

bateau/ (Fr. m.), a boat, *not batt-; pl.* **-x**

Bath: et Well:, sig. of Bp. of Bath and Wells (two colons)

Bath/ brick, — bun, — chair, — chap (*not* chop), **— Oliver, — stone** (cap. B) (two words)

baton, music conductor's stick

battalion, abbr. **battn.** *or* **bn.**

battels, account for provisions etc., at Oxford University

batter (typ.), a damaged letter

batterie de cuisine (Fr. f.), a set of cooking utensils

battery, abbr. **batt.** *or* **bty.**

battleaxe (one word)

battle-cry (hyphen)

battle/dress, -field, -ground, -ship (one word)

battue (Fr. f.), shooting

Batum, Georgia, USSR, *not* -oum

Baudelaire (Charles), 1821-67, French poet, *not* Beau-

Bauhaus, 20th-c. Ger. archit. school

baulk, of timber; area on billiard-table; as verb *use* **balk**

Bavaria/, -n, abbr. **Bav.;** in Ger. **Bayern**

bawbee (Sc.), halfpenny, *not* bau-

bayadère, Hindu dancing girl

Bayard (Pierre de Terrail, Chevalier de), 1475–1524, 'The Knight without fear and without reproach'

Bayern, Ger. for **Bavaria**

bayonet/, -ed, -ing; — fitting, plug (elec.), two words

bayou, tributary creek

Bayreuth, Bavaria, *not* Bai-. *See also* **Beirut**

bazaar

BB (pencils), double black; *also* **2B**

BBB (pencils), treble black; *also* **3B**

BBC, British Broadcasting Corporation

BC, Bachelor of Surgery, Board of Control, British Columbia (Canada)

BC, before Christ (s. caps.), should always be placed *after* the numerals

b.c. (meteor.), partly cloudy sky

BCA, Bureau of Current Affairs

BCC, British Copyright Council, — Council of Churches, — Crown Colony

BCD (comp.), binary coded decimal

BCE, Bachelor of Civil Engineering; before Christian (*or* Common) Era (*see* **ACE**)

B.Ch., B.Chir., Bachelor of Surgery

B.Ch.D., BDS, Bachelor of Dental Surgery

BCG, bacillus of Calmette and Guérin (anti-TB inoculation)

BCL, Bachelor of Civil Law

B.Com., B.Comm., B.Com.Sc., Bachelor of Commerce

BCP, Book of Common Prayer

BD, Bachelor of Divinity

Bd. (Ger.), *Band* (a volume)

bd., board, bond, bound

BDA, British Dental Association

Bde. (Ger.), *Bände* (volumes)

Bde. Maj., Brigade Major

bdg., binding

bdl., bdls., bundle, -s

Bdr., Bombardier

BDS, Bachelor of Dental Surgery, Bomb Disposal Squad

bds., (bound in) boards

BE, B.Eng., Bachelor of Engineering

Be, beryllium (no point)

b.e., bill of exchange

BEA, British Epilepsy Association, — Esperanto Association, — European Airways (*see* **BA**)

beach-la-mar, pidgin English of W. Pacific

bear (Stock Exchange), speculator for a fall

beard (typ.), the space between the bottom of the x-height (q.v.) and the upper edge of the shank

beargarden (one word)

béarnaise (Fr. cook.), rich white sauce

bearskin, military cap (one word)

beastings, *use* bee-

beasts of the chase (law), buck, doe, fox, marten, roc; ditto **forest** (law), boar, hare, hart, hind, wolf; ditto **warren** (law), cony and hare; ditto **of venery** (law) are 'beasts of the forest'

beatae memoriae (Lat.), of blessed memory; abbr. **BM**

Beata/ Maria, or — Virgo (Lat.), the Blessed Virgin; abbr. **BM** *or* **BV** (no points). *See also* **BMV, BVM**

Beatles (the), pop group, *fl.* 1960s

beatnik, member of 'beat' generation

beau (*pl.* **beaux**), not ital.

Beauclerc, sobriquet of Henry I

Beauclerk, fam. name of Duke of St. Albans

Beaudelaire, *use* **Bau-**

beau-fils (Fr. m.), son-in-law, stepson; without hyphen, beautiful son

Beaufort scale, of wind force (one cap.); *for details see under* **wind**

beau ideal, model of excellence (not ital.)

beauidealize, *not* -ise (one word)

Beaujolais, a burgundy

Beaulieu, Hants

beau monde, the fashionable world

Beaune, a burgundy

beau-père (Fr. m.), father-in-law, stepfather; without hyphen, beautiful father

beaux-arts (Fr. pl. m.), fine arts

beaux/ esprits (Fr. pl. m.), brilliant wits, *sing.* **bel esprit**; *— yeux,* good looks

beccafico/, small edible Italian bird, *not* beca-; *pl.* **-s**

béchamel (Fr. cook.), white sauce named after Béchamel, steward of Louis XIV

bêche-de-mer, sea-slug, a Chinese delicacy

Bechstein (Friedrich Wilhelm Carl), 1826–1900, German pianoforte maker; — (**Johann Matthäus**), 1757–1822, German naturalist

Bechuana/, -land, S. Africa. *See* **Botswana**

Becket (à), *see* **à Becket**

Beckett (Samuel), b. 1906, Irish writer

becquerel, unit of radioactivity, abbr. **Bq**

Becquerel, French family of physicists: — (**Antoine César**), 1788–1878, father of — (**Alexandre Edmond**), 1820–91, father of — (**Antoine Henri**), 1852–1908, discoverer of radioactive **- rays,** *now* **gamma rays**

bed (typ.), the part of the press on which the forme lies

B.Ed., Bachelor of Education

bedaw/een, -i, -in, *use* **bedouin**

bed/bug, -clothes (one word)

Beddgelert, Gwynedd, *not* Beth-

bedeguar, mosslike excrescence on roses

bedel, Oxford official form of beadle; at Cambridge and London, **bedell**

Bedfordshire, abbr. **Beds.**

bedouin, a desert Arab, *not* the many variants; *pl. same*

bed/owy, *use* **-ouin**

bed/pan, -post, -ridden, -room (one word)

Beds., Bedfordshire

bed/-sitter, -sitting-room (hyphens)

bed/sore, -time (one word)

bed/uin, *use* **-ouin**

beefsteak (one word)

beehive (one word)

beerhouse (law), where beer is sold to be drunk *on* or *off* the premises

beer-mat (hyphen)

beershop (law), where beer is sold to be drunk *off* the premises

beestings, first milk from a mammal, *not* bea-, bie-

beeswax (noun *and* verb)

Beethoven (**Ludwig van,** *not* von), 1770–1827, German composer; divide Beet-hoven

BEF, British Expeditionary Force

be/fall, *not* -fal; past **-fell**

Beggar's Opera, not -s'

behemoth, biblical animal (book of Job), thought to be the hippopotamus

behoof (noun), benefit

behove (verb), to be incumbent on

Behring (**Emil von**), 1854–1917, German bacteriologist

Behring Isle, Sea, and **Strait,** *use* **Bering** —

beignet (Fr. m.), a fritter

Beijing, Pinyin form of **Peking**

Beinn Bhuidhe, near Inveraray, Strathclyde. *See also* **Ben Buie**

Beirut, Lebanon, *not* Ba-, Bai-, Beyrout(h). *See also* **Bayreuth**

bel, unit of ten decibels (**decibel** is more often used); abbr. **B**

Belalp, Switz. (one word)

Bel and the Dragon (Apocr.), abbr. **Bel & Dr.**

beldam (arch.), a hag, *not* -e

Belém, Brazil, *formerly* **Pará**

bel/ esprit (Fr. m.), a brilliant wit, *pl.* **beaux esprits;** —
étage, the first floor, *not belle* —

Belgique (Fr. f.), Belgium;
belge, Belgian (not cap., unless used as a noun)

Belg/ium, -ian, -ic, abbr. **Belg.**

Belgrade, Yugoslavia

believable, *not* -eable

Belitung, Indonesia, *not* Billiton

Belize/, Cent. America, *formerly* British Honduras; adj. **-an**

Bell (**Acton, Currer, Ellis**), *see* **Brontë**

belladonna (bot.), deadly nightshade (one word)

belle (not ital.)

belle/ amie (Fr. f.), female friend; — *-de-jour* (bot.), convolvulus; — *-de-nuit* (bot.), marvel of Peru; — *époque,* before 1914; — *-fille,* daughter-in-law, stepdaughter; without hyphen, beautiful girl; — *laide,* fascinatingly ugly woman

Belle-Île-en-Mer, dép. Morbihan, France (hyphens)

Belle Isle (**Straits of**), between Newfoundland and Labrador

Belleisle, Co. Fermanagh

belle-mère (Fr. f.), mother-in-law, stepmother; without hyphen, beautiful mother

belles-lettres, literature, is pl. (hyphen)

belletrist/, one devoted to *belles-lettres*; adj. **-ic** (*not* -ettr-)

Bell Island, off NE coast of Newfoundland

Bell Rock, North Sea. *See also* **Inchcape Rock**

Bel/oochee, -uchi, *use* **Baluchi**

Belorussia, White Russia, a Soviet Socialist Republic, *not* Bye-. *See also* **USSR**

belvedere (It.), a raised building, *not* belvi- (not ital.)

Belvedere, Kent; Calif., US

Belvidere, Ill., US

Belvoir, castle, Leics.

BEM, British Empire Medal

Benares, India, off. **Varanasi;** in Hindi **Banaras**

Ben Buie, Mull, Strathclyde. *See also* **Beinn Bhuidhe**

bench-mark (hyphen)

Ben Day (typ.), mechanical method for producing shading and stippling effects

Benedic, canticle from Psalm 103

Benedicite, the 'Song of the Three Children', from Apocr., as canticle

Benedick, confirmed bachelor newly married, *not* -ict (from Shakespeare's *Much Ado about Nothing*)

Benedictus, the Canticle of

Zacharias (Luke 1 : 68 ff.), or a choral passage in the Mass, beginning *Benedictus qui venit* (Matt. 21 : 9)

benefactor, *not* -er

benefit/, -ed, -ing

Benelux, Belgium, the Netherlands, Luxemburg

Beneš (Eduard), 1884–1948, president of Czechoslovakia, 1935–8, 1939–48

Benet, English form of the name Benedict; used of St. Benedict and institutions bearing his name, e.g. St. Benet's Church, — — Hall

Bene't Street, Cambridge

Benét (Stephen Vincent), 1898–1943, and — **(William Rose),** 1886–1950, his brother, US poets

B.Eng., Bachelor of Engineering

Bengali/, person born in, language of, (adj.) of, Bengal, *not* -ee, *pl.* **-s;** abbr. **Beng.**

Benghazi, Libya, *not* -gasi

Ben-Gurion (David), 1886–1973, Israeli statesman

Ben/Lawers, Tayside; — **Macdhui,** Grampian

Bennet, family name of Earl of Tankerville; family in Austen's *Pride & Prejudice;* — **(John),** *c.*1600, English madrigalist

Bennett (Enoch Arnold), 1867–1931, English writer; — **(Richard Rodney),** b. 1936, and — **(William Sterndale),** 1816–75, English composers

benth/al, of ocean depths exceeding 6,000 feet; **-os,** the flora and fauna of the sea-bottom, as opp. to **plankton**

Bentham (George), 1800–84, botanist; — **(Jeremy),** 1748–1832, English polit. philosopher

ben trovato (It.), well invented

Ben Venue, Central (two words)

ben venuto (It.), welcome

Ben/ Vorlich, — **Vrackie,** Tayside

Benzedrine (med.), propr. term

for an inhalant or nerve stimulant, *not* -in (cap.)

benzene, a substance obtained from coal-tar

benzine, a spirit obtained from petroleum

benzoin, aromatic resin of Japanese tree

benzol, crude benzene, *not* -ole

benzoline, benzine

Beowulf, Anglo-Saxon epic

bequeather, a testator

Béranger (Pierre Jean de), 1780–1857, French poet

berberry, *use* **bar-**

berceau (Fr. m.), cradle

berceaunette, *use* **bassinet**

berceuse, a cradle song

Berg/ (Ger. m.), a mountain; *pl.* **-e** (cap.)

berg (Afrik.), a mountain

Bergamask, native of Bergamo, Lombardy; **bergamasque,** rustic dance attributed to them

bergamot, tree of citrus family; type of pear, *not* burg-

Bergerac, dép. Dordogne, France; wine made there; — **(Savinien Cyrano de),** 1619–55, French writer

Bergman (Ingmar), b. 1918, Swedish film and theatre director; — **(Ingrid),** 1917–82, Swedish actress

Bergson (Henri),1859–1941, French philosopher

beriberi, tropical disease, *not* -ria (one word)

Bering (Vitus Jonassen), 1680–1741, Danish navigator

Bering Isle, Sea, and **Strait,** *not* Beh-, -ings

Berkeleian, of the philosopher Berkeley

Berkeley, Calif., US; — **(George, Bishop),** 1685–1753, Irish philosopher; — **(Sir Lennox),** b. 1903, English composer

berkelium, symbol **Bk**

Berkhamsted, Herts.

Berks., Berkshire

Bernard (Claude), 1813–78, French physiologist

Bernardine, Cistercian, from St. Bernard of Clairvaux, 1091–1153

Berne, Swiss canton and town, in Ger. **Bern; — Convention** (1886 and later revisions) deals with copyright: there are now over sixty members (incl. UK but not US or USSR). Basically each member-nation extends benefit of its own copyright laws to works by citizens of other member-nations, and no registration of the book or claim to copyright is required. The author also has the 'moral right' during his lifetime to object to any alteration of his work, regardless of copyright-ownership. *See also* **Universal Copyright Convention**

Berners (Gerald Hugh Tyrwhitt-Wilson, Baron), 1883–1950, English composer; — **(John Bourchier, Baron),** 1467–1533, English translator; — **(Dame Juliana),** wrote *Boke of St. Albans*, first ed. 1486

Bernhardt (Rosine, *called* **'Sarah'),** 1845–1923, actress

Bernoulli, family of Swiss mathematicians, *not* -illi

bernouse, Arab cloak, *use* **burnous**

bersaglier/e (It.), a rifleman; *pl.* *-i*

Berthelot (Pierre Eugène Marcelin), 1827–1907, French chemist and statesman

Berthollet (comte Claude Louis), 1748–1822, French chemist

Berwickshire, former county of Scotland

beryllium, symbol **Be**

Berzelius (Jöns Jakob, baron), 1779–1848, Swedish chemist

B. ès A., *now B. ès L.*

Besançon, dép. Doubs, France

B. ès L. (Fr.), *Bachelier ès lettres* (Bachelor of Letters)

beso/las manos (Sp.), 'I kiss the hands' (frequently said or written); — *los pies,* 'I kiss the feet'

B. ès S. (Fr.), *Bachelier ès sciences* (Bachelor of Science)

Bessarabia, district in Moldavia and Ukraine, ceded by Romania to USSR, 1940

Bessbrook, Co. Armagh

Bessemer process, in steel mfr. (one cap.)

Besses o' th' Barn, Greater Manchester

bestialize, *not* -ise

beta/ particle, — **ray** (two words)

betatron, apparatus for accelerating electrons

betel, leaf of betel-pepper, chewed in East

betel-nut, misnomer for areca--nut, chewed with betel

Betelgeuse (astr.), red star in Orion, *not* -x

bête noire (Fr. f.), one's pet aversion, *not* — *noir; pl.* *-es -es*

bethel, Nonconformist chapel (*not* cap.)

bêtise (Fr. f.), stupidity

Betjeman (Sir John), b. 1906, English poet and writer on architecture; appointed Poet Laureate 1972

betony (bot.), *not* bett-

better, one who bets, *not* -or

Betws-y-coed, Gwynedd (hyphens)

Bevan (Aneurin), 1897–1960, Welsh politician. *See also* **Bevin**

bevel/, -led, -ling

Beveridge (William Henry, Baron), 1879–1963, English economist

Beverley, Humberside

Beverly, Mass., US

Beverly Hills, Calif., US

Bevin (Ernest), 1881–1951, English politician. *See also* **Bevan**

bevy, proper word for a company of ladies, larks, maidens, quails, or roes; also in gen. use

Bewick (Thomas), 1753–1828, English engraver

Bexleyheath, London (one word)

bey (Turk.), a governor; **Bey of Tunis,** formerly ruler of Tunisia (cap.)

beylic, jurisdiction of a bey, *not* -ik

Beyrout(h), Lebanon, *use* **Beirut**

bezant, gold coin first struck at Byzantium, *not* by-

b.f., bloody fool, brought forward, (typ.) bold face

BFBS, British and Foreign Bible Society

BFI, British Film Institute

BFPO, British Forces' Post Office

BGC, bank giro credit

Bhagavadgita, philosophical dialogue in the Hindu epic *Mahabharata*

B'ham, Birmingham

bhang, Indian hemp, *not* ba-

BHC, British High Commissioner

b.h.p., brake horsepower

BHS, British Home Stores

Bhutan/, protectorate of Rep. of India since 1949; adj. **-ese**

Bi, bismuth (no point)

Biafra, portion of Nigeria, seceded 1967–70

biannual, half-yearly, *but* **biennial,** two-yearly

bias/, -ed, -ing

Bib., Bible

bib., biblical

bibl., *bibliotheca* (library)

Bible, abbr. **Bib.** (cap., *but* three bibles = three copies of the Bible). *See also* **authorities, quotations**

Biblia Pauperum, medieval 'Bible of the poor', in pictures (caps.)

biblical (not cap.), abbr. **bib.**

bibliograph/er, -ic, -ical, -y; abbr. **bibliog.**

bibliopeg/y, bookbinding as a fine art; **-ist**

biblio/theca (Lat.), a library; *-thécaire* (Fr. m.f.), a librarian; *-thèque* (Fr. f.), a library

Bibliothek/ (Ger. f.), a library; *-ar* (Ger. m.), a librarian (caps.)

bicameral, with two legislative chambers

Bickleigh, Devon

Bickley, Cheshire, London, N. Yorks.

bid, past **bade,** partic. **bidden,** but **bid** is used for both in auction sense (I bid £10 for it; £10 was bid) and for partic. in some phrases, e.g. do as you are bid

Biedermeier, Ger. 19th-c. conventional artistic and literary style

biennial, two-yearly, *but* **biannual,** half-yearly

bienni/um, a two-year period; *pl.* **-ums**

bienséance (Fr. f.), propriety

bienvenu/ (Fr.), *fem.* **-e**, welcome; **-e** (f.), a welcome

BIF, British Industries Fair

biff/é (Fr.), *fem.* **-ée,** cancelled

Bigelow (**Erastus Brigham**), 1814–79, US inventor and industrialist; — (**John**), 1817–1911, US journalist and diplomat

Biglow Papers (*The*), anti-slavery poems, 1848 and 1857–61, by James Russell Lowell

bigot/, -ed

bijou/, a 'gem'; *pl.* **-x**

Bilbao, Spain

bilberry, *not* bill-

bilbo/, Spanish sword or rapier; *pl.* **-s**

bilboes, *pl.*, fetters

Bildungsroman (Ger. m.), novel of early life

bile-duct (hyphen)

bill (parliamentary) (cap. when part of title)

billet/, -ed, -ing

billet-doux, a love letter; *pl.* **billets-doux** (hyphens)

billiard/-ball, -cue, etc., not billiards- (hyphens)

billion, in Britain, France (since 1948), and Germany, a million millions; in USA, and increas-

billion (*cont.*)
ingly in British usage, a thousand millions; abbr. **bn.** *See also* **trillion**

Billiton, Indonesia, *use* **Belitung**

bill of/ exchange, abbr. **b.e.**; — — **lading,** abbr. **b.l.**; — — **sale,** abbr. **b.s.**

biltong (Afrik.), sun-dried meat

Biluchi, *use* **Bal-**

BIM, British Institute of Management

bimetall/ic, -ism, -ist

bimillenary (adj. and noun), (designating) a period of 2,000 years; a two-thousandth anniversary

bi-monthly, avoid as ambiguous; *use* (adj. or adv.) **semi-monthly** or (adv.) twice a month, every two months, as appropriate

bindery, a bookbinder's establishment

binnacle (naut.), a compass-stand

binocle, a field-glass

binocular/ (adj.), for two eyes; **-s** (pl. noun), field- or opera-glasses

Binstead, IW

Binsted, Hants, W. Sussex

Binyon (Laurence), 1869–1943, English poet

biochemist/, -ry (one word)

biodegradable (one word)

biograph/er, -ic, -ical, -y; abbr. **biog.**

biological nomenclature: see *Hart's Rules*, pp. 6–7

biolog/y, -ical; abbr. **biol.**

bio/mathematics, -physics (one word)

Bipont/, -ine, books printed at Bipontium (Zweibrücken), W. Germany

bird-bath (hyphen)

bird/cage, -seed (one word)

bird's-eye view (one hyphen)

bird/-song, -table (hyphens)

Birmingham, abbr. **Birm., B'ham;** — **Small Arms, BSA**

Birnam, Tayside, immortalized in Shakespeare's *Macbeth*

Birstall, Leics., W. Yorks.

birth certificate (two words)

birth/-control, -rate (hyphens)

birthmark (one word)

birthplace, abbr. **bpl.**

birthright (one word)

BIS, Bank for International Settlements

bis (Fr., It., Lat.), twice, encore
bis., bissextile

Biscayan, abbr. **Bisc.,** pertaining to Biscay, Spain; Basque

bis dat qui cito dat (Lat.), he gives twice who gives quickly

bise (Fr. f.), cold N. wind in S. France

BISF, British Iron and Steel Federation

Bishop, abbr. **Bp.,** (chess) **B**

Bishopbriggs, Strathclyde

Bishop *in partibus infidelium,* RC bishop taking title from ancient see situated 'in territory of unbelievers', now termed 'Titular Bishop'

Bishop Rock, lighthouse, Scilly Is.

Bishop's/ Castle, Shropshire; — **Cleeve,** Glos.; — **Frome,** Hereford & Worc.; — **Lydeard,** Som. (apos.)

bishop's signature, usually consists of Christian name and diocese (John Birmingham). For special spellings and punctuation of certain signatures refer to individual entries in this book

Bishop's Stortford, Herts. (apos.)

Bishopsteignton, Devon (one word)

Bishopston, Avon, W. Glam.

Bishopstone, Bucks., Hereford & Worc., Kent, E. Sussex, Wilts.

Bishop's Waltham, Hants (apos.)

Bishopthorpe, residence of Abp. of York (one word)

Bishopton, Co. Durham, N. Yorks., Strathclyde

bisk, *use* **bisque**

Bismarck (Karl Otto Eduard Leopold, Fürst von), 1815–98, German statesman, *not* -ark

bismillah (Arab.), 'in the name of God', *not* biz-

bismuth, symbol **Bi**

bisque, in tennis, croquet, golf; unglazed white porcelain; rich soup (not ital.)

bissextile (noun and adj.), Latin leap year, abbr. **bis.**

bistoury, surgeon's scalpel

bit (US), 12½ cents; (comp.) binary digit (*see also* **byte**)

bitts (naut., pl.), pair of posts on deck for fastening cables, etc.

bitumin/ize, -ous

bivouac/, -ked, -king

bi-weekly, avoid as ambiguous; *use* (adj. or adv.) **semi-weekly** or fortnightly, (adv.) twice a week or every two weeks, as appropriate

bizarre (not ital.)

Bizet (Georges), 1838–75, French composer

Bjørnson (Bjørnstjerne), 1832–1910, Norwegian writer

Bk, berkelium (no point)

bk., bank, book

bkg., banking

bkt., bkts., basket, -s

BL, Bachelor of Law, ditto Letters, ditto Literature, British Leyland, ditto Library

bl., bls., bale, -s, barrel, -s

b.l., bill of lading

Black, a Negro (cap.)

black (typ.), a space inked in

Blackburn:, sig. of Bp. of Blackburn (colon)

blackcock, *see* **black game**

Blackfoot (N. Amer. Indian), *pl.* **Blackfeet**

Blackfriars, London

black game (two words), black grouse: *male*, **blackcock** (one word), *female* **grey-hen** (hyphen)

black letter (typ.), general term

for Gothic or Old English (q.v.) type. *See also* **Fraktur**

black list (noun, two words)

blacklist (verb, one word)

Black Maria, prison van (two words, caps.)

Blackmoor, Hants

Blackmore, Essex, Shropshire; — **(Richard Doddridge),** 1825–1900, English writer

black-out (noun, hyphen)

black out (verb, two words)

Blackrock, Co. Cork, Cornwall, Co. Dublin, Gwent, Co. Louth

Blackrod, Gr. Manchester

Black Rod, abbr. for 'Gentleman Usher of the Black Rod'

blackshirt/, a Fascist; **-ed** (one word)

blaeberry, Sc. and N. Eng. for **bilberry**

Blaenau Ffestiniog, Gwynedd (no hyphen)

blague/ (Fr. f.), humbug; *-ur* (m.), hoaxer

Blairadam, Fife (one word)

Blair Atholl, Tayside (caps., no hyphen)

Blairgowrie, Tayside (one word)

Blairs College, Aberdeen (no apos.)

blameable, *see* **-able**

blancmange (cook.) (one word, not ital.)

Bland (Mrs), *see* **Nesbit**

blanket/, -ed, -ing

blanquette (Fr. cook. f.), white meat in white sauce

Blantyre, Strathclyde, Malawi

blas/é, bored with enjoyment (not ital.) ; *fem.* **-ée**

Blatherwycke, Northants, *not* -wyck, -wick

bldg., bldgs., building, -s

bled, bleed (typ.), an illustration going to the trimmed edge

bleed (bind.), to overcut the margins and mutilate the printing

blesbok (Afrik.), an antelope, *not* -buck

blind/ blocking, *or* — **tooling** (bind.), *see* **blocking**

blind ¶, paragraph mark with solid loop

B.Lit., Bachelor of Literature

B.Litt., *Baccalaureus Litterarum* (Bachelor of Letters)

blitzkrieg (Ger.), violent attack; Eng. colloq. **blitz**

blk. (typ.), block, -s

B LL, Bachelor of Laws

bloc (Fr.), combination of nations or parties

Bloch (Ernest), 1880–1959, Swiss-American composer

block (typ.), letterpress printing plate for an illustration, (brass) stamp to impress book--cover

blocking (bind.), impressing lettering or a design into the case of a book; **blind —,** without foil or colour; **gold —,** with gold foil

Block (Maurice), 1816–1901, French economist

Blok (Alexander Alexandrovich), 1880–1921, Russian poet

blond/, fair-complexioned; light-coloured (of hair); *fem.* **-e** (not ital.)

blood/ bank; — cell, — count, — group (two words)

blood/-pressure, -supply (hyphens)

blood sport (two words)

bloodstain/, -ed (one word)

bloodstream (one word)

blottesque, painting with blotted touches (not ital.)

blow, partic. **blown,** but **blowed** in sense 'cursed' (well I'm blowed)

blow-up (noun), an enlarged picture (hyphen)

blow up (verb), to enlarge (two words)

blowzy, coarse-looking, *not* blou-, -sy

Blubberhouses, N. Yorks. (one word)

Blücher (Gebhard Leberecht von), 1742–1819, German field marshal

bluebell (one word)

blue book, Parliamentary or Privy-Council report (two words)

blueing, *not* blui-

blueish, *use* **bluish**

bluejacket, RN sailor (one word)

blueprint (one word)

blue-printing, a photog. process (hyphen)

bluestocking, *but* **Blue Stocking Society**

bluey, bluelike

bluing, *use* **blueing**

bluish, *not* blue-

Blundellsands, Merseyside (one word)

blurb, eulogy or résumé of book, printed on jacket

Blut und Eisen (Ger.), blood and iron

Blyth, Northumberland

Blyth Bridge, Borders

Blythebridge, Staffs.

BM, Bachelor of Medicine, ditto Music; *Beata Maria* (the Blessed Virgin), *beatae memoriae* (of blessed memory); bench-mark; British Museum

BMA, British Medical Association

BME, Bachelor of Mining Engineering

B.Met., Bachelor of Metallurgy

BMJ, British Medical Journal

BMR, basal metabolic rate

B.Mus., Bachelor of Music

BMV, *Beata Maria Virgo* (Blessed Virgin Mary)

bn., battalion, billion

B'nai B'rith, sons of covenant, Jewish fraternity (two apos.)

BNB, British National Bibliography

BNC, Brasenose College, Oxford

BO, body odour

b.o., box-office, branch office, broker's order, brought over, buyer's option

BOA, British Optical Association, — Olympic Association

Boabdil, last Moorish king of Granada, 1482–92; d. *c.*1533

BOAC, British Overseas Airways Corporation (*see* **BA**)

Boadicea, use **Boudicca**

Boanerges, *sing.* or *pl.*, loud preacher(s)

boarding/-house, -school (hyphens)

Board of Trade, abbr. **B.o.T.** *or* **B. of T.,** absorbed in **Department of Trade and Industry,** 1970

boatswain, contr. **bos'n**

boat-yard (hyphen)

Bobadil, braggart, from Bobadill in Jonson's *Every Man in His Humour*

bobolink, N. American songbird

bob-white, American quail

Boccaccio (Giovanni), 1313–75, Italian writer (four *c*s)

Boccherini (Luigi), 1743–1805, Italian composer

Bodhisattva, in Buddhism and Hinduism, person able to reach nirvana

Bodicote, Oxon.

Bodleian Library, abbr. **Bod. Lib., Bodl., Bodley**

Bodoni (Giambattista), 1740–1813, Italian printer and punch-cutter; designer of Bodoni type

body (typ.), the measurement from front to back of the shank of a piece of type; in the Anglo--American system 12 pt. = 0.166 in. *See also* **Didot**

body-blow (hyphen)

bodyguard (one word)

body-weight (hyphen)

Boeotia/, district of cent. Greece; **-n,** stupid

Boerhaave (Herman), 1668–1738, Dutch naturalist

Boethius (Anicius), AD 470–525, Roman statesman and philosopher

bœuf (Fr. m.), beef

B. of T., Board of Trade, q.v.

bogey, in golf

bogie, wheeled undercarriage

Bogota, NJ, US

Bogotá, Colombia, S. America

bog/y, ghost, *not* -ey; *pl.* **-ies**

Bohemian (noun and adj.), socially unconventional (person); **bohemianism** (not cap.)

Bohr (Niels Hendrik David), 1885–1962, and his son — **(Aage),** b. 1922, Danish physicists

Boïeldieu (François Adrien), 1775–1834, French composer

boiler-room (hyphen)

boiler suit (two words)

boiling-point (hyphen), abbr. **BP** *or* **b.p.**

Bokhara, river, NSW, Australia. *See* **Bukhara** for the town in Uzbekistan

Bokmål, *see* **Riksmål**

bold-face type, as this; indicated in copy by wavy underline; abbr. **b.f.**

bolero/, Spanish dance, woman's short jacket; *pl.* **-s**

Boleyn (Anne), 1507–36, 2nd wife of Henry VIII of England

Bolívar (Simón), 1783–1830, S. American patriot

bolívar, Venezuelan monetary unit

Bolivia/, -n, abbr. **Bol.**

Böll (Heinrich), b. 1917, German writer

Bologna, Italy, in Lat. **Bononia;** adj. **Bolognese,** but **spaghetti bolognese**

bolometer, radiation measurer

boloney, humbug, *not* bal-

Bolshevik, advocate of proletarian dictatorship by soviets; member of majority group of Russian Social Democratic Party, later the Communist Party, opp. to **Menshevik**

bolts (bind.), the three outer edges of a folded sheet before trimming

Boltzmann constant (phys.), symbol *k*

bolus/, large pill; *pl.* **-es**

Bombardier, cap. as title; abbr. **Bdr.**

bombazine, a fabric, *not* -basine, -bazeen, -bycine

bombe (Fr. cook. f.), conical confection

bona/ fide (Lat.), genuinely; — **fides,** good faith (no accent)

Bonaparte, *not* Buonaparte, Corsican family

bona/ peritura (Lat.), perishable goods; — *vacantia,* unclaimed goods

bon-bon, a kind of sweet; **bonbonnière,** a box for sweets

bon-chrétien, a pear; *pl.* **bons--chrétiens**

bond, abbr. **bd.**

Boney, sobriquet of Napoleon. *See also* **bony**

bondholder (one word)

bond paper, a hard, strong paper

Bonduca, *use* **Boudicca**

Bo'ness, Central (apos., being orig. Borrowstounness). *See also* **Bowness**

bon/ gout (Fr. m.), good taste; — *gré, mal gré,* whether one likes or not

Bonheur (Rosa), 1822–99, French painter

bonhomie (Fr. f.), good nature

bonhomme (Fr. m.), a pleasant fellow, a friar

Bonhomme (Jacques), the French peasant

bonjour (Fr. m.), good-day (one word)

bon/ marché (Fr.), a cheap shop, cheap; — *mot,* a witticism, *pl.* **bons mots** (two words)

bonne (Fr. f.), a maid

bonne-bouche, a dainty morsel, *pl.* **bonnes-bouches;** *bonne/ foi,* good faith; — *fortune,* success, *pl.* **bonnes fortunes;** — *grâce,* gracefulness, *pl.* *bonnes grâces*

bonnet/, -ed, -ing

bonnet rouge (Fr. m.), Republican's cap

Bononia, Lat. for Bologna, Italy

bonsai (Jap.), dwarfed plant

bonsoir (Fr. m.), good-evening, good-night (one word)

bontebok (Afrik.), pied antelope, *not* -buck

bon ton (Fr. m.), good style

bon vivant/, one fond of luxurious food (no hyphen, not ital.); *pl.* **-s.** The fem., **bonne vivante,** is not in French usage

bon viveur, one who lives luxuriously (not in French usage)

bon voyage! (Fr.), a pleasant journey!

bony, *not* -ey. *See also* **Boney**

Booerhave, *see* **Boerhaave**

book/, -s, abbr. **bk., bks.**

bookbind/er, -ing (one word)

book-end (hyphen)

bookkeep/er, -ing (one word)

booklet (typ.), usually an inset, saddle-stitched piece of printing, with limp cover

bookmak/er, -ing (one word)

bookmark(er), (one word)

book-plate (hyphen)

books (cited titles of), in italic; *but* series titles roman

book/seller, -stall (one word)

book sizes (typ.): the following are the standard untrimmed book sizes in mm., 4to and 8vo: **metric crown,** 252 × 192, 192 × 126; **metric large crown,** 264 × 204, 204 × 132; **metric demy,** 282 × 222, 222 × 141; **metric royal,** 318 × 240, 240 × 159; **A4,** 303 × 213; **A5,** 216 × 151. For the former sizes *see* **crown, demy, foolscap, imperial, medium, pott, royal, small royal**

books of Scripture (typ.), not to be italic, nor quoted; for abbrs. *see* separate titles

Booksellers Association, abbr. **BA**

book trim (bind.): standard

trim is 3 mm. off head, foredge, and tail margins

bookwork (one word)

Boole/ (George), 1815–64, English logician and mathematician; adj. **-an**

Boone (Daniel), 1735–1820, American explorer and colonizer

Boötes (astr.), the constellation

booze, (to) drink, *not* -se, bouse, bowze

Bophuthatswana, S. Africa, indep. 1977

bor., borough

Borak (Al), the winged horse on which Muhammad, in a vision, ascended to heaven

Bordeaux, dép. Gironde, France; a claret or any wine of Bordeaux

Bordeaux mixture, of lime and copper sulphate, to kill fungus and insect parasites on plants

border (typ.), an ornament

bordereau (Fr. m.), a memorandum, docket

borderland (one word)

Borders, region of Scotland

Boreas (Gr. myth.), god of the north wind

Borehamwood, Herts.

Borghese, Italian family

Borgia (Cesare), 1476–1507; — **(Lucrezia),** 1480–1519

born/, of child, with reference to birth, abbr. **b.**; **-e,** carried, also of birth when 'by' (with name of mother) follows

born/é (Fr.), *fem.* **-ée,** narrow-minded

Borodin (Alexander Porfiryevich), 1834–87, Russian composer and chemist

boron, symbol **B**

borough, abbr. **bor.**

Borstal, institution for reformative training (cap.)

bortsch, Russian soup

borzoi/, Russian wolfhound; *pl.* **-s**

bosbok, Afrik. for **bushbuck**

Bosch (Hieronymus), ?1450–1516, Dutch painter

bos'n, contr. of **boatswain**

Bosporus, strait separating European and Asian Turkey, *not* Bosph-

Bossuet (Jacques Bénigne), 1627–1704, French bishop and writer

Boswellize, to write laudatory biography (cap.)

B.o.T., Board of Trade, q.v.

bot/, *not* -tt; **-fly** (hyphen)

botanize, *not* -ise

botan/y, -ical, -ist, abbr. **bot.**; genera, species, and varieties to be ital.; all other divisions roman: Rosales (order); Rosaceae (family); *Rosa* (genus); *Rosa moschata* (species); *Rosa moschata* var. *nastarana* (variety); *Rosa moschata* cv. Plena (cultivar). Specific and infraspecific epithets, even when derived from names of persons, should be l.c.: *Lilium wardii, Camellia* × *williamsii, not Wardii, Williamsii.* Also l.c. are names of flowers used non-technically, as geranium, lobelia. See *Hart's Rules,* pp. 6–7

bothy (Sc.), hut, *not* -ie

bo-tree, sacred tree of India

Botswana, S. African independent republic 1966, *formerly* Bechuanaland

Botticelli (Sandro), 1447–1515, Florentine painter

Boucicault (Dion), 1822–90, Irish playwright

Boudicca, d. AD 61, Queen of Iceni

boudoir (not ital.)

bouffant, puffed out, as a dress (not ital.)

bougainvillaea, trop. plant with rosy bracts

Bouguereau (William Adolphe), 1825–1905, French painter

bouillabaisse, thick fish soup (not ital.)

bouilli (Fr. m.), stewed meat (not ital.)

bouillon, broth (not ital.)

boul., boulevard

boule, a French game akin to bowls (not ital.)

boule (inlay), *usually spelt* **buhl,** q.v.

Boulē, legislative council in ancient Greek city-state

boulevard/ (Fr. m.), abbr. **boul.; -ier,** a 'man about town'

boulevers/é (Fr.), overturned; *-ement* (m.), a violent inversion

Boulez (Pierre), b. 1925, French composer

Boulogne, déps. Haute-Garonne and Pas-de-Calais, France

bounceable, *not* -cable

bouncy, *not* bouncey

bound, abbr. **bd.**

bouquetin, the ibex

bouquinist, second-hand bookseller: in Fr. m. *bouquiniste*

bourgeois/ (Fr.), *fem.* **-e,** one of the middle class; **-ie,** the middle class (not ital.)

bourgeois (typ.), name for a former size of type, about 9 pt.; *pron.* berjoyce (stress on second syllable)

Bourn, Cambs.

Bourne, Lincs.

bourn (arch.), a limit, *not* -ne

Bournemouth, Dorset

Bournville, Birmingham

bourrée (Fr.), old lively dance, music for this (not ital.)

bourse, foreign money-market, esp. Paris

bouse (naut.), to haul, *not* -wse. *See also* **booze**

boustrophedon, written from left to right and right to left on alternate lines, esp. in Gr. inscriptions

boutique, small shop (often within a shop) for individual clothes and accessories

boutonnière (Fr. f.), a buttonhole

bouts-rimés, rhymed endings

Bouvet I. (Norw.), Antarctic Ocean

bowdlerize, to expurgate, from Dr T. Bowdler (not cap.)

bowled (cricket), abbr. **b.**

Bowness, Cumbria. *See also* **Bo'ness**

Boxing Day, first weekday after Christmas Day (caps.)

box number (two words)

box-office (hyphen), abbr. **b.o.**

boycott, to exclude from society or business, from Capt. Boycott (*not* cap.)

Boy Scout, *now* **Scout**

BP, British Petroleum, — Pharmacopoeia; boiling-point

BP, before present (s. caps.), year to precede

Bp., Bishop

b.p., below proof, bills payable, blood-pressure, boiling-point

BPC, British Pharmaceutical Codex

B.Phil., Bachelor of Philosophy

b.p.i. (comp.), bits (*see* **bit**) per inch

BPIF, British Printing Industries Federation

bpl., birthplace

Bq, becquerel

BR, British Rail (*formerly* Railways)

Br, bromine (no point)

Br., British, Brother (monastic title)

bra, colloq. for brassière (no point)

'Brabançonne (La)', Belgian national anthem

brace (typ.), *see* **punctuation** XIV (⁓)

bracket/, -ed, -ing

brackets (typ.), square [ ], angle < > *or* ⟨ ⟩; not the parentheses or round form ( ). *See* **punctuation** XI

Bradford:, sig. of Bp. of Bradford (colon)

Braggadochio, in Spenser's *Faerie Queene*

braggadocio, empty boasting (not ital.)

Brahe (Tycho), 1546–1601, Danish astronomer

Brahma, supreme Hindu god (not ital.)

brahma, brahmaputra, Asian domestic fowl (not cap.)

Brahmaputra, Indian river

brahmin/, member of Hindu priestly caste; *pl.* **-s**; adj. **-ical**

Brahmoism, reformed theistic Hinduism

brail, to haul up

braille, signs composed of raised dots replacing letters, for the blind

Braine (John), b. 1922, English novelist

braise (cook.), *not* -ze. *See also* **braze**

brake/, for wheel, a large vehicle, etc., *not* break; — **horsepower** (two words)

Br. Am., British America

Bramah (Ernest), 1867–1942, English writer; — **(Joseph),** 1749–1814, English inventor of — **lock** etc.

brand-new, *not* bran-

Brandt (Willy), b. 1913, German statesman, Chancellor of W. Germany 1969–74

Brandywine, battle, 1777, of American War of Independence

brant-goose, *use* **brent-**

Brantôme (Pierre de Bourdeille, Seigneur de), 1540–1614, French writer

Brasilia, new cap. of Brazil, 1960

brass/ (binder's), die for blocking book-cases, *pl.* **-es**; — **rule** (typ.), for printing simple borders or lines between columns of type

brassard, badge worn on arm

brasserie, beer garden, licensed restaurant; *brasserie* (Fr. f.), *also* brewery

brassie, golf-club with brass sole, *not* -y

brassière, woman's undergarment supporting breasts. *See also* **bra**

bratticing, (archit.) open carved work; (coal-mining) wooden shaft-lining or screen

Brauneberger, a white wine

Braunschweig, Ger. for **Brunswick**

brav/a! (It.), 'well done!' to a woman; *-o!* to a man; **brav/o** (noun), a cry of 'well done!'; *pl.* **-os**

bravo/, a desperado; *pl.* **-es**

braze, to solder, *not* -ise, -ize. *See also* **braise**

brazier, worker in brass, pan for holding lighted coal, *not* -sier

Brazil/, -ian, abbr. **Braz.**

Brazzaville, cap. of Congo, q.v., Cent. Africa

BRCS, British Red Cross Society

BRD (Ger.), *Bundesrepublik Deutschland*, Federal Republic of Germany

bread/board, -crumb (one word)

break/ (typ.), the division into a fresh paragraph; — **-line,** last one of a paragraph: never to begin next page, and should have more than five letters, except in narrow measures. *See also* **brake**

breakdown (noun), a collapse, analysis of statistics (one word)

Breakspear (Nicholas), Pope Adrian IV from 1154 to 1159, *not* -eare

breakthrough (noun, one word)

breakup (noun, one word)

breathalyser, device for testing amount of alcohol in the breath, *not* -iser, -izer; **breathalyse,** to carry out test

breathings, *see* **accents**

breccia, rock composed of angular fragments, *not* -cchia

Brechin, Tayside

Brecht (Bertolt, *not* -th-, -d), 1898–1956, German playwright

Brecon, Powys

breech birth (two words)

breeches-buoy, life-saving apparatus (hyphen)

bren-gun, light machine-gun (hyphen, not cap.)

brent-goose, *not* bra-

Brescia, Lombardy, Italy

Breslau, Ger. for **Wrocław** (Poland)

Bretagne, Fr. for **Brittany; Grande —,** Great Britain

Breton, (inhabitant) of Brittany, abbr. **Bret.**

Bretton Woods, New Hampshire, US (UN Monetary Conf., 1944)

Breughel, *use* **Brueghel** *or* **Bruegel,** q.v.

brev., brevet, -ed

breve, a curved mark (˘) to indicate a short vowel or syllable

brevet d'invention (Fr. m.), a patent

breveté s.g.d.g. (Fr.), patented without government guarantee (*sans garantie du gouvernement*)

breviary, book containing RC daily service

brevier (typ.), name for a former size of type, about 8 pt.; *pron.* brĕveer (stress on second syllable)

briar, *use* **-er**

Briarean, many-handed, like **Briareus,** of Gr. myth., *not* -ian

bribable, *not* -eable

bric-à-brac, curiosities in furniture and furnishings (hyphens, accent)

brick/bat, -work (one word)

Bridge End, Cumbria, Devon, Lincs., Shetland

Bridgend, Cornwall, Cumbria (near Patterdale), Donegal, Glos., Lothian, Mid Glam., Strathclyde, Tayside

Bridge of Allan, Central

Bridgeport, Ala., Calif., Conn., Nebr., Penn., Tex. (US)

Bridges Creek, Va. (US), bpl. of George Washington

Bridges (Robert Seymour), 1844–1930, English poet, Poet Laureate 1913–30

Bridgeton, Glasgow; New Jersey, US

Bridgetown, Corn., Som., Staffs., Co. Wexford, Barbados, NS (Canada), W. Austral.

Bridgewater, NS (Canada); — **(Duke of),** 1736–1803, projector of the — **Canal;** — **(Earl of),** 1756–1829, founder of the — **Treatises**

Bridgnorth, Shropshire

Bridgwater, Somerset, *not* Bridge-

bridle/-path, -road, -way (hyphens)

brier/, -root, -rose (hyphens), *not* briar, brere

Brig., brigad/e, -ier

Brillat-Savarin (Anthelme), 1755–1826, French gastronomist

brilliant (typ.), name for a former size of type, about $3\frac{1}{2}$ pt.

bring up (typ.), to underlay or interlay a block to bring it to printing height

Bri-Nylon, propr. term (caps., hyphen)

brio (mus.), fire, life, vigour

briquette, block of compressed coal-dust, *not* -et

brisling, Norwegian sprat, *not* brist-

Bristol:, sig. of Bp. of Bristol (colon)

Bristol-board, cardboard used by artists

Brit., Britain, Britann/ia, -icus, -ica, British

Briticism, idiom current in Britain, *not* Britishism

British/, abbr. **Brit.;** — **Academy, Association, BA;** — **America, Br. Am.;** — **Columbia, BC;** — **Honduras,** *now* **Belize;** — **Library, BL;** — **Museum, BM;** — **Rail, BR;** — **Standard, BS** (*see* **BSI**)

britschka, acceptable form for **bryczka,** q.v.

Britt. (on coins), Britanniarum, 'of the Britains' (Roman *Britannia Inferior* and *Britannia Superior*)

Brittany, in Fr. **Bretagne**

Britten (Edward Benjamin, Baron), OM, CH, 1913–76, British composer
Bro., Brother (monastic title)
bro., bros., brother, -s
Broad-Churchman (caps., hyphen)
Broad Haven, Dyfed, Mayo
Broad Oak, Cumbria, Dorset, Hereford & Worc., E. Sussex
Broadoak, Cornwall, Dorset, Glos., Kent, Shropshire
Brobdingnag, land of giants in Swift's *Gulliver's Travels,* not -dignag
broccoli, a hardy cauliflower, It. *pl.,* but Eng. *sing.* or *pl.*
Broderers, livery company
Bromberg, Ger. for **Bydgoszcz**
bromine, symbol **Br**
bronch/i, *pl.,* the main forks of the windpipe (*sing.* -**us**); -**ia,** *pl.,* ramifications of the above; -**ioles,** their minute branches
bronco/, Mexican horse, *not* -cho; *pl.* -**s**
Brontë (Anne), 1820–49 ('Acton Bell'); — **(Charlotte)**, 1816–55 ('Currer Bell'); — **(Emily)**, 1818–48 ('Ellis Bell'), English writers
Bronx (the), borough in New York City
bronzing (typ.), printing in yellow and then dusting with bronze powder to imitate gold
brooch, a dress-fastening, *not* -ach
Brooke, Leics., Norfolk; — **(Baron)**; — **(Rupert)**, 1887–1915, English poet
Brookline, Mass., US
Brooklyn, New York
Brooks's Club, London, *not* -es's
broomstick (one word)
Brother (monastic title), abbr. **Bro.**
brother/, abbr. **b.** *or* **bro.,** *pl.* -**s, brethren,** abbr. **bros.;** — **-german,** having same parents; — -**in-law,** *pl.* **brothers-in-law;** — -**uterine,** having same mother only (hyphens)

brougham, one-horse or electric closed carriage (not cap.)
brouhaha (Fr. m.), a commotion
Brown (Ford Madox), 1821–93, English painter; — **(John)**, 1800–59, US abolitionist; — **(Robert)**, 1773–1858, Scottish botanist
Browne (Charles Farrar), *see* **Ward;** — **(Hablot Knight)**, 1815–82, English artist, pseud. Phiz, illustrated Dickens; — **(Sir Thomas)**, 1605–82, English physician and author
Brownian motion of particles (observed by Robert Brown)
Brownie, junior Guide (cap.); **brownie,** Sc. house-spirit (not cap.)
browse, to eat, read desultorily, *not* -ze
BRS, British Road Services
Bruckner (Anton), 1824–96, Austrian composer
Bruegel (Pieter), *c.*1520–69, Flemish painter; **Brueghel (Pieter,** 1564–1638, and **Jan,** 1568–1625), his sons, also painters; *not* Breu-
Bruges, Belgium, in Fl. **Brugge**
Brummagem, abbr. **Brum,** derog. form of Birmingham; (adj.) tawdry
Brummell (George Bryan, 'Beau'), 1778–1840
Brunei, NW Borneo
Brunel (Marc Isambard), 1769–1849, and — **(Isambard Kingdom)**, 1806–59, his son, British engineers; — **University,** Uxbridge, 1966
Brunelleschi (Filippo), 1377–1446, Italian architect and sculptor
Brunhild (Norse myth.)
Brünnhilde, Wagnerian heroine
Brunonian, noun, an alumnus of Brown University, US; adj., of the system of medicine of Dr John Brown of Edinburgh, 1736–88

Brunswick, normal Eng. for
　Braunschweig, Germany
Brussels, cap. of Belgium, in Fl.
　Brussel, in Fr. **Bruxelles**
Brussels sprouts (cap., no
　apos.)
brut (Fr.), unsweetened
brutalize, *not* -ise
Bruxelles, Fr. for **Brussels**
bryczka, a Polish carriage (the
　Polish spelling; for Eng. form
　use **britschka**)
Bryn Mawr College, Philadel-
　phia, US
Brynmawr, Gwent
bryology, study of mosses; abbr.
　bryol.
Brython, Welsh for Briton, a S.
　British Celt
Brythonic, (language) of S.
　British Celts
BS, Bachelor of Surgery *or* (US)
　Science; Blessed Sacrament;
　British Standard
b.s., balance sheet, bill of sale
BSA, Birmingham Small Arms,
　British School at Athens
B.Sc., Bachelor of Science;
　B.Sc.(Econ.), ditto in faculty of
　Economics; **B.Sc.(Eng.),** ditto
　Engineering, Glas.;
　B.Sc.Tech., Bachelor of Tech-
　nical Science
b.s.g.d.g., see breveté
BSI, British Standards Institu-
　tion. A complete list of British
　Standards is in BS Yearbook.
　Standards in fields relevant to
　this dictionary are BS 1219
　(Proof correction), BS 1629
　(Bibliographical references), BS
　1749 (Alphabetical arrange-
　ment), BS 1991, Part I (Letter
　symbols, signs, abbreviations),
　BS 2961 (Typeface nomencla-
　ture), BS 2979 (Transliteration
　of Cyrillic and Greek charac-
　ters), BS 3700 (Index prepara-
　tion), BS 3763 (SI units), BS
　3814 (Glossary of rotary press
　terms), BS 3862 (Symbols for
　languages, geographical areas,
　and authorities), BS 4148 (Ab-

breviation of periodical titles),
BS 4149 (Glossary of paper and
ink terms), BS 4277 (Terms used
in offset lithography), BS 4280
(Transliteration of Arabic), BS
4812 (Romanization of Japan-
ese), BS 5261 (Proof correction
and copy preparation), BS 5775
(Quantities, units and symbols)
BSJA, British Show Jumping
　Association
BSR, British School at Rome
BST, British Summer Time (be-
　fore 1968 and from 1972); —
　Standard Time (1968–71) (both
　1 hr. ahead of GMT)
Bt., Baronet, *not* Bart.
BTA, British Tourist Authority
B.Th., Bachelor of Theology
BTU (*or* **B.Th.U.**), British Ther-
　mal Unit
bty., battery
BU, Brown University, Rhode
　Island, US; Baptist Union
bu., bushel(s)
buccaneer, piratical adven-
　turer esp. in seventeenth cen-
　tury, *not* -ier
Buccleuch (Duke of), *not* -gh
Bucellas, white wine from B.,
　near Lisbon
Buch (Ger. n.), book; *pl. Bücher*
Bucharest, Romania, *properly*
　Bucureşti, *not* Buka-
Buchausgabe (Ger. f.), edition
　of a book
Buckinghamshire, abbr.
　Bucks.
Bucknall, Lincs., Staffs.
Bucknell, Oxon., Shropshire;
　also American university at
　Lewisburg, Pa.
buckram, coarse fabric, (bind.)
　a good quality cloth
Budapest, Hungary, *not* -pesth
　(one word). Collective name for
　two towns, Buda and Pest
Buddha (Skt.), Gautama Sid-
　dhārtha, d. *c.*483 BC
Buddh/ism, -ist; if necessary
　divide at stroke; abbr. **Budd.**
budgerigar, a parakeet
budget/, -ed, -ing

buen/as noches (Sp.), good-night; **-as tardes,** good-afternoon; **-os días,** good-day, good-morning

Buenos/ Aires, cap. of Argentine Rep.; also in Panama. *Use* **— Ayres** in titles of Arg. railway companies

buffalo/, *pl.* **-es**

buffer state (two words)

buffet/, -ed, -ing

Buffon (Georges Louis Leclerc, comte de), 1707–88, French naturalist

buhl, (inlaid with) brass, tortoiseshell, etc., German form of **boule,** from **Boule (André),** cabinet-maker to Louis XIV

building, abbr. **bldg.,** *pl.* **bldgs.**

BUJ (*Baccalaureus utriusque juris*), Bachelor of Canon and Civil Law

Bukarest, *use* **Buch-**

Bukhara, town, Uzbekistan, USSR, *not* Bo-. *See also* **Bokhara**

buksheesh, *use* **bak-**

bulbul (Pers.), Asian song-thrush; singer or poet

Bulgaria/, -n, abbr. **Bulg.**

bull (Stock Exchange), speculator for a rise

bulletin (not ital.); abbr. **bull.**

bullfight/, -er, -ing (one word)

bullring (one word)

bull's-eye, centre of target, lantern, peppermint sweetmeat (apos. and hyphen)

bull-terrier (hyphen)

bulrush, *not* bull-

bulwark, defence, *not* bull-, -work

Bulwer-Lytton (Edward George Earle Lytton), first **Lord Lytton,** 1803–73, English novelist

Bulwer-Lytton (Edward Robert), first **Earl of Lytton,** 1831–91, son of above, English poet and diplomat, pseud. **Owen Meredith**

bumf, papers

bumkin (naut.), projection from bows or stern of ship

bumpkin, yokel

bump out (typ.), increase word or letter spacing to fill a line, or line spacing to fill a page

buncombe, *use* **bunkum**

Bundesanstalt, German off. institution since 1945. *See* **Reichsanstalt**

Bundesrat, Bundestag (Ger.), upper and lower houses of W. German Parliament

bundle/, -s, abbr. **bdl/., -s.**

Bunker Hill, Mass., battle of American War of Independence, 1775, actually fought at neighbouring Breed's Hill, *not* -er's

bunkum, *not* -combe

Buñuel (Luis), b. 1900, Spanish-born film director

Buonaparte, *use* **Bonaparte**

buona sera (It.), good-evening; **buon giorno,** good-day

buoyan/t, -cy

BUPA, British United Provident Association

bur, clinging seed-vessel or catkin. *See also* **burr**

Burckhardt (Jacob Christoph), 1818–97, Swiss historian

Burdett-Coutts (Angela Georgina, Baroness), 1814–1906, English philanthropist

bureau/, *pl.* **-x** (not ital.); **-cracy, -crat,** adj. **-cratic**

Bureau Véritas, maritime underwriters' association at Brussels (two caps.)

burg., burgess, burgomaster

burgamot, *use* **ber-**

burgeois, *see* **bourgeois** (typ.)

burgess, abbr. **burg.**

burgh (Sc.), a chartered town, *not* borough

burgomaster (one word); abbr. **burg.**

burgrave, a governor, *not* burgg-

burgundy, a wine (not cap.)

burial/ ground, — place, — service (two words, no caps.)

burl, a lump in cloth

burl., burlesque

Burlington House, London, headquarters of Royal Academy, British Association, and other artistic and learned bodies

Burm/a, *not* -ah; **-an,** native of Burma, *pl.* **-ans; -ese,** native or language of Burma

Burmah Oil Co., *not* Burma

Burne-Jones (Sir Edward Coley), 1833–98, English painter (hyphen)

Burnett (Frances Eliza Hodgson), 1849–1924, US writer

Burnham scale, of teachers' salaries

burnous, Arab cloak, *not* -e, bernouse

Burntisland, Fife (one word)

burnt sienna, orange-red pigment, *not* — siena

burr, a rough edge, rough sounding of letter *r*; kind of limestone. *See also* **bur**

burro (Sp.), donkey

Burroughs (Edgar Rice), 1875–1950, US novelist; — **(John),** 1837–1921, US naturalist and poet; — **(William),** b. 1914, US novelist

bursar, treasurer of a college, *not* -er

Burundi, Cent. Africa, indep. 1962

Bury St. Edmunds, Suffolk (no apos., no hyphen)

bus/, omnibus, *not* 'bus, *pl.* **-es;** as verb, past **-ed,** partic. **-ing**

bushbuck (S. Afr.), antelope, in Afrik. **bosbok**

bushel, 36.4 litres; in US 35.3; abbr. **bu.** (*sing.* and *pl.*)

Bushey, Herts., *not* -hy

bushido, code of the samurai or Japanese mil. caste

Bushire, Iran

Bushy Park, London

businesslike (one word)

business/ man, — **woman** (two words)

busing, *see* **bus**

busybody, *not* busi-

busyness, state of being busy, distinct from **business**

Butler (Samuel), 1612–80, English poet (*Hudibras*); — (—), 1835–1902, English writer (*Erewhon*)

Butte, Montana, US

butter-fingers (hyphen)

buttermilk (one word)

buttonhole (one word)

buyer's option, abbr. **b.o.**

Buys Ballot (Christoph Hendrik Didericus), 1817–90, Dutch meteorologist

BV, *Beata Virgo* (the Blessed Virgin); (points) pseudonym of James Thomson, q.v.

BVM, *Beata Virgo Maria,* the Blessed Virgin Mary

BWI, British West Indies, *now* **WI**

BWV, *Bach Werke Verzeichnis*, prefix to enumeration of J. S. Bach's compositions

by- (as prefix) is tending to form one word with the following noun, but a hyphen is still frequently found, as in ten cases given below

by and by (no hyphens)

by-blow, side-blow, bastard (hyphen)

Bydgoszcz, Poland, in Ger. **Bromberg**

bye/, -s (cricket), abbr. **b.**

bye-bye, familiar form of goodbye (hyphen)

by-election, by-form (hyphens)

Byelorussia, *use* **Be-**

bygone (one word)

by-lane (hyphen)

by-law, *not* bye-

byline, in football (one word)

byname, a sobriquet (one word)

bypass (one word)

bypath (one word)

by-play (hyphen)

by-plot, by-product, *not* bye-

by-road (hyphen)

bystander (one word)

by-street (hyphen); *but* Masefield's *The Widow in the Bye Street*

by the by (no hyphens)
byte (comp.), group of 8 bits (*see* **bit**)
byway (one word)

byword (one word)
byzant, *use* **be-**
bzw. (Ger.), *beziehungsweise* (respectively)

C

C, carbon, (elec.) coulomb, centum (a hundred), Celsius or centigrade, (mus.) common time, the third in a series

C., Cape, Catholic, century, chairman, chief, Church, Command paper (q.v.), congress(ional), Conservative, consul, contralto, Council, counter-tenor, Court, Gaius

ⓒ, copyright

c (as prefix), centi-

c., cent, -s, centime, century, chapter, city, colt, conductor, constable, copeck, cubic, (cricket) caught, (meteorol.) cloudy

C (ital. cap.), symbol for capacitance

c., circa, about

Ɔ (inverted C), 500

CA, California (off. postal abbr.), Central Africa, Central America, M. Inst. Chartered Accountants of Scotland, Chief Accountant, chronological age, Civil Aviation, College of Arms, commercial agent, Consumers' Association, Controller of Accounts, Court of Appeal

C/A, capital account, commercial account, credit account, current account

Ca, calcium (no point)

ca., (law) cases

ca., abbr. of *circa, use* **c.**

CAA, Civil Aviation Authority

Caaba, *use* **Kaaba**

caaing-whale, *use* **ca'ing--whale**

CAB, Citizens' Advice Bureau, (US) Civil Aeronautics Board

cabal, secret faction

caballero/, Spanish gentleman, *pl.* **-s**

cabbala, Heb. tradition, *not* k-, cabala, -alla

Cabell (James Branch), 1879–1958, US novelist

Cable (George Washington), 1844–1925, US writer

Cabul, *use* **Kabul**

CAC, Central Advisory Committee

ca'canny (Sc.), go carefully!

cacao/, trop. American tree from whose seeds cocoa and chocolate are made; **-tree** (hyphen). *See* **coco**

cachemere, -mire, *use* **cashmere**

cachet, distinctive mark, capsule containing medicine

cachou, lozenge to sweeten breath, *not* cashew

cachucha, Spanish dance

caciqu/e, W. Indian and S. American chief, political boss; **-ism;** *not* caz-

cacodemon, an evil spirit, *not* -daemon

cacoethes/ (Lat. fr. Gr.), an evil habit; — *loquendi,* an itch for speaking; — *scribendi,* ditto for writing

CACTM, Central Advisory Council for the Ministry, *now* **ACCM**

cact/us (bot.), *pl.* **-i**

c.-à-d. (Fr.), *c'est-à-dire* (that is to say)

caddie, golf-attendant

caddis/, mayfly larva, *not* -ice; **-worm** (hyphen)

caddy, tea-box

cadi, Muslim judge, *not* k-

Cadmean/, of Cadmus of Thebes, *not* -ian; — **victory,** a ruinous one

cadmium, symbol **Cd**

cadre, nucleus of personnel (not ital.)

caduce/us, the wand of Hermes; *pl.* **-i**

caducous (bot.), tending to fall

Cædmon, d. 670, English poet, *not* Ce-

Caenozoic, *use* **Cai-**

Caermarthen, *use* **Car-**
Caernarfon, Gwynedd, *not*
 Car-, -von
Caerphilly, Mid Glam.; also the
 cheese
caerulean, *use* **ce-**
Caesar/, -ean, *not* Ces-, -ian
Caesarea Philippi, in ancient
 Palestine
caesium, symbol **Cs**
caesura, division of a metrical
 line, *not* ce-
café (Fr. m.), coffee-house (not
 ital.)
café/ au lait (Fr. m.), coffee with
 milk; — *noir,* strong coffee
 without milk
Caffre, *use* **Kaffir** *or* **Kafir,**
 qq.v.
caftan, long-sleeved Turkish or
 Persian garment, *not* k-
cag/ey, shrewd, *not* -gy; adv.
 -ily, noun **-eyness**
cagoule, waterproof outer garment, *not* ka-
cahier (Fr. m.), a paper book,
 sheets of MS, exercise-book
Caiaphas, Jewish high priest,
 Luke 3: 2
caiman, *use* **cay-**
Cainan, grandson of Seth. *See
 also* **Canaan**
**Caine (Sir Thomas Henry
 Hall),** 1853-1931, Manx
 novelist
ca'ing-whale, round-headed
 porpoise, *not* caa-
Cainozoic (geol.), *not* Caeno-,
 Ceno-, Kaino- (cap.)
caique, Bosporan skiff, Levan-
 tine sailing ship
'Ça ira', 'It will be!', French
 revolutionary song
Caius, Roman praenomen, *use*
 Gaius
Caius (*properly* **Gonville and
 Caius**) **College,** Cambridge,
 pron. as keys; abbr. **Cai.,** *or*
 CC
cajole/, persuade by flattery;
 noun **-ry** (one *l*)
caky, *not* -ey
Cal. (California), *use* **Calif.**

cal, calorie (no point)
cal., calendar
cal. (mus.), calando
calamanco/, a woollen stuff, *pl.*
 -es
calamary, the squid
calando (mus.), volume and rate
 diminished; abbr. *cal.*
calcareous, chalky, *not* -ious
calcedony, *use* **chal-**
calceolaria, slipper-like flower,
 not calci-
calcium, symbol **Ca**
calculator, *not* -er
calcul/us, stony concretion in
 the body, *pl.* **-i,** adj. **-ous;** a
 system of computation, *pl.*
 -uses, adj. **-ar**
**Calderón de la Barca
 (Pedro),** 1600-81, Spanish
 playwright
caldron, *use* **caul-**
calecannon, *use* **colcannon**
calendae, calends, *use* **k-**
calendar, an almanac, *not* k-;
 abbr. **cal.**
calend/er, to make smooth
 (cloth, paper, etc.); a machine
 that does this; **-erer,** one who
 does this, *not* k-
calender (Pers., Turk.), a der-
 vish, *not* k-
calends, the first day of a Roman
 month; the **Greek** —, a time
 that never comes; *not* k-
calf/, abbr. **cf.; -leather, -love**
 (hyphens)
calfskin (one word)
Caliban/, in Shakespeare's *The
 Tempest;* '— **upon Setebos',** by
 R. Browning, 1864
Caliburn, King Arthur's sword,
 not Cala-, Cale-; cf. **Excalibur**
calico/, *pl.* **-es**
California, off. abbr. **Calif.** or
 (postal) **CA**
californium, symbol **Cf**
caligraphy, *use* **calli-**
calipers, *use* **callipers**
caliph/, Muslim chief civil and
 religious ruler; **-ate;** *not* ka-,
 kha-, -if
calisthenics, *use* **call-**

cal/ix (physiol. and biol.), cup-like cavity or organ; *pl.* **-ices**

calk, *use* **caulk**

Callaghan (Leonard James), b. 1912, British statesman, Prime Minister 1976–9

Callander, Central

called, abbr. **cld.**

calligraphy, *not* cali-

Calliope (Gr.), muse of eloquence and epic poetry; **calliope**, steam-organ on roundabout

calliper, metal leg-support, (*pl.*) compasses for measuring inside or outside diameters of objects, *not* cali-

Callirrhoe, wife of Alcmaeon

callisthenics, gymnastic exercises, *not* cali-

call/us, thick skin; *pl.* **-uses**; adj. **-ous**

Calmann-Lévy, publishers, Paris

calmia, Calmuck, *use* k-, K-

Calor gas, propr. term (cap., two words)

calori/e, unit of heat or energy (not now in scientific use), abbr. **cal**; **-meter**, measurer of heat; **-motor**, voltaic battery (one word)

calotte, skull-cap (RC clergy)

Caltech, California Institute of Technology

caltrop (mil.), horse-maiming iron ball with spikes, *not* -throp, -trap

calumniator, *not* -er

Calvé (Emma), 1866–1942, French operatic soprano

Calverleigh, Devon

Calverley, Kent, W. Yorks.; — **(Charles Stuart)**, 1831–84, English writer

calves-foot jelly, *not* -f's, -ves', -feet

cal/x, residue from burning; *pl.* **-ces**

Calypso, sea-nymph in the *Odyssey*; **calypso**, W. Indian

Negro song, usu. about current affairs

caly/x (bot.), whorl of leaves forming outer case of bud; *pl.* **-ces**

camaraderie, good-fellowship

Camargue (the), S. France

Camb., Cambridge

Cambodia, Indo-China, sovereign state 1953, *not* Kamboj-, -oja; in Fr. **Cambodge**; officially **Kampuchea (Democratic)**

cambric, fine white linen

Cambridge/, abbr. **Camb.** or **Cantab.**; — **University, CU**; — **University Press, CUP**

Cambridgeshire, abbr. **Cambs.**

camellia, flowering shrub, *not* -elia

camelopard, giraffe

Camembert, a French cheese

cameo/, striated stone carved in relief; *pl.* **-s**. *See also* **intaglio**

camera/, *pl.* **-s**; — **obscura**, *not* — oscura; —**ready copy** (typ.), abbr. **CRC**

Camera Stellata, the Star Chamber

Cameroon/ (United Republic of), Cent. Africa, indep. 1960; adj. **-ian**

Cammaerts (Emile), 1878–1953, Belgian poet

Camoens (Luis Vaz de), 1524–80, Portuguese poet, *not* Camö-; in Port. **Camões**

camomile, medicinal plant, *not* cha-

Camorra, Neapolitan secret society, practising violence and extortion. *See also* **Mafia**

Campagna di Roma, the plain round Rome

campanile/, bell-tower, usu. detached; *pl.* **-s**

camp/-bed, -chair, -fire (hyphens)

Campbell-Bannerman (Sir Henry), 1836–1908, British statesman, Prime Minister 1905–8 (hyphen)

Campbeltown, Strathclyde

Campeche Bay, City, *or* State, Mexico, *not* -peachy

camp/o (It.), open ground, *pl.* -i; *campo santo* (It., Sp.), a cemetery

campsite (one word)

Campus Martius, field of Mars, anc. Rome

Can., Canad/a, -ian, Cantoris

can., canon, canto

Canaan, son of Ham. *See also* Cainan

Canad/a, -ian, abbr. Can.

Canadian River, Okla., US

canaille (Fr. f.), the rabble

canaker, *use* kanaka

Canaletto (Giovanni Antonio), real surname Canal), 1697–1768, Italian painter; also used by his nephew Bernardo Bellotto, 1720–80, Italian painter

canalize, *not* -ise

canapé/, caviare, anchovies, etc., on fried bread or toast; *pl.* -s

Canar/i, -ese, *use* K-

canard (Fr. m.), an absurd story, a duck; *canard sauvage*, wild duck

canasta, a card-game

canaster, a tobacco, *not* k-, -ister

cancan, high-kicking Parisian dance (one word)

cancel (typ.), incorrect matter, new-printed leaf correcting that (its signature preceded by *)

cancel/, -lation, -led, -ler, -ling

cancellate (biol.), porous, marked with cross-lines

cancelled/, abbr. cld.

Candahar, *use* K-

candela, standard unit of light intensity; abbr. cd

candelabr/um, large candle-stick; *pl.* -a; also *sing.* -a, *pl.* -as

Candide, by Voltaire, 1759

c. and l.c. (typ.), capital(s) and lower case

candle-light (hyphen)

Candlemas Day, 2 Feb.

candle/power, -stick, -wick (one word)

canephor/us (archit.), sculptured youth or maiden with basket on head; *pl.* -i

caneton (Fr. m.), duckling

canister, small metal box, *not* -aster

cannabis, Indian hemp, *not* cana- (not cap., except for the bot. genus)

cannelloni, pasta rolls

Cannes, dép. Alpes-Maritimes, S. France

cannibalize, *not* -ise

Canning (George), 1770–1827, and his son — (Charles John, Earl), 1812–62, English statesmen

cannon/-ball, -bone, -fodder (hyphens)

canoe/, -d, -ing, -ist

canon, abbr. can.

canon (typ.), name for a former size of type, about 48 pt.

cañon (Sp.), a gorge

canonize, *not* -ise

'Canon's Yeoman's Tale', by Chaucer

Canosa, Apulia, Italy

Canossa, Emilia, Italy, scene of penance of Emperor Henry IV before Pope Gregory VII, 1077; hence to go to —, 'to eat humble pie'

Canova (Antonio), 1757–1822, Italian sculptor

canst (no apos.)

Cant., Canterbury, Canticles

can't, to be printed close up

cantabile (mus.), in a singing, graceful style

Cantabrigian, of Cambridge, *not* -dgian

Cantabrigiensis (Lat.), of Cambridge University; abbr. Cantab.

cantaloup, a musk-melon, *not* cante-, -loupe

Cantate, Psalm 98 as canticle

cantatrice (It. and Fr.), professional woman singer

55

cante hondo (Sp.), mournful song

Canterbury, abbr. **Cant.**

Canterbury bell, flower (one cap.)

canticle, a hymn; abbr. **cant.** *See also* next

Canticum Canticorum, 'The Song of Songs', *also* **Canticles,** abbr. **Cant.,** same as **Song of Solomon**

cantilever, a form of bridge or support, *not* canta-, -liver

Cantire, Strathclyde, *use* **Kintyre**

canto/, division of long poem, a song; *pl.* **-s,** abbr. **can.**

cantonment, mil. quarters

Cantoris, of the precentor, that is on north side of choir; abbr. **Can.**

Cantor Lectures (Royal College of Physicians, *and* Society of Arts)

Cantuar., Cantuarius, Cantuariensis (of Canterbury)

Cantuar:, sig. of Abp. of Canterbury (colon)

Cantuarian, of Canterbury

cantus firm/us (mus.), basic melody in counterpoint, *pl.* **-i;** in It. *canto fermo, pl.* **canti fermi**

Canute, accepted variant of **Cnut,** 995–1035, king of Norway, Denmark, and England

canvas/, coarse linen; **-ed, -es**

canvass/, to solicit votes; **-ed, -er, -es**

canyon, a gorge

caoutchouc, unvulcanized rubber

CAP, Common Agricultural Policy (EEC)

Cap (Fr. geog. m.), Cape

cap., capital letter, chapter, foolscap

cap-à-pie, from head to foot (accent, hyphens, not ital.)

Cap d'Ail, town, dép. Alpes--Maritimes, France

Cape/, cap. when with name, abbr. **C.; — Breton,** Nova

Scotia, *not* Briton, Britton, Britun, abbr. **CB; — of Good Hope,** abbr. **CGH; — Province** (of S. Africa), abbr. **CP**

Čapek (Josef), 1887–1945, an his brother — (**Karel**), 1890–1938, Czech writers

Cape Town, S. Africa (two words)

Cape Verde/ Islands, off W. Africa, indep. 1975; *not* — Verde — de Verd, adj. **-an**

Cap Haitien, town in Haiti

capias (law), writ of arrest

Capital Gains Tax, abbr. **CG**

capitalization (see *Hart's Rules* pp. 8–14, 91–5 for French, 106–7 for German, 121–3 for Russian). When in doubt use lower case. Use initial capitals for:

Abstract qualities personified: O Fame!

Acts of Parliament and **Bills** (titles of).

Adjectives derived from proper names, when connection with the name is still felt to be alive, as Christian, Homeric, etc., but not when the connection is remote or conventional, as roman (numerals), or when the sense is an attribute, as chauvinistic, gargantuan, quixotic.

Aircraft (types of), as a Concorde, a Starfighter.

Archaeological eras, *see* **Historical eras** below.

Associations, as Charity Organization Society.

Bible (and books of the Bible: *but see* s.v.), **Book of Common Prayer, Old Testament,** etc.

Botany, all divisions higher than species, that is, genera, families, orders, classes, etc. *See also* **botany** (as main entry).

Churches, as Baptist Church.

Compass point abbreviations, N., N. b E., NNE, NE b N., etc.

Compass points when denoting a region (the mysterious East, unemployment in the North).

Compound titles, as Assistant Adjutant-General, Chief Justice, Ex-President, Major-General, Vice-President.

Days, as Christmas Day, Lady Day, Monday, New Year's Day.

Deity (the), and related words, as Allah, Almighty, Christ, *Dominus*, Father, God, Jehovah, Lord, the Deity, the Holy Trinity, Yahveh; pronouns to have lower case except at author's request, *but always* who, whom.

Festivals, as Easter, Whitsuntide.

Full point, cap. follows unless the point marks an abbreviation.

Geographical names with recognized status, as Northern Ireland (but northern England), East Africa, New England (USA), River Plate (but the river Thames).

Geological formations, as Devonian, and **eras,** as Palaeozoic.

German nouns: see *Hart's Rules*, pp. 106–7.

Government (the).

Historical eras and events, as Dark Ages, Early Minoan, Perpendicular, the Renaissance, First World War (but not archaeological eras, as neolithic, palaeolithic).

House of Lords, Commons, etc.

Institutions and movements, as Christianity, Islam, Marxism; the Crown, Parliament, the State.

MS, MSS, manuscript, -s.

Nicknames, as the Iron Duke, the Admirable Crichton.

O! and **Oh!** (the interjections), and **O** (vocative).

Palaeontology, all divisions higher than species.

Period, *see* **Full point** above.

Poetry, generally at beginning of each line in classical English and French; not necessarily in German, Latin, or Greek.

Political bodies and parties, as Assembly, Senate; Conservative, Republican.

Post-codes, as BR6 8JU (even full caps., no points, lining numerals; but when the fount has non-lining numerals, or in a s.-cap. context (e.g. publisher's imprint), even s. caps. may give a better effect).

Proper names, including English Christian names, surnames, and names of an individual, family, place, locality, and the like; also attributes used as proper names, as a Black (Negro), a Blue (university), a Red (a communist).

Proprietary terms: Aspro, Kodak, Persil.

PS, postscript.

Question, the next word following a question mark (?) should generally begin with a capital letter.

Quotations, when syntactically complete: Consider this 'Thou art the man', *but* Do not follow 'the madding crowd'.

Rank, when individuals are referred to or addressed by their rank, as 'the Squire said', 'Good evening, General'.

Religious denominations, Nonconformist, Orthodox.

Roads, Gardens, Gates, Groves, Hills, Parks, Squares, Streets, Terraces, etc., when with name.

Ships, as the *Cutty Sark*, HMS *Dreadnought*.

Sovereign (the), in proclamations, all personal pronouns referring to, as Her, Him, His, etc.

capitalization (*cont.*)

Titles and Subtitles of books, films, newspapers, periodicals, and plays, the first word and important words only, as 'I Forbid the Banns', *Farmer and Stockbreeder, The Merchant of Venice*; but scientific and technical bibliographies often capitalize first word and proper names only. For French titles see *Hart's Rules*, p. 92; in Italian titles capitalize the first word only; for Russian titles see *Hart's Rules*, p. 122.

Titles of corporations, as Department of Trade.

Titles of courtesy, honour, and rank: HRH, TRH, the Prince of Wales, the Bishop of Oxford, President Roosevelt, His Grace, Sir John Smith, J. Smith, Esq., Mr J. Smith, etc.; note compound titles: Assistant Adjutant-General, Vice-President.

Titles of distinction, as FRS, LL D, are usu. put in large caps. Even s. caps. often improve general effect.

Titles of pictures, as for books; *but* identificatory descriptions given by cataloguers etc. where no artist's title exists take initial capital only.

Titles of poems and songs, when formed from the first line, to be capitalized as book-titles, as 'I Fear no Foe', 'Where the Bee Sucks'.

Trade names, *see* **Proprietary terms** above.

Vehicles (types of), as a Dormobile, a Jaguar.

Verbs derived from proper names when there is direct reference to the name, as Americanize, Hellenize.

Zoology, all divisions higher than species, that is, genera, families, orders, classes, etc. See also **zoology** (as main entry)

capitalize, *not* -ise

capital letter, abbr. **cap.**

capitals (large), indicated in copy by three lines underneath the letter; abbr. **cap.** *See also* **capitalization**

capitals (small), USUALLY ABOUT TWO-THIRDS SIZE OF LARGE CAPS., AS THIS TYPE, indicated in copy by two lines underneath; may be used for subsidiary headings (with large-cap. initials) for the signatures of printed correspondence, and for any even large-cap. matter that would otherwise gain excessive prominence on the page. They are not used in German, Greek, or Russian. Abbr. (typ.) **s.c.** (without points as a proof-correction mark); *also* **s. cap.,** *pl.* **s. caps.** *See also* **even s. caps.**

Capitol, temple of Jupiter on Capitoline Hill, anc. Rome; Congress building, Washington DC, US

capitulum (bot., anat.), small head, knob

cappuccino (It.), coffee with milk

capriccio/ (mus.), fanciful work; *pl.* **-s,** adj. **-so**

caps. and smalls (typ.), set in small capitals with the initial letters in large capitals; abbr. **caps** (*or* **c) and sc**

capsiz/e, *not* -ise; noun **-al**

Captain, abbr. **Capt.**

caption (typ.), descriptive matter printed above or below an illustration; (US) heading of chapter, page, or section

Captivity (the) of the Jews (cap. C)

caput/ (lat.), head; **—** *mortuum,* worthless residue after distillation

Car., Carolus (Charles)

car., carat

carabineer, soldier with carbine, *not* carb-, -ier; *but* **Carabiniers & Greys**

carabinier/e, Italian soldier serving as policeman; *pl. -i*

Caracci, family of painters, 1550–1619, *not* Carr-

Caractacus, accepted form of **Caratacus,** *fl.* AD 50, king of Silures, anc. Britain

caracul, *use* **karakul**

carafe, glass container for water or wine, *not* -ff, -ffe

caramba! (Sp.), wonderful!, how strange!

carat, a weight, *not* caract, carrat, karat; abbr. **K.** *or* **car.**

caratch; use **kharaj**

Caravaggio (Michelangelo Merisi da), 1569–1609, Italian painter

caravanserai, inn with courtyard used as halting-place for caravans, *not* -sary, -sery, -sera. *See also* **khan**

caravel, Spanish ship

caraway, plant with aromatic fruits ('seeds'), *not* carra-

carbineer, *use* **carbineer**

carbohydrate (one word)

carbon, symbol **C**

carbon copy, abbr. **c.c.**

carbon dating (two words)

carbonize, *not* -ise

carbon paper (two words)

carburett/ed, -or, *not* -t-, -er

carcass/, *not* -ase; *pl.* **-es**

Carcassonne, dép. Aude, France

carcinom/a, a cancer; *pl.* **-ata**

card., cardinal

cardamom, aromatic capsules of ginger plants, *not* -mon, -mum

card-game (hyphen)

Cardigan, Dyfed. *See also* **Ceredigion**

card index (noun), **card-index** (verb)

Carducci (Giosué), 1835–1907, Italian poet

carefree (one word)

carême (Fr. m.), Lent

care of, abbr. **c/o**

caret, insertion mark (ʌ)

caretaker (one word)

Carew (Richard), 1555–1620, English poet; — **(Thomas),** 1595–?1639, English poet and courtier

car/ex, sedge; *pl.* **-ices**

Carey (Henry), ?1690–1743, English poet and composer; — **(Henry Charles),** 1793–1879, US economist, son of next: — **(Mathew,** one *t*), 1760–1839, US publicist; — **Street,** London, syn. with the Bankruptcy Court. *See also* **Cary**

cargo/, *pl.* **-es**

Caribbean Sea, W. Indies, *not* Carr-

Cariboo Mountains, BC, Canada

caribou, N. American reindeer, *pl. same* (*not* -boo)

Caribou Highway, BC, Canada

Caribou Mountain, Idaho, US

carillon, (chime of) bells

Carington, *see* **Carrington**

Carinthia, Austria, in Ger. **Kärnten**

cariole, *use* **carr-**

Carisbrooke, IW

carità (It. art), representation of maternal love

Carleton (William), 1794–1869, Irish novelist

Carlile/ (Richard), 1790–1843, English reformer and politician; — **(Revd Preb. William),** 1847–1942, founder of Church Army. *See also* **Carlisle, Carlyle**

Carliol:, sig. of Bp. of Carlisle (colon)

Carlisle, Cumbria; — **(Earl of).** *See also* **Carlile, Carlyle**

Carlist, supporter of Spanish pretender Don Carlos (1788–1855)

Carlovingian, *use* **Carolingian**

Carlsbad, Czechoslovakia, *not* K-; in Czech **Karlovy Vary**

Carls/krona, Sweden, **-ruhe,** W. Germany, *use* **K-**

Carlton Club, London

Carlyle (Thomas), 1795–1881, Scottish writer. *See also* **Carlile, Carlisle**

Carlylean, *not* -eian, -ian

Carmagnola, Piedmont, Italy

'Carmagnole (La)', French revolutionary song and dance

Carmarthen, Dyfed, *not* Caer-

Carnarvon, Gwynedd, *use* **Caernarfon,** *but* — **(Earl of)**

Carnatic, Madras, India, *not* K-

Carnaval/, suite of piano pieces by Schumann, 1834–5; — *des animaux* **(le),** suite for small orchestra by Saint-Saëns, 1886; — *romain* **(le),** overture by Berlioz, 1844. *See also* **Carnival**

'Carnaval de Venise, (le)', (mus.), popular type of air and variations, early nineteenth century

Carnegie (Andrew), 1835–1919, Scottish-born US millionaire and philanthropist

carnelian, *use* **cor-**

carnival, in Fr. m. *carnaval,* in Ger. n. *Carneval*

Carnival, overture by Dvořák, 1891. *See also* **Carnaval**

Carnoustie, Tayside

carol/, -led, -ler, -ling

Carolean, of the time of Charles I or II, *not* -ine

Carolingian, of the Frankish dynasty of Charlemagne, *not* Carlovingian, Karl-

Carolus, Lat. for Charles; abbr. **Car.**

carous/e, drink heavily; **-al**

carousel, tournament, rotating conveyor system. *See* **Carrousel**

car-park (hyphen)

Carpathian Mts., E. Europe, *not* K-

carpe diem (Lat.), enjoy the day

carpet/, -ed, -ing

carport (one word)

carp/us, the wrist; *pl.* -i

Carracci, *use* **Caracci**

carrageen, Irish moss; *not* — moss, -gheen; from Carragheen, near Waterford

Carrantuohill, Kerry, highest mt. in Ireland

Carrara, N. Italy, source of marble

carrat, *use* **carat**

carraway, *use* **cara-**

Carribbean, *use* **Cari-**

Carrigtohill, Cork (one word)

Carrington (Baron), family name **Carington** (one *r*)

carriole, a carriage, *not* cariole

Carroll (Lewis) (Charles Lutwidge Dodgson), 1832–98, English author

Carrousel (place du), Paris, from Fr. *carrousel* (m.), a tilting-match

Carrutherstown, Dumfries & Galloway (one word)

carry-cot (hyphen)

carsick (one word)

Carte (Richard D'Oyly), 1844–1901, *see* **D'Oyly Carte**

carte (fencing), *use* **quart**

carte blanche (Fr. f.), full discretion (ital.)

carte-de-visite (not ital.)

Carter, Paterson & Co., former carriers, London

Cartesian, *see* **Descartes**

cart-horse (hyphen)

cartouche (archit.), scroll ornament, (archaeol.) ring enclosing hieroglyphic royal name, *not* -ch

cartridge paper, a firm, strong paper

cart/-wheel, -wright (hyphens)

carvel-built (naut.), with smooth planking

Cary (Henry Francis), 1772–1844, English translator of Dante; — **(Arthur Joyce Lunel),** 1888–1957, English novelist; — **(Lucius),** Lord Falkland, 1610–43; — **(Alice),** 1820–71, — **(Phoebe),** 1824–71, US sisters, poets. *See also* **Carey**

caryatid/, female figure as a column; *pl.* -s

CAS, Chief of the Air Staff

Casabianca (Louis de), ?1755–98, French seaman

Casablanca, Morocco

Casa Grande, prehistoric ruin in Arizona, US

'Casa Guidi Windows', by Mrs Browning, 1851

Casals (Pablo), 1876–1973, Catalan cellist

Casanova de Seingalt (Giovanni Giacomo), 1725–1803, Italian adventurer

Casas Grandes, Mexican village with famous ruins

Casaubon (Isaac), 1559–1614, French scholar and theologian; — **(Méric),** 1599–1671, his son, clergyman; — **(Mr),** in G. Eliot's *Middlemarch*

casava, *use* cassava

casbah, *use* k-

case (bind.), bookbinding consisting of two boards joined by a flexible spine, and usually covered with cloth etc.; **upper** — (typ.), capitals; **lower** —, small letters

casein, milk protein, *not* -ine

cases (law), abbr. **ca.**

casework (one word)

cashew, trop. American tree with edible nuts

Cashmer/, -e, *use* **Kashmir**

cashmere, soft wool fabric, *not* -meer, cachemere, -ire. *See also* **kerseymere**

Casimir-Perier (Jean Pierre Paul), 1847–1907, French Pres. 1894–5 (no accent)

casine, *use* casein

casino/, *pl.* **-s** (not ital.)

Caslon (William), 1693–1766, typefounder; cutter of Caslon type

cassata, It. ice-cream (not ital.)

Cassation (Cour de), highest court of appeal in France

cassava, the manioc, *not* cas-, -ave

Cassel, W. Germany, *use* **K-**

Cassell & Co., Ltd., division of **Cassell & Collier Macmillan Publishers, Ltd.**

cassimere, *use* **kerseymere.** *See also* **cashmere**

Cassiopeia (astr.), a constellation

Cassivellaunus, *fl.* 50 BC, British prince

cast, actors in a play

caste, hereditary class

Castellammare, Campagna and Sicily, Italy

Castelnuovo-Tedesco (Mario), 1895–1968, Italian composer

caster, one who casts (*see also* **-or**); (typ.) a casting machine

Castil/e, N. Spain, *not* -ille; **-ian**

cast iron (noun); **cast-iron** (adj.)

Castle Cary, Somerset

Castlecary, Central, Donegal

Castlerea, Co. Roscommon, Ireland

Castlereagh (Robert Stewart, Viscount), 1769–1822, British statesman

Castleton, Borders, Derby., Gr. Manchester, Gwent, N. Yorks.

Castletown, Cumbria, Dorset, Highland, I.o.M., Staffs., Tyne & Wear

cast off (typ.), to estimate amount of printed matter copy would make

castor, a beaver (hence **castor oil**) or its fur; a small wheel for furniture; a small pot with perforated lid (hence **castor sugar**). *See also* **caster**

Castor and Pollux, stars, also patrons of sailors

castrat/o (mus.), castrated male soprano; *pl.* **-i**

casual, incidental, not regular. *See also* **causal**

casus/ belli (lat.), the cause of war; — *foederis,* case stipulated by treaty; — *omissus,* case unprovided for by statute

CAT, College of Advanced Technology, computer-assisted typesetting

Cat., Catalan, Catullus

cat., catalogue, catechism

cataclasm, violent disruption

cataclysm/, deluge, violent event; **-al, -ic, -ist,** *not* -atist

Catalan, of Catalonia; abbr. **Cat.**

Cataline, *use* **Catiline**

catalogu/e, abbr. **cat.;** **-ed, -er, -ing**

catalogue raisonné (Fr. m.), an explanatory catalogue

Catalonia, in Sp. **Cataluña**

catal/yst, a substance facilitating chem. change; **-yse,** *not* -ize; **-ysis; -ytic**

catamaran, a raft; *also* two-hulled boat

catarrh, inflammation of mucous membrane

catarrhine, narrow-nosed (of monkeys), *not* -arhine

Catch (Jack), *use* **Ke-**

catch/-all, -as-catch-can (hyphens)

catchline (typ.), a line containing catchword(s), signature letter, etc.

catch-phrase (hyphen)

catchpole, a sheriff's officer, *not* -poll

catch-22, inescapable dilemma, after novel by Joseph Heller, 1961

catchword (typ.), the first word of next page printed at bottom of preceding page; word(s) printed at head of dictionary page as guide to its contents

Câteau Cambrésis (Peace of), 1559

catechism, teaching by question and answer; abbr. **cat.**

catechiz/e, -er, -ing, *not* -is-

catechumen, convert under instruction before baptism

categorize, to classify, not -ise

caten/a, a chain; *pl.* **-ae**

cater-cousin, a good friend, *not* qua-

caters (campan., *pl.*), changes on nine bells

Cath., Catherine, Catholic

cath., cathedral

Catharine, Kans., NY, Pa., US

cathars/is, purging; *pl.* **-es**

Cathay, poetical for China, *not* K-

cathedral, cap. when with name, as Ely Cathedral; abbr. **cath.**

Catherine, abbr. **Cath.; St. —,** hence **— pear** and **— wheel; — I** and **II,** empresses of Russia; **— of Aragon, — Howard, — Parr,** first, fifth, and sixth wives of Henry VIII of England; **— de Medici,** 1519–89, Queen of France

cathisma, use **ka-**

cathod/e, negative electrode; **—e ray tube,** abbr. **CRT; -ic, -ograph;** *not* k-

Catholic, abbr. **C.** *or* **Cath.**

catholic/ism, -ize, *not* -ise

Catiline (Catilina, Lucius Sergius), 108–62 BC, Roman conspirator, *not* Cata-

cation (elec.), ion carrying positive charge of electricity, *not* k-

cat-o'-nine-tails (hyphens)

cat's/-cradle, a children's game; **-eye,** precious stone, **(C-,** propr. term) stud in road

Catskill Mountains, New York, US

cat's-paw, person used as tool, (naut.) a light air, *not* catspaw

catsuit (one word)

Cattegat, *use* Dan. **Kattegat** (Swed. **Kattegatt**)

Cattleya, an orchid genus

Catullus (Gaius Valerius), ?84–?54 BC, Roman poet, abbr. **Cat.**

Cauchy (Augustin Louis), 1789–1857, French mathematician

caucus, political meeting

Caudillo, Spanish 'leader', title assumed by Gen. Franco after Spanish Civil War

caught (cricket), abbr. **c.**

cauldron, *not* cal-

caulk, to make watertight, *not* calk

caus., causa/tion, -tive

causal, having to do with a cause. *See also* **casual**

causa/tion, -tive, abbr. **caus.**

cause célèbre (Fr. f.), famous law case, *pl. -es -es*

causerie (Fr. f.), chat, usu. as informal essay

cauterize, *not* -ise

Cavafy (Constantine), 1863–1933, Greek poet, *not* K-, -is

Cavalleria Rusticana (Rustic Gallantry), opera by Mascagni, 1890

cavass, *use* k-

caveat, a formal warning (not ital.)

caveat/ actor (Lat.), let the doer beware; — *emptor,* ditto buyer; — *viator,* ditto traveller

cave canem (Lat.), beware of the dog

caviare, pickled sturgeon-roe, *not* -iar, -ier

cavil/, to make trifling objections, **-led, -ler, -ling**

Cavour (Count Camillo), 1810–61, Italian statesman

Cawnpore, usual form in English history of the now correct **Kanpur**

Caxton (William), *c.*1422–91, first English printer

Cayenne, capital of Fr. Guiana

cayman/, American alligator, *not* cai-, kai-; *pl.* **-s**

Cayman Islands, W. Indies

cazique, *use* **cacique**

CB, Cape Breton, Cavalry Brigade, Chief Baron, citizens' band (radio), Coal Board, Common Bench, Companion (of the Order) of the Bath, (naval) Confidential Book, confined to barracks, cost benefit, County Borough, (mus.) *contrabasso* (double bass)

CBA, cost benefit analysis, Council for British Archaeology

CBC, Canadian Broadcasting Corporation

CBE, Commander (of the Order) of the British Empire

CBEL, Cambridge Bibliography of English Literature

CBI, Confederation of British Industry; (US) Cumulative Book Index

CC, 200, Caius College, Central Committee, Chamber of Commerce, Chess Club, Circuit Court, City Council, -lor, Civil Court, Common Councilman, Companion of the Order of Canada, Consular Corps, Countryside Commission, County Clerk, — Commissioner, — Council, -lor, — court, Cricket Club, Cycling Club

cc (no points), cubic centimetre (in engines, etc.), *but use* **cm³** in scientific and technical work

cc., *capita* (chapters)

c.c., carbon copy

CCC, 300, Corpus Christi College, Oxf. and Camb.

CCCC, 400

CCF, Combined Cadet Force

C.Chem., chartered chemist

CCP, Code of Civil Procedure, Court of Common Pleas, credit card purchase

C.Cr.P., Code of Criminal Procedure

CD, 400, Civil Defence

Cd, cadmium (no point)

cd, candela (no point)

Cd., Command paper, q.v. (with number following)

c.d., cum dividend (with dividend)

CDC, Commonwealth Development Corporation (till 1963 Colonial — —)

Cdr., Commander

Cdre., Commodore (naval)

CE, Chief Engineer, Civil Engineer

CE, Christian (*or* Common) Era (non-Christian usage in place of AD)

Ce, cerium (no point)

cease-fire (noun, hyphen)

Ceauşescu (Nicolae), b. 1918, president of Romania 1967–

cedar, a tree

ceder, one who cedes

cedilla, mark under a letter, as ç

Cedron, *see* **Kidron**

CEGB, Central Electricity Generating Board

Celanese, an artificial silk (propr. term)

cela/ va sans dire (Fr.), needless to say; — *viendra,* that will come

Celebes, *see* **Sulawesi**

celebrator, *not* -er

celiac, *use* **coeliac**

cellar, *not* -er

Cellini (Benvenuto), 1500–71, Italian goldsmith and sculptor

cell/o, violoncello; *pl.* -os; -ist, the player

Cellophane, transparent wrapping material (propr. term)

Celsius, temperature scale same as centigrade, abbr. **C**

Celt/, -ic, -icism, *not* K-; abbr. **Celt.**

cembal/o, harpsichord, *pl.* -os; -ist, the player

CEMS, Church of England Men's Society

C.Eng., chartered engineer

cenobite, *use* **coe-**

Cenozoic, *use* **Caino-**

censer, incense vessel

censor, Roman magistrate, official with power to suppress (parts of) books, etc.

cent., central, century; **cent** (no point), *see* **per cent**

cent/, -s, US etc. coin; abbr. **c., ct.,** *or* **cts.**

cental/, corn measure of 100 lb.; *pl.* -s; abbr. **ctl.**

centauromachy, (representation of) centaurs fighting

centaury, a plant

centenarian, one who is a hundred years old

centen/ary, -nial

centering, framing for an arch, *not* -reing. *See also* **centre**

centi, prefix meaning one-hundredth, abbr. **c**

centigrade (not cap.); abbr. **C**

centigram, *not* -gramme; abbr. **cg** (no point in scientific and technical work)

centilitre, abbr. **cl** (no point in scientific and technical work)

centime/, one-hundredth of French franc; *pl.* **-s;** abbr. **c.** *or* **cts.**

centimetre, 0.394 inch, *not* -er; abbr. *sing.* and *pl.* **cm.** (*but* no point in scientific and technical work)

Cento, Central Treaty Organization

cent/o, writing composed of scraps from various authors; *pl.* **-os**

Central, region of Scotland

central, abbr. **cent.**

centralize, *not* -ise

Central Provinces (India), *now* **Madhya Pradesh**

centr/e, -ed, -ing; (typ.), set matter in middle of line, measure, etc. *See also* **centering**

centre-notes (typ.), those between columns

cents, abbr. **c.,** *or* **cts.**

centum, a hundred; abbr. **C** (no point). *See also* **per cent**

centumvir/, Roman commissioner; *pl.* **-i**

centurion, commander of company in Roman army, *not* -ian

century (typ.), spell out in bookwork; abbr. **C.** *or* **cent.**

cephalic, of the head, *not* k-

Cephalonia, Gr. island, anc. name **Cephallenia**

ceramic, etc., *not* k-; abbr. **ceram.**

Cerberus (Gr. myth.), three--headed dog at gate of Hades

cerebell/um, the hinder brain; *pl.* **-a;** adj. **-ar**

cerebro/-spinal, -vascular (hyphens)

cerebr/um, the brain proper; *pl.* **-a;** adj. **-al**

cere/cloths, -ments, grave--clothes, orig. dipped in wax, *not* sere-

64

Ceredigion, district of Wales, formerly Cardiganshire

cerge, *use* **cie-**

ceriph (typ.), *use* **serif**

cerise (Fr. f.), cherry

cerise, a colour

cerium, symbol **Ce**

CERN, *Conseil Européen* (now *Organisation Européenne*) *de Recherches Nucléaires,* European Council for Nuclear Research

cert, colloq. abbr. of certainty (no point)

cert., certificate, certify

Cert. Ed., Certificate in Education

certiorari, a writ removing a case to a higher court

cerulean, blue, *not* cae-, coe-

Cervantes Saavedra (Miguel de), 1547–1616, author of *Don Quixote*

cervi/x (anat.), *pl.* **-ces,** adj. **-cal**

Cesarean, *use* **Cae-**

cesarevitch, Tsar's eldest son, *use* **tsarevich**

Cesarewitch, horse-race

cesser (law), the coming to an end

c'est/-à-dire (Fr.), that is to say; **— la guerre,** it is according to the customs of war; **— le premier pas qui coûte,** it is the first step that is difficult

Cestr:, sig. of Bp. of Chester (colon)

Cestrian, of Chester

cestui que/ trust (law), a beneficiary, *pl.* **cestuis — —** (*not* trustent); **— — vie,** he on whose life land is held, *pl.* **cestuis — —**

cesura, *use* **cae-**

CET, Central European Time

cetera/ desiderantur or **— desunt** (Lat.), the rest are missing

ceteris paribus (Lat.), other things being equal; abbr. *cet. par.*

Cetinje, Yugoslavia, formerly cap. of Montenegro, *not* Cett-, Zet-, -ingé

Ceylon, republic within Commonwealth, 1972, officially **Sri Lanka (Ceylon)**; adj. **Ceylonese.** *See also* **Sinhalese**

Cézanne (Paul), 1839–1906, French painter

CF, Chaplain to the Forces

Cf, californium (no point)

cf., calf, *confer* (compare)

c.f., carried forward

CFA, *Communauté Financière d'Afrique,* African Financial Community

CG, Captain of the Guard, coastguard, Coldstream Guards

C.-G., Captain-General, Commissary-General, Consul-General

cg, centigram, -s (no point in scientific and technical work)

c.g., centre of gravity

CGM, Conspicuous Gallantry Medal

CGPM, *Conférence Générale des Poids et des Mesures,* General Conference of Weights and Measures

CGS, Chief of General Staff

c.g.s., centimetre-gram-second system

CGT, Capital Gains Tax; *Confédération Générale du Travail,* General Federation of Labour

CH, clearing-house, Companion of Honour, court-house, custom-house

Ch., chaplain, Chin/a, -ese, Church, (Lat.) *Chirurgiae* (of surgery), (mus.) choir organ

ch, in Spanish and Welsh, a separate letter, not to be divided

ch., chapter, chief, child, -ren

Chablis, a white burgundy

cha-cha-cha, ballroom dance (hyphens)

chaconne, dance

chacun à son goût (Fr.), everyone to his taste

Chad/, Cent. Africa, independent republic 1960, adj. **-ian; Lake —,** in Sudan, *not* Tchad

chadar (Ind.), *use* **chuddar**

Chagall (**Marc**), b. 1889, Russian painter

chain/-gang, -smoker (hyphens)

chain reaction (two words)

chairman (m. *and* f.), abbr. **C.** *or* **chm.**

chair/person, -woman (one word)

chaise-longue/, a couch, day-bed; *pl. -es -es*

chal, a gypsy; *fem.* **chai**

chalaz/a (biol.), string attached to yolk of egg; *pl. -ae*

chalcedony, a form of quartz, *not* calce-, chalci-

Chald/ea, Babylonia (*not* -aea); **-aic, -aism** (*not* -ism), **-ean, -ee**; abbr. **Chald.**

chalet, a Swiss cottage, *not* châ- (not ital.)

Chaliapin (**Feodor Ivanovich**), 1873–1938, Russian bass singer

Châlons-sur-Marne, dép. Marne, France, battle 451

Chalon-sur-Saône, dép. Saône-et-Loire, France

chamb., chamberlain (cap. as a title of office)

chamber-maid (hyphen)

chamber music (two words)

chamber-pot (hyphen)

Chambers's/(Edinburgh) Journal; — *Encyclopaedia*

Chambertin, a red burgundy

chameleon, lizard of change-able colour, *not* chamae-

chamfer, to bevel

chamois-leather, *not* shammy-

chamomile, *use* **cam-**

Chamonix, dép. Haute-Savoie, France, *not* -ouni, -ounix, -ouny

champagne, a sparkling wine

champaign, flat open country

champerty (law), an illegal agreement, *not* -arty

champignon (Fr. m.), a mushroom, *not* -pinion

Champlain (**Lake**), between NY and Vt., US

Champlain (**Samuel de**), *c.* 1570–1635, French founder of Quebec, Can.

champlevé, a kind of enamel

Champollion (**Jean François**), 1790–1832, French Egyptologist

Champs-Élysées, Paris

chancellery, *not* -ory

chancellor (cap. as title), abbr. **chanc.**

chance-medley (law), a form of homicide (hyphen)

Chancery, abbr. **Chanc.**

chancre, venereal ulcer

chandelier, branched hanging support for lights

Chandler's Ford, Hants

Chanel (**Gabrielle**), 1883–1971, French perfumer

change/able, -ability

change-over (noun, hyphen)

channel/, -led, -ling

Channel Islands, *not* — Isles; abbr. **CI**

chanson/ (Fr. f.), a song; **-nette** (f.), a little song; **-nier** (m.), singer or song-writer in cabaret

chant/, singing; **-er**, of bagpipe, *not* chau-

Chantilly, dép. Oise, France

Chantrey/ (**Sir Francis Legatt**), 1781–1841, sculptor; — **Fund**, Royal Academy of Arts, London

chantry, endowed chapel

chanty, *use* **shanty**

chap., chapel, chaplain, chapter

chaparejos (*pl.*), cowboy's leather riding-breeches, *not* -ajos

chapeau/ (Fr. m.), a hat, *pl.* **-x**; **-x bas!** hats off!; — **-bras**, three-cornered hat carried under arm

chapel, abbr. **chap.**; (typ.), smallest organized union group in printing works, in journal and in some bookwork editorial offices; **father of** —, its chief officer

Chapel-en-le-Frith, Derby. (hyphens)

chapelle ardente (Fr. f.), a

66

chapel lighted for a lying-in-
-state (no hyphen)
Chapel Royal, *pl.* **Chapels —**
chaperon, *not* -one, -onne
Chap.-Gen., Chaplain-General
chaplain, abbr. **chap.**
Chapman (Geoffrey), publisher
Chapman & Hall Ltd., publishers
Chappell (William), 1809–88,
music publisher
chapter/, -s, abbr. **c., cap., ch.,**
or **chap.**
chapter-headings (typ.),
choice of type is a matter of
taste; no general rule can be
stated
char/, -red, -ring
char, small trout, *not* -rr
charabanc/, a motor coach; *pl.*
-s; *not* char-à-banc
characteriz/e, *not* -ise; **-able,**
-ation, -er
chargé, *more fully* **chargé**
d'affaires; *pl.* **chargés** (not
ital.)
charisma/, spiritual power of
person or office; adj. **-tic**
charivari, a mock serenade;
London Charivari (The), former
subtitle of *Punch*
Charlemagne, 742–814, king
of the Franks from 768
Charles, abbr. **C.** *or* **Chas.**
Charles's Wain (astr.), Ursa
Major
Charleston, Ill., SC (source of
the dance), W. Va., US
Charlestown, Mass., US
Charlottenburg, Berlin
charlotte russe (cook.), cus-
tard or cream enclosed in
sponge-cake (not ital.)
Charollais, breed of cattle, *not*
-olais
Charon (class. myth.), the
ferryman of the Styx
charpoy, Indian bedstead
charr, *use* char
charter flight (two words)
charter-member (hyphen)
charter-party, written evi-

dence of agreement, esp. naut.
(hyphen)
Charters Towers, Queens-
land, Australia (no apos.)
chart paper, machine-made
from best rags, suitable for lithographic map-printing
chartreuse, a liqueur first made
at Carthusian monastery near
Grenoble
Chartreux, a Carthusian monk
Charybdis, personified whirlpool in Str. of Messina. *See also*
Scylla
Chas., Charles
chase (typ.), metal frame holding composed type
Chasid/, -ic, *use* **H-**
Chasles (Michel), 1793–1880,
French mathematician; —
(Philarète), 1798–1873,
French lit. critic
chassé(Fr. m.), dance step; —
croisé, double *chassé*
chassis, window sash, a frame-
work as of motor car, *pl.* same
(not ital.)
chastis/e, *not* -ize; **-ement, -ing**
château/ (Fr. m.), a castle; *pl* **-x**
(not ital.)
Chateaubriand (François
René, vicomte de), 1768–
1848, French writer and statesman (no accent)
Châteaubriand (Fr. cook.), fillet of beef, cut thick and grilled
Châteaubriant, dép. Loire-
-Inférieure, France
Châteaudun, dép. Eure-et-Loir,
France
Châteauguay, Quebec, Can.
Château/-Lafite, -Latour,
-Margaux, clarets (hyphens)
châteaux en Espagne (Fr.),
'castles in the air'
chatelain/, lord of the manor;
fem. **-e**
Chattanooga, Tenn., US, battle
in Civil War
Chatto & Windus, Ltd., pub-
lishers
Chaucer/ (Geoffrey), ?1345–
1400, English poet; adj. **-ian**

chauffeu/r, driver of motor car, *fem.* **-se**

chaunt/, -er, *use* **chant**

chaussée (Fr. f.), a causeway, the ground level

chaussures (Fr. f. pl.), boots, shoes, etc.

Chautauqua, New York State, celebrated resort

chauvin/ism, -ist (not cap.)

Ch.B., *Chirurgiae Baccalaureus* (Bachelor of Surgery)

Ch.Ch., Christ Church, Oxford

Chebyshev (Pafnuti Lvov-ich), 1821–94, Russian mathematician, *not* Tch-

check, to stop. *See also* **cheque**

check/-in, -list, -out, -up (nouns, hyphens)

Cheddar cheese, *not* -er

Cheeryble Brothers, in Dickens's *Nicholas Nickleby*

cheese/board, -cake, -cloth (one word)

cheesy, *not* -ey

cheetah, hunting leopard, *not* chet-

chef (Fr. m.), a cook; *chef/ de cuisine,* head cook; — *d'orchestre,* leader of the orchestra; — *-d'œuvre,* a masterpiece, *pl.* **chefs-**

cheir-, *use* **chir-**

Cheka, USSR 'Extraordinary Commission (against counter--revolution)', 1917–22. *See also* **OGPU**

Chekhov (Anton Pavlovich), 1861-1904, Russian playwright and story-writer, *not* the many variants

CHEL, Cambridge History of English Literature

Chelmsford:, sig. of Bp. of Chelmsford (colon)

chem/ical, -ist, *not* chy-; abbr. **chem.**

chemico-physical (hyphen)

chemin de fer (Fr. m.), railway; a form of baccarat

chemistry (typ.), caps. for initial letters of symbols, no point at end. Names of chemical compounds to be in roman, not caps. Some prefixes are ital. or s. caps.; abbr. **chem.** *See also* **elements (chemical)**

Chemnitz, E. Germany, *now* **Karl-Marx-Stadt**

cheque, a written order (US **check**)

chequer, (noun) pattern of squares; (verb) to variegate, interrupt (*not* -ck-) ; **Chequers,** Prime Minister's official country house in Bucks.

Cherbourg, dép. Manche, France

cherchez la femme (Fr.), look for the woman

chère amie (Fr. f.), a sweetheart (*è*)

Cherenkov (Pavel Alekse-evich), b. 1904, Russian physicist, discoverer of — **radiation**

chér/i (Fr.), *fem.* **-ie,** darling (*é*)

chernozem, Russian black soil

Cherokee, N. American Indian

chersonese, name applied to various peninsulas; cap. when used as name (esp. of Thracian peninsula W. of Hellespont)

cherub/, an angelic being; *pl.* **-s**; Heb. *pl.* **-im**

Cherubini (Maria Luigi), 1760–1842, Italian composer

Cherwell, Oxford river

che sarà sarà (It.), what will be, will be

Cheshire, abbr. **Ches.**

Chesil Beach, Dorset

chess/-board, -man (hyphens)

Chester, formal name for Cheshire, now only hist. esp. in 'County palatine of Chester'

Chester-le-Street, Co. Durham (hyphens)

chestnut, *not* chesnut

chetah, *use* **chee-**

Chetniks, guerrilla forces raised in Yugoslavia during Second World War by Gen. Mihailovich; in Serbo-Croat *četnici*

cheval/ (Fr. m.), a horse, *pl.* *chevaux;* — *de frise* (mil.), obstacle to cavalry advance, *usually in the pl.* **chevaux de frise**

cheval-glass, tall mirror on frame

Chevalier/ (Albert), 1861–1923, English music-hall artiste; **— (Maurice),** 1888–1972, French entertainer

chevalier d'industrie (Fr. m.), a swindler

Chevallier (Gabriel), 1895–1969, French novelist

cheville, violin peg, stopgap word in sentence or verse

chevrotain, deerlike animal, *not* -tin

chevy, *use* **chivvy**

Chevy Chase (Ballad of)

Cheyenne, Wyoming, US; *also* an Indian tribe

Cheyne Walk, Chelsea

chez (Fr.), at the house of

chi/, Gr. letter X, χ; **— -square,** method of comparing statistical experiment with theory

Chi., Chicago

Chiang Kai-shek, 1887–1975, Nationalist Chinese statesman

Chianti, Italian wine

chiaroscuro/, light and shade, *pl.* **-s**

chibouk, a Turkish pipe, *not* chibouque

chic (noun and adj.), stylish(ness) (not ital.)

Chicago, Ill., abbr. **Chi.**

chiccory, *use* **chicory**

Chichele, Oxford professorships

Chichen Itzá, Mexico, site of antiquities

Chickahominy (battles of the), in SE Virginia, American Civil War

Chickamauga (battles of), 1863, N. Georgia

chicken-pox (hyphen)

chicory, *not* chiccory

chief, abbr. **C.** *or* **ch.; Chief/ Accountant, CA; — Baron, CB; — Justice** (caps., no hyphen), abbr. **CJ; — of General Staff,** abbr. **CGS**

chield (Sc.), a young man, *not* chiel

chiffon, dress fabric or ornaments

chiffonier, a sideboard, *not* -nnier (not ital.)

chiffre (Fr. m.), figure, numeral, monogram

chignon, a coil of hair

chigoe, W. Indian parasite, *not* jigger

child/bed, -birth (one word)

Childe Harold's Pilgrimage, by Byron, 1811–17

Childermas, 28 Dec.

childlike (one word)

child/-minder, -proof (hyphens)

Chile, S. America, *not* Chili; adj. **Chilean**

Chilianwala, Pakistan, *not* Chilianwallah, Killianwala

chilli/, red, or Guinea, pepper, *not* chile, chili, chilly; *pl.* **-es**

chimaera, *use* **chimera**

Chimborazo, Ecuador

chimer, bishop's upper robe, *not* -ere

chimera, a creation of the imagination, (biol.) an organism with two cells of two or more genetically distinct types, *not* -aera

chincapin, dwarf chestnut, *not* -kapin, -quapin

Chindits, British glider-borne troops in Burma in Second World War

chiné (Fr.), variegated (of stuffs)

Chinese/, abbr. **Ch.** *or* **Chin.;** (typ.) printed in characters numbering several thousand, which are phonetically transcribed into the Latin alphabet according to various systems, notably (1) Pinyin, e.g. Hua Guofeng, Tianjin, sometimes with accents indicating tone of spoken syllables, (2) Wade-Giles, e.g. Hua Kuo-feng, Tientsin, sometimes with superior figures indicating tone;

Chinese (*cont.*)
— **Classics,** the sacred books of Confucianism
chinoiserie, Chinese decorative objects
chipmunk, a N. American ground-squirrel, *not* -uck
Chipping Campden, Glos.
chi-rho, monogram of *XP*, first letters of Greek 'Christos'
chiromancy, palmistry, *not* cheir-
Chiron, the centaur, *not* Cheir-
chiropodist, one who treats hands and feet, *not* cheir-, -pedist
chiropract/ic (sing.), removal of nerve interference by manipulation of spinal column; **-or,** one who does this
Chiroptera, the bat order
chirosophy, palmistry, *not* cheir-
chirrup/, -ed, -ing
chirurgia (Lat.), of surgery; abbr. **Ch.**
chisel/, -led, -ling
chit, chitty (Anglo-Ind.), a letter
Chittagong, Bangladesh
chivvy, to harass, *not* che-, chivy
chlorine, symbol **Cl**
chlorodyne, medicine containing chloroform, *not* -ine
chlorophyll (bot.), green colouring matter of plants, *not* -il, -yl
Ch.M., *Chirurgiae Magister* (Master of Surgery)
chm., chairman (m. *and* f.)
choc., chocolate
chock/-a-block, -full (hyphens)
Choctaw, tribe of N. American Indians
choir, part of a church, *or* singers, *not* quire
cholera morbus, acute gastro--enteritis
cholesterol, fatty alcohol found in parts of the body, *not* -in
choosy, *not* -ey
Chopin (**Frédéric François**), 1810–49, Polish composer

'Chops of the Channel', west entrance to English Channel
chop-suey, Chinese dish
choreograph/y, ballet design; **-er,** the arranger
chorography, description of districts (intermediate between geography and topography)
chorus/, pl. -es; -ed
chose (law), a thing
Chose (**Monsieur**) (Fr.), Mr So-and-so
chose jugée (Fr. f.), a matter already decided
chota hazri (Anglo-Ind.), early breakfast
Chota Nagpur, Bihar, India
chou/ (Fr. cook. m.), cabbage, puff (pastry), a dress rosette; *—-fleur,* cauliflower (hyphen), *pl. choux-fleurs; choux de Bruxelles,* Brussels sprouts; *chou marin,* sea-kale
Chou En-lai, *use* **Zhou Enlai**
choux, rich light pastry
chow mein, Chinese dish (two words)
Chr., Christ, Christian, Chronicles
Chr. Coll. Cam., Christ's College, Cambridge
Chrétien de Troyes, 12th c., French poet
Christ Church, Oxford, *not* — — College; abbr. **Ch.Ch.**
Christchurch, Hants, *and* NZ
Christe eleison (eccl.), 'Christ, have mercy'
Christian/, -ity (cap.)
Christiania, former name of Oslo, capital of Norway; a type of turn in skiing, abbr. **Christie**
Christianize, *not* -ise (cap.)
Christie, Manson, & Woods, auctioneers, London, ('Christie's')
Christmas Day (caps.); **Old** — —, 7 January
Christoff (**Boris**), b. 1919, Bulgarian bass singer
Christ's College, Cambridge; abbr. **Chr. Coll. Cam.**

Christy-minstrels, coloured entertainers, with bones, banjos, etc. (after George Christy, of New York)

chromatography (chem.), separation by passing over adsorbent

chrome yellow, a pigment

chromium, symbol **Cr**

chromo/ (typ.), (no point), abbr. of **chromolithograph,** picture printed in colours from stone or plates; *pl.* **-s**

chromolithography (one word)

chromo paper, one with a special heavy coating for lithographic printing

chromosome (biol.), structure carrying genes in cell

chromosphere, gaseous envelope of sun

chron., chronolog/y, -ical, -ically, chronometry

Chronicles, First, Second Book of, abbr. **1 Chr., 2 Chr.**

chronological age, abbr. **CA**

chrysal/is, *pl.* **-ides**

Chrysaor, son of Neptune

Chryseis, daughter of Chryses, a Trojan priest

chryselephantine (Gr. sculpture), overlaid with gold and ivory

Chrysler/, US make of car; — **Building,** New York

chrysoprase, green chalcedony, *not* -phrase

Chrysostom (St. John), Greek Father, 347–407

chthon/ian, -ic, (Gr. myth.), dwelling in the underworld

chuddar (Anglo-Ind.), overgarment, *not* the many variants

Chur, Grisons, Switzerland; in Fr. **Coire**

Church, initial cap. when referring to a body of people, as the Methodist Church; lower case for building; abbr. **C.** *or* **Ch.**

Church Assembly (*properly* National Assembly of the Church

of England), set up 1920, absorbed into General Synod, 1970

Church Commissioners, *formerly* Ecclesiastical Commissioners and Queen Anne's Bounty

church/-goer, -going (hyphens)

Churchill (Sir Winston Leonard Spencer), 1874–1965, British statesman; — (**Winston**), 1871–1947, US writer

church mouse (two words)

church/warden, -yard (one word)

Churrigueresque, lavishly ornate Spanish baroque

chute, slide etc.

chutney, condiment, *not* -nee, -ny

chymist, *use* che-

CI, Channel Islands, Chief Inspector, — Instructor, Commonwealth Institute, Communist International

Ci, curie (no point)

CIA, (US) Central Intelligence Agency

ciao (It.), word of greeting

Cibber (Colley), 1671–1757, Poet Laureate 1730–57

Cic., Cicero

cicad/a, the tree cricket, *not* cig-, -ala; *pl.* **-ae** (biol.), **-as** (general)

cicatrice/, scar; *pl.* **-s**

cicatri/x (in surgery), scar; *pl.* **-ces**

cicatrize, *not* -ise

Cicero (Marcus Tullius, 'Tully'), 106–43 BC, Roman orator; abbr. **Cic.**

cicero (typ.), European type-measure of 12 Didot (q.v.) points

ciceron/e, a guide, *pl.* **-i**

Cicestr:, sig. of Bp. of Chichester (colon)

cicisbe/o (It.), a married woman's gallant; *pl.* **-i**

CID, Criminal Investigation Department

Cid Campeador (El), *c.* 1040–99, Spanish warrior

cider, *not* cy-
ci-devant (Fr.), formerly
CIE, Companion of the Order of the Indian Empire
Cie (Fr.), *compagnie* (company) (no point)
cierge, large candle, *not* cer-, ser-
cigala, *use* **cicada**
ci-gît (Fr.), here lies
CIGS, Chief of Imperial General Staff
CII, Chartered Insurance Institute
cili/um, a hairlike appendage; *pl.* **-a; -ary, -ate, -ation**
cill, *use* **sill**
Cimabue (Giovanni), *c.*1240–1302, Italian painter
cimbalom, dulcimer
cim/ex, bedbug; *pl.* **-ices**
Cimmerian, intensely dark; (*pl.*) nomadic people, 7th c. BC (cap.)
C.-in-C., Commander-in-Chief
Cincinnati, Ohio, US
Cincinnatus, 5th c. BC, Roman hero
cine-camera (hyphen)
cinerari/a, plant with ash--coloured foliage; *pl.* **-as**
cinerari/um, niche for urn with ashes of deceased; *pl.* **-a**
Cingalese, *use* **Sinhalese**
cingul/um, a girdle or zone; *pl.* **-a**
cinquecento, 1500–99, and Renaissance art of that century
Cinque Ports, Hastings, New Romney, Hythe, Dover, and Sandwich, plus Winchelsea and Rye
cinques (campan., *pl.*), changes on eleven bells
cion, *use* **scion**
cipher, *not* cy-
cipolin (one *l*), Anglicized form of It. *cipollino*, veined white and green marble
Cipriani (Giambattista), 1727–85, Italian painter and engraver
circ., *circiter* (about)

circa (Lat. prep.), about; in Eng used mainly with dates and quantities; abbr. *c.*
Circean, of Circe, *not* -aean
circiter (Lat. adv.), about, approximately; abbr. *circ.*
circle, sign ○; **arc of circle,** ⌒
Circuit Court, abbr. **CC**
circuit edges (bind.), limp covers turned over to protect the leaves
circulariz/e, to issue circulars, *not* -ise; **-ing**
circum (Lat.), about; in Eng. used as prefix meaning around, surrounding
circumcise, *not* -ize
circumflex (typ.), the accent ^ *See also* **accents**
circumflexion, *not* -flect-
Cirencester, Glos.
cirque, natural amphitheatre (not ital.)
cirrhos/is, a liver disease; *pl.* **-es**
cirriped, marine crustacean, *no* -de
cirr/us, (bot.) tendril, (zool.) slender appendage, (meteor.) tuftlike cloud formation; *pl.* **-i**
cisalpine, on the Roman side o the Alps (not cap. except in Cisalpine Gaul)
cispontine, on the north side o the Thames. *See also* **transpon tine**
cist (archaeol.), tomb or coffin, *not* k-. *See also* **cyst**
cit., citation, cited, citizen
citations, *see* **authorities, quotations**
cities (typ.), names of, not to be abbreviated if avoidable
citizen, abbr. **cit.**
citizens' band (radio), abbr. **CB**
citoyen/ (Fr.), *fem.* **-ne,** citizen
citrine (noun and adj.), lemon--colour(ed)
Citroën, Fr. make of car
cittadin/o (It.), a citizen, *pl.* **-i;** *fem.* **-a,** *pl.* **-e**

city, abbr. **c.; the City,** London's financial centre; **City Editor,** the supervisor of financial reports, (N. Amer.) editor of local news. *See also* **cities**

ciudad (Sp.), city, not *ciu-*

Ciudad Trujillo, recent name for Santo Domingo, cap. of Dominican Republic

civ., civil, -ian

civics, the study of citizenship

Civil/ Court, abbr. **CC;** — **Engineer,** abbr. **CE**

civiliz/e, *not* -ise; **-able, -er**

Civil/ Servants, — Service (caps.), abbr. **CS**

Civitavecchia, Italy (one word)

CJ, Chief Justice

Cl, chlorine (no point)

cl, centilitre, -s (no point in scientific and technical work)

cl., class, clause, cloth

Clackmannan, district and former county of Scotland (three ns)

Clairaut (Alexis Claude), 1713–65, French mathematician, *not* -ault

claire-cole, *use* **clearcole**

clairvoyant/, *fem.* **-e** (not ital.)

clam/, to clog, smear; **-miness, -my**

clamjamphrie (Sc.), rubbish, *also* a mob, *not* the many variants

clam/our, *but* **-orous**

clamp-down (noun, hyphen)

clang/our, *but* **-orous**

Claparède (Jean Louis René Antoine Édouard), 1832–70, Swiss naturalist

claptrap, empty words (one word)

claque/, hired applauders; **-ur,** member of claque (not ital.)

clar., (mus.) clarinet; (typ.) clarendon

clarabella (mus.), an organ stop, *not* clari-

Clarenceux, the second King-of-Arms, *not* -cieux

clarendon (typ.), a thickened modern type, *not* a generic term for bold-face type (q.v.); abbr. **clar.**

Clarendon/ Press, named after — **Building,** a former printing house of the University Press at Oxford; hence still the imprint on certain learned books of the OUP

claret, red wine of Bordeaux

clarinet/, -tist, *not* -ionet; abbr. **clar.**

Clarissa Harlowe, by Richardson, 1748

class, abbr. **cl.**

class., classic, -al, classification

classes (bot., zool.), (typ.) to be in roman with capital initials

clause so-and-so (typ.), to have lower-case *c*; abbr **cl.,** *pl.* **cll.**

Clausewitz (Karl von), 1780–1831, Prussian mil. writer

clayey, of the substance of clay

CLB, Church Lads' Brigade

cld., called, cancelled, cleared (goods or shipping), coloured

clean (typ.), said of proofs or revises with few errors, or pulled after matter has been corrected

clean disc, *see* **disc**

clearcole, (noun and verb), (to paint with) a coating of size, *not* claire-

clear days, time to be reckoned exclusive of the first and last

cleared (goods or shipping), abbr. **cld.**

clear/-eyed, -headed, -sighted (hyphens)

clearing bank (two words)

clearing-house, abbr. **CH**

clearstory, *use* **clere-**

clearway (one word)

clef (mus.). Three are now in use:

the C clef, 𝄡 (movable),

the treble clef, 𝄞, and the

bass clef, 𝄢 .

cleistogam/ic, -ous (bot.), permanently closed and self--fertilizing, *not* k-

Clemenceau (Georges Eugène Benjamin), 1841–1929, French politician, *not* Clé-

Clemens (Samuel Lang-horne), 1835–1910 ('Mark Twain'), US writer
clench, to secure (a nail or rivet), to grasp firmly; the action or result of these. *See also* **clinch**
clepsydra/, an ancient water--clock, *not* k-; *pl.* **-s**
cleptomania, *use* k-
clerestory (archit.), upper part of wall with row of windows, *not* clear-
clerk, abbr. **clk.**
Clerke (Agnes Mary), 1842–1907, English astronomer
Clerk-Maxwell (James), 1831–79, *see* **Maxwell (James Clerk)**
Clerk/ of Parliaments (caps.); **— of the Peace,** abbr. **CP; — of the Privy Council,** abbr. **CPC**
Clermont-Ferrand, dép. Puy--de-Dôme, France (hyphen)
Cleveland, county of England
clevis/, U-shaped iron at end of beam for attaching tackle, *pl.* **-es**
clew, ball of thread; (naut., verb or noun) (to draw up) lower corner of sail. *See also* **clue**
cl. gt. (bookbinding), cloth gilt
cliché (typ.), European term for a block or stereotype; a hack-neyed phrase
clicker (typ.), former title of foreman of a companionship (q.v.)
Clicquot (Veuve), a brand of champagne
clientele, clients collectively (no accent, not ital.)
climacteric (noun or adj.), (pertaining to) a critical period; **grand —,** the age of 63
climactic, pertaining to a climax
climatic, pertaining to climate
clinch, to make fast a rope in a special way, to embrace, to settle conclusively; the action or result of these. *See also* **clench**

Clinker (Humphry, *not* -ey), by Smollett, 1771
clinometer, instrument for measuring slopes, *not* k-
Clio (Gr.), muse of history
cliometrics (*pl.* used as *sing.*), statistical method of historical research
cliqu/e, -ish, -ism, -y, *not* -eish, -eism, -ey (not ital.)
clish-ma-claver (Sc.), gossip (hyphens)
Clitheroe, Lancs.
clitor/is (anat.), female organ; adj. **-al**
clk., clerk
cll., clauses
Cllr., Councillor
Cloaca Maxima, sewer of anc. Rome (caps.)
cloche (hort.), bell-glass *or* glass cover for plants; woman's close--fitting hat (not ital.)
clock-face (hyphen)
cloff, an allowance on commod-ities, *not* clough
cloisonné (Fr.), a kind of enamel
close (typ.), the second member of any pair of punctuation marks, as ']); *pron.* kloze
close up (typ.), to push together, to remove spacing-out leads; **close matter,** unleaded, or thinly spaced
closure, a stopping, esp. of de-bate in Parliament; in Fr. f. *clôture*
Clos Vougeot, a burgundy
clot-bur (bot.), burdock, *not* clote-, cloth-
clote (bot.), burdock and similar burry plants, *not* — -bur
cloth/, abbr. **cl.; — of Bruges,** gold brocade
Clotho (Gr. myth.), the Fate that draws the thread of life
clôture (Fr. f.), closure
clou (Fr. m.), point of chief interest
Cloud-cuckoo-land (hyphens, one cap.)
cloudy (meteor.), abbr. **c.**

clough, a ravine

clove hitch, a knot (two words)

clubbable, sociable

club-foot, congenital disorder (hyphen)

clubhouse (one word)

clue, information to be followed up. *See also* **clew**

Cluniac, (monk or nun) of branch of Benedictine order stemming from Cluny, France

Clwyd, river and county, N. Wales

Clydebank, Strathclyde, a town

Clytemnestra, wife of Agamemnon, *not* Clytae-

CM, Certified Master *or* Mistress, *Chirurgiae Magister* (Master of Surgery), (mus.) common metre, Corresponding Member, Member of the Order of Canada

Cm, curium (no point)

c.m., *causa mortis* (by reason of death)

cm., centimetre(s) (*but no point in scientific and technical work*)

Cmd., Command paper (q.v.) (old series), with number following

cmdg., commanding

Cmdr., Commodore (naval)

CMG, Companion of (the Order of) St. Michael and St. George

Cmnd., Command paper (q.v.) (new series), with number following

CMP, Commissioner of the Metropolitan Police

CMS, Church Missionary Society

CNAA, Council for National Academic Awards

CND, Campaign for Nuclear Disarmament

Cnossus, Crete, *use* **Knossos**

Cnut, more correct form of **Canute,** *not* Knut

CO, Colorado (off. postal abbr.), Commanding Officer, conscientious objector, criminal offence, Crown Office

Co, cobalt (no point)

Co., Colón, Company, County

c/o, care of

coad., coadjutor

coaetaneous, *use* **coe-**

coagul/um, a clot; *pl.* **-a**

Coalbrookdale, Shropshire

coal/-dust, -face (hyphens)

coal/field, -man (one word)

coal-mine (hyphen)

coalmouse (ornith.), the coal titmouse, *not* cole-

coal-pit (hyphen)

Coalville, Leics.

coastguard, one man or the body of men (one word), abbr. **CG**

coat of arms (three words)

co-author, noun and verb (hyphen)

coaxial (one word)

cobalt, symbol **Co**

Cobbe (**Frances Power**), 1822–1904, British social writer

coble, boat, *not* cobble

Coblen/ce, -z, W. Germany, *use* **Koblenz**

COBOL (Common Business Oriented Language), a computer-programming language; caps. or even s. caps.

cobra de capello, Indian snake, *not* da, di

Coburg, Iowa, US; **Saxe- —,** Germany, *not* -ourg, -urgh

coca, S. Amer. shrub, source of cocaine

Coca Cola, abbr. **Coke,** carbonated soft drink from cola nut and cacao (propr. terms)

Cocagne, *use* **Cockaigne**

cocaine, *not* -ain; pop. abbr. **coke**

cocco (bot.), Jamaica plant tuber, *not* cocoa, coco

coccy/x (anat.), end of spinal column; *pl.* **-ges**

Cochin-China, formerly part of Fr. Indo-China, now of Vietnam (hyphen)

cochle/a, the ear cavity; *pl.* **-ae**

cochon de lait (Fr. m.), sucking pig

Cockaigne, London, *not* Cocagne, Cockayne

cock-a-leekie (Sc.), soup, *not* cockie-leekie, cocky-leeky

cockatiel, small crested Australian parrot, *not* -teel

cockatoo/, large crested Australian parrot; *pl.* **-s**

cockatrice, fabulous serpent

Cockcroft (Sir John Douglas), 1897–1967, British nuclear scientist

Cocker (Edward), 1631–75, reputed author of *Arithmetic*; '**according to Cocker**', accurate

cockney, native of E. London (not cap.)

cockscomb, a plant. *See also* **coxcomb**

cockswain, *use* **coxswain**

Cocles (Horatius), who kept the Sublician Bridge

coco, trop. palm-tree bearing coconut. *See also* **cacao, cocco**

coconut, *not* cokernut

Cocos Islands (Keeling Islands), Bay of Bengal

cocotte, small dish, (arch.) a prostitute

Cocytus, a river in Hades

COD, cash, *or* collect, on delivery

COD, Concise Oxford Dictionary

cod., codex

Code Napoléon (Fr. m.), civil code promulgated 1804–11

cod/ex, ancient MS, abbr. **cod.**; *pl.* **-ices,** abbr. **codd.**

codfish (one word)

codling, small cod, elongated apple

cod-liver oil (one hyphen)

Cody (William Frederick), 1846–1917, 'Buffalo Bill', US showman

coed, coeducational, girl in co-educational school (no point)

coeducation/, -al (one word)

coefficient (one word)

Coelebs in Search of a Wife, by Hannah More, 1709

Coelenterata, a zool. phylum (not ital.)

coeliac, abdominal, *not* ce-

coenobite, member of monastic community, *not* cen-

coequal (one word)

coerulean, *use* **ce-**

coetaneous, of the same age, *not* coae-

coeternal (one word)

Cœur de Lion, Richard I of England, or Louis VIII of France (no hyphens)

coeval, of the same age (one word)

coexist/, -ence (one word)

coextensive (one word)

C. of E., Church of England

coffee bar (two words)

coffee/-bean, -cup, -house, -pot (hyphens)

coffer-dam, a watertight enclosure

cogito, ergo sum (Lat.), I think, and therefore I exist (motto of Descartes)

cognate, abbr. **cog.**

cogniz/e, to become conscious of, *not* -ise; **-able, -ance, -ant**

cognoscent/e (It.), a connoisseur; *pl.* **-i,** *not* con-

cognovit (*actionem*) (law), an acknowledgement that the action is just

coheir/, -ess (one word)

COHSE, Confederation of Health Service Employees

COI, Central Office of Information

COID, Council of Industrial Design, *now* Design Council

coiff/eur (Fr. m.), **-euse** (f.), hairdresser; **-ure** (f.), head-dress, hair-style

Coire, Grisons, Switzerland; *use* **Chur**

coits, *use* **quoits**

coke, pop. abbr. for **cocaine**

Coke, abbr. for **Coca Cola** (propr. term)

Coke (Sir Edward), 1552–1634, Lord Chief Justice of England

cokernut, *use* **coconut**

Col., Colonel, Colossians

col., colonial, column

cola, W. Afr. tree and its seed, *not* k-

colander, a strainer, *not* colla-, culle-

colcannon, an Irish dish, *not* cale-, cole-

cold composition, (typ.) *see* **hand-setting;** (comp.) *see* **strike-on**

Cold Harbor, Va., US, scene of three Civil War battles

Coldharbour, Dorset, Glos., Surrey

Cold Harbour, Lincs., Oxon.

cold/ metal, — type, (typ.) sorts cast up for hand composition or correction; (comp.) *see* **strike-on**

Coldstream/, Borders, a town; **— Guards,** *not* The Coldstreams, abbr. **CG**

cold war (the) (not caps.)

colemouse, *use* **coalmouse**

Coleoptera (Zool.), the beetles

cole-pixy, *use* **colt-pixie**

Colerain, NC, US

Coleraine, Co. Londonderry

Coleridge (Hartley), 1796–1849, English minor poet and critic, son of next; **— (Samuel Taylor),** 1772–1834, English poet and critic; **— (Sara,** *not* -ah), 1802–52, English writer

Coleridge-Taylor (Samuel), 1875–1912, English composer

coleslaw, salad of chopped cabbage (one word)

Colette, pen-name of **Sidonie Gabrielle Claudine Colette,** 1873–1954, French writer

colic/, -ky

Coligny (Gaspard de), 1519–72, French Huguenot leader

Coliseum, London theatre. *See also* **Coloss-**

coll., colleague, collect/ed, -ion, -or, college

collage, art composition with components pasted on to a surface

colla/ parte *or* **— voce** (mus.), adapt to principal part *or* voice; abbr. **col. p.,** or **col. vo.**

collander, *use* **colander**

collapsible, *not* -able

coll' arco (mus.), with the bow

collar of SS, *see* **SS (collar of)**

collat., collateral, -ly

collat/e, (bind.) to check sequence of gathered sections; (typ.) to put the sheets of a book in right sequence; (bibliog.) to compare minutely two copies of same book, to provide information for such a comparison; **-or,** *not* -er

colla voce, *see* **colla parte**

colleague, abbr. **coll.**

collect., collectively

collectable, *not* -ible

collectanea (Lat. pl.), collected notes

collect/ion, -or, abbr. **coll.**

collections, an Oxford college examination

collective nouns, *see* **nouns (collective)**

colleen (Anglo-Irish), a girl

college, abbr. **coll.** (cap. when used as part of name)

Collège de France (m.)

College of Arms, abbr. **CA**

College of Justice (Sc.), supreme civil courts

collegi/an, a member of a college; **-ate,** belonging to a college

collegi/um, an ecclesiastical body uncontrolled by the State; *pl.* **-a**

col legno (mus.), with the wood of the bow

Colles fracture (med., of wrist)

collie, a dog, *not* -y

Collins, a letter of thanks for hospitality (from the character in Jane Austen's *Pride and Prejudice*)

Collins (William) Sons & Co., Ltd., publishers

collogue, to talk confidentially

collop, a piece of meat

colloq., colloquial, -ly, -ism

colloqui/um, conference; *pl.* **-a**

colloqu/y, talk; *pl.* **-ies; -ize,** *not* -ise

collotype (typ.), a planographic (q.v.) process used mainly for the finest reproduction work

Colo., Colorado, US

Cologne, W. Germany, in Ger. **Köln**

Colombia (Republic of), S. America. *See also* **Colu-**

Colombo, cap. of Sri Lanka, *not* Colu-

Colón, Cent. Amer.; abbr. **Co.** *See also* **Columbus**

colon, *see* **punctuation** IV

Colonel/, abbr. **Col.;** — **Bogey** (golf), the imaginary good player

colonial, abbr. **col.**

coloniz/e, *not* -ise; **-able, -ation**

colonnade, row of columns

colophon (typ.), an inscription at the end of a book, giving the sort of details now usually found on the title-page; incorrect name for publisher's device or trade mark

colophony, dark resin

Colorado, off. abbr. **Colo.**

coloration, *not* colour-

Colosseum, Rome. *See also* **Colis-**

Colossians, abbr. **Col.**

coloss/us, *pl.* -i

Colossus of Rhodes (cap.)

colourant, *not* colorant

Coloured/ (S. Afr.), of Asian or mixed ancestry (adj.); **-s,** Cape Coloured people (cap.).

colourist, *not* colorist

Colour-Sergeant, *not* -jeant; abbr. **Col.-Sgt., Col.-Sergt.**

Colour (Trooping the), *not* Colours

col. p. (mus.), colla parte (adapt to principal part)

colporteur, hawker of books, esp. bibles (not ital.)

Colquhoun, *pron.* ko-hoon (stress on second syllable)

colt, abbr. **c.**

colter, *use* **cou-**

colt-pixie, a mischievous fairy, *not* cole-pixy

columbari/um, a dovecot, place for cremation urns; *pl.* -a

Columbia (British), Canada, abbr. **BC**

Columbia (District of), US, abbr. **DC** (no points). *See also* **Colo-**

columbium, *now called* **niobium**

Columbo, *use* **Colombo**

Columbus (Christopher), 1451–1506, Genoese navigator; in Sp. **Cristóbal Colón**

column, abbr. **col.**

colure (astr.), each of two great circles of the celestial sphere

Colville of Culross (Viscount)

col. vo. (mus.), colla voce (adapt to principal voice)

Com., Commander, Communist

com., comedy, comic, commercial, commission, -er, committee, common, -er, -ly, commun/e, -ity, communicat/e, -ed, -ion

com/a (med.), stupor; *pl.* **-as**

com/a (astr.), nebulous envelope round head of comet, (bot.) tuft of hair; *pl.* **-ae**

comb., combin/e, -ed, -ing

combat/, -ed, -ing, -ive

Combermere (Viscount)

combin/e, -ed, -ing, abbr. **comb.**

Comdr., Commander

Comdt., Commandant

come-at-able, accessible (hyphens)

Comecon, economic association of Communist countries

Comédie-Française (La), off. name of Le Théâtre Français (caps.)

Comédie Humaine (La), series of novels by Balzac, 1842–8

comédien/ (Fr.), actor; *fem.* **-ne**

comedy, abbr. **com.**

come/ prima (mus.), as at first; — **sopra,** as above, abbr. **co. sa.**

comfit/, sweetmeat, *not* con-

comfrey (bot.), ditch plant, *not* cum-

comic, abbr. **com.**

Comintern, the Communist

(Third) International, 1919-43,
not Kom-

:omitadji, band of irregulars
(in the Balkans), *not* k-

omitatus (Lat.), a retinue, a
county or shire; *pl. same*

omitia (Lat. pl.), Roman assembly of the people

Comitia (*pl.*), a meeting of the
Senate of Dublin University;
— *Aestivalia,* ditto in summer;
— *Hiemalia,* ditto in winter;
— *Vernalia,* ditto in spring

omity of nations, international courtesy

Comm. (naval), Commodore

omm., commentary, commerce, commonwealth

omma/, *pl.* **-s;** *see* **punctuation** II

Commandant, abbr. **Comdt.**

ommandeer, to seize for mil.
or other service

Commander/, abbr. **Cdr.** (Admiralty), **Com.** *or* **Comdr.;**
— **-in-Chief** (hyphens), abbr.
C.-in-C.

:ommanding, abbr. **cmdg.;**
Commanding Officer, abbr.
CO

:ommando/ (member of) body
of troops, esp. shock troops; *pl.*
-s

Command paper, abbr. **C.**
(1870-99) *or* **Cd.** (1900-18) *or*
Cmd. (1919-56) (old series);
Cmnd. (1956-   ) (new series),
followed by number

omme ci, comme ça (Fr.),
indifferently, so-so

:ommedia dell'arte, Italian
Renaissance comedy, mainly
improvised by guild of professional actors

omme il faut (Fr.), as it should
be

Commemoration, at Oxford,
annual ceremony in memory of
founders and benefactors

Commencement, ceremonial
conferring of university degrees,
as at Cambridge; (US) a speech-
-day

commendam, ecclesiastical
benefice held *in commendam,* i.e.
without duties

commentary, abbr. **comm.**

commentator, *not* -er

commerce, abbr. **comm.**

Commissary-General (hyphen); abbr. **CG**

commission/, -er, abbr. **com.**

commissionaire, uniformed
door-attendant (not ital.); in Fr.
m. *commissionnaire,* less specific in meaning

commit/, -ment, -table (*not*
-ible), **-tal, -ted, -ting**

committee, representative
group from a larger body (coll.
sing., with sing. verb); abbr.
com. See also **committor**

committ/er, one who commits;
-or (law), a judge who commits
a lunatic to the care of a **committee** (stress on final syllable)

Commodore (naval), abbr.
Cdre, Cmdr. *See also* **Air Commodore**

common/, -er, -ly, abbr. **com.**

Common Bench, abbr. **CB**

common metre (mus.), abbr.
CM

commonplace (adj.) (one
word)

Common Pleas, abbr. **CP**

common sense, adj. with noun
(hyphen when attrib.)

commonsensical (adj.) (one
word)

Common Serjeant, *not* -geant;
abbr. **CS** *or* **Com.-Serj.**

common time (mus.), abbr. **C**

Common Version (of the Bible); abbr. **CV** *or* **Com. Ver.**

commonwealth (cap. when
part of name); abbr. **comm.**

Commonwealth (Australian),
abbr. **Cwlth**

commune, communal group
(esp. sharing accommodation),
small territorial division
(France and elsewhere); abbr.
com.

communicat/e, -ed, -ion, abbr.
com.

communications satellite, *not* -ation (two words)

communiqué, an official report

Communist, abbr. **Com.;** — **Party,** abbr. **CP**

communize, to make common; *not* -ise

commutator, apparatus for reversing electric current

commuter, one who travels regularly to work in a town

Comor/o Islands, Indian Ocean, indep. 1975, adj. **-an**

comp., comparative, comparison, compil/e, -ed, -er, -ation, compos/er, -ite, -ition, -itor, compound, -ed

compagnie (Fr. f.), company; abbr. *Cie*

compagnon de voyage (Fr. m.), travelling companion

Companies Act (no apos.)

companionship (typ.), former name of a group of compositors working together under a clicker (q.v.); abbr. **'ship**

company, abbr. **Co.,** (mil.) **Coy.** (point). *See also compagnie*

comparative, abbr. **comp.**

compare, abbr. **cf.** (*confer*), or **cp.**

comparison, abbr. **comp.**

compass (typ.), the compound points, when printed in full, to be hyphened without caps., as north-east, north-by-north-east, north-north-east, north-east-by-north. *See also* **abbreviations, capitalization**

compendi/um, *pl.* **-a**

competit/or, *fem.* **-ress**

Compiègne, dép. Oise, France

compil/e, -ed, -er, -ation, abbr. **comp.**

complacen/t, self-satisfied; **-cy**

complaisan/t, obliging; **-ce**

Compleat Angler (*The*), by Izaak Walton, 1653

complement (verb and noun), to make complete, that which makes complete

completori/um (Lat.), compline; *pl.* **-a**

complexion, *not* -ction

compliment (verb and noun) (to) praise, flatter(y)

compline, last service of the da (RCC), *not* complin

compos/er, -ite, -ition, -itor, abbr. **comp.**

compos mentis (Lat.), in one' right mind

compote (Fr. f.), stewed fruit

compound/, -ed, abbr. **comp**

compound ranks *or* **titles,** each word to have cap., as Assistant Adjutant-General, Vice-President

comprehensible, *not* -able

compris/e, *not* -ize; **-ed, -ing**

compromise, *not* -ize

compte/ (Fr. m.), an account; — *rendu,* official report, *pl. comptes rendus*

'Comptes Rendus', reports of the French Academy (caps.)

Compton-Burnett (Dame Ivy), 1884–1969, English novelist (hyphen)

comptroller, retain when citing official title which has it, otherwise *use* **controller**

Comptroller-General of the National Debt Office, ditto **of the Patent Office** (caps., hyphen)

computer, calculating and data-processing machine

computerize, *not* -ise

Com.-Serj., Common Serjeant

Comsomol, *use* K-

Comt/e (Auguste), 1798–185 French philosopher; **-ian, -ism, -ist**

comte (Fr.), Count; *fem.* **comtesse** (not cap.)

Con., Consul

con/, to direct a ship's course, t examine, to swindle, *not* -nn, -un; **-ning**

con., conclusion, conics, conver sation

con., contra (against, in oppositio to)

Conakry, Guinea

on amore (It.), with affection (not ital. in mus.)

on brio (mus.), with vigour

onc., concentrat/ed, -ion

Conceição, name of many places in Portugal and Brazil

oncensus, *use* **conse-**

oncentration camp (two words)

Concepción, name of many Cent. and S. American places (but not in Brazil)

concert/-goer, -hall, -master (hyphens)

Concert/meister, -stück (Ger. mus.), *now* **Konzert-** (cap.)

Concertgebouw Orchestra, Amsterdam

concertin/o (mus.), group of soloists, short concerto, *pl.* **-os**

concerto/ (mus.), *pl.* **-s** (but **concerti grossi**); abbr. **cto.**

concessionaire, holder of a concession, usu. of a monopoly from a foreign government (not ital.), in Fr. m. **concession-naire**

conch/, a shell; *pl.* **-s**

conch., conchology

conchy, *see* **conscientious objector**

oncierg/e (Fr. m.f.), house--porter, -portress; **-erie** (f., no accent), porter's lodge

Conciergerie (La), former Paris prison

onclusion, abbr. **con.**

Concord, Calif., Mass., NC, NH, US

oncordance, index of words or phrases used in book or by author

oncordat, an agreement (not ital.)

Concorde, make of aircraft

concours (Fr. m.), a competition; *— d'élégance,* parade of vehicles

concur/, -red, -ring

onde (Sp. & Port.), a count, *not* **-dee**

Condé, Louis II de Bourbon,

Prince de ('The Great'), 1621–86, French general

condensed type, a narrow typeface

condottier/e (It.), a captain of mercenaries; *pl.* **-i**

conductor, *not* -er; abbr. **c.**

con espressione (mus.), with expression, *not* ex-; abbr. **con esp.**

coney, *use* **cony**

Coney Island, NY, US

conf., conference

confection/ery (noun), **-ary** (adj.)

Confederacy (US hist.), the league of seceding states in the Civil War

confer (Lat.), compare; abbr. **cf.**

conference, abbr. **conf.**

conférence, (Fr. f.), lecture

confer/, -red, -ring, -rable, -rer, *but* **-ee, -ence, -ment**

Confessio Amantis, by Gower, 1393

confetti, bits of coloured paper, etc.

confidant/, a trusted friend, *fem.* **-e**

confrère (Fr. m.), colleague (not ital.)

Confucius, 551–479 BC, Chinese sage

con fuoco (mus.), with passion

cong., congregation, -al, -alist -ist, congress, -ional

congé/ (Fr. m.), leave; *— d'élire,* leave to elect, *not — de lire*

Congo (Democratic Repub-lic of), Cent. Africa, *now* **Zaïre**

Congo (People's Republic of the), Cent. Africa, indep. 1960, *formerly* Fr. Middle Congo; adj. **Congolese**

congou, a black tea, *not* -o, kongo

congregation/, -al, -alist (cap. as relig. denomination); abbr. **cong.**

Congresbury, Avon

Congress, India, the political party; US, the legislative body

Congress (*cont.*)
(cap.); spell out number, as
fifty-fourth Congress, not 54th
congress/, -ional, abbr. **C.** *or*
cong.
conics, geometry of the cone
and its sections; abbr. **con.**
conjugation, inflexions of a
verb; abbr. **conj.**
conjunction, connection; abbr.
conj.; (astr.) apparent proxi-
mity of two heavenly bodies,
sign ☌
conjunctiv/a, mucous membrane between eyelid and eye-
ball; *pl.* **-ae** (not ital.); **-itis,**
inflammation of this
conjuror, a magician; one
bound by oath, *not* -er
con moto (mus.), with (more)
rapid motion
Conn., Connecticut, US
Conna, Co. Cork
Connacht, province of Ireland
Connah's Quay, Clwyd
connect/er, one who connects;
-or, that which connects
Connecticut, US; abbr. **Conn.**
or (postal) **CT**
connection, *not* connexion
connivance, tacit assent, *not*
-ence
connoisseur, well-informed
judge in matters of taste (not
ital.)
conoscente, use **cogno-**
conquistador/ (Sp.), a conqueror, esp. of Mexico and Peru;
pl. **-es**
Conrad (Joseph), real name
Teodor Józef Konrad Korzen-
iowski, 1857–1924, Polish-Eng-
lish novelist, *not* K-
cons., consecrate(d), conserva-
tion, consonant, constable, constitution, -al
Conscience (Hendrik), 1812–
83, Belgian novelist
conscience' sake
conscientious objector, abbr.
CO; (First World War colloq.)
conchy
consenescence, general decay

consensus, *not* -census
conservat/oire (Fr. m.); *-orio*
(It., Port., Sp.); *Konserva-*
torium (Ger. n., cap.)
consol., consolidated
console, (archit.) bracket; key
-desk of an organ; cabinet for
radio(gram), etc.
Consolidated Annuities,
abbr. **Consols**
consommé, clear soup
consonant, abbr. **cons.**
consonantize, to make conso-
antal; *not* -ise
con sordino (mus.), with a mu
conspectus (Lat.), a general
view (not ital.)
con spirito (mus.), spiritedly
constable, abbr. **c.** *or* **cons.**
Constance, lake and town,
Switzerland
Constans, Roman emperor
337–50
Constans, mythical King of
Britain
Constant (Benjamin; in full,
**Henri Benjamin Constant
de Rebecque**), 1767–1830,
French Swiss political novelist
— (**Benjamin Jean Joseph**)
1845–1902, French painter
Constantino/ple, *now* usu. **Istanbul; -politan**
Constitution of a country (ca
C)
constitution/, -al, abbr. **cons**
construction, abbr. **constr.**
Consuelo, by George Sand,
1842
Consul, abbr. **C.** *or* **Con.;** Lat.
pl. **consules,** abbr. **coss.**
Consul-General, abbr. **C.-G.**
Consumers' Association,
abbr. **CA**
consummat/e (verb), to perfec
or complete; (adj.), perfect,
complete; noun **-ion**
consummatum est (Lat.), it i
finished
cont., containing, contents, con
tinent, continue, -d
contadin/o, Italian peasant, *pl.*
-i; *fem.* **-a,** *pl.* **-e**

contagi/um, contagious matter;
  pl. -a
containing, abbr. **cont.**
contakion, use **kontakion**
contang/o (Stock Exchange),
  charge for carrying over; *pl.* **-os**
conte (Fr. m.), short story
conte (It. m.), a count
contemptible, *not* -able
contents, abbr. **cont.**
conterminous, with a common
  end or boundary, *not* cot-
Contes drolatiques, by Balzac,
  1832-7
contessa (It.), a countess
Continent (the) (cap.); **con-**
  tinent of Europe, etc. (not
  cap.); abbr. **cont.**
continu/e, -ed, abbr. **cont.**
continu/o (mus.), an improvised
  keyboard accompaniment; *pl.*
  -os
continu/um (philos., phys.), a
  continuous quantity; *pl.* **-a**
cont-line (naut.), spiral space
  between rope strands, space between stowed casks
contr., contract, -ed, -ion, -s,
  contrary
contra (Lat.), against, in opposition to; abbr. **con.**
contrabasso (It. mus.), the
  double-bass, abbr. **CB**
contract/, -ed, -ion, -s, abbr.
  contr.
contracting-out (noun, hyphen)
contractions, *see* **abbrevi-**
  ations
contractor, *not* -er
contrafagotto (It. mus.), the
  double bassoon (one word)
contra jus gentium (Lat.),
  against the law of nations
contralt/o (mus.), lowest female
  voice; *pl.* **-os;** abbr. **C.**
contra/ mundum (Lat.),
  against the world; — *pacem*
  (law), against the peace
contrariwise
contrary, abbr. **contr.**
Contrat social (*Le*), by J. J.
  Rousseau, 1762

contretemps, a mishap (not
  ital.)
contributor, *not* -er
control/, -led, -ling
controller, *see* **comptroller**
Controller-General, *but*
  Comptroller-General of the
  National Debt Office *and* of
  the Patent Office
Controller of Accounts, abbr.
  CA
conundrum/, riddle; *pl.* **-s**
convector, apparatus for heating by convection
convenances (*les*) (Fr. f. pl.),
  the proprieties
convener, one who calls a meeting
convention/, -al, abbr. **conv.**
conventionalize, *not* -ise
conversation, abbr. **con.**
conversazione/, a meeting for
  learned conversation; *pl.* **-s**
converter, *not* -or
convertible, *not* -able
convey/er (person); **-or** (thing)
  esp. in **conveyor belt**
Convocation, a provincial assembly of the Church of England; a legislative assembly of
  certain universities
convolvulus/, flower; *pl.* **-es**
Conwy, river and town, Gwynedd
cony/, a rabbit, *not* coney, *pl.*
  conies; — **-garth,** a rabbit-
  -warren
Conyngham (Marquess)
Cooch Behar, India, *not* Kuch
cooee, (esp. Australian bush)
  call, *not* -ey, -hee, -ie
Cook (Captain James), 1728-
  79, British explorer; —
  (**Thomas**), 1808-92, English
  travel agent
cookie, (Sc.) a bun, (US) a small
  cake or biscuit, *not* -ey, -y
coolabah, Austral. gum-tree, *not*
  -ibah
coolie (Ind., Chin.), native hired
  labourer, *not* -y
Coomassie, Ghana, *use* **Ku-**
  masi

Cooper (**James Fenimore**, one *n*), 1789–1851, US novelist

co-operat/e, -ion, -ive (hyphen), *but* **uncooperative** (no hyphen)

Co-operative Society (hyphen)

co-opt (hyphen)

co-ordinate (hyphen); *but* in math. usu. **coordinate**; *and* **uncoordinated**

Cop., Copernican

cop., copper, copulative

copaiba (med.), a balsam from a S. American tree, *not* -va

coparcen/ary (law), joint heirship to an undivided property, *not* -ery; **-er,** a joint heir

COPEC, Conference on Christian Politics, Economics, and Citizenship

copeck, in Russ. **kopeĭka,** a hundredth of a rouble, *not* -ec, -ek, ko-; abbr. **c.**

Copenhagen, in Dan. **København**

Copernican, abbr. **Cop.**

copier, *not* copyer

copper, abbr. **cop.**; symbol **Cu** (*cuprum*)

copperas, ferrous sulphate, green vitriol

copperize, to impregnate with copper, *not* -ise

copro- (in compounds), dung-, *not* k-

Copt., Coptic, liturgical language of Egyptian Christians

copul/a (gram., logic, anat., mus.), that which connects; *pl.* **-ae**

copulative, abbr. **cop.**

copy (typ.), matter to be reproduced in type. *See also* **manuscript**

copy-book (hyphen)

copyer, *use* **-ier**

copyholder (typ.), (formerly) a proof-reader's assistant

copy preparation, *see* **manuscript**

copyright notice, the symbol ©, name of copyright-owner, and date of publication must appear (usu. on the verso of the title-page) in every copy of a book claiming copyright protection under the Universal Copyright Convention (q.v.). *See also* **Berne Convention**

copy/-typist, -writer (hyphens)

coq/ (Fr. m.), cock; — *au vin,* chicken cooked in wine; — *de bruyère,* black game

coquet/, *fem.* **-te,** a flirt; **-ry, -ting, -tish**

coquille (Fr. typ. f.), a misprint

Cor., 1 and 2 Corinthians

cor., corpus; (mus.) cornet, corno (horn); coroner

coram (Lat.), in presence of; — *judice* (law), before a judge; — *nobis,* before us; — *paribus,* before one's equals; — *populo,* before the people

Coran, *use* **Koran**

cor anglais (mus.), instrument of oboe family

coranto, *use* **courante**

Corcyra, anc. name of **Corfu**

Corday (**Charlotte**), 1768–93, French revolutionist

cordelier, Franciscan friar, wearing knotted waist-cord

cordillera/ (Sp.), a mountain chain; *pl.* **-s**

cordite, an explosive

Cordoba, Spain and Argentina *not* -va; in Sp. **Córdoba**

córdoba, monetary unit of Nicaragua

cordon bleu, blue ribbon, a first-rate cook

cordon sanitaire, sanitary cordon (of men) to prevent spread of contagious disease

cordwain, Spanish leather

Corea/, -n, *use* **K-**

co-respondent, in divorce case (hyphen)

corf, coal-miner's basket, tub, or trolley; lobster-cage; *pl.* **corves**

Corfe Castle, Dorset, castle and village

Corflambo, the giant in Spenser's *Faerie Queene*

Corfu, Gr. island, anc. name **Corcyra,** in mod. Gr. **Kerkyra**

Corinthians, (First, Second, Epistle to the), abbr. **1 Cor., 2 Cor.**

Corn., Cornish, Cornwall

corncrake, bird (one word)

corne/a, the eyeball covering; *pl.* **-as**

Corneille (Pierre), 1606–84, French playwright

cornelian, red quartz, *not* carnelian, -ion

cornemuse, French bagpipe

corner-stone (hyphen)

cornet (mus.), abbr. **cor.**

cornett/o (mus.), old woodwind instrument; *pl.* **-i**

cornfield (one word)

cornflour, finely ground maize

cornflower, blue-flowered plant

Corniche (La), coast road from Nice to Genoa; in It. **La Cornice**

Cornish/, abbr. **Corn.;** — **gillyflower,** a variety of apple

Corn Laws (caps.)

corn/o (mus.), a horn, *pl.* **-i;** *-o inglese* (It.), cor anglais

cornu/ (anat.), hornlike process; *pl.* **-a**

Cornubia (Lat.), Cornwall

cornucopia/, horn of plenty; *pl.* **-s;** adj. **-n**

Cornwall, abbr. **Corn.;** in Fr. f. **Cornouailles**

coroll., corollary, inference

corolla/ (bot.), *pl.* **-s**

coron/a (archit., bot., anat., astr.), *pl.* **-ae**

coroner, cap. as title; abbr. **cor.**

coronis, mark of contraction (crasis, ') in Greek

Corot (Jean Baptiste Camille), 1796–1875, French painter

Corporal, abbr. **Cpl.**

corpor/al, of the human body;

-eal, physical, mortal, opp. to spiritual

corporealize, to materialize, *not* -ise

corposant, St. Elmo's fire

corps, *sing.* and *pl.*

corps/ d'armée (Fr m.), army corps, — *de ballet,* company of ballet-dancers; — *de bataille,* central part of an army; — *d'élite,* a picked body; — *des lettres* (Fr. typ.), the body of the type; — *diplomatique,* diplomatic body

corp/us, body; *pl.* **-ora,** abbr. **cor.**

Corpus Christi, festival of institution of the Eucharist, on the Thursday after Trinity Sunday

Corpus Christi College, Oxf. and Camb., abbr. **CCC**

corpus delicti (law), the facts constituting an alleged offence

corp/us lut/eum (physiol.), ovarian body; *pl.* **-ora -ea**

corp/us vil/e, worthless substance; *pl.* **-ora -ia**

corr., correct/ion, -ive, -or; correspond, -ence, -ent, -ing; corrupt, -ed

corral (noun and verb), pen (for) cattle (two *r*s)

correcteur (Fr. typ. m.), corrector of the press

correct/ion, -ive, -or, abbr. **corr.**

correction of proofs, *see* **proof-correction marks**

Correggio (Antonio Allegri da), 1494–1534, Italian painter

corregidor (Sp.), a magistrate

correlate, bring into relation

correlative, abbr. **correl.**

correspond/, -ence, -ent, -ing, abbr. **corr.**

Corresponding Member, abbr. **CM** *or* **Corr. Mem.**

Corrèze, dép. France

corrida, a bullfight (not ital.); *corrida de toros* (Sp.), a bullfight (itals.)

corrigend/um, thing to be corrected; *pl.* **-a**

Corr./ Mem., Corresponding Member; — **Sec.,** Corresponding Secretary

corroboree, Australian aboriginal dance, *not* -bery, -borie, -bory

corrupt/, -ed, -ion, abbr. **corr.**

corrupter, *not* -or

corset, close fitting inner bodice

Corsica, abbr. **Cors.;** in Fr. f. **Corse**

corslet, (armour) cuirass; combined waistband and brassière; *not* -elet

corso (It.), (horse-)race, street for it

Corstorphine, Edinburgh

cort., cortex

cortège, (funeral) procession

Cortes, the legislative assembly of Spain

Cortés (Hernando), 1485–1547, conqueror of Mexico

cort/ex, bark; *pl.* **-ices,** abbr. **cort.**

Corunna, in Sp. **La Coruña**

corvée (Fr. f.), feudal forced labour, drudgery

corvette, a small frigate

Coryate (Thomas), ?1577–1617, English traveller

corybant/, a Phrygian priest; *pl.* **-es** (not ital.); adj. **-ic,** frenzied

Corycian nymphs, the Muses

coryphae/us, a chorus leader, *not* -eus; *pl.* **-i**

coryphée (Fr. m.), chief ballet-dancer

COS, Chief of Staff

Cos, Gr. island, *not* Kos; hence **cos lettuce**

cos (math.), cosine (no point)

co. sa. (mus.), come sopra, as above

cosec (math.), cosecant (no point)

cosh (math.), hyperbolic cosine (no point)

cosh/ar, -er (Heb.), *use* **kosher**

Così fan tutte ('All women are like that'), opera by Mozart, 1790

co-signatory (hyphen)

cosine (math., no hyphen), abbr. **cos**

cosmogony, cosmography, abbr. **cosmog.**

COSPAR, Committee on Space Research

coss., consules (consuls)

cosseted, pampered, *not* -etted

co-star (hyphen)

cotangent (math., no hyphen), abbr. **cot**

cote (Fr. f.), market quotation, figure, mark, share

côte (Fr. f.), hillside, shore

côté (Fr. m.), side

Côte d'Azure, eastern Mediterranean coast of France

Côte-d'Or, dép. France

côtelette (Fr. f.), a cutlet

coterie, a 'set' of persons (not ital.)

coterminous, *use* **cont-**

Côte-rôtie (Fr. f.), a red wine

côtes de bœuf (Fr. f.), ribs of beef

Côtes-du-Nord, dép. France

cotillion, a dance; in Fr. m. *cotillon*

cottar, peasant, *not* -er

cotter, pin, wedge, etc.

Cotter's Saturday Night (The), by Burns, 1786

Cottian Alps, France and Italy

cottier, a cottager

cotton/, in Fr. m. *coton; -tail* (one word), American rabbit; — **wool** (two words), raw cotton

Cottonian Library, in British Museum (British Library, Reference Division)

cottonize, to make cotton-like, *not* -ise

couch, a kind of grass. *See also* **cutch**

coudé (astr.), telescope with light-path bent at an angle

Couéism, psychotherapy by auto-suggestion (Emile Coué, 1857–1926)

couldn't, to be printed close up

couldst (no apos.)

coulé (mus.), a slur

oulée, a lava-flow

ouleur (Fr. f.), colour; — *de
rose,* roseate (figurative)

oulisses (pl.), the wings in a
theatre

ouloir, a gully

**oulomb (Charles Augustin
de),** 1736–1806, French physicist; **coulomb** (elec.), unit of
charge, abbr. **C**

oulter, a plough blade,
not col-

uncil/, assembly; **-lor,** member of a council, abbr. **Cllr.**

**ouncil/ estate, — flat, —
house** (two words), properties
owned by council; **-house** (hyphen), in which council meets.

ounsel/, advice, barrister; **-led,
ling; -lor,** one who counsels

unt/, -ess (cap. as title); abbr.
Ct.

ountdown (noun, one word)

ounter (typ.), space wholly or
mainly enclosed within a letter,
e.g. the centre of O

ounter/balance, -charge
(one word)

ounter/-cheer, -claim (hyphens)

ounter-clockwise (one hyphen)

ounter-quarte (fencing) (hyphen), *not* -carte

ounter/-reformation (caps.
with hist. reference); **-revolu-
tion** (hyphens)

ounter-tenor (mus.), abbr. **C.**

ountesthorpe, Leics.

ounties palatine, Cheshire
and Lancs. (cap. C only)

ountrif/y, -ied, *not* country-

ountry/ dance, — house (two
words)

ountryside (one word)

ountry-wide (hyphen)

ounty, abbr. **Co.**

ounty Council/, -lor, abbr.
CC

ounty court (one cap.), abbr.
CC

oup, a stroke, esp. a political
revolution

Coupar Angus, Tayside (no
hyphen), *not* Cupar —

coup/ de foudre (Fr. m.), stroke
of lightning; — *de fouet* (fencing), a 'beat'; — *de grâce,* a
finishing stroke; — *de main,*
sudden attack to gain a position;
— *de maître,* a master-stroke;
— *de pied,* a kick; — *de poing,*
blow with the fist, — *de soleil,*
sunstroke; — *d'essai,* first attempt; — *d'état,* sudden or
violent change in government;
— *de théâtre,* sudden sensational act; — *d'œil,* a glance,
wink

coupe, a shallow dish

coupé, a covered motor car for
two (accent, not ital.)

coupee, in dancing, a salute to
partner

couper (Sc.), a dealer

Couperin (François), 1688–
1733, French composer

coup manqué (Fr. m.), a
failure

courante, (music for) dance
with gliding step, *not* coranto

Cour de Cassation, see **Cassation**

courgette, small marrow (not
ital.)

courier, travelling attendant, in
Fr. m. *courrier*

Court, abbr. **C.** or **Ct.**

Courtauld Institute of Art,
London

court bouillon (Fr. cook. m.),
fish stock

Courtelle, propr. term for an
acrylic fibre (cap.)

Courtenay, fam. name of Earl
of Devon

courtesan, prostitute with
wealthy clients

court-house, abbr. **CH**

Courtmantle, *use* Cu-

court martial (noun, two
words), *pl.* **courts —; court-
-martial** (verb, hyphen)

Court of/ Appeal, abbr. **CA;
— — Common Pleas,** abbr.
CCP; — — Probate, abbr.

Court of (*cont.*)
CP; — **St. James's** (apos.); — — **Session,** abbr. **CS**
Courtrai, Belgium, *not* -ay; in Fl., **Kortrijk**
Courts of Justice (caps.)
courtyard (one word)
couscous, NW African dish of granulated flour steamed over broth. *See also* **cuscus**
cousin-german, *not* -aine, -ane; *pl.* **cousins-german**
coûte que coûte (Fr.), at any cost, *not* — qui —
Coutts & Co., bankers
couturière (Fr. f.), a dressmaker
Covenanter (Sc. hist.)
covenantor (law)
Coventry:, sig. of Bp. of Coventry (colon)
Coverley (**Sir Roger de**)
cover-up (noun, hyphen)
Cowling, Lancs., N. Yorks.
Cowlinge, Suffolk
Cowper (**William**), 1731–1800, English poet
cowrie, the shell, used as money in Africa and S. Asia, *not* -ry. *See also* **kauri**
coxcomb, a fop. *See also* **cockscomb**
coxswain (naut.), *not* cocks-; colloq. **cox**
Coy. (mil.), company
coyote, N. American prairie wolf
coypu, S. American aquatic rodent, source of nutria fur, *not* -ou
CP, Cape Province (of S. Africa), Central Provinces, *now* Madhya Pradesh (India), Chief Patriarch, Civil Power, — Procedure, Clarendon Press, Clerk of the Peace, Code of Procedure, College of Preceptors, Common Pleas, — Prayer, Communist Party, *Congregatio Passionis* (Passionist Fathers), Court of Probate
cp., compare
c.p., candlepower, carriage paid

CPC, Clerk of the Privy Council
c.p./i., characters per inch; — ditto per line; — **m.,** ditto per minute; — **s.,** ditto per second cycles per second
Cpl., Corporal
CPO, Chief Petty Officer, Compulsory Purchase Order
CPR, Canadian Pacific Railway
CPRE, Council for the Protection of Rural England
CPS, *Custos Privati Sigilli* (Keeper of the Privy Seal)
CPSA, Civil and Public Service Association
CPSU, Communist Party of the Soviet Union
CPU (comp.), central processing unit
CR, *Carolina Regina* (Queen Caroline), *Carolus Rex* (King Charles), *Civis Romanus* (a Roman citizen), Community of the Resurrection, *Custos Rotulorum* (Keeper of the Rolls), credit rating, current rate
Cr, chromium (no point)
Cr., credit, -or, Crown
cr., created
Crabb (**George**), 1778–1854, English philologist
Crabbe (**George**), 1754–1832 English poet
Cracow, acceptable Anglicized spelling of **Kraków,** Poland
Craigton, Glasgow
Crane (**Harold Hart**), 1899–1932, US poet; — (**Stephen**), 1871–1900, US novelist and poet
crane's-bill, plant of the genus *Geranium,* *not* cranesbill
crani/um, skull; *pl.* **-a**
crap/e, gauzelike fabric, usu. black, for mourning; adj. **-y.** *See also* **crêpe**
crassa negligentia (law), criminal negligence
crawfish, large marine lobster
crayfish, freshwater crustacean
CRC, camera-ready copy
cream/-laid, a writing paper

with wire marks; — **-wove,** ditto without

reated, abbr. **cr.**

Creation (the) (cap.)

rèche, a day nursery

Crécy (battle of), 1346, *not* Cressy, Créci

redit/, -or, abbr. **Cr.**

redit card (two words)

rédit/ foncier (Fr. m.), society for loans on real estate; — *mobilier,* ditto on personal estate

redo/, creed; *pl.* **-s**

reese, *use* **kris**

refeld, W. Germany, *use* **Krefeld**

reighton (Mandell), 1843–1901, Bp. and historian

rematori/um, *pl.* **-a**

rème de la crème (Fr. f.), the very best, *not* cre, crê-

remona, a violin made there (cap.)

renate (bot.), notched (of leaves)

renellate(d) (mil.), furnish(ed) with battlements

renulate (bot.), finely notched

reole, (strict use, usu. cap.) pure-blooded descendant of French, Spanish, Portuguese settlers in W. Indies, Louisiana, Mauritius, Africa, and E. Indies; (loose use, not cap.), a Negro born in America; a Creole–Negro half-breed speaking a Spanish or French dialect; dialect thus imposed by a dominant group

reosote, *not* kreosote

rêpe/ (Fr. m.), crape fabric other than black, (f.) pancake; — **de Chine,** raw silk crêpe; — **lisse,** smooth crêpe; — **Suzette,** small pancake. *See also* **crape**

rescendo/ (mus.), growing in force; abbr. **cres., cresc.**; (nontechn.), gradually increasing noise; *pl.* **-s**

ressy, *use* **Crécy**

retaceous, chalky, *not* -ious

Cretan, of Crete

cretin, type of mental defective; in Fr. m. *crétin,* f. *crétine*

cretonne, a cotton cloth

Creusot (Le), dép. Saône-et--Loire, France, *not* -zot

crevasse, large fissure, esp. in the ice of a glacier

Crèvecœur (Michel Guillaume Jean de), 1735–1813, French-born US writer

crève-cœur (Fr. m.), heart--break; *pl. same*

crevette (Fr. f.), prawn

crevice, small fissure

Criccieth, Gwynedd

Crichton (James), 1560–?85, Scottish scholar and soldier, 'the Admirable —'

cricket/, -er, -ing

cri de cœur (Fr. m.), passionate appeal

crim. con. (law), criminal conversation, adultery

crime passionnel (Fr. m.), crime caused by sexual passion (two *n*s)

Crimplene, propr. term for bulked Terylene (cap.)

cring/e, -ing

crinkum-crankum, intricate, crooked, *not* -cum -cum

cris/is, *pl.* **-es**

crispbread, thin rye etc. biscuit (one word)

crit., critic/al, -ized

criteri/on, standard of judgement; *pl.* **-a**

criticaster, a petty critic

criticize, *not* -ise

critique, a review (not ital.)

Critique of Pure Reason, in German *Kritik der reinen Vernunft,* by Kant, 1781

CRMP, Corps of Royal Military Police

Croat, a native of Croatia; Serbo-Croat **Hrvat**

crochet/, hooked-needle work; **-ed, -ing**

Crockford, in full *Crockford's Clerical Directory,* the 'Who's Who' of the clergy, from 1858

Crockford's, London club
crocus/, *pl.* **-es**
Croesus, 6th c. BC, rich king (and the last) of Lydia
croissant, crescent-shaped roll, *not* croisant
Croix de Guerre, French decoration
Cro-Magnon (anthrop.), of a prehistoric race, remains found at —, dép. Dordogne, France
Cromartie (Earl of)
cromesquis (Fr. cook. m.), *see* **kromesky**
Crome Yellow, by Aldous Huxley, 1921
Cronos, *use* K-
Crookback, sobriquet of Richard III, *not* Crouchback
Crookes (Sir William), 1832–1919, English physicist
Croonian Lecture, of Royal Society
croquet/, game (not ital.); as verb, **-ed, -ing**
croquette (Fr. f.), rissole (not ital.)
crore (Ind.), ten millions, point thus: 1,00,00,000. *See also* **lakh**
crosette, *use* **crossette**
crosier, Bp.'s or Abp.'s staff, *not* -zier
croslet, *use* **crosslet**
cross (typ.), proof-correction sign for a battered sort
crossbar (one word)
cross-bench(er) (Parl.) (hyphen)
cross-bill (law), a promissory note given in exchange; a bill brought by defendant against plaintiff in a Chancery suit (hyphen)
crossbill, a passerine bird
crossette (archit.), a ledge, *not* crose-
Crossgates, Cumbria, Fife, Powys, N. Yorks.
Cross Gates, W. Yorks.
Crosshill, Fife, Strathclyde
Cross Hill, Derby., Shropshire
Crosskeys, Co. Antrim, Co. Cavan, Gwent

Cross Keys, Kent, Wilts.
crosslet (her.), small cross, *not* croslet
cross, Maltese ✠; **Latin** †; **Greek** +; **tau cross,** of St. Anthony, ⊤. *See also* **crux** and **ecclesiastical signs**
crossroad(s) (one word)
cross/-section, -wind (hyphens)
crossways, *not* -way, -wide
crossword (puzzle)
crotchet/, a mus. note; a whin -ed, -ing, -y
Crouchback, *use* **Crookback**
croupier, gaming-table attendant
croûton (Fr. m.), a bit of crust c toast
Crowland, Lincs., Suffolk, *not* Croy-
Crown (the) (cap. C); abbr. **C.**
crown, former size of paper, 15 × 20 in.; — 4to, 10 × 7½ in. — 8vo, 7½ × 5 in. (untrimmed basis for size of metric crown, 768 × 1008 mm. *See also* **book sizes**
Crowner's quest, dialectal fo **Coroner's inquest** (cap. C only)
Crown Office, abbr. **CO**
Crowther-Hunt (Baron) (hy phen)
crozier, *use* **cros-**
CRP, *Calendarium Rotulorum Pate tium* (Calendar of the Patent Rolls)
CRR, Curia Regis Roll
CRT, cathode ray tube
cru (wine), growth (no accent)
Crucifixion (the) (cap. C)
Cruft's Dog Show, *not* -s'
Cruikshank (George), 1792–1878, English caricaturist and illustrator
crumhorn, *use* **krummhorn**
cruse, a jar, *not* cruise
crush bar (two words)
crush-barrier (hyphen)
Cruso, NC, US
Crusoe (Robinson), by Defoe 1719

crux, of an argument; *pl.* **cruces** (not ital.)

crux/ ansata, the cross with a handle, ♀; — **commissa,** the tau cross, ⊤; — **decussata,** cross of St. Andrew, or St. Patrick, ×; — **stellata,** the cross with arms ending in stars

cryogenics, the study of low--temperature refrigeration

cryptogam/ (bot.), any member of the Cryptogamia, flowerless plants; adj. **-ous**

crypto/gram, anything written in code; **-grapher, -graphic, -graphy**

crypton (chem.), *use* **k-**

cryptonym, a private name

crystal., crystallography

crystalliz/ation, -e, *not* -is-; **-ed, -ing**

CS, Chemical Society (*now* part of Royal Society of Chemistry), Civil Service, Clerk to the Signet, College of Science, Common Serjeant, Court of Session, *Custos Sigilli* (Keeper of the Seal)

Cs, caesium (no point)

c/s, cycles per second

Csar etc., *use* **Ts-**

csárdás, Hungarian dance, *pl. same*, *not* cz-

CSC, Civil Service Commission, Conspicuous Service Cross (*now* **DSC**)

CSE, Certificate of Secondary Education

CSI, Companion of the Order of the Star of India

CSM, Company Sergeant-Major

CSU, Civil Service Union

CT, Connecticut (off. postal abbr.)

Ct., Count, Court

ct., cent

CTC, Cyclists' Touring Club

Ctesiphon, city of anc. Mesopotamia

ctl., central, -s

cto. (mus.), concerto

cts., centimes, cents

CU, Cambridge University

Cu, *cuprum* (copper) (no point)

cu., cubic

cube root (two words)

cubic/, abbr. **c., cu.,** *or* **cub.**; — **metre** etc., abbr. **cu.m.** or in scientific work usu. **m³**

cubicul/um (archaeol.), burial--chamber; *pl.* **-a**

cudgel/, -led, -ling

CUDS, Cambridge University Dramatic Society

cue/, -ing. *See also* **queue**

Cufic, *use* **Kufic**

cui bono? (Lat.), who gains by it?

cuidad, use *ciudad*

Cuillins, mountains in Skye

cuirass/, body armour; **-ier,** soldier wearing it

cuisine, cookery (not ital.)

culch, oyster spawn, *not* cultch

cul-de-sac, a blind alley, a trap, *pl.* **culs-de-sac**

cul/ex, a gnat; *pl. -ices*

cullender, *use* **colan-**

Cullinan, famous diamond

Culpeper, Va., US, from Governor — (**Thomas, Lord**), 1578–1662

cultivar (bot.), variety made by cultivation, abbr. **cv.**

CUM, Cambridge University Mission

cum (Lat.), with (not ital.)

Cumaean, of Cumae, near Naples

Cumbernauld, Strathclyde, 'new town', 1956

Cumbria, county of England

cum dividend, with dividend; abbr. **c.d.**

cumfrey, *use* **co-**

cum grano salis (Lat.), with a grain of salt

cumin, plant with aromatic seeds, *not* cummin

cummerbund, a waistbelt, *not* ku-, -band

Cummings (Edward Estlin), 1894–1962, US poet who renounced use of capital letters

cum multis aliis (Lat.), with many others

cum. pref., cumulative preference

cumquat, *use* k-

cumul/us (meteor.), a cloud form; *pl.* **-i**; abbr. **k.**

cuneiform, wedge-shaped, *not* cunif-, cunef-

Cunninghame Graham (Robert Bontine), 1852–1936, British writer

CUP, Cambridge University Press

Cupar/, Fife; **— Angus,** Tayside, *use* **Coupar Angus**

cupbearer (one word)

cupro-nickel, alloy of copper and nickel

cuprum, copper, symbol **Cu**

cur., currency, current

curable, *not* -eable

curaçao, liqueur, *not* -oa

curare, a drug, *not* -a, -i, urari

curb, for verb, and curb of bridle. *See also* **kerb**

curbstone, *use* **kerb-**

curé, French priest; *petit* **—,** French curate

cure-all, a universal remedy (hyphen)

curfuffle, use **ker-**

Curia/, the papal court (cap.); **— *advisare vult,*** the court desires to consider; abbr. *c.a.v.*

Curie (Marie Sklodowska), 1867–1934, and her husband **— (Pierre),** 1859–1906, French scientists

curie, unit of radioactivity, abbr. **Ci** (cap.); now replaced by **becquerel**

curio/, an object of art; *pl.* **-s** (not ital.)

curium, symbol **Cm**

curlicue, a decorative curl, *not* -eque, -ycue

currach, a coracle, *not* -agh

Curragh (The), Co. Kildare

curren/t, -cy, abbr. **cur.;** **electric current,** symbol *I* or *i* (ital.)

currente calamo (Lat.), easily, fluently

curricul/um, course of study,

pl. **-a; -um vitae,** summary of a career, *pl.* **-a vitae;** abbr. **c.v**

Currie, Lothian

Curry, Co. Sligo

cursor (comp.), movable spot o light on VDU (q.v.) screen, indicating item being edited

curtain-raiser (theat.), Fr. *leve de rideau*

Curtiss (Glenn Hammond), 1878–1930, US pioneer aviato

Curtmantle, sobriquet of Henry II, *not* Cou-

curtsy/, *not* -sey; **-ing**

curvilinear (one word)

Curwen (John), 1816-80, pioneer of tonic sol-fa

cuscus/, a marsupial; **—-grass** of India. *See also* **couscous**

custodia legis (Lat.), in the custody of the law

custom-house (hyphen); abbr **CH**

cust/os (Lat. m.), a custodian, *pl* **-odes;** *Custos/ Privati Sigill* Keeper of the Privy Seal, abbr **CPS;** **— *Rotulorum,*** Keeper of the Rolls, abbr. **CR;** **— *Sigilli,*** ditto Seal, abbr. **CS**

cut and dried (adj., no hyphens)

cut-back (noun, hyphen)

Cutch, India, *use* **K-**

cutch, catechu, extract of India plants, tough paper sheets used by gold-beaters, *not* k-. *See also* **couch**

cut/-off, -out (nouns), **-price** (adj.) (hyphens)

cuts (typ.), illustrations

Cutty Sark, famous clipper, now in dry dock at Greenwich

cutty-stool, stool of repentance

cuvée (Fr. f.), a vatful, or sort, o wine (ital.)

Cuyp, Dutch artists, *not* K-

cv. (bot.), cultivar

c.v., curriculum vitae

CVO, Commander of the Royal Victorian Order

Cwlth (no point), Commonwealth (Australian)

Cwmbran, Gwent, 'new town', 1949

Cwmcarn, Gwent

CWS, Co-operative Wholesale Society

cwt., hundredweight (*but* no point in scientific and technical work)

cybernetics, study of communication processes in animals and machines

cyc., cyclopaed/ia, -ic

Cycle of the Saros (astr.), 6,585½ days

cyclopaed/ia, -ic; but use **-ped-** in quoting titles using that form; abbr. **cyc.**

Cyclop/s, a giant with one eye; *pl.* **-es;** adj. **-ean**

cyclotron, apparatus for accelerating particles, used in nuclear disintegration, artificial radioactivity, etc.

cyder, *use* **cider**

Cym., Cymric

cyma (archit.), a moulding of the cornice

cymbalist, a cymbal-player. *See* **cembalist**

cymbal/o, a dulcimer, *pl.* **-os**

cymbiform, boat-shaped, *not* cymbae-

Cymmrodorion (Honourable Society of) (two *m*s)

Cymric, Welsh, *not* K-

Cymru, Wales

Cymry, the Welsh nation

CYMS, Catholic Young Men's Society

Cynewulf, eighth century, Anglo-Saxon poet

cynocephalus, a dog-headed creature

Cynthia, the moon

Cynthius, epithet of Apollo

cypher, *use* **cipher**

cy pres (law), as near as possible to a testator's intentions

Cyprian, of Cyprus; —, in Lat. **Thascius Caecilius Cyprianus,** *c.* 200–58, Christian writer and martyr

Cypriot, inhabitant or language of Cyprus

Cyprus, independent republic 1960

Cyrenaic, pertaining to Cyrene or its school of philosophy

Cyrillic, (pertaining to) the alphabet used by Slavonic peoples of the Eastern Church (two *l*s)

cyst/ (biol.), sac; **-ic;** *not* ci-

Cytherean, pertaining to Aphrodite, *not* -ian

CZ, Canal Zone (Panama)

Czar etc., *use* **Ts-**

czárdás, *use* **cs-**

Czech, native or language of Bohemia

Czechoslovak, *not* -ian, (person) belonging to **Czechoslovakia**

Czerny (Karl), 1791–1857, Austrian composer

D

D, 500, deuterium (= heavy hydrogen), the fourth in a series

D., Deputy, (US) Democrat, Duke, *Deus* (God), *Dominus* (Lord), prefix to enumeration of Schubert's works (*see* **Deutsch**)

Ð ð, OE and Icelandic (lower case ð) and phonetic letter, *pron.* dh (as th in *then*); *eth*, *see* **eth**

d, (prefix) deci-

d., date, daughter, day, dead, degree, departs, desert/ed, -er, died, dime, dioptre, dose, (Fr.) *douane* (customs), *droite* (the right hand), (It.) *destra* (right), (Lat.) *decretum* (a decree), *denarii* (pence), *denarius* (penny), (meteor.) drizzling

d', as prefix to an un-Anglicized proper name should, in accordance with continental practice, be lower case and *not* cap., as d'Arsonval, except at beginning of sentence. Signatures to be copied. *See also* **accents**

∂ (math.), sign of partial derivative

𝔡 (typ.), delete

DA, deposit account, (US) District Attorney

Da., Danish

da, (prefix) deca-

DAB, *Dictionary of American Biography*

da/ ballo (mus.), dance style; — **capo,** *or* — **capo al fine,** repeat to the word *fine,* abbr. **DC;** — **capo dal segno,** repeat from the sign '$., abbr. **DS;** — **capella,** *or* — **chiesa,** in church style

d'accord (Fr.), I agree

Dachau, Nazi concentration camp, 1933–45

dachshund, badger-dog

dacoit, Indian robber, *not* dak-, dec-

dactyl/, foot of three syllables (‾ ˘ ˘), adj. **-ic**

Dada/, unconventional art *c.*1920; **-ism**

daddy-long-legs, the crane-fly (hyphens)

dado/, wooden skirting; *pl.* **-s**

Daedalus, builder of the Cretan labyrinth

daemon (Gr. myth.), supernatural being, indwelling spirit, *not* daimon, demon

daffadowndilly, a daffodil, *not* daffi-, daffo-, daffy- (one word)

Dagapur, Ethiopia

dagger (†), **double —** (‡), (typ.), *see* **reference marks.** †, in Eng. before, in Ger. after, a person's name, signifies 'dead' or 'died'

daggle-tail, *use* **draggle-**

dago/, Spaniard, Portuguese, *or* Italian (derog.), *pl.* **-s**

Dagonet, Sir, King Arthur's fool

Daguerre (Louis Jacques Mandé), 1789–1851, French pioneer of photography

daguerreotype (not cap.)

dahabeeyah, Nile sailing-boat

Dahomey, W. Africa, *now* **Benin**

Dáil Éireann, Lower House of Irish Parliament

d'ailleurs (Fr.), besides

daimio/, Japanese noble; *pl.* **-s**

Daimler motor car

daiquiri, a cocktail

dais, small platform, *not* daïs

dakoit, *use* **dac-**

Dakota (North and South), off. abbrs. **N. Dak.** *or* **ND, S. Dak.** *or* **SD**

Dalai Lama, the Grand Lama of Tibet

Dali (Salvador), b. 1904, Spanish painter

Dalila, *see* **Delilah**

Dallapiccola (Luigi), 1904– 75, Italian composer

Dalmatian, spotted dog, *not* -ion

94

al segno (It. mus.), repeat from the sign '$., abbr. **DS**

amageable, *not* -gable

amara/, -land, SW Africa

amascen/e, -er, ornament(er) in metal, *not* -keen, -kin

ame aux Camélias (*La*), play by Dumas fils, 1848

amien de Veuster (**Joseph**), 1840–89, Belgian priest to leper colony in Hawaiian Islands, 'Father Damien'

amnosa hereditas (Lat.), a legacy involving loss

amnum absque injuria (Lat.), damage without wrong

amon and Pythias (Gr. myth.), model friends

an., (Book of) Daniel

an/aë, mother of Perseus, *also* asteroid; *Danaea,* fern genus; **-aïd,** dau. of Danaus; **-aus,** son of Belus

andie Dinmont, breed of Scottish terrier, from **Andrew Dinmont,** farmer in Scott's *Guy Mannering*

andruff, scurf, *not* -riff

anegeld, land-tax, *not* -lt

aniel (**Book of**), abbr. **Dan.**

aniell's battery (elec.), usu. called a **Daniell cell**

anish/, abbr. **Da.;** **— alphabet** (typ.), *see* **accents**

anke schön (Ger.), many thanks

annebrog, Danish national standard, *also* order of knighthood, *not* Dane-

'Annunzio (**Gabriele**), 1864–1938, Italian writer

anse macabre (Fr. f.), dance of death

anseuse, a female dancer (not ital.)

ant/e Alighieri, 1265–1321, Italian poet; **-ean, -esque, -ist** (caps.)

anzig, Baltic free port, *now* **Gdańsk**

aou, *use* **dhow**

DAR, Daughters of the American Revolution

d'Arblay (**Mme**), 1752–1840, the English novelist **Fanny Burney**

Darby and Joan, devoted old married couple

d'Arc (**Jeanne**), Joan of Arc, 1412–31

Darcy de Knayth (**Baroness**)

daredevil (one word)

daren't, to be printed close up

dare say (two words)

Dar es Salaam, cap. of Tanzania (no hyphens)

Darjeeling, *not* Darji-

Dark Ages (**the**) (caps.)

dark-room (hyphen)

darshan (Hindi), seeing a revered person

Darwen, Lancs., *not* Over —

Darwin/ (**Charles Robert**), 1809–82, author of *Origin of Species*; **—** (**Erasmus**), 1731–1802, English physician and poet, grandfather of Charles; **-ian, -ism, -ist**

das (Ger.), the (*neut. sing. nominative and accusative*); *also* that. *See also* **das heisst**

dash, *see* **punctuation** XII

dashboard (one word)

das/ heisst (Ger.), that is to say, abbr. **d.h.;** **— ist,** that is, abbr. **d.i.**

dass (Ger.), that (conjunction)

dat., dative

data, *see* **datum**

data/ bank, — base (two words)

datable, capable of being dated, *not* -eable

data processing (hyphen when attrib.)

date, abbr. **d.** The order is day, month, year, as 5 June 1903; in Fr. le 5 juin 1903; in Ger. den 5. Juni 1903. Names of days and months generally in full, but may be abbreviated in footnotes. All-figure form in Britain 5.6.03; in US 6.5.03 (month first); international dating system (*see* BS 4795) is year, month, day, e.g. 1971–04–20. No

date (*cont.*)
comma in four-figure years, but 10,000 BC, etc. For periods use least number of figures, as 1904–7, 1920–1, 1926–8; *but* 1890–1905, 1913–15, and always in full in display matter

date-line (hyphen)

dative, indirect object; abbr. **dat.**

dat/um, thing known or granted, *pl.* **-a**

daube, braised meat stew

Daudet (Alphonse), 1840–97, French novelist; — **(Léon),** 1867–1942, French writer and politician

daughter/, abbr. **d.** *or* **dau.**; — **-in-law,** *pl.* **daughters- - - —** (hyphens)

d'Aumale (duc)

Daumier (Honoré), 1808–79, French painter

Dauntsey's School, Devizes

dauphin, eldest son of King of France, 1349–1830; **dauphiness,** his wife, in Fr. *dauphine*

Dauphiné, region of SE France

Dav., David

Davenant (Sir William), 1606–68, English playwright and poet, Poet Laureate 1660–8

Davies (Sir Henry Walford), 1869–1941, musician; — **(William Henry),** 1871–1940, poet

da Vinci (Leonardo), *see* **Leonardo da Vinci**

Davis (Jefferson), 1808–89, Pres. Confederate States

Davis/ apparatus, for escape from submarine; — **Cup,** tennis trophy

Davy (Sir Humphry), *not* -ey, -ey, 1778–1829, English chemist, inventor of Davy lamp

Davy Jones's locker, the sea- -bed

Dawley, Shropshire

day, abbr. **d.**; (typ.), of the week, and of fasts, feasts, festivals, holidays, to have initial caps.

Abbr., when necessary, **Sun., Mon., Tue., Wed., Thur., Fri., Sat.** In Fr. the names of the days do not take caps., as **lundi**

Dayak, use **Dyak**

day-bed (hyphen)

daybook (one word); abbr. **d.b**

day/-break, -dream (hyphens)

Day-Lewis, *see* **Lewis**

daylight (one word)

day/-long, -room (hyphens)

day nursery (two words)

day's journey (Heb.), 16.95 miles

day/time, -work (one word)

Dayton, Ohio, US

Daytona Beach, Fla., US

DB, Domesday Book

dB, decibel (no point)

d.b., daybook, double bed, double-breasted

DBE, Dame Commander of the Order of the British Empire

dbk., drawback

dbl., double

DC, 600, deputy-consul, District of Columbia (US), (mus.) da capo (from the beginning)

d.c., direct current (elec.)

D.Ch., Doctor of Surgery

DCL, Doctor of Civil Law

DCM, Distinguished Conduct Medal

DCMG, Dame Commander of the Order of St. Michael and St George

DCVO, Dame Commander of the Royal Victorian Order

DD, direct debit, *Divinitatis Doctor* (doctor of Divinity)

D.d., *Deo dedit* (gave to God)

dd, in Welsh, a separate letter, not to be divided

dd., delivered

d.d., *dono dedit* (gave as a gift)

D-Day, 6 June 1944

DDA, Dangerous Drugs Act

DDD, *dat, dicat, dedicat* (gives, devotes, and dedicates); *dono dedit dedicavit* (gave and consecrated as a gift)

DDR (Ger.), *Deutsche Demokra-*

tische Republik (German Democratic Republic)

DDS, Doctor of Dental Surgery

DDT, dichloro-diphenyl-trichloro-ethane (an insecticide)

DE, Delaware (off. postal abbr.), Department of Employment

de, as prefix to a proper name, in accordance with continental practice, should *not* have initial cap., as de Candolle; except when Anglicized, as De Vinne, or at beginning of sentence. Signatures to be copied

deaconate, *use* **diac-**

dead, abbr. **d.** *See also* **dagger**

dead heat (noun), **dead-heat** (verb)

Dead Letter Office, *now* **Returned** ditto; **dead/line, -lock** (one word); — **reckoning** (naut.), abbr. **DR;** — **reprint** (typ.), a reprint with no correction; — **weight,** abbr. **d.w.**

dean, wooded valley, *use* **dene**

Dean of Faculty (Sc.), (*not of* the), president of the Faculty of Advocates; abbr. **DF**

Dear/Madam, — **Sir,** in printed letter full left (comma, no dash)

deasil (Sc.), clockwise, opp. to **withershins;** *not* the many variants

death/-bed, -rate, -roll, -trap, -warrant, -watch beetle (hyphens)

deb, colloq. for **débutante** (no point)

deb., debenture

débâcle, a downfall

debarkation, disembarkation

debatable, *not* -eable

debauchee, a libertine

debenture/, abbr. **deb.;** — **-holder** (hyphen)

debonair, gaily elegant, *not* -aire; in Fr. **débonnaire**

Debrett (*Peerage*)

debris, ruins (no accent); in Fr. m. **débris**

début/, -ant, *fem.* **-ante** (not ital.). *See also* **deb**

Dec., Decani, December, (archit.) Decorated

dec., deceased, declaration, declared, declension, declination, decorative

dec. (mus.) decrescendo

déc. (Fr.), *décéd/é,* fem. *-ée* (deceased), *décembre* (not cap.) (December)

deca-, prefix meaning ten, abbr. **da**

Decalogue, the Ten Commandments

Decameron (*The*), by Boccaccio, 1352

Decan, India, *use* **Decc-**

Decani, dean's or south side of a choir (north side in Durham Cathedral), abbr. **Dec.;** opp. to **Cantoris**

decanter, bottle to hold decanted liquor

Deccan, India, *not* Decan

deceased, abbr. **dec.**

décéd/é (Fr.), *fem. -ée,* deceased; abbr. *déc.*

December, abbr. **Dec.;** in Fr. m. *décembre,* abbr. *déc.* (not cap.), *also* X^{bre}

decenni/um, a decade; *pl.* **-a**

decentralize, *not* -ise

decern (Sc. law), to judge. *See also* **discern**

deci-, prefix meaning one-tenth, abbr. **d**

decibar, one-tenth of a bar, q.v.

decibel, unit for comparing intensity of noises, power, etc.; abbr. **dB**

Decies (Baron)

decigram, *not* -mme, abbr. **dg** (no point in scientific and technical work)

decilitre, abbr. **dl** (no point in scientific and technical work)

decimal currency (British, as from 15 February 1971). For amounts less than a pound express as (new) pence: 54p (roman, no point); for amounts of one pound or more express as pounds and decimal fractions of a pound (point in medial

decimal currency (*cont.*) position): £54, £54·65, £54·07 (omitting p); the (new) halfpenny should be expressed as a fraction: £54·07½

decimal fractions. No decimal can be plural, or take verb in pl., however many numerals it contains; (typ.) print in numerals. The decimal point should be on the line, as 1.5, except in **decimal currency**. In languages other than English, a decimal comma is often used (1,5)

decimalize, *not* -ise. But Decimal Currency Board used -ise, -isation

decimal point (two words)

decimator, one who decimates, takes every tenth part or man, *not* -er

decimetre, one-tenth of a metre, 3.937 in., *not* -er; abbr. **dm** (no point in scientific and technical work)

decimo-octavo (typ.), 18mo (not ital.); *see* **eighteenmo**

deckle edge, ragged edge of hand-made paper, *not* -el

declar/ation, -ed, abbr. **dec.**

Declaration of Independence, US, 4 July 1776

déclass/é (Fr.) *fem.* -ée, who has fallen to an inferior status

declension (gram.), system of case-endings; abbr. **dec.**

declinable, *not* -eable

declination (astr.), angular distance from the celestial equator; abbr. **dec.**

decoit, *use* dac-

decollate, to behead; **Decollation** of St. John the Baptist, 29 August

décollet/age (Fr. m.), (wearing of) low neck of dress; *-é, fem. -ée,* with low-necked dress

decolorize, *not* -ise

decolour, to render colourless

décor, stage or room furnishings and fittings

decorat/e, -or

Decoration Day (US), usu. last Mon. in May

decorative, abbr. **dec.**

decorum, propriety of conduct

decree nisi (law), the first stage in the dissolution of a marriage

decrepit, decayed, *not* -id

decrescendo/ (mus.), decreasing in loudness; abbr. *dec. or decres.*; as noun, *pl.* -s

decret/um (Lat.), a decree; *pl. -a*; abbr. **d.**

Dedlock (**Sir Leicester** *and* **Lady**), in Dickens's *Bleak House*

deducible, able to be inferred, *not* -eable

deductible, able to be subtracted, *not* -able

deemster, a judge (Isle of Man) *not* dempster

de-escalate (hyphen)

def., defective, defendant, deferred (shares), defined, definite definition, defunct

de facto (Lat.), in actual fact

defecat/e, -or, *not* defae-

defective, abbr. **def.**

defector, *not* -er

defence, *not* -se (except US)

defendant, abbr. **def.**

défense/ d'afficher (Fr.), stick no bills; — *d'entrer,* no admittance; — *de fumer,* no smoking

defensible, *not* -ceable, -sable

defensor fidei (Lat.), Defender of the Faith; abbr. **DF**

defer/, -ence, -rable, -red, -rer, -ring

deferred (shares), abbr. **def.**

de fide (Lat.), authentic, to be believed as part of the (Christian) faith

definable, *not* -eable

defin/ed, -ite, -ition; abbr. **def.**

definite article, *see* **the**

deflate, to remove air, to reverse economic inflation

defle/ct, to bend downwards; **-xion,** correct, *but* **-ction** is increasingly common, and usual in US

Defoe (**Daniel**), 1661–1731, English writer, *not* de Foe

defunct, abbr. **def.**

deg., degree(s)

dégag/é (Fr. m.), *fem.* *-ée,* unconstrained

Degas (Hilaire Germain Edgar), 1834–1917, French painter (no accent)

de Gaulle (General Charles André Joseph Marie), 1890–1970, French President 1944–5, 1959–69

degauss/, -ing, anti-magnetic-mine device or method

degradable, *not* -eable

degree, 1/360th part of circle; sign °; — **of latitude,** 69 statute or 60 geog. miles; abbr. **d.** or **deg.**

degrees of/ inclination (typ.), to be in words, as 'an angle of forty-five degrees', except in scientific and technical work; —— **temperature,** to be in figures, as 70°F (no point in scientific and technical work)

dégringolade (Fr. f.), rapid deterioration

de/ haut en bas (Fr.), contemptuously; — *haute lutte,* with a high hand

Deïaneira, wife of Hercules; *also* an asteroid

de-ice (hyphen)

Dei/ gratia (Lat.), by the grace of God; abbr. **DG**; — *judicium,* the judgement of God (*see also* **Deo**)

Deity (the), see **capitalization**

déjà vu (Fr.), seen before

déjeuner/ (Fr. m.), breakfast *or* lunch; *petit* —, coffee and rolls on rising; — *à la fourchette,* meat breakfast, early lunch

de jure (Lat.), by right

Dekker (Thomas), *c.*1570–1640, English playwright

del (math.), the symbol ∇

Del., Delaware (off. abbr.)

del., delegate, delete

del., delineavit (drew this)

Delacroix (Ferdinand Victor Eugène), 1799–1863, French painter

de la Mare (Walter), 1873–1956, English poet

de la Ramée (Louise), pseud. **Ouida**

De la Roche (Mazo), 1885–1961, Canadian novelist

Delaroche (Paul), 1797–1856, French painter

Delaware, off. abbr. **Del.** or (postal) **DE**

De La Warr (Earl)

dele (typ.), delete (proof-correction sign ◌)

deleatur (Lat.), let it be deleted

delegate, abbr. **del.**

delenda (Lat.), things to be deleted

delft, glazed earthenware made at Delft (*formerly* Delf) in Holland

Delibes (Léo), 1836–91, French composer

delicatesse, delicacy (not ital.); in Fr. f. *délicatesse*

delicatessen, (Germanic pl. of above), prepared foods, shop or department selling them

Delilah (OT), *but* **Dalila** in Milton's *Samson Agonistes*

delineable, that can be delineated

delineavit (Lat.), drew this; abbr. *del.*

delirium/, disordered state of mind, with hallucinations, *pl.* **-s;** — **tremens,** delirium with trembling induced by heavy drinking, abbr. **d.t., d.t.'s**

De L'Isle (Viscount)

deliverer, *not* -or

Della-Cruscan (noun and adj.), (member) of Florentine *Accademia della Crusca,* concerned with purity of Italian, sixteenth century; or of English poetical group in Florence, eighteenth century

Della-Robbia, an enamelled terracotta invented by Luca della Robbia. *See also* **Robbia**

Delphi, Greece, site of ancient oracle, in Gr. **Delphoi**

de luxe (Fr.), luxurious

dem (Ger.), to *or* for the (*m. and neut. sing.*, *dative*)

dem. (typ.), demy

demagog/ue, -y, -uery, -ic, (of, action of) a leader of the people

dem/ain, *use* **-esne**

de mal en pis (Fr.), from bad to worse

demarcate, to mark the limits of, *not* -kate

demarch, chief officer of anc. Attic deme; modern Greek mayor

démarche (Fr. f.), political procedure

demean, to lower in dignity, to conduct (oneself)

demeanour, bearing towards another

démenti (Fr. m.), official denial

dementia, mental enfeeblement (not ital.)

demesne, a landed estate, *not* -ain

demi/god, -john (one word)

demilitarize, *not* -ise

demi/-monde (Fr. f.), prostitutes (ital., hyphen); —*-pension*, accommodation with one main meal per day; —*-saison*, spring or autumn fabric

demise, death; to bequeath, *not* -ize

demitasse, small coffee-cup; in Fr. f. *demi-tasse*

demo/, colloq. for demonstration, *pl.* **-s**

demobilize, to discharge from the army; *not* -ise; colloq. **demob**

Democrat/, -ic, abbr. (US) **D.**

democratize, *not* -ise

Democrit/us, 5th c. BC, Greek philosopher; adj. **-ean**

démod/é (Fr.), *fem.* **-ée,** out of fashion

demoiselle, young lady (not ital.)

Demoivre (Abraham), 1667–1754, French mathematician (one word)

demon, *see* **daemon**

demon., demonstrative

demonetize, to divest of value as currency, *not* -ise

demon/ic, -ize, *not* dae-

demonstrable, *not* -atable

demonstrator, *not* -er

demoralize, *not* -ise

De Morgan (Augustus), 1806–71, English mathematician; — **(William Frend),** 1839–1917, his son, English novelist

de mortuis nil nisi bonum or *bene* (Lat.), speak nothing but good of the dead

demos, the people (not ital.)

dempster, use **deem-**

demurrable, that may be demurred to

demurrage, undue detention of ship, railway wagon, etc.; compensation for this

demurrer (law), plea that opponent's facts, though true, do not support his case

demy, a scholar at Magdalen Coll., Oxford; *pl.* **demies**

demy, *pron.* dĕmeye, with stress on second syllable (typ.), former size of paper, $17\frac{1}{2} \times 22\frac{1}{2}$ in.; — 4to, $11\frac{1}{4} \times 8\frac{3}{4}$ in.; — 8vo, $8\frac{3}{4} \times 5\frac{5}{8}$ in. (untrimmed); basis for size of metric demy, 888×1128 mm. *See also* **book sizes**

D.En., Department of Energy

den (Ger.), the (*m. sing.*, *accusative*); to the (*pl.*, *dative*)

Den., Denmark

denar, *use* **dinar**

denar/ius (Lat.), penny; *pl.* **-ii,** pence; abbr. **d.**

denationalize, *not* -ise

dene, wooded valley, *not* dean

D.Eng., Doctor of Engineering

dengue, *not* — fever, denga, -gey

Denholm, Borders

Denholme, W. Yorks.

deniable etc., *see* **deny**

denier, one-twelfth of French sou; unit of silk, rayon, nylon yarn weight

denim, twilled cotton fabric for overalls, etc.

Denmark, abbr. **Den.** *See also* **Assemblies**

de nos jours (Fr.), of our time

denouement, an unravelling (no accent, not ital.)

denounce, give notice to terminate (treaty)

de nouveau (Fr.), afresh

de novo (Lat.), afresh

dent., dent/al, -ist, -istry

dentelle (Fr. f.), lace-work

dentil, one of the toothlike blocks under the bed-moulding of a cornice, *not* -el, -ile

den/y, -iable, -ial, -ier

deoch an doris (Sc.), a parting-cup, *not* the many variants

deodand, personal chattel that has caused death

deodar, E. Indian cedar

deodoriz/e, *not* -ise; **-er**

Deo/ favente (Lat.), with God's favour; — *gratias,* thanks to God; — *volente,* God willing, abbr. **DV** (*see also Dei*)

dep., departs, deposed, deputy

dép. (Fr.), *département* (shire), *député* (deputy)

DEP, Department of Employment and Productivity, *now* **DE**

département (Fr. m.), shire; abbr. **dép.**

department, abbr. **dept.**

departmentalize, *not* -ise

Department of Education and Science, abbr. **DES**

Department of Employment, abbr. **DE**

Department of Energy, abbr. **D.En.**

Department of Health and Social Security, abbr. **DHSS**

Department of Industry, abbr. **DI**

Department of the Environment, abbr. **D.o.E.**

Department of Trade, abbr. **D.o.T.**

Department of Trade and Industry, *see now* **Department/ of Trade,** — **Industry**

dépays/é (Fr.), *fem.* **-ée,** out of one's usual element

dependable, *not* -ible

depend/ant (noun), **-ence, -ent** (adj.)

depilatory, removing, or remover of, hair

de pis en pis (Fr.), from bad to worse

de plano (law), clearly

depolarize, *not* -ise

deposed, abbr. **dep.**

depositary, a person

depositor, *not* -er

depository, a place

depot, in Fr. m, *dépôt*

Deputy-Lieutenant, abbr. **DL**

depressible, *not* -able

de profundis (Lat.), out of the depths

de proprio motu (Lat.), of his, *or* her, own accord

dept., department

député, a member of the lower French Chamber

deputize, *not* -ise

deputy, abbr. **dep.**

De Quincey (Thomas), 1785–1859, English essayist

der (Ger.), the (*m. sing. nominative*); of *or* to *or* for the (*f. sing., genitive and dative, or m., f.,* and *neut. pl., genitive*)

der., deriv/ation, -ative, -ed

Derbyshire, abbr. **Derby.**

de règle (Fr.), in order

de rigueur (Fr.), according to etiquette

derisible, laughable, *not* -able

derisory, derisive, scoffing

deriv/ation, -ative, -ed; abbr. **der.**

derm, the true skin; *also* **derm/a, -is**

dernier/ (Fr.), last; — *cri,* the very latest; — *ressort,* a last resource

derring-do, daring action, *not* — doe

Derry, postally acceptable abbr. of **Londonderry** (town or county)

der Tag (Ger.), the (great) day

derv, fuel oil for heavy vehicles

DES, Department of Education and Science

des (Fr.), of the (*pl.*); as prefix to a proper name, treat as **de**, q.v.

dès (Fr.), since

des (Ger.), of the (*m. and neut. sing., genitive*)

Descartes (**René**), 1596–1650, French mathematician and philosopher; adj. **Cartesian**

descend/ant (noun), person or thing descended; **-ent** (adj.), descending

descender (typ.), lower part of letters such as g, j, p, q, y

descendible, that may descend or be descended, *not* -able

desert, a wilderness; to abandon. *See also* **dessert**

desert/ed, -er, abbr. **d.**

déshabillé (Fr. m.), undress, dishabille

desiccate, to dry, *not* dessicate

desiderat/um, something desired; *pl.* **-a** (not ital.)

desirable, *not* -eable

Des Moines, Iowa, US

desorb, desorption, release from adsorbed state

despatch, *use* **dis-**

desperado/, a desperate man; *pl.* **-es** (not ital.)

despicable

despise, *not* dis-, -ize

despotize, to act the despot, *not* -ise

dessert, a dinner course. *See also* **desert**

dessertspoonful/, *pl.* **-s,** two drams (one word)

dessicate, *use* **desiccate**

destra/ (It.), right-hand side, abbr. **d.**; — **mano** (mus.), the right hand, abbr. **DM**

destructible, *not* -able

destructor, a refuse-burning furnace, *not* -er

desuetude, disuse

desunt/ cetera (Lat.), the rest are missing; — **multa,** many things are wanting

Detaille (**Jean Baptiste Édouard**), 1848–1912, French painter (no accent)

detector (person or thing), *not* -er

de te fabula narratur (Lat.), of thee is the story told

detent, a catch (mechanical)

détente (Fr. f.), cessation of strained relations between States

détenu/ (Fr.), *fem.* **-e,** one detained in custody

deterrent, *not* -ant

detonat/e, -or

detour, a circuitous way; in Fr. m. **détour**

detract/or, *fem.* **-ress**

detritus, debris (not ital.)

de trop (Fr.), not wanted; superfluous

Deus/ (Lat.), God, abbr. **D.**; — **avertat!** God forbid! — **det,** God grant; — **ex machina,** 'a god from a machine' (device in anc. Gr. theatre), a solution or intervention in the nick of time; — **misereatur,** God be merciful

deuterium, symbol **D** (no point); — **oxide,** heavy water

Deuteronomy, abbr. **Deut.**

Deutsch/ (**André**), **Ltd.,** publishers; — (**Otto Erich**), 1883–1967, compiler of thematic catalogue of Schubert's works, *see* **D.**

Deutsche Mark (W. Ger. currency), usu. written and spoken **Deutschmark** *or* **D-Mark,** abbr. **DM** (no point). *See also* **mark**

Deutschland, Deutsches Reich, Ger. names for Germany

de Valera (**Éamon**), 1882–1975, Prime Minister of Ireland, 1937–48, 1951–4, 1957–9, President, 1959–73

Devanagari, Ind. alphabet

develop/, -able, -ment, *not* -pe

devest, *see* **divest**

deviat/e, -or, *not* -er

deviling, a young devil

devilling, working as a hack

devilry, *not* -try

devil's advocate, official at papal court who proposed objections to a canonization, an adverse critic

Devil's Island, penal settlement, Fr. Guiana

devis/e, *not* -ize; **-er,** one who devises (non-legal)

devis/ee, one who is bequeathed real estate; **-or,** one who bequeaths it

devitalize, to render lifeless, *not* -ise

devoir, an act of civility (not ital.)

Devon, off. name of the county

Devonshire, abbr. **Devon.**

De Vries (Hugo), 1848–1935, Dutch botanist

DEW, distant early warning

Dewalee, *use* **Diwali**

dewar, vacuum flask (not cap.)

De Wet (Christian Rudolph), 1854–1922, Boer general and statesman

Dewey (George), 1837–1917, US admiral; — **(John),** 1859–1952, US philosopher; — **(Melvil),** 1851–1931, US librarian; — **(Thomas Edmund),** 1902–71, US politician

De Witt (Jan), 1625–72, Dutch statesman

dexter (her.), the shield-bearer's right, the observer's left, opp. **sinister**

dextrous/, *not* -erous; **-ly**

DF, *defensor fidei* (Defender of the Faith), Dean of Faculty, direction-find/er, -ing

DFC, Distinguished Flying Cross

DFM, Distinguished Flying Medal

dft., draft

DG, *Dei gratia* (by the grace of God), *Deo gratias* (thanks to God), Director-General, Dragoon Guards

dg, decigram(s) (no point in scientific and technical work)

d.h. (Ger.), *das heisst* (that is to say)

dhobi, Indian washerman

dhooly, *use* **doolie**

dhoti, nether garment of male Hindu, *not* -ee, -ootie

dhow, an Arab vessel (accepted misspelling of **dow**), *not* daou, daw

DHSS, Department of Health and Social Security

Dhuleep, *use* **Du-**

dhurrie, Indian cotton fabric, *not* durrie

DI, Defence Intelligence, Department of Industry

d.i. (Ger.), *das ist* (that is)

dia., diameter

diablerie, devilry, *not* -ry (not ital.)

diachylon, a plaster, *not* -um, -culum

diaconate, office of deacon, *not* de-

diacritical signs, *see* **accents**

diaeresis (typ.), sign (¨) directing the second of two vowels to be pronounced separately, as in naïve; *not* die-

Diaghilev (Sergei Pavlovich), 1872–1929, Russian ballet impresario

diagnos/is, *pl.* **-es**

dial/, -led, -ling

dialect/, -al, -ic, -ical; abbr. **dial.**

diallage (rhet.), figure of speech in which various arguments are brought to bear on one point, (mineral.) a brown, grey, or green mineral similar to augite

dialling/ code, — tone (two words)

dialogue, conversation in drama, novels, etc., not necessarily between two persons only

dialyse (chem.), to separate by filtration through a membrane, *not* -ize, -yze

dialys/is, *pl.* **-es**

diamanté (Fr.), sparkling with powdered crystal

diameter, abbr. **dia.** *or* **diam.**

diamond (typ.), name for a former size of type, about $4\frac{1}{2}$ pt.

diarchy, rule by two authorities, *not* dy-

diarrhoea, *not* -œa

diaspora, a dispersal; (cap.) that of the Jews after the Babylonian captivity of 597–538 BC

diatessaron, a harmony of the four gospels (not cap.)

diathes/is, a habit of body, *pl.* **-es**

DIC, Diploma of Membership of Imperial College, London

dic/ey, risky, *not* -cy; **-ier, -iest**

dichotomize, to divide into two parts, *not* -ise

Dickens House, London, headquarters of the Dickens Fellowship (no apos.)

dickey, *use* **dicky**

Dicksee (Sir Frank), 1853–1928, painter, PRA 1924–8

dicky, rear seat (usu. folding type), false shirt-front, small bird, *not* -ey

dict., dictator, dictionary

Dictaphone (propr. term)

dictionnaire (Fr. m., two *n*s), dictionary

dict/um, a saying; *pl.* **-a**

dicy, *use* **dicey**

didactyl, two-fingered, *not* -le

didn't, to be printed close up

Didot (typ.), European system for type measurement; 12 pt. Didot = 4.512 mm. *See also* **body**

didst (no apos.)

die (Ger.), the (*f. sing., and m., f., and neut. pl.*, nominative and accusative)

diecious, *use* **dioe-**

died, abbr. **d.** *See also* **dagger**

die-hard (hyphen)

dieresis, *use* **diae-**

dies (Lat.), day(s)

diesel, compression-ignition engine

dies/ fausti (Lat.), auspicious days; — *infausti,* inauspicious days; — *irae,* day of wrath

dies/is (typ.), rare name for double dagger ‡; *pl.* **-es**

dies/ juridicus (Lat.), a day on which courts sit; — *nefasti,*

blank days; — *non* (l* *m*) a day on which no business is done

die-stamping (typ.), intaglio (q.v.) process leaving raised impression

dietitian, *not* -cian

Dieu et mon droit, God and my right (English royal motto)

Die Wacht am Rhein (Ger.), the Watch on the Rhine, famous German patriotic song

differ/, -ence, abbr. **diff.**

differenti/a, a distinguishing mark; *pl.* **-ae**

digamma, archaic Gr. letter (*Ϝ, ϝ*)

digester, person or instrument *not* -or

digestible, *not* -able

digitiz/ation (comp.), electronic reduction of characters, by scanning, to a series of digital signals; **-ed fount,** fount so prepared stored electronically in photosetter (also called **ECM**)

dignitary, *not* -atory

digraph, two letters representing one sound, as *ph* in *digraph. See also* **diphthongs** and *Hart's Rules,* p. 62

dike, *not* dyke

diktat, categorical statement (not ital.)

dil., dilute

dilapidat/e, -ed, -ion, *not* de-

dilatable, *not* -eable

dilatation, expansion, 'more correct than dilation' (*OED*)

dilator, muscle or instrument that expands, incorrectly formed but more common than **dilatator**

dilemma/, position involving choice between two unsatisfactory lines of argument or action; *pl.* **-s**

dilettant/e (It.), a lover of or dabbler in the fine arts; *pl.* **-es** or **-i**

Dilhorne (Viscount), fam. name Manningham-Buller

diligence, a stage-coach (not ital.)

Dillon (Viscount)

diluvi/um (geol.), aqueous deposit; *pl.* **-a** (not ital.)

dim., *dimidium* (one half), diminutive

dim., diminuendo

dime (US), ten cents; abbr. **d.**

diminuendo (It. mus.), getting softer; abbr. *dim.*

DIN, Deutsche Industrie-Norm; the international paper-size system (some trimmed sizes in mm.): A0, 841 × 1,189; A4, 210 × 297; A5, 148 × 210; B0, 1,000 × 1,414; B1, 707 × 1,000. For some untrimmed sizes in this system, *see* **book sizes**

Dinan, dép. Côtes-du-Nord, France

Dinant, Belgium

dinar, Byzantine gold coin (denarius); unit of currency in various Asian and African countries and Yugoslavia, *not* de-

Dinard, dép. Côtes-du-Nord, France

d'Indy (Paul Marie Théodore Vincent), 1851–1931, French composer (small *d*)

Ding an sich (Ger. philos. n.), thing in itself

ding-dong (hyphen)

dinghy, small boat, *not* -gey, -gy

dingo/, Australian native dog; *pl.* **-es**

dingy, grimy. *See also* **dinghy**

dining/-car, -room, -table (hyphens)

dinner-jacket, abbr. **DJ**

dioces/e, -an; abbr. **dioc.**

dioecious (bot.), *not* die-

Diogenes, 412–323 BC, Cynic philosopher

Diogenes Laertius, biographer of philosophers, about AD 200

dionym, a binomial name, as *Homo sapiens*

Dionysia (noun pl.), orgiastic and dramatic festival(s) of Dionysus

Dionysi/ac, of Dionysus; **-an,** of Dionysus or of Dionysius

Dionysius, 430–367 BC, and his son, *fl.* 350 BC, tyrants of Syracuse

Dionysus, Greek god Bacchus

dioptre, lens measure, abbr. **d.**

Dioscuri (the), Castor and Pollux

Dip., Diploma

Dip. A.D., Diploma in Art and Design

Dip. Ed., Diploma in Education

Dip. H.E., Diploma of Higher Education

diphtheria, *not* dipth-

diphthongs (typ.), *Æ, æ, Œ, œ,* for *single sounds*, are employed in Danish, French, Icelandic, Norwegian, and Old-English words, and as phonetic symbols, instead of the separate letters *ae, oe. ae* and *oe* should be used in Latin and Greek words. Generally no ligatures are used in US. *See also* **German**

diplomat, *not now* diplomatist; abbr. **dipl.**

diplomate, one holding a diploma; in Fr. *diplôm/é, fem.* **-ée**

Diplomatic, Lectureship in, *not* Diplomatics

dipsomania/, -c

Directoire, French Directory of 1795–9; adj., of the dress or furniture of the period (cap.)

directress, female director

directrix (geom.)

Dirichlet series (math.)

dirigible

dirigisme (Fr. m.), economic control by the State

dirndl, Alpine peasant bodice and full skirt

dis-, *see* **dys-**

dis., discipline, discontinue, -ed, discount, (typ.) distribute

dis aliter visum (Lat.), the gods have thought otherwise

disappoint (one *s*, two *p*s)

disassemble, take (machine) apart. *See also* **dissemble**

disassociate, *use* **dissociate**

disburden, *not* -then

disbursement, *not* -sment

disc, now accepted for **disk,** esp.
in compds., disc harrow, disc
jockey, slipped disc; (in photosetter) of glass or plastic, bearing
master-images of fount(s) to be
set; (in comp. input) of limp
plastic, storing input-material
and driving photosetter

disc., discover/ed, -er; discount

disc jockey, abbr. **DJ**

discern/, to see; **-ible.** *See also*
decern

disciplinary, *not* -ery

discipline, abbr. **dis.**

discipular, disciple-like

disco/, dancing-party with
records, *pl.* **-s**

discobol/us, a discus-thrower,
not -*ulus*; *pl.* **-i** (not cap.). The
Discobolus (cap.), the lost statue
by Myron, of which *discoboli* are
copies

discoloration, *not* discolour-

discolour/, -ed, -ment, *not* dis-
color

discomfit/, -ed, -ing, to thwart;
noun, **-ure**

discomfort, (noun) lack of
ease; (verb) to make uneasy

disconnection

discontinu/e, -ed, abbr. **dis.**

discothèque, room or small hall
used for dancing to recorded
music

discount, abbr. **dis.**

discover/ed, -er; abbr. **disc.**

discreet, judicious

discrete, separate

disect, *use* **dissect**

disenfranchise, *use* **disfranchise**

disenthral/, *not* -enthrall,
-inthral(l); **-led, -ment**

disenthrone, *use* **dethrone**

diseu/r (Fr. m.), artist entertain-
ing with monologues, *fem.* **-se**

disfavour, *not* -vor

disfranchise, *not* -ize

disguis/e, -er

dishabille, untidy dress

dishevelled, with hair or dress
in disorder, *not* -eled

disinterested, impartial (not
the same as uninterested)

disjecta membra (Lat.), scat-
tered remains, *but* 'disjecti membra poetae' (Horace)

dismissible, *not* -able

disorganiz/e, -er, *not* -ise

disorient/ed, -ing, and **dis-
orientat/ed, -ing,** are equally
established in modern usage. *See
also* **orient**

dispatch/, -er, *not* des-

dispensable, *not* -ible

dispensary, abbr. **disp.**

dispise, *use* **de-**

display/ (typ.), the setting and
leading of titles, advertisements,
etc., direction to set in this style;
— **type,** of size or cut suited to
this, *or* any type cast in an
especially durable metal

Disraeli (Benjamin), 1804–81
Earl of Beaconsfield

d'Israeli (Isaac), 1766–1848,
writer, father of the foregoing

dissect/, -ion, -or, *not* dise-

disseis/e, to dispossess wrong-
fully, *not* -ze; **-ee, -in, -or,
-oress**

dissemble, to conceal. *See also*
disassemble

disseminat/e, to scatter
abroad; **-or**

Dissenter, from Church of England (cap.)

dissertation, abbr. **diss.**

dissociate, to separate, *not* disassociate

dissoluble, *not* -uable

dissolvable, *not* -ible

dissyllable, *use* **disy-**

dissymmetry, *not* disy-

dist., distance, distilled, distinguish, -ed, district

distension, state of being
stretched, *not* -tion

distich/, couplet; *pl.* **-s**

distil/, *but* **-lation; -led,** abbr.
dist.

distingu/é (Fr.), *fem.* **-ée,** distin-
guished-looking

distinguish/, -ed; abbr. **dist.**

distrait/, (Fr.), *fem.* *-e*, absent-
-minded

distributary, one of the streams
of a river delta

distribute (typ.), to put type
back into case, or to melt down
type; abbr. **dis.**

distributor, one who *or* that
which distributes

district, abbr. **dist.**

District/ Attorney (US), abbr.
DA; — **Court,** — **of Colum-
bia,** abbr. **DC**; — **Registry,
DR**

disyllab/lc, -ic, -ize, of two
syllables, *not* diss-; in Fr. m.
dissyllabe

disymmetry, *use* **diss-**

ditheism, belief in two gods, *not*
dy-

dithyramb, wild hymn of
Bacchic revellers, *not* dythi-

ditto, abbr. **do.**

div., divide, -d, dividend, divine,
division, divisor, divorce, (Fr.)
divers (diverse)

divers, sundry

diverse, different

diverticul/um, a byway, esp. of
the intestines; *pl.* **-a**

divertiment/o (It.), a kind of
ballet, (mus.) entertainment-
-piece, *pl.* **-i**; also (Fr. m.)
divertissement

divest, *not* de- (except in law, as
devest out of)

**divid/e, -ed, -end, division,
divisor,** abbr. **div.**; sign for
divide ÷

divide et impera (Lat.), divide
and rule

Divina Commedia (*La*), 1300-
18, by Dante, *not* Come-

Divis (Ger. typ. n.), the hyphen

divisi (mus.), in several parts

divisible, *not* -able

divisim (Lat.), separately

division mark, ÷

division of words. Usually
divide a word after a vowel,
taking over the consonant. In
present participles take over

-ing, as sound-ing, divid-ing;
but trick-ling, chuck-ling, and
similar words. Generally, when
two consonants or vowels come
together divide between them,
as splen-dour, appreci-ate. Ter-
minations such as -cian, -sion,
-tion should not be divided when
forming one sound, as Gre-cian,
ascen-sion, subtrac-tion. Hyphenated words should be
divided at the hyphen, and in
lexical work etc. a second hyphen may be used (as in this
dictionary) to clarify their spelling, e.g. second-/-hand. Avoid
divisions which might confuse:
leg-ends (*le-gends*), reap-pear
(*re-appear*), ex-acting (*exact-ing*).
American printers divide
strictly according to pronunciation. For a fuller treatment and
for word-division in foreign languages see *Hart's Rules,* pp. 14–
16, 96-7, etc.

divisor (math.), a factor

divorc/é (Fr.), *fem.* *-ée*, a divorced person; Eng. **divorcee**
is common gender

Diwali, Hindu festival, *not*
Dewalee

Dixie/ *or* **-land** (one word), the
Southern States, US

DIY, do it yourself

DJ, dinner-jacket, disc jockey

Djakarta, Indonesia, *use* **Jak-**

djellaba, djibbah, *use* j-

Djibouti, *formerly* French Territory of Afars and Issas, *use* **Jibuti**

djinn, *see* **jinnee**

DL, Deputy-Lieutenant

dl, decilitre(s) (no point in
scientific and technical work)

D.Lit., Doctor of Literature;
D.Litt., Doctor of Letters. *See
also* **Lit.D., Litt.D.**

DLM (mus.), double long metre

DLO, Dead Letter Office, *now*
Returned — —

DM, Deputy Master, Deutsche
Mark (W. Ger. currency), Doc-
tor of Medicine (Oxford), (Fr.)
Docteur en Médecine (Doctor of

DM (*cont.*)
Medicine), (It. mus.) *destra mano* (the right hand)

dm, decimetre(s) (no point in scientific and technical work)

D-Mark, *see* **Deutsche Mark**

D.Mus., Doctor of Music

DN, *Dominus noster* (our Lord)

DNA, deoxyribonucleic acid

DNB, Dictionary of National Biography

Dnieper River, White Russia–Ukraine; in Russ. **Dnepr**

Dniester River, Ukraine; in Russ. **Dnestr**

D-notice, official request to news editors not to publish item

do (noun), *pl.* **dos**

do (mus.), *use* **doh**

do., ditto, the same

DOA, dead on arrival

doat, *use* **dote**

Dobermann pinscher, German breed of hound

doc., document(s)

Docent (Ger. m.) a university teacher, *now Doz-*

doch-an-doris, *use* **deoch an doris**

docket/, -ed, -ing

dockyard (one word)

Doctor, abbr. **D.,** (*but* **Dr** before name); **Doctor of/ Canon Law** *or* **Civil Law, DCL;** — — **Dental Surgery, DDS;** — — **Divinity, DD;** — — **Laws, LL D;** — — **Letters, D.Litt.** *or* **Lit(t).D.;** — — **Literature, Lit.D.;** — — **Medicine, MD** *or* **DM;** — — **Music, D.Mus.** *or* **Mus.D.;** — — **Philosophy, Ph.D** *or* **D.Phil.;** — — **Science, D.Sc.** *or* **Sc.D.;** — — **Veterinary Science** *or* **Surgery, DVS**

Doctors' Commons, London, where marriage licences were formerly issued by the Bishop of London's Registry (apos.)

doctrinaire, an unpractical theorist; theoretical and unpractical (not ital.)

documentary (noun), factual report or film

documents (typ.), to be an exact reprint; abbr. **doc.**

dodecaphonic (mus.), of the twelve-note scale

dodgem (car at fun-fair)

dodo/, extinct flightless bird; *pl.* **-s**

D.o.E., Department of the Environment

doesn't, to be printed close up

dogana (It. f.), custom-house

dogate, office of doge, *not* -eate (not ital.)

dogcart (one word)

dog-days, 3 July–11 August (hyphen)

doge, chief magistrate of Venice (not ital.)

dogfight, dogfish (one word)

doggerel, unpoetic verse, *not* -grel

Doggett's Coat and Badge, trophies of Thames Watermen's championship

doggie, noun; **doggy,** adj.

doghouse (one word)

dog-Latin, barbarous Latin (hyphen)

dogma/, authoritative doctrine; *pl.* **-s**

dogmatize, *not* -ise

dogsbody, (naut. slang) junior officer; general drudge

dog-star (hyphen)

doh (mus.), *not* do

Dohnányi (Ernö), 1877–1960, Hungarian composer

doily, a napkin, *not* -ey, doyley, -ly, d'oyley, -ie

dolce (mus.), sweetly

dolce/ far niente (It.), delightful idleness; — *vita,* sweet life

doleful, *not* -ll

Dolgellau, Gwynedd, *not* -ey

dollar mark (typ.), $ or $, to be *before,* and close up to, the figures, as $50. Various dollars differentiated as $A (Australian), $HK (Hong Kong), $US, etc.

Dollfuss (Engelbert), 1892–

1934, Austrian politician, Chancellor 1932–4

doll's house, not -s' —

dolman, Turkish robe, hussar's jacket, woman's mantle

dolmen, prehistoric stone table

dolorous, doleful, but **dolour**

DOM, *Deo optimo maximo* (To God the best and greatest); *Dominus omnium magister* (God the Master, or Lord, of all)

Dom (Ger.), cathedral; (Russ.) house; (Port.) title of nobility; *also* title of certain monks, as Dom Gasquet; *not* Don

dom., domestic, dominion

domaine (Fr. m.), a vineyard

Domenichino, real name **Domenico Zampieri,** 1581–1641, Italian painter

'Domesday Book', not Dooms-; abbr. **DB**

domestic, abbr. **dom.**

domicile, not -cil

Domine dirige nos (Lat.), O Lord, direct us (motto of the City of London)

Dominica, one of the Windward Islands, W. Indies

Dominican Republic, W. Indies, *formerly* **Santo Domingo,** the eastern portion of the island of Hispaniola

dominie (old Sc.), teacher, preacher; (current locally) headmaster

dominion, abbr. **dom.**

Dominion Day, Canada, 1 July

domino/, cloak, mask, piece in game; *pl.* **-es**

Dominus/ (Lat.), Lord; abbr. **D.;** — *noster,* our Lord, **DN**

Domus Procerum (law), the House of Lords; abbr. **DP** or **Dom. Proc.**

Don, Spanish title; *see* **Dom**

Donegal, Ireland

Donegall (Marquess of)

Donizetti (Gaetano), 1797–1848, Italian composer

donna (It.), a lady (ital.)

Donne (John), 1573–1631, English poet

donnée (Fr. f.), basic fact

don't, to be printed close up

doolie (Ind.), a litter, palanquin, *not* dhooley, -lie, -ly, dooly

Doolittle (Eliza), heroine of Shaw's *Pygmalion*; — **(Hilda),** 1886–1961, US poet

'Doomsday Book', *use* **Domesday** —

door-keeper (hyphen)

doormouse, *use* dor-

Doornik, Fl. for **Tournai**

dop (Afrik.), brandy, tot

doppelgänger (Ger. m.), a wraith

Doppelpunkt (Ger. typ m.), the colon

doppio (movimento) (It.), double (speed)

Doppler effect (phys.), apparent change of frequency when source of vibrations is approaching or receding, from **Christian Johann Doppler,** 1803–53, Austrian physicist; *not* ö

Dor., Doric

DORA, Defence of the Realm Act, 1914

Doré (Paul Gustave), 1833–83, French painter and engraver

dormer window (two words)

dormeuse (Fr. f.), a settee, nightcap, travelling sleeping carriage

dor/mouse, *not* door-; *pl.* **-mice**

dormy (golf), as many up as there are holes to play, *not* -ie

Dorneywood, Bucks., country house used as off. residence by any Minister designated by the Prime Minister

dorp (Afrik.), village, country town

d'Orsay (Alfred Guillaume Gabriel, Count), 1801–52, 'the last dandy'

Dorset, *not* Dorsetshire

dory, John Dory, fish, *not* -ey

dos-à-dos (Fr.), back to back, a sofa made for sitting so

dosage (med.), *not* -eage

dose (med.), abbr. **d.**

dosimeter, *not* dosemeter

Dos Passos (**John Roderigo**), 1896–1970, US writer

dossier, papers referring to some matter (not ital.)

Dostoevsky (**Fyodor Mik-hailovich**), 1821–81, Russian novelist

D.o.T., Department of Trade

dot (Fr. f.), dowry

dot/e, to show great fondness, *not* doat; **-age, -ard**

Douai, dép. Nord, France, *not* -y; **Douai School,** *but* **Douay Bible**

douane (Fr. f.), custom-house; abbr. **d.**

double (typ.), a word, etc., erroneously repeated

double-barrelled (hyphen), *not* -eled

double-bass (mus.), (hyphen)

double entendre, word or phrase of two meanings, one of them usu. indecent (obs. Fr., now Anglicized; the mod Fr. is *double entente* (*un mot à*))

double pica (typ.), name for a former size of type, about 22 pt.

doublure (Fr. typ. f.), ornamental inner lining to a book-cover

doucement (Fr. mus.), sweetly

douceur (Fr. f.), a gratuity

douche, jet of water directed at the body (not ital.)

doughboy (US), a dumpling, an American infantryman (one word)

doughnut (one word)

Doukhobors, *use* **Duk-**

Dounreay, Highland, atomic research station

douse, to drench with water, extinguish. *See also* **dowse**

douzaine (Fr. f.), dozen; abbr. **dzne.**

dovecot, *not* -cote

dow, *use* **dhow**

dowager, a widow with a dower or retaining her title; abbr. **dow.**

dowel/, headless pin usu. of wood, to fasten with this; **-led, -ling**

dower, widow's life-interest in her husband's property

Dow-Jones, NY Stock Exchange index (hyphen)

Down, N. Ireland

downbeat (one word)

Downe (**Viscount**)

down/fall, -hill (one word)

down-market (adj., hyphen)

down/pour, -right, -stairs, -stream, -trodden, -wind (one word)

dowry, money or property brought by a bride to her husband

dowse, to use a divining rod (*see also* **douse**)

doyen/, the senior member of a society (not ital.); *fem.* **-ne**

d'oyl/ey, -ie, *use* **doily**

D'Oyly Carte (**Richard**), 1844–1901, theatrical impresario

doze, to sleep

dozen/, -s, abbr. **doz.**

Dozent (Ger. m.), a university teacher, not now *Doc-* (cap.)

DP, displaced person (refugee); (law) *Domus Procerum* (the House of Lords)

DPAS, Discharged Prisoners' Aid Society

DPH, Diploma in Public Health

D.Phil., Doctor of Philosophy (Oxford, Sussex, York); **Ph.D.** more usual

DPP, Director of Public Prosecutions

DR, (naut.) dead reckoning, dispatch rider, District/ Railway, — Registry, (Ger.) *Deutsches Reich* (German Empire)

Dr, doctor (before name)

dr., drachm, -s, drachma (dram *or* coin), -s, dram, -s, drawer

drachm, *use* **dram,** except for apothecaries' measure and weight; abbr. **dr.,** sign ℈

drachma/, a dram, various coins; *pl.* **-s,** abbr. **dr.**

draft, a deduction in weighing, a mil. party, a money order, a rough sketch; to draw off, to sketch; abbr. **dft**. *See also* **draught**

draftsman, one who drafts documents. *See also* **draughtsman**

dragée, a sweetmeat enclosing drug or nut or fruit

draggle-tail, a slut, *not* daggle-

dragoman/, Arabic interpreter; *pl*. **-s,** *not* -men

dragon-fly (hyphen)

dragonnade, French persecution of Protestants in 1681, *not* -onade, -oonade

Dragoon Guards, abbr. **DG;** these are not 'Guardsmen'

dram, sixty grains, one teaspoonful, sixty minims, small drink, *see* **drachm;** *pl*. **-s,** abbr. **dr.**

dramat/ic, -ist, abbr. **dram.**

dramatis personae, (list of) characters in a play, abbr. **dram. pers.**

dramatize, *not* -ise

Drang nach Osten (Ger.), desire for expansion eastwards

Drapier's Letters, by Swift, 1724

draught, act of drawing, a take of fish, 20 lb. of eels, act of drinking, a dose, a vessel's depth in water, a current of air, liquor 'on draught'. *See also* **draft**

draughtsman, one who makes drawings, plans, etc., piece in game of draughts. *See also* **draftsman**

drawback, in tech. meaning of excise duty remitted, abbr. **dbk.**

drawer, abbr. **dr.**

drawing-room (hyphen)

draw-on cover (bind.), limp cover glued to back of book

Drdla (Franz), 1868–1944, Czech composer

dreadnought, thick cloth, battleship, *not* -naught

Dred Scott, Negro slave, subject of US law case

Dreiser (Theodore), 1871–1945, US novelist

driblet, a trickle, *not* dribb-

dri/er, -est, -ly; *also* **drier,** one who or that which dries; but **dry/ish, -ness**

driftwood (one word)

drip/-dry, -proof (hyphens)

drivable, *not* -eable

drivell/ed, -er, -ing (two *l*s)

drive out (typ.), to set matter with wide word spacing, such setting

drizzling (meteor.), abbr. **d.**

droit (Fr. m.), moral and legal right

droit de seigneur (Fr. m.), lord's right to enjoy vassal's bride

droite (Fr. f.), the right hand; abbr. **d.**

Drontheim, *use* **Trondheim**

dropped head (typ.), opening title of chapter etc. set lower than first line of text on other pages

droshky (Russ. *pl*. **drozhki**), a four-wheeled vehicle, *not* -sky

drought/, -y, arid(ity), *not* drouth

DRP, Deutsches Reichspatent (German patent)

drum/beat, -fire, -head, -stick (one word)

drunkenness (three *n*s)

Druse, one of a Syrian sect, *not* Druz, -e

Dryasdust, a dull pedant (character in prefaces of Scott's novels) (one word, cap.)

dry/er, -est, -ly, *use* **dri-**

dry goods (US), textiles

dry/ish, -ness

dry-point, etching needle, the work produced by it (hyphen)

DS, (mus.) dal segno (from the sign); disseminated sclerosis

Ds., *dominus*; at Cambridge, a Bachelor of Arts

DSC, Distinguished Service Cross

D.Sc., Doctor of Science

DSIR, Department of Scientific and Industrial Research

DSM, Distinguished Service

DSM *(cont.)*
Medal; (mus.) double short metre

DSO, (Companion of the) Distinguished Service Order

d.s.p., *decessit sine prole* (died without issue)

d.t., d.t.'s, delirium tremens

Du., Dutch

du, as prefix to a proper name, in accordance with continental practice, should *not* have initial cap., as du Châtelet, except when Anglicized or at beginning of sentence. Signatures to be copied

Dual Monarchy (caps.)

dub., *dubitans* (doubting), *dubius* (dubious)

Du Barry (Marie Jeanne Bécu, comtesse), 1746–93, favourite of Louis XV of France

dubbin, grease for leather, *not* -ing

Dublin, abbr. **Dubl.**

Dubois-Raymond (Emil), 1818–96, German physiologist, *not* du Bois

duc (Fr.), Duke; *fem.* **duchesse** (not cap.)

duces tecum (Lat.), a subpoena

du Chaillu (Paul Belloni), 1837–1904, French-American explorer in Africa

duckbill (one word)

ductus (Lat. med.), a duct; *pl.* same

Dudevant, *see* **Sand (George)**

duell/er, -ing, -ist (two *l*s)

duenna, Spanish chaperon (not ital.)

duffel/, coarse woollen cloth; — **bag,** — **coat** (two words)

Dufy (Raoul), 1877–1953, French painter

dug-out, canoe made from single tree-trunk (hyphen)

Duguesclin (Bernard), 1314–80, Constable of France

duiker (Afrik.), a small antelope, *not* duy-

duke, cap. as title, abbr. **D.;** in Fr. *duc* (not cap.)

Dukhobors, Russian sect, *not* Douk-

Duleep Singh, not Dhuleep, Dulip

dullness, *not* dulness

Dulong and Petit (law of) (physics)

Dumas (Davy de la Pailleterie, Alexandre), 1802–70, French writer; — (fils, **Alexandre**), his son, 1824–95, also writer; — (**Jean Baptiste André**), 1800–84, chemist

du Maurier (George Louis Palmella Busson), 1834–96 English artist and author; — (**Sir Gerald**), 1873–1934, his son, actor-manager; — (**Dam Daphne**), b. 1907, his daughter author

Dumbarton, Strathclyde

Dumbarton Oaks, estate in Washington, DC, at which —. — Conference was held, 1944

dumb-bell (hyphen)

dumbfound/, -ed, -er, *not* dumb

Dumfries & Galloway, regio of Scotland

dum sola (*et casta*) (law), whi unmarried (and chaste)

Dunbarton, former Scottish county. *See also* **Dumbarton**

Dunblane, Central, *not* Dum-

Duncan-Sandys (Baron) (hy phen)

Dunelm:, sig. of Bp. of Durham (colon); of Durham University (no point, no colon)

dungaree, coarse calico; *pl.* overalls made of it; *not* -eree(s)

duniwassal (Sc.), a Highland gentleman, *not* dunni-

Dunkirk, dép. Nord, France; i Fr. **Dunkerque**

Dun Laoghaire, port of Dubli

Dunnottar Castle, Grampian (two *n*s, two *t*s)

Duns Scotus (Johannes), 1265 or 1274–1308, metaphysi cian

duo/, pair of artistes; *pl.* **-s**

duodecimo/ *or* **twelvemo/** (typ.), a book based on 12 leave

24 pages, to the sheet; *pl.* **-s;**
abbr. **12mo**

uologue, (stage) conversation
between two

uplex/ apartment (US), one
on two floors; — **house** (US),
one for two families

uplication of points, *see*
punctuation XV

u Pont, family of American
industrialists

**uquesne/ (Abraham, mar-
quis),** 1610–88, French
admiral; —, **Fort,** French fort,
1754, renamed Fort Pitt (Pitts-
burg, Pa.), 1758, on its capture
by the English

ur (Ger. mus.), major

urab/le, -ility

ürer (Albrecht), 1471–1528,
German painter; **Dureresque,**
in his style

uress, constraint, *not* -e

'Urfey (Thomas), 1653–
1723, English playwright

urham, abbr. **Dur.**

ürrenmatt (Friedrich), b.
1921, Swiss playwright

urrie, *use* **dhurrie**

ushanbe, *formerly* Stalinabad,
cap. of Tadjikistan, USSR

ussek (Jan Ladislav), 1760–
1812, Bohemian composer

üsseldorf, W. Germany

ust-bowl, area denuded of ve-
getation by drought (hyphen);
Dust Bowl (caps., no hyphen),
the region along the western
edge of the Great Plains, US

utch/, abbr. **Du.;** — **alphabet**
(typ.), same as English but *q* and
x are used only in borrowed
foreign words. Accented letters
are often used in stressed syllables. Marked letters *ë* and *ö*,
diaeresis. *ch* must never be separated; for *y* use *ij*, cap. *IJ*

utiable, subject to duty

uumvir/, one of a pair of
Roman officials; Eng. *pl.* **-s**

dux, top pupil (not ital.)

dux (Lat.), a leader, *pl.*
duces; — *gregis,* leader of the
flock

duyker, *use* **dui-**

DV, *Deo volente* (God willing)

Dvořák (Antonin), 1841–1904,
Czech composer

DVS, Doctor of Veterinary Science (*or* Surgery)

d.w., dead weight, delivered
weight

dwarf/, *pl.* **-s**

dwt., pennyweight, -s, 24 troy
grains, *not* pwt.

d.w.t., dead-weight tonnage

Dy, dysprosium (no point)

Dyak, of Borneo, *not* Day-

dyarchy, *use* **diarchy**

Dyce Airport, Aberdeen

dyeing (cloth, etc.)

dye-line, diazo print (hyphen)

Dyfed, Welsh county

dying (ceasing to live)

dyke, *use* **dike**

dynamics, science of matter and
motion (sing.); abbr. **dyn.**

dynamo/, *pl.* **-s**

dyne, unit of force; abbr.
dyn (no point); in SI units 10
μN

dyotheism, *use* **ditheism**

dys-, prefix meaning difficult or
defective; distinguish it from
dis-, meaning apart

dysentery

dyslex/ia, word-blindness; adj.
-ic

dyspep/sia, chronic indigestion,
not dis-; adj. **-tic**

dysprosium, symbol **Dy**

dytheism, *use* **di-**

dythiramb, *use* **dithy-**

dzho, cow–yak hybrid, *not* the
many variants

dziggetai, Mongolian wild ass,
not the many variants

dzne. (Fr.), *douzaine* (dozen)

D-Zug (Ger. m.), *Durchgangszug*
(an express train)

E

E, Egyptian (in £E), einsteinium₁ (no point), (prefix) exa-, the fifth in a series. *See also* **Lloyd's**

E., Earl, east, -ern, English

e, eccentricity of ellipse, (dyn.) coefficient of elasticity, (elec.) electromotive force of cell

e, **e,** or ϵ (math.), base of natural or Napierian logarithms

e (It., Port.), and

é (Port.), is

è (It.), is

è (**e grave accent**), to be used for the last syllable of past tenses and participles when that otherwise mute syllable is to be separately pronounced, as, 'Hence, loathèd Melancholy!'. *See also* **accents**

e. (meteor.), wet air but no rain

each (gram.) must be followed by verb and pronoun in sing., as 'each person *knows his* own property'; abbr. **ea.**

EAEC, European Atomic Energy Community (*now* **Euratom**)

eagre, tidal wave, *not* -er

E. & O. E., errors and omissions excepted

earache (one word)

ear-drum (hyphen)

earl (cap. as title); abbr. **E.**

Earls Court, London (no apos.)

ear/mark, -phone (one word)

ear/-piece, -plug, -ring (hyphens)

earth (**the**), cap. only in astronomical contexts and in a list of planets; sign ⊕

earthwork (one word)

east/, -ern, abbr. **E.** *See also* **capitalization, compass**

eastbound (one word)

East Bridgford, Notts., *not* Bridge-

East End, London (caps.)

Easter Day, first Sunday after the calendar full moon on, or next after, 21 March

East Kilbride, Strathclyde, 'new town', 1947

easy chair (two words)

easygoing (adj., one word)

eau/-de-Cologne, — -de-Nil —-de-vie (hyphens); — **for** nitric acid, *also* an etching; — **-fortiste,** an etcher; — **sucrée,** sugar and water (not ita

EB, Encyclopaedia Britannica

Ebbw Vale, Gwent, a town

ebd. (Ger.), *ebenda,* ibid.

Eblis, chief of fallen angels in Muslim myth., *not* I-

ebonize, *not* -ise

Ebor., *Eboracum* (York)

Ebor:, sig. of Abp. of York (colon)

EBU, European Broadcasting Union

eburnean, like ivory, *not* -ian

EC, East-Central (postal distri of London), Education(al) Committee, Electricity Counc Engineer Captain, Episcopal Church, Established Church, Executive Committee

écarté, a game of cards

Ecce Homo, 'Behold the Man

eccentric/, -ally, -ity. For mat sense *see also* **excentric**

Ecclefechan, Dumfries & Gal loway, birthplace of Carlyle, 4 Dec. 1795

Ecclesiastes, in OT, abbr. **E** cles.

ecclesiastical/, abbr. **eccles.** — **signs:** Greek Cross ✠ used in service books to notify 'mak the sign of the cross', also befor signatures of certain Church dignitaries; in service books: ℞ response, ℣ versicle, ✱ word to be intoned. *See also* **cross, crux**

Ecclesiasticus, in Apocr., abbr. **Ecclus.**

ECG, electrocardiogram

Echegaray (José), 1833–1916, Spanish playwright

echelon (not ital., no accent)

echo/, *pl.* **-es,** *not* -os

echt (Ger.), genuine

éclair, small pastry filled with cream

éclaircissement (Fr. m.), explanation

éclat, renown (not ital.)

eclectic, borrowing freely from various sources

eclogue, a pastoral poem

ECM, electronic character master. *See also* **digitized fount**

ecology, the study of organisms in their environment, *not* oe-; abbr. **ecol.**

Economist (The), (cap. *T*)

economize, *not* -ise

econom/y, -ical, -ics, -ist, abbr. **econ.**

écossais (Fr.), *fem.* **-e,** Scottish (not cap.); *also* *écossaise,* a dance

ecraseur, surgical instrument (no accent)

écrevisse (Fr. f.), crayfish

ecru, unbleached linen colour (no accent, not ital.)

ECSC, European Coal and Steel Community

ecstasy, rapture, *not* ex-, -cy

ECT, electroconvulsive therapy

ECU, European Currency Unit

Ecuador, *not* Eq-; abbr. **Ecua.,** adj. **-ean**

ecumenic/, -al, belonging to the entire Christian Church, *not* oec- except in special uses

eczema, a skin inflammation

ed., edited, editor

ed. cit., the edition cited

EDD, English Dialect Dictionary

Edda/, a collection of Icelandic legends; *pl.* **-s; The Elder —,** a collection of poems, eleventh century and earlier; **The Younger —,** thirteenth--century stories, prosody, and commentary

Eddy (Mary Morse Baker), 1821–1910, founder of Christian Science

Eddystone Lighthouse

edelweiss, Alpine plant

edema/, -tous, *use* **oed-**

Edgbaston, Birmingham, *not* Edge-

edge-tool, *not* edged —, but 'with edged tools' (fig. sense)

edgeways, *not* -way, -wise

Edgeworth (Maria), 1767–1849, English novelist; — **(Richard Lovell),** 1744–1817, her father, English educationist

Edgware, London; — **Road**

edh, *use* **eth**

edidit (Lat.), edited this

edile, *use* **aedile**

Edinburgh, abbr. **Edin.**

Edipus, *use* **Oe-**

Edirne, formerly **Adrianople,** Turkey

Edison (Thomas Alva), 1847–1931, US inventor

édit/é, fem. **-ée, par** (Fr.), published by

éditeur (Fr. m.), a publisher, *not* editor

edition, the state of a book (also the copies, or any one copy, so printed) at its first publication, and after each revision, enlargement, abridgement, or change of format (2nd, 3rd, etc./revised/enlarged/abridged/ paperback, etc., edition); *not* any reprint containing no substantial alteration (*see* **impression**); abbrev. **edn.;** in trade practice **1/e, 2/e,** etc. = 1st, 2nd, etc., edn. *See also* **reprint**

édition de luxe (Fr.), a sumptuous edition (ital.)

editio/ princeps, a first printed edition; *pl.* **-nes principes**

editor, abbr. **ed.,** *pl.* **eds.** or **edd.**

edn., edition

EDP, electronic data processing

EDS, English Dialect Society

educ/ate, -able

educationist, *not* -alist

educ/e, -ible

Eduskunta, Parliament of Finland

EE, Early English

ee (Sc.), eye; *pl.* **een**

E.E. & M.P., Envoy Extraordinary and Minister Plenipotentiary

EEC, European Economic Community (the Common Market)

EEG, electroencephalogram

eerie, weird, *not* -y; **eerily**

EETS, Early English Text Society

effect/er, one who effects; **-or** (biol.), an organ that effects a response to a stimulus

Effendi, Turkish title of respect, *not* -dee

effluvi/um, vapour from decaying matter; *pl.* **-a**

effluxion, that which flows out, *not* -ction

EFTA, European Free Trade Association

e.g., *exempli gratia* (for example) (l.c., not ital., comma before). Use 'for example' rather than e.g.

egg/head, -shell (one word)

ego/, the self which is conscious and thinks; *pl.* **-s**

egotize, to act egotistically, *not* -ise

egret, the lesser white heron (*see also* **aigrette**)

Egypt/, -ian. *See also* **United Arab Republic**

egyptian (typ.), a typeface with thick stems and slab serifs

Egyptian pound, abbr. **£E**

Egyptolog/ist, -y, abbr. **Egyptol.** (cap.)

eh, when exclamation, to be followed by exclamation mark; when question, by question mark

Ehrenburg (Ilya Grigorevich), 1891–1967, Russian writer

-ei, *not* -ie-, in: beige, ceiling, conceit, conceive, counterfeit, cuneiform, deceit, deceive, forfeit, geisha, heifer, heinous, in-

veigh, inveigle, kaleidoscope, leisure, meiosis, neigh, -bour, -bourhood, neither, nonpareil, obeisance, perceive, plebeian, receipt, receive, Seidlitz, seigneur, seigniorage, seise, seize, seizure, sleight, surfeit, their, weir, weird. *See also* **-ie-**

eidol/on, a phantom; *pl.* **-a**

Eifel Mountains, W. German

Eiffel/ (Alexandre Gustave) 1832–1923, French engineer; **— Tower,** Paris

eigen/frequency, -function, -value (math. and phys., one word)

eighteenmo/ (typ.), a book based on 18 leaves, 36 pages, to the sheet; also called octodecimo, decimo-octavo; *pl.* **-s;** abbr. **18mo**

Eighteenth Amendment to the Constitution of the US, 1920 introducing prohibition of intoxicating liquor; repealed by Twenty-first Amendment, 193

eigret/, -te, *use* **aigrette.** *See als* **egret**

eikon, *use* **icon**

Eikon Basilike, pamphlet reputedly by Charles I, more probably by John Gauden, 1605–52 bishop and writer; another, written by Titus Oates, ded. to William III and attacking James II

Einstein (Albert), 1879–1955 German physicist (naturalized American 1940); **— (Alfred),** 1880–1952, German musicologist

einsteinium, symbol **E**

Eire, in Gaelic means 'Ireland'; former off. name of the Republic of Ireland (or 'the Irish Republic'), i.e. the 26 counties, formerly Irish Free State

eirenicon, a peace proposal, *not* ir-

Eisenhower (Dwight David) 1890–1969, President of US, 1953–61; American general, Supreme Commander of Allied

Expeditionary Force in Second World War

Eisenstein (Sergei Mikhailovich), 1898–1948, Russian film-director

eisteddfod/, a congress of Welsh bards; *pl.* **-s**

either (gram.), correlative *or*, not *nor*; for number of following verb, *see* **neither**

ejector, one who, that which, ejects, *not* -er

ejusdem/ (Lat.), of the same, abbr. *ejusd.*; — **generis,** of the same kind

El (Heb.), God

el (Sp. m.), the

él (Sp.), he

Elagabalus, 204–22, emperor of Rome 218–22, *not* Elio-, Helio-

El Al, Israel Airlines, Ltd.

El Alamein, Egypt; battle, 1942

élan, dash, spirit; — **vital,** life-force (ital.)

Elbrus (Mount), Caucasus, *not* -ruz, -urz

Elburz, mts., Iran, *not* -bruz

Elchee, Anglicized version of *elchi*, Turkish ambassador

Elder Brethren of Trinity House (caps.)

El Dorado, the golden land (two words, *not* ital.)

Eldorado/, place of great abundance, *pl.* **-s**

elec., electricity, electrical, electuary

elector, *not* -er; (hist.) German prince electing emperor (cap. as title)

Electra complex, exaggerated attachment of daughter to father, the fem. analogue of **Oedipus complex**

electress, *not* -toress

electro/ (no point), abbr. of **electrotype;** *pl.* **-s**

electrocute, to put to death by electricity, *not* -icute

electroencephalogram (one word), abbr. **EEG**

electrolyse, to break up by electric means, *not* -yze

electrolyte, solution able to conduct electric current

electromagnet/ic, -ism, (pertaining to) electricity and magnetism (one word)

electrometer, electricity measurer (one word)

electromotive, producing electricity (one word)

electromotor, an electric motor (one word)

electron-volt, unit of energy (hyphen); abbr. **eV** (one cap.)

electroplate, coat with metal by electrolysis, objects thus made (one word)

electrostatic, of electricity at rest (one word)

electrotype, duplicate printing plate made by copper electrolysis (one word); abbr. **electro**

electuary, a sweetened medicine; abbr. **elec.**

eleemosynary, charitable

eleg/y, funeral song, pensive poem; adj. **-iac; -ist,** writer of elegy, *not* -iast; **-ize,** write elegy, *not* -ise. *See also* **elogy, eulogy**

elementary, *not* -ery; abbr. **elem.**

elements (chemical), no point after symbols. *See also under each name*

elench/us (logic), a refutation; *pl.* **-i;** adj. **elenctic,** *not* -chtic

elephant, name for a former size of paper, 20 × 27 in.

Elephantine, Egypt, famed for archaeological yields

elevator, *not* -er

Elgin Marbles, transported to British Museum from Athens, 1805–12, by Earl of Elgin

El Giza, Egypt, site of the pyramids, *use* **Giza**

Elia, pseud. of Charles Lamb

Elias, Gr. form of Elijah used in NT (AV)

Elien:, sig. of Bp. of Ely (colon)

eligible, *not* -able

Eliogabalus, *use* **Ela-**

Eliot, fam. name of Earl of St.

Eliot (*cont.*)
Germans; — (**George**), 1819–
90, pseud. of Mary Ann, or
Marian, Evans, English novelist; — (**Thomas Stearns**),
1888–1965, US (naturalized
British) poet

Élisabethville, Zaïre, *now* **Lubumbashi**

elision (typ.), suppression of
letters or syllables in such contractions as e'en, there'll, I'd,
you've, it's (it is, has), William's
(William is, has). In Eng. and
Ger. set close up. In Fr. put a
space after an apostrophe following a word of two or more
syllables (e.g. bonn' petite *but*
j'ai). In It. set close up where the
apostrophe follows a consonant
(e.g. dall'aver *but* a' miei). In
Gr., Lat., Sp. to be spaced. *See
also* **ellipsis**

élite, the chosen ones; size of
letters on some typewriters (not
ital.)

elixir, alchemist's preparation to
change metals into gold or to
prolong life indefinitely

Elizabeth, abbr. **Eliz.**

Ellice Islands, W. Pacific, *now*
Tuvalu

Elliot, in fam. name of Earl of
Minto; family in Jane Austen's
Persuasion

Elliott (**Ebenezer**), 1781–1849,
English poet

ellips/is, the omission of words;
pl. **-es;** (typ.) three points (not
asterisks) separated by normal
space of line are sufficient. When
three points are used at the end
of an incomplete sentence a
fourth point should not be
added; normal space before the
first point. Where the sentence
is complete, the closing point is
set close up, followed by the
three points for omission. See
Hart's Rules for rules in foreign
languages. *See also* **elision**

Ellis Island, New York harbour

éloge (Fr. m.), an oration of

praise; Anglicized as **eloge** (n
accent)

elogy; elogi/um, *pl.* **-a; eloge**
(*see above*); (funeral) oration of
praise. *See also* **elegy, eulogy**

Elohim (Heb.), the Deity

eloin (law), to abscond, *not* -gn

Elois/a, -e, *see* **Héloïse**

E. long., east longitude

El Paso, Texas, US, *not* -ss-

El Salvador, republic of Cent.
America, cap. San Salvador

Elsass-Lothringen, *use*
Alsace-Lorraine

Elsevier, *see* **Elzevier**

Elsinore, Eng. for Dan. **Helsingør**

elucidator, one who makes
clear, *not* -er

elusive, difficult to grasp, mentally or physically. *See* **illusive**

elver, a young eel

Elyot (**Sir Thomas**), ?1499–
1546, English writer

Elysée palace, Paris

Elysium (Gr. myth.), abode of
dead heroes (cap., not ital.)

elytr/on, hard wing-case of
beetle; *pl.* **-a** (not ital.)

Elzevier, Dutch family of book
sellers and printers, in business
?1583–1712, *not* Elzevir, although the form is found in
some bibliographical sources.
The name of the modern publisher is **Elsevier**

EM, Earl Marshal, Edward
Medal, *Equitum Magister* (Master of the Horse)

Em., Emmanuel, Emily, Emma

em (typ.), the square of the body
of any size of type; *also* the
standard unit of typographic
measurement, equal to 12 pt.,
4.2 mm., the 'pica' em

'em, = them

EMA, European Monetary
Agreement

Emanuel School, Wandswort

embalmment, preservation of
dead body by aromatic drugs
(three *m*s); **embalming** is the
more usual noun

embargo/, temporary ban on trade; *pl.* **-es**

embarkation, *not* -cation

embarras/ (Fr. m.); — *de richesse(s),* a superfluity of good things; — *du choix,* difficulty of choosing, not *de*

embarrass/, -ment

embassy, *not* am-

embathe, *not* im-

embed/, -ded, -ding, *not* im-

embezzlement, fraudulent appropriation of money, *not* -lment

emblaz/e, -onry, *not* im-

embod/y, -ied, -ier, -iment, *not* im-

embolden, *not* im-

embonpoint, plumpness (not ital.)

embosom, to receive into one's affections, *not* im-

emboss, to raise in relief, *not* im-

embouchement, a river-mouth (not ital.)

embouchure (mus.), mouthpiece, shaping of the mouth (not ital.)

embower, to shelter with trees, *not* im-

embrangle, to entangle, confuse, *not* im-

embrasure, door or window recess, widening inwards, *not* -zure

embroglio, use im-

embroil, to confuse, involve in hostility, *not* im-

embrue, use im-

embryo/, *pl.* **-s**

embryology, abbr. **embryol.**

embue, use im-

em dash (typ.), a long dash (—); *see* **punctuation** XII

emend, to alter or remove errors in a text etc. *See also* **amend**

emerald (typ.), name for a former size of type, about 6½ pt.

emerit/us, honourably retired; *pl.* **-i**; *fem.* **-a**

emerods, use **haemorrhoids**

Emerson (Ralph Waldo), 1803–82, US poet, essayist, and philosopher

emeu (bird), use **emu**

émeute (Fr. f.), insurrection

e.m.f., electromotive force

EMI, Electrical and Musical Industries (Ltd.)

émigr/é (Fr.), *fem.* **-ée,** an emigrant, esp. political exile

éminence grise (Fr. f.), power behind the throne

Emir, in Turk. and Indian titles. *See also* **Amir**

Emmanuel/ College, Cambridge; — **(I–III),** Kings of Italy, *properly* **Emanuele**

Emmental, Swiss cheese, *not* -thal

emmesh, use enm-

Emmet (Robert), 1778–1803, Irish patriot

Emp., Emperor, Empress

empaestic, connected with embossing, *not* -estic

empanel/, to enrol, *not* im-; **-led, -ling**

empassion, use im-

Emperor, abbr. **Emp.**

emphas/is, *pl.* **-es**

emphasize, *not* -ise

emplane, *not* en-

employ/é (Fr.), *fem.* **-ée;** *pl.* **-és,** *fem.* **-ées** (not ital.), use **employee/s** (m. and f., no accent)

empolder, use im-

empori/um, centre of commerce, large shop; *pl.* **-a**

emprise (arch.), chivalrous enterprise, *not* -ize

emptor, a purchaser

empyema, pus in a cavity, esp. of a lung

empyre/an, the highest heaven; adj. **-al**

empyreum/a, the 'burnt' smell of organic matter; *not* -ruma; *pl.* **-ata**

em rule (typ.), a long dash (—); *see* **punctuation** XII

EMS, European Monetary System

emu, Australian running bird, *not* emeu

e.m.u., electromagnetic unit(s)

emulator, rival, *not* -er

en (typ.), half the width of an em

enactor, *not* -er

enamel/led, -ler, -ling

enamorato, *use* in-

enamoured, inspired with love

en/ arrière (Fr.), behind; — *attendant,* meanwhile; — *avant,* forward; — *bloc,* in the mass; — *brosse,* bristly-cut

enc., enclos/ed, -ure

Encaenia, Commemoration at Oxford, *not* -cenia (not ital.)

encage, encapsulate, encase, *not* in-

enceinte, pregnant, a fortified enclosure (not ital.)

enchase, to put in a setting, engrave, *not* in-

enchiridi/on, handbook or manual; *pl.* **-a** (not ital.)

en clair (Fr.), in ordinary language (not in cipher)

enclave, territory surrounded by foreign dominions

enclos/e, -ure, *not* in- (*but* Inclosure Acts)

encomium/, a eulogy; *pl.* **-s** (not ital.)

encore, again (not ital.)

encroach, to intrude usurpingly, *not* in-

encrust, *not* in-; *but* **incrustation**

encumber, to hamper, burden, *not* in-

encyclopaed/ia, -ic, -ical, -ism, -ist, -ize; *but use* -ped- in quoting titles where that spelling was preferred; abbr. **ency.**

Encyclopaedia Britannica, abbr. *EB, Ency. Brit.*

Encyclopédie, the French encyclopaedia ed. by Diderot, 1751–72; *encyclopédiste,* a writer for it (not cap.)

en dash (typ.), a short dash as in '1914–18', 'the Fischer–Spassky match'; contrast hyphen. *See also* **punctuation** XII

endemic, regularly found in a specified place, opp. to **epidemic**

endemn/ify, -ity, *use* in-

endent/, -ure, *use* in-

en/ dernier ressort (Fr.), as a last resource; — *déshabillé,* in undress

end leaves, *see* endpapers

endors/e, to write on the back of, *not* in-; **-able,** *not* -eable, -ible

endpapers (bind.), the sheets at the beginning and end of a book, half of each pasted to inside of cover, half forming flyleaf

endue, to clothe, *not* in-

endur/e, -able, -ed, -er, -ing, *not* in-

endways, *not* -wise

Endymion, Greek legend, and Keats's poem; *also* Disraeli's last novel

ENEA, European Nuclear Energy Agency

enema/, (syringe for making) injection into rectum; *pl.* **-s** (not ital.)

energize, *not* -ise

en/ face (Fr.), facing; — *famille,* with one's family

enfant/ gâté (Fr. m.), spoilt child; — *prodigue (l'),* the prodigal son; — *terrible,* indiscreet person

enfeoff (hist.), to transfer land to possession of a subordinate tenant. *See also* **fee, feoff/ee, -ment**

en fête (Fr.), in festivity

enfin (Fr.), finally (one word)

enfold, *not* in-

enforce/, -able, *not* in-

enfranchis/e, -able, -ement, -ing, *not* -ize

Eng., England, English

eng., engineer, -ing, engrav/ed, -er, -ing

Engadine, Switzerland; in Ger. **Engadin**

engag/é (Fr.), *fem.* **-ée,** morally committed

en garçon (Fr.), as a bachelor

Engels (Friedrich), 1820–95,

German socialist, associate of Karl Marx

engineer/, -ing, abbr. **eng.**

engine-room (hyphen)

Engl/and, -ish, abbr. **Eng.**

English (typ.), name for former size of type, about 14 pt.

engraft, *not* in-

engrain, to dye in the raw state, *not* in-. *See also* **ingrain**

en grande/ tenue, or — — *toilette* (Fr.), in full dress; *en grand seigneur,* magnificently

engross, to buy wholesale, copy in a large hand, absorb attention of, *not* in-

engulf, *not* in-, -gulph

enigma/, riddle; *pl.* **-s**

enjambment, in verse, the continuation of sense beyond the end of a line; in Fr. m. *enjambement*

en masse (Fr.), in a body

enmesh, to entangle, *not* emm-, imm-

Enniskillen, Co. Fermanagh; — (**Earl of**) *but* **Inniskilling Dragoons** and **Fusiliers**

ennui, boredom (not ital.)

ennuy/é, *fem.*-**ée,** bored (not ital.)

ENO, English National Opera

enoculate, *use* in-

enology, *use* oen-

en/ pantoufles (Fr.), relaxed; — *passant,* by the way, in passing (also of a pawn capture in chess); — *pension,* as a boarder

enplane, use **em-**

en/ poste (Fr.), in official position; — *prince,* in princely style; — *prise* (chess), in a position to be taken

enquire, to ask. *See also* **inquire**

enquiry, a question, the general word, **inquiry** being reserved for an official investigation

en/ rapport (Fr.), in sympathy; — *règle,* as it should be, in order; — *revanche,* in revenge

enrol/, -ment, *not* enroll-; *but* **-led, -ler, -ling**

en route (Fr.), on the way

en rule (typ.), *see* **en dash, punctuation** XII

ens, an entity; Lat. *pl.* **entia**

En Saga (Swed.), A Saga (title of tone-poem by Sibelius, 1892)

Enschede, Netherlands, *but* **Enschedé en Zonen,** printing house at Haarlem

ensconce, to protect with an earthwork, establish snugly, *not* ins-, es-

en secondes noces (Fr.), by second marriage

ensemble, general effect, group of players (not ital.)

ensheath, *not* -the

ensilage, storage of green fodder in pits or silos

ensnare/, -ment, *not* in-

en somme (Fr.), in short

ensuing, *not* -eing

en suite (Fr.), to match, forming a unit

ensuite (Fr.), after, following (one word)

ensure, to make safe. *See also* **assurance, insurance**

enswathe, to wrap, *not* in-

ENT, ear, nose, and throat

entailed estate, one settled on a series of heirs, *not* in-

entente/ (Fr. f.), meaning; — *cordiale,* cordial understanding, (caps.) that between Britain and France, 1904 onwards; — (*un mot à double*), a word or phrase with two meanings, *pl.* **mots à double entente.** *See* **double entendre**

enterprise, *not* -ize

enthral/, -led, -ler, -ling, -ment

enthron/e, -ization, *not* in-

entitle, *not* in-. *See also* **intitule**

entomology, the study of insects; abbr. **entom.**

entourage, surroundings, attendants (not ital.)

en-tout-cas (Fr.), umbrella--cum-parasol; *En-Tout-Cas*

en-tout-cas (*cont.*)
(caps.) propr. term for type of hard tennis-court

entozo/on, internal parasite; *pl.* **-a** (not ital.)

entr'acte/ (performance in) the interval between two acts; *pl.* **-s**

en train (Fr.), in progress

entrain, to board a train, to put aboard a train

entrain (Fr. m.), heartiness

entrammel/, to entangle; **-led, -ling**

entrap/, -ped, -ping, *not* in-

entreat, to beseech, *not* in-

entrechat, a jump in which a dancer strikes his heels together several times (not ital.)

entrecôte (Fr. cook. m.), the 'undercut'

entrée, a dish served between courses, right of admission (not ital.)

entremets, side dishes (not ital.)

entrench, *not* in-

entre nous (Fr.), confidentially

entrepôt, a market (not ital.)

entrepreneur/, person in control of business enterprise, a contractor (not ital.); adj. **-ial**

entresol, a storey between ground floor and first floor (not ital.)

entrust, *not* in-

Ent. Sta. Hall, Entered at Stationers' Hall

en/twine, -twist, *not* in-

enunciat/e, -or

enure, *use* **inure**

envelop/ (verb), **-ed, -ment**

envelope (noun)

Env. Extr., Envoy Extraordinary

enweave, *use* **in-**

enwrap/, -ped, -ping, *not* in-

enwreath, *not* in-

enzootic, endemic in animals

enzyme (chem.), protein with catalytic properties

EO, Education Officer

Eocene (geol.) (cap., not ital.)

e.o.d., every other day

EOL (comp.), end of line

Eolian etc., *use* **Aeo-**

eon, *use* **aeon**

Eōthen ('from the east'), by Kinglake, 1844

EP, electroplate

Ep., Epistle

e.p. (chess), *en passant*

epact, age of the moon on 1 Jan.

épat/ant (Fr.), shocking; **-er les bourgeois,** to shake up the hidebound

epaulette, shoulder-piece, *not* -et

EPDA, Emergency Powers Defence Act

épée, sword used in fencing

epergne, table ornament to hold flowers or fruit (not ital., no accent)

Épernay, French white wine

Eph., Ephesians, Ephraim

ephah, Hebrew measure, *not* epha

ephedrine, an alkaloid used as cardiac depressive or for asthma, *not* -in

ephemer/a, *same as* **-on;** *pl.* **-as** (not ital.)

ephemer/is, an astronomical almanac; *pl.* **-ides**

ephemer/on, insect that lives for a day, *pl.* **-ons** (not ital.); (bibliog.) printed item intended for transient use (e.g. theatre programme, bottle label), *pl.* **-a**

Ephesians, abbr. **Eph.**

ephod, Jewish priestly vestment

ephor/, Spartan magistrate, *pl.* **-s**

ephphatha (Aram.), 'Be opened'

epicedi/um, funeral ode; *pl.* **-a**

epicene, sexless (contemptuous); (gram.) referring to either sex

epicentre, point over centre of earthquake

Epicoene, by Ben Jonson, 1609

epicure/, one who indulges in a refined way in the pleasures of the table; adj. **-an**

Epicurean, a philosopher who identifies pleasure with virtue, following the Greek **Epicurus,** 342–270 BC

pideictic, adapted for display, *not* -ktic

pidemic, (disease) breaking out locally and lasting only for a time

pidermis, the outer skin or cuticle

pilogue, concluding part of a lit. work

piornis, *use* aepy-

piphany, the manifestation of a god; (cap.) 6 Jan., the manifestation of Christ to the 'wise men'

piphyte, a plant that lives on the surface of another

Epirot, inhabitant of Epirus, *not* -te

piscopal, belonging to bishops; abbr. **episc.**; (cap.) title of Anglican Churches in Scotland and US

'Epithalamion', by Spenser, 1595

epithalami/um, a nuptial song; *pl.* **-a**

epitheli/um, surface tissue; *pl.* **-a**

epitomize, to shorten, summarize, *not* -ise

epizo/on (zool.), an animal that lives on the surface of another; *pl.* **-a**; **-otic,** epidemic in animals

e pluribus unum (Lat.), many made one (motto of US)

EPNS, electroplated nickel silver

eppur si muove (It.), and yet it does move (ascribed to Galileo, after his recantation)

épreuve (Fr. typ. f.), a proof

épris/ (Fr.), *fem.* **-e,** enamoured

Epstein (Sir Jacob), 1880–1959, English sculptor

épuis/é (Fr.), *fem.* **-ée,** exhausted; out of print

epyornis, *use* ae-

eq., equal

equable, uniform, not easily disturbed. *See also* **equatable**

Equador, *use* **Ecu-**

equal/ *to* (no longer *with*); **-led, -ling**

equaliz/e, -ation, *not* -ise

equal mark (typ.), = (a space before and after)

equanimity, evenness of mind or temper

equatable, able to be regarded as equal. *See also* **equable**

Equator/, -ial, abbr. **Eq.**

equerry, off. attendant on royal personage, *not* -ery

equestri/an, horseman; *fem.* **-enne**

equilibri/um, balance, *pl.* **-a**

equinoctial, *not* -xial

equinox/, period of year when night and day are equal; *pl.* **-es**

equivalent, abbr. **equiv.**

equivocator, user of misleading words, *not* -er

equivoque, a quibble (not ital.)

ER, Eastern Region, *Edwardus Rex* (King Edward), *Elizabetha Regina* (Queen Elizabeth)

Er, erbium (no point)

eraser, one who, *or* that which, rubs out, *not* -or

Erasmus (Desiderius), 1466–1536, Dutch scholar

Erato (Gr.), the lyric muse

erbium, symbol **Er**

Erckmann–Chatrian, joint pseud. of two French writers collaborating from 1848 to 1870

Erdgeist (Ger. myth.), an earth-spirit

Erechtheum, temple at Athens

erector, one who, *or* that which, erects, *not* -er

erethism (path.), excitement, *not* ery-

Erevan, Armenia, *use* **Yerevan**

Erewhon, by Butler, 1872

erf (Afrik.), plot of ground; *pl.* **erven**

erg, unit of work, in SI units 10^{-1} μJ

ergo (Lat.), therefore (ital.)

Erie, one of the Great Lakes (US and Canada)

Erin go bragh (Ir.), Erin for ever

Eriny/s (Gr. myth.), any of the three Furies, Eumenides, or avenging deities; *pl.* **-es**

Eris, Greek goddess of discord
erl-king, in Germanic myth.; in Ger. *Erlkönig* (rom. quoted as title of poem by Goethe and of songs by Loewe, Schubert, and others)
Ernie, device for selecting weekly and monthly winners in Premium Savings Bonds (Electronic Random Number Indicator Equipment)
'Eroica' Symphony, by Beethoven, 1804
Eros, Greek god of love
erpetolog/y, -ist, *use* h-
errat/um, error in writing or printing; *pl.* **-a** (not ital.)
erroneous, wrong, *not* -ious
ersatz (Ger.), substitute
Erse, the Highland or Irish Gaelic language
erstens (Ger.), in the first place; *erstgeboren,* first-born
erven, *see* **erf**
erysipelas, a skin disease, St. Anthony's fire, *not* -us
erythism, *use* ere-
Erzerum, Turkish Armenia, *not* -oum, -om
ES (paper), engine-sized
ESA, European Space Agency
escalade, to mount and enter by a ladder
escalat/e, -ion, to rise, rising, steadily and inevitably, like an escalator
escallop, shellfish, *use* **scallop**
escalope (Fr. f.), collop, slice of meat
escargot (Fr. m.), edible snail
escarp/, -ment, a steep slope, *not* -pe
Escaut, Fr. for **Schelde**
eschalot, *use* **shallot**
eschatology, (theol.) doctrine of last things
escheator, official who watches over forfeited property, *not* -er
eschscholtzia, a yellow-flowered plant
esconce, *use* **ens-**
Escorial, Spain, *not* Escu-
escritoire, *not* -oir (not ital.)

Esculap/ius, -ian, *use* **Ae-**
escutcheon, heraldic shield, plate for keyhole, etc., *not* scut-
Esdras, in the Apocrypha, First Second, Book of, abbr. **1, 2 Esd**
Eskimo/, *not* Esquimau; *pl.* **-s**
ESN, educationally subnormal
esophag/us, -eal, *use* **oes-**
esp., especially
ESP, extra-sensory perception
espagnol/, fem. -e (Fr.), Spanish
espagnolette, fastening for french window
espalier, a lattice-work, a trained fruit-tree (not ital.)
especially, abbr. **esp.**
Esperanto, an artificial universal language, invented by L. L. Zamenhof, 1887
espièglerie, roguishness (not ital.)
espionage, spying (not ital.)
espressivo (mus.), with expression, *not* ex-
espresso, apparatus for making coffee under pressure, the coffee produced, *not* exp- (not ital.)
esprit/ (Fr. m.), genius, wit; — *de corps,* members' respect for a society; — *de l'escalier,* inspiration too late; — *fort,* a strong-minded person, *pl. esprits forts* (ital.)
Esquimalt, Vancouver Island, Canada, *not* -ault
Esquimau, *use* **Eskimo**
Esquire, abbr. **Esq.**; J. Smith, **jun., Esq.,** *not* Esq., jun.
ess, name of the letter *s*; *pl.* **esses**
ess., essences
essays (typ.), cited titles to be roman quoted; caps. as in title
Essouan, *use* **Aswan**
established, abbr. **est.**
Established Church (caps.); abbr. **EC**
Establishment (the), (the values held by) the established sector of society (cap.)
estamin, a woollen fabric, *not* ét-
estaminet (Fr. m.), a café

stanci/a (Sp.-Amer.), cattle farm; *-ero*, its keeper

Estate/s of the Realm, the Parliament of Scotland before union with England in 1707; **—s (the Three),** Lords Spiritual (bishops), Lords Temporal, Commons; — **(the Third),** French bourgeoisie before the Revolution; — **(the Fourth),** the Press

Esther (OT), not to be abbr., but **Rest of Esth.** (Apocr.)

esthet/e, -ic, *use* aes-

estimator, *not* -er

estiv/al, -ation, *use* aes-

Estonia, *not* Esth-, a Soviet Socialist Republic. *See also* **USSR**

estoppel (law), a conclusive admission, *not* -ple, -pal

estr/ogen, -us, *use* oe-

ESU, English-Speaking Union

e.s.u., electrostatic unit(s)

eszett (Ger.), the character ß

ET, English translation

ETA, estimated time of arrival

et alibi (Lat.), and elsewhere; abbr. *et al.*

et alii (Lat.), and others; abbr. *et al.,* *not* et als.

etalon (phys.), interference-device (no accent, not ital.)

etamin, *use* **estamin**

état-major (Fr. m.), a staff of military officers

et cetera (not ital.), abbr. **etc.;** comma before 'etc.' if more than one term precedes; comma after 'etc.' if it would be needed after an equivalent phrase such as 'and the like'.

etceteras, extras, sundries (one word)

Eternal City (the), Rome

eth, the Old-English Ð, ð and Icelandic Ð, ð (distinguish from **thorn** and **wyn,** qq.v.), also used in phonetic script; *not* edh

ether, medium filling all space, *not* ae-

ethereal/, -ity, -ly, *not* -ial

Etherege (Sir George), ?1635–91, English playwright

ethics, the science of morals

Ethiopia, modern name of Abyssinia

ethn/ic, concerning races; **-ology, -ological,** abbr. **ethnol.**

et hoc genus omne (Lat.), and all this kind of thing

ethology, the study of character formation or animal behaviour

etiology, *use* **ae-**

etiquette, conventions of courtesy (not ital.)

Etna, Sicily, *not* Ae-

étrier (Fr. m.), short rope-ladder

et sequen/s, and the following, abbr. *et seq.* or *et sq.; pl. -tes,* n. *-tia,* abbr. *et sqq.*

étude/ (Fr. f.), a study; *— de concert* (mus.), concert study

étui, case for small articles, *not* etwee (not ital.)

etymolog/y, the study of derivations; **-ical, -ically, -ist, -ize;** abbr. **etym.**

etym/on, a root-word, *pl.* **-a**

Eu, europium (no point)

eucalypt/us (bot.), *pl.* **-uses**

euchre, a card-game, *not* eucher, eucre

Euclid/, *c.*300 BC, Athenian geometer, his treatise (cap.); adj. **-ean** (cap.)

eudemon, a good angel

Euer (Ger. m.), your; abbr. *Ew.*

Eugène, French Christian name (è); — **(Prince, of Savoy),** 1663–1736, Austrian general

Eugénie (Empress), 1826–1920, wife of Napoleon III (é)

eulogium/, eulogy; *pl.* **-s**

eulogize, *not* -ise

eulogy (a, *not* an), spoken or written praise. *See also* **elegy, elogy**

Eumenides (Gr. myth.), the avenging Furies: Alecto, Tisiphone, and Megaera; (ital.) play by Aeschylus

euonymus (a, *not* an), plant of genus *Euonymus,* e.g. spindle-tree

euphem/ism (a, *not* an), a mild term for something unpleasant; **-istic(al)**; abbr. **euphem.**

euphemize, *not* -ise

euphonize, *not* -ise

euphony (a, *not* an), agreeable sound

euphorbia (bot.), spurge

euphor/ia, a feeling of well-being; **-ic**

Euphues, a prose romance in affected style, 1579, by John Lyly; hence **euphuism**

Euratom, European Atomic Energy Community

Eure-et-Loir, Fr. dép.; *not* Loire; not to be confused with Fr. dép. **Eure**

eureka, I have found (it)!, *not* heu- (not ital.). *See also* **heuristic**

eurhythmic, *not* eury-

Europe, abbr. **Eur.**

European (a, *not* an); abbr. **Eur.**

Europeanize, *not* -ise (cap.)

europium, symbol **Eu**

Euterpe (Gr.), muse of music

euthanasia, peaceful death, mercy killing

eV, electron-volt

evangelize, to inform about the Gospel, *not* -ise

evaporimeter, instrument to measure evaporation, *not* -ometer

evening, abbr. **evng.**

even pages (typ.), the left-hand, or verso, pages

even s. caps. (typ.), word(s) set entirely in small capitals

ever, joined to interrogatives used universally (whatever is, is right) but not to those used emphatically (what ever do you mean?)

everglades (US), tracts of low swampy ground; (cap.) a large area like this in Florida

evermore (one word)

every must be followed by verb and pronoun in sing., as 'every bird *tries* to protect *its* young'

every/body, -day (adj.), **Everyman, -one** (but — **one** when each word to retain its meaning; cf. *Hart's Rules*, p. 77 **-thing, -way** (adverb), **-wher** (one word)

Ew. (Ger.), *Euer* etc. (your)

ewe, female sheep (a, *not* an)

ewer, jug (a, *not* an)

Ewigkeit (Ger. f.), eternity

ex (Lat.), out of, excluding (not ital.)

ex., example, executed, executiv

exa-, prefix meaning 10^{18} times abbr. **E**

exactor, one who exacts, *not* -e

exaggeration

exalter, one who exalts, *not* -or

examination/, colloq. **exam;** — **paper**, the questions; — **script**, the answers

examplar, *use* **exemplar**

example, abbr. **ex.**, *pl.* **exx.**; i Fr. m. *exemple*

exanimate, lifeless, depressed

ex animo (Lat.), from the minc earnestly

exasperate, to irritate, *not* -irat

Exc., Excellency

exc., excellent, except, -ed, -ion

exc., *excudit* (engraved this)

Excalibur, King Arthur's sword, *not* -bar, -bour (not ital.) *See also* **Caliburn**

ex cathedra (Lat.), from the Pope's, professor's, chair, with authority

excavator, *not* -er

excellence/, superiority; *pl.* **-s,** *not* -cies

Excellenc/y, *pl.* **-ies** (persons), abbr. **Exc.**

excellent, abbr. **exc.**

excentric, used in math. sense instead of **eccentric** to avoid sense of 'odd'

except/, -ed, -ion, abbr. **exc.**

exceptionable, open to criticism

exceptional, out of the ordinary

exceptis excipiendis (Lat.), with the necessary exceptions

excerpt, an extract

exch., exchange, exchequer

exchange, abbr. **exch.; Stock —** (caps.), abbr. **St. Ex.**

exchangeable, *not* -gable

exchequer, abbr. **exch.**

excisable, liable to excise duty, removable by excision, *not* -eable

excise (verb), *not* -ize

excit/able, *not* -eable; **-ability,** *not* -ibility

exciter, one who, *or* that which, excites, (elec.) auxiliary machine supplying current for another, *not* -or

excitor (anat.), a nerve stimulating part of the body, *not* -er

exclamat/ion, -ory, abbr. **excl.,** or **exclam.**

exclamation mark (!), *see* **punctuation** VII

excreta, *pl.,* excreted matter (not ital.)

excudit (Lat.), engraved this; abbr. *exc.*

ex curia (Lat.), out of court, *not* -â

excursus/, a digression; *pl.* **-es** (not ital.)

excusable, *not* -eable

ex-directory (hyphen)

ex dividend, without next dividend; abbr. **ex div.** or **x.d.**

exeat/, 'let him depart', a formal leave of temporary absence; *pl.* **-s** (not ital.) *See also* **exit,** *exit*

execut/ed, -ive, abbr. **ex.**

execut/or, one who carries out a plan, etc.; (law) person appointed to execute will, *not* -er, abbr. **exor.;** *fem.* **-rix,** abbr. **exrx.,** *pl.* **-rices**

exege/sis, explanation; *pl.* **-ses** (not ital.); adj. **-tic**

exemplaire (Fr. m.), a specimen, a copy; in Ger. n. *Exemplar*

exemplar, pattern, *not* exa-

exemple (Fr. m.), example

exempli gratia (Lat.), for example; abbr. **e.g.** (l.c. not ital., comma before)

exequatur/, an off. recognition by a foreign government of a consul or other agent; *pl.* **-s** (not ital.)

exequies (*pl.*), funeral ceremony (not ital.)

exercise, *not* -ize

Exeter, in Lat. *Exonia,* abbr. **Exon.** *See also* **Exon:**

exeunt/, see **exit;** — *omnes* (Lat. stage direction), they all leave

ex gratia (Lat.), voluntary

exhibitor, *not* -er

exhilarat/e, to enliven; **-ing, -ion**

ex hypothesi (Lat.), by hypothesis

exigency, urgency

exigent, urgent

exiguous, small, scanty

ex interest, without next interest; abbr. **ex int.** or **x.i.**

existence, *not* -ance

existentialism, the philosophy that the individual is a free and responsible agent determining his own development

exit/, a way out; *pl.* **-s**

exit (Lat. stage direction), he or she goes out; *pl. exeunt,* they go out (ital.). *See also* **exeat**

ex lege (Lat.), arising from law

ex-libris (*sing.* and *pl.*), 'from the library of', a book-plate (not ital.); abbr. **ex lib.**

ex new, without the right to new shares; abbr. **ex n.** or **x.n.**

ex nihilo nihil fit (Lat.), out of nothing, nothing comes

exodi/um, conclusion of a drama, a farce following it; *pl.* **-a;** Anglicized as **exode.** *See also* **exordium**

exodus, a departure, esp. of a body of people; (cap., OT) departure of Israelites from Egypt, biblical book, abbr. **Exod.**

ex officiis (Lat.), by virtue of official positions

ex officio (Lat.), by virtue of one's official position; **ex-officio** (adj.)

exon, officer of the Yeomen of the Guard

Exon.:, sig. of Bp. of Exeter (colon)

exor., executor

exorcize, to expel an evil spirit from, *not* -ise

exordium/, introductory part of a discourse; *pl.* **-s.** *See also* **exodium**

exotic type, non-roman faces, e.g. Arabic, Hebrew

exp., export, -ation, -ed, express

expanded type (typ.), a type with unusually wide face

ex parte (Lat.), one-sided(ly); **ex-parte** (law adj., hyphen, not ital.)

ex pede Herculem (Lat.), judge from the sample

expendable, able to be sacrificed for some end

expense, *not* -ce

experimenter, *not* -or

expertise, expert knowledge

expertize, give expert opinion

experto crede (Lat.), believe one who has tried it

expi/ate, -able

explanation, abbr. **expl.**

explicit (Lat.) (here) ends; as noun, the conclusion (of a book)

export/, -ation, -ed, abbr. **exp.**

exposé, explanation, exposure (not ital.)

expositor, one who expounds, *not* -er

ex post facto (Lat.), after the fact

expostulator, one who remonstrates, *not* -er

Ex-President (caps.)

express, abbr. **exp.**

expressible, *not* -able

expressivo (mus.), with expression, *use* **es-**

expresso, *use* **es-**

expressway (one word)

exrx., executrix

ext., extension, exterior, external, -ly, extinct, extra, extract

extasy, extatic, *use* **ecs-**

extempore (adj. and adv.), un-prepared, without preparation (not ital.)

extemporize, *not* -ise

extender, *not* -or

extendible, *not* -able, *but* in gen. use next

extensible, *not* -able

extension, abbr. **ext.**

extensor, a muscle, *not* -er

extent (typ.), the amount of printed space, expressed in ens (*see* **en**) or pages, filled by a given batch of copy

exterior, abbr. **ext.**

external/, -ly, abbr. **ext.**

externalize, *not* -ise

externe, non-resident hospital physician, day-pupil

exterritorial, *use* **extra-**

extinct, abbr. **ext.**

extirpator, one who destroys completely, *not* -er

extol/, -led, -ler, -ling

extra, extract, abbr. **ext.**

extractable, *not* -ible

extractor, *not* -er

extracts, *see* **authorities, quotations**

extra/marital, outside marriage, **-mural,** outside the scope of ordinary teaching (one word); *but* **Extra-Mural Studies** (of Universities)

extraneous, brought in from outside, *not* -ious

extra-sensory perception, abbr. **ESP**

extraterritorial, not under local jurisdiction, *not* exterr-

extravaganza, a fantastic composition (not ital.)

extrovert, *not* extra-

ex ungue leonem (Lat.), judge from the sample

exuviae, cast coverings of animals

ex-voto/, an offering made in pursuance of a vow; *pl.* **-s** (not ital.)

exx., examples

Exzellenz (Ger. f.), Excellency; abbrev. *Exz.,* *not* Exc-

Eyck, *see* **Van Eyck**

**eye/ball, -bath, -bright,
-brow, -ful, -glass** (one word)
eyeing, *not* eying
eye/lash, -lid (one word)
eye-muscles (hyphen)
eye/piece (optics), **-sight,
-sore, -wash, -witness** (one
word)

eyot, islet, use **ait**
**Eyre & Spottiswoode (Pub-
lishers), Ltd.**
eyrie, *not* aerie
eyrir, coin of Iceland, one-hundredth of krona; *pl.* **aurar**
Ezekiel, abbr. **Ezek.**
Ezra, not to be abbr.

F

F, Fahrenheit, fail, farad, fluor-ine, (photog.) focal length, (math.) function, (pencils) fine, (on timepiece regulator) fast, the sixth in a series

F., fair, Fellow, felon, formul/a, -ae, Friday, all proper names with this initial except those beginning ff

f, femto-, (math.) function

f., farthing, fathom, female, fem-inine, filly, (photog.) focal length, (meteor.) fog, following, franc, free, from, furlong

f (mus.), *forte* (loud)

f. (Lat.), *fortasse* (perhaps); (Ger.) *für* (for)

FA, Football Association; Fédér-ation Aéronautique

Fa., Florida, *use* **Fla.**

fa (mus.), *use* **fah**

Faber & Faber, Ltd., publishers

fabliau/, a metrical tale; *pl.* -**x**

fac., facsimile

façade, front face of a building (not ital.)

face (typ.), the printing surface of type; the design of a particu-lar fount

facet/, one side of a cut gem; -**ed,** -**ing,** *not* -tt-

facetiae, humorous anecdotes (not ital.)

facia, tablet over shop, with occupier's name; dashboard of a motor car. *See also* **fascia**

facies (nat. hist.), general aspect (ital.)

facile princeps (Lat.), easily best (ital.)

facilis descensus Averni (*not* Averno) (Lat.), easy is the descent to Avernus

façon de parler (Fr.), mere form of words

facsimile/, exact copy (one word, not ital.); *pl.* -**s;** abbr. **fac.**

facta, non verba (Lat.), deeds, not words

factious, characterized by fac-tion(s)

factitious, not natural

factorize, *not* -ise

factotum/, servant who attends to everything; *pl.* -**s** (not ital.)

factum est (Lat.), it is done

facul/a, bright solar spot; *pl.* -**ae;** *not* fae-

fado/, Portuguese folk-song; *pl.* -**s**

faec/es, *pl.*, excrement; adj. -**al**

faerie (arch. and poet.), Fairy-land, *not* -y

Faerie Queene (*The*), by Spen-ser, 1590-6

Faeroe Isles, Danish, in N. Atlantic, *use* **The Faeroes;** adj. **Faeroese**

fag-end (hyphen)

faggot, bundle of sticks, seasoned meatball, derog. name for woman

fah (mus.), *not* fa

Fahrenheit, temperature scale; abbr. **F**

FAIA, Fellow of the Association of International Accountants

faience, glazed pottery, *not* faï-, fay-

fainéant, idle, idler (not ital.)

faint ruled (paper), *use* **feint**

fair, abbr. **F.**

fair and square (no hyphens except when attrib.)

fair copy, transcript free from corrections; abbr. **f. co.**

Fair Isle, Shetland, and knit-ting-design

fair play (two words)

fairway, *not* fare- (one word)

fairyland (one word)

fairy-tale (hyphen)

fait accompli (*un*) (Fr. m.), an accomplished fact

faith-heal/er, -ing (hyphen)

aizabad, Uttar Pradesh (India), *use* **Fyz-**

akir, Muslim or Hindu religious mendicant, *not* the many variants

alang/e, -ist, (member of) political party in Spain or Syria

aldstool, *not* fold- (one word)

alernian, wine of ancient Campania

alkland/ Islands, S. Atlantic; **— (Viscount)**

alk Laws, 1874–5, German anti-Catholic laws introduced by politician Adalbert Falk

alla (Manuel de), 1876–1946, Spanish composer

al-lal, piece of finery, *not* fallol; collect. noun **fallallery**

allible, liable to err, *not* -able

allodon (Edward, Viscount Grey), 1862–1933, English statesman, *not* -en

all-out (noun), from nuclear explosion (hyphen)

falsa lectio (Lat.), a false reading; abbr. *f. l.*

alucca, *use* **felucca**

alutin, *see* **high-falutin**

FAM, Free and Accepted Masons

fam., familiar, family

familiarize, *not* -ise

family, abbr. **fam.** *See also* **botany, zoology**

fan/ belt, — club, — dance (two words)

fandango/ (Sp.), a dance or its music; *pl.* **-es** (not ital.)

Faneuil Hall, historic building, Boston, Mass., US

fanfaronade, arrogant talk, *not* -nnade, -faronade

fanlight (one word)

fantas/ia (mus.), a free composition (not ital.); *pl.* **-ias**

fantasize, *not* -ise

fantasmagoria, *use* **phantas-**

fantast, a dreamer, *not* ph-

fantasy, *not* ph-

Fanti, tribe in Ghana, *not* -te, -tee

Fantin-Latour (Henri), 1836–1904, French painter

fantoccini (It.), marionettes

fantom, *use* **ph-**

FAO, Food and Agriculture Organization (of UN)

FAP, First Aid Post

faqu/eer, -ir, *use* **fakir**

far., farriery, farthing

farad/ (elec.), unit of capacitance; abbr. **F** (no point); **-ic current** (med.), *not* -aic

Faraday (Michael), 1791–1867, English chemist

fareway, *use* **fair-**

farewell (one word)

far/-fetched, -flung (hyphens)

farinaceous, starchy, *not* -ious

Faringdon, Oxon., *not* Farr-

farm-hand (hyphen)

farm/house, -stead, -yard (one word)

Farne Islands, N. Sea, *not* Farn, Fearne, Ferne

far niente (It.), doing nothing. *See also* **dolce — —**

Faroe, *use* **Faeroe**

farouche, sullen from shyness (not ital.)

Farquhar (George), 1678–1707, Irish dramatist

farrago/, a hotchpotch, *not* fara-; *pl.* **-s**

Farrar (Frederick William, Dean), 1831–1903, English divine and writer

farriery, horse-shoeing; abbr. **far.**

Farringdon Within, Without, wards of the City of London

Farsi, language of Iran

farth/er, -est, *use* **furth/er, -est**

farthing, abbr. **f.** *or* **far.**

FAS, Fellow of the Anthropological Society; ditto Antiquarian Society

fasces (*pl.*), a bundle of rods, the symbol of power (ital.)

fasci/a (archit.), long flat surface under eaves or cornice; *pl.* **-ae.** *See also* **facia**

fascicle, fascicule, fascicul/us (*pl.* **-i**), a bundle, bunch; one part of a book published in instalments

Fasc/ism, -ist (noun and adj.);
  -istic (adj.)
Fastens/-een, -eve, -even
  (Sc.), Shrove Tuesday, *not*
  Feastings-
fata Morgana, mirage seen in
  Str. of Messina (ital.)
fat face type, display type (q.v.)
  with thick stems and fine hair-
  -lines
Fates (the Three), Atropos,
  Clotho, and Lachesis
Father (relig.), abbr. **Fr.;** —
  (the), as Deity (cap.)
father-in-law, *pl.* **fathers-** —-
  -— (hyphens)
father of the chapel, *see*
  chapel
fathom, abbr. **f.** *or* **fm.**
fatstock (one word)
faubourg (Fr. m.), a suburb,
  cap. F when with name (not
  ital.)
Faulkland, in *The Rivals,* by
  Sheridan
Faulkner (William), 1897–
  1962, US novelist
faultfind/er, -ing (one word)
faun/, Latin rural deity; *pl.* **-s**
fauna/ (*collect. sing.*), the animals
  of a region or epoch; *pl.* **-s** (not
  ital.)
Faure (François Félix), 1841–
  99, French statesman
Fauré (Gabriel), 1845–1924,
  French composer
faute de mieux (Fr.), for want
  of better
fauteuil/ (Fr.), armchair, membership of French Academy; *pl.*
  -s
Fauves (Les), group of painters
  led by Matisse; **fauvism,** style
  of painting associated with them
  (not cap.)
faux pas (Fr. m.), a blunder (two
  words, *pl. same*)
favour/, -able, -ite, -itism
Fawkes (Guy), 1570–1606, conspirator
fawn, young deer, its colour; as
  verb, to grovel
fayence, *use* **faience**

FBA, Fellow of the British Aca-
  demy, *not* FRBA
FBAA, Fellow of the British
  Association of Accountants an(
  Auditors
FBCS, Fellow of the British Com
  puter Society
FBI, (US) Federal Bureau of
  Investigation
FBIM, Fellow of the British
  Institute of Management
FBOA, Fellow of the British
  Optical Association
FC, Football Club, Free Church
  (of Scotland)
FCA, Fellow of the Institute of
  Chartered Accountants in Eng
  land and Wales (or Ireland)
  (off.)
FCGI, Fellow of the City and
  Guilds of London Institute
FCIA, Fellow of the Corporatio(
  of Insurance Agents
FCIB, Fellow of the Corporatio(
  of Insurance Brokers
FCII, Fellow of the Chartered
  Insurance Institute
FCMA, Fellow of the Institute o
  Cost and Management Accoun
  tants
FCO, Foreign and Common-
  wealth Office
f. co., fair copy
FCP, Fellow of the College of
  Preceptors
fcp., foolscap
FCS, Fellow of the Chemical
  Society, *now* **FRSC**
FCSA, Fellow of the Institute of
  Chartered Secretaries and Ad-
  ministrators
FCSP, Fellow of the Chartered
  Society of Physiotherapy
FCST, Fellow of the College of
  Speech Therapists
FD, *fidei defensor* (Defender of the
  Faith)
Fe, *ferrum* (iron) (no point)
Fearne Islands, *use* **Farne** —
feasible, practicable, *not* -able
Feastings-, *see* **Fastens-**
featherfew, *use* **fever-**

featherweight paper, light but bulky book paper

February, abbr. Feb.

fecal, feces, *use* fae-

fecerunt (Lat. pl.), made this; abbr. *ff.*

fecial, *use* fetial

fecit (Lat.), made this; abbr. *fec.*

fedayeen, Arab guerrillas (pl.)

Federal Assemblies, *see* Assemblies

Federalist, abbr. Fed.

federalize, *not* -ise

fee/ (law), an inherited estate of land; — simple, estate unlimited as to class of heir (two words); — -tail, estate so limited (hyphen). *See also* fief

feedback, return of part of output to input (one word)

feedstock (one word)

feeoff, *use* fief

feeoffee, *use* feoffee

feet, *see* foot

feff/, *use* fief; -ment, *use* feoffment

fehmgericht (Ger.), *use* Femgericht, q.v.

feilding, fam. name of Earl of Denbigh. *See also* Fie-

feint, (noun) sham attack, (verb) to pretend

feint ruled (paper), *not* faint

feld/spar, a rock-forming mineral; adj. -spathic; *not* felspar

Félibre, member of *Félibrige*, school of Provençal writers

Felixstowe, Suffolk

fellah, Egyptian peasant; pl. -in, *not* -s, -een. *See also* Fulahs

fellmonger, hide-dealer

felloe, wheel rim, *not* felly

Fellow, abbr. F.; in Lat., *socius*; or, in the Royal Society, *sodalis*

fellow/ citizen, — men (two words)

fellow/-feeling, -traveller (hyphens)

felly, *use* felloe

felo de se (Anglo-Lat.), suicide; pl. *felos de se*

felon, one who commits a grave crime; abbr. F.

felspar, *use* feldspar

felucca, small Mediterranean vessel, *not* fal-, fil-

fem., feminine

female (bot., zool., sociol.), abbr. f., sign ♀

feme/ (law), wife, *not* femme; — covert, married woman, — sole, woman without husband (two words). *See also* femme

Femgericht/ (Ger.), pl. *-e*, medieval German tribunal, Anglicized as Vehmgericht, q.v.

feminine, abbr. f. *or* fem.

feminize, to make feminine, *not* -ise

femme/ *de chambre* (Fr. f.), chambermaid, lady's maid, pl. *femmes — —*; — *fatale*, a dangerously attractive woman; — *galante*, a prostitute; — *incomprise*, an unappreciated woman; — *savante*, a learned woman. *See also* feme

femto-, prefix meaning 10^{-15}, abbr. f

fem/ur, thigh bone, pl. -ora (not ital.); -oral

fencible, able to be fenced; (hist., noun) soldier liable only for home service; *not* -able

Fénelon (François de Salignac de La Mothe), 1651–1715, French ecclesiastic and writer, *not* Féné-

fenestr/a (anat.), a small hole in a bone; pl. -ae (not ital.)

fenugreek, leguminous plant, *not* foenu-

feoff, *use* fief

feoff/ee (hist.), one enfeoffed, *not* feeo-; -ment, act of enfeoffing. *See also* enfeoff, fee, fief

ferae naturae (law), adj., wild (of animals), literally 'of a wild nature'

Ferd., Ferdinand

Ferdausi, Persian poet, *use* Fir-

fer de lance, venomous snake (trop. S. Amer.)

Ferghana, cent. Asia, *not* -gana

Feringhee, in the Orient, a European, esp. a Portuguese; *not* the many variants

Fermanagh County, N. Ireland, abbr. **Ferm.**

fermium, symbol **Fm**

Ferne Islands, *use* **Farne**

Ferrara, Italy; — **(Andrea),** 16th c., Italian swordsmith

Ferrari (Paolo), 1822–89, Italian dramatist

Ferrero (Guglielmo), 1871–1942, Italian historian

ferret/, -ed, -er, -ing

ferrule, metal cap on end of stick (two *r*s), *not* ferrel. *See also* **ferule**

ferrum, iron, symbol **Fe**

ferryman (one word)

fertilize, *not* -ise

ferule, a cane or rod for punishment (one *r*). *See also* **ferrule**

fess (her.), *not* fesse

festa (It.), a festival

Festschrift/ (Ger. f.), writings presented to scholar; *pl.* **-en**

fêt/e, entertainment, **-ed** (not ital.); *fête-champêtre,* outdoor entertainment; *Fête--Dieu,* feast of Corpus Christi, *pl. Fêtes—* (ital., hyphen)

fetial, ambassadorial, *not* fecial

fetid, ill-smelling, *not* foe-

fetish/, inanimate object worshipped by savages, (psych.) abnormal cause of sexual desire, *not* -ich, -iche; **-eer, -ism, -ist** (not ital.)

fetor, bad smell, *not* foe-

fetus, *use* **foe-**

feu/ (Sc.), ground-rent, *pl.* **-s**

feu/ (Fr.), late, deceased; *fem.* **-e**

Feuchtwanger (Lion), 1884–1958, German novelist

feud., feudal

feudalize, to make feudal, *not* -ise

feu/ d'artifice (Fr. m.), firework, *pl.* **feux —;** **—** *de joie,* a salute, *pl.* **feux — —**

feu-duty (hyphen)

feuille (Fr. typ. f.), a sheet

Feuillet (Octave), 1821–90, French novelist and playwrigh

feuillet/ (Fr. typ. m.), a leaf; — *blanc,* blank leaf

feuilletage (Fr. cook. m.), puff--pastry

feuilleton (Fr. m.), light literature; in Fr. part of newspaper etc. devoted to this

Feulhs, *use* **Fulahs**

feverfew (bot.), *Chrysanthemum parthenium,* *not* feather-, fetter--foe

février (Fr. m.), February, abbr *fév.* (not cap.)

fez/, a cap; *pl.* **-zes** (not ital.); adj. **-zed**

FF, *Felicissimi Fratres* (Most Fortunate Brothers)

ff (typ.), as initials for proper name, *not* Ff; in Welsh, a separate letter, not to be divided

ff., folios, following (*pl.,* preferre to *et seqq.*)

ff (mus.), fortissimo (very loud)

ff., fecerunt, q.v.

FFA, Fellow of the Faculty of Actuaries (Scotland)

FFAS, Fellow of the Faculty of Architects and Surveyors

fff (mus.), fortississimo (as loud as possible)

F.F.Hom., Fellow of the Facult of Homoeopathy

FFPS, Fellow of the Faculty of Physicians and Surgeons

FFR, Fellow of the Faculty of Radiologists

FGS, Fellow of the Geological Society

FH, fire-hydrant

FHA, Fellow of the Institute of Health Service Administrators

FHS, Fellow of the Heraldry Society

FIA, Fellow of the Institute of Actuaries

F.I.A.A. & S., Fellow Architect Member of the Incorporated Association of Architects and Surveyors

FIAC, Fellow of the Institute of Company Accountants

fiacre (Fr. m.), a four-wheeled cab

FIAI, Fellow of the Institute of Industrial and Commercial Accountants

fiancé/e, *fem.* **-ée,** one betrothed (not ital.)

fianchetto/ (chess), placing bishop on long diagonal (*pl.* **-es**; not ital.)

fianna, the militia of Finn and other legendary Irish kings; **Fianna/,** *pl.,* the Fenians; — **Eireann,** the Fenians of Ireland; — **Fáil,** the Irish Republican party

F.I.Arb., Fellow of the Institute of Arbitrators

FIAS, Fellow Surveyor Member of the Incorporated Association of Architects and Surveyors

fiasco/, a failure; *pl.* **-s**

fiat, formal authorization (not ital.)

Fiat, Fabbrica Italiana Automobile Torino (It. motor car, company, and factory)

fiat/ justitia (Lat.), let justice be done; — **lux,** let there be light

FIB, Fellow of the Institute of Bankers

F.I.Biol., Fellow of the Institute of Biology

fibrin, blood protein appearing as network of fibres, *not* -ine

fibul/a, leg-bone, brooch; *pl.* **-ae**

FICE, Fellow of the Institute of Civil Engineers

fiche, microfiche; *pl. same*

fichu, woman's neckerchief (not ital.)

FICS, Fellow of the Institute of Chartered Shipbrokers; ditto International College of Surgeons

fic/tion, -titious, abbr. **fict.**

fictionalize, *not* -ise

FID, Fellow of the Institute of Directors

fidalgo (Port.), a noble

fiddle-de-dee, nonsense (hyphens)

Fidei Defensor (Lat.), Defender of the Faith; abbr. **FD** *or* **Fid. Def.**

fides Punica (Lat.), Punic faith, bad faith

fidget/, -ed, -ily, -ing, -y (one *t* only)

FIDO, Fog Investigation Dispersal Operation

fidus Achates (Lat.), a trusty friend

FIEE, Fellow of the Institution of Electrical Engineers

fief (hist.), land held by tenant of a superior, *not* feoff. *See also* **enfeoff, fee, feoff/ee, -ment**

field-glasses (hyphen)

Fielding (Henry), 1707–54, English novelist. *See also* **Fei-**

Field Marshal (two words, caps. as title); abbr. **FM**

Field Officer (two words, caps. as title); abbr. **FO**

fieri facias (Lat.), 'see that it is done', a writ; abbr. **fi. fa.**

fiesta (Sp.), festivity, holiday (ital.)

FIFST, Fellow of the Institute of Food Science and Technology

fift/y, -ieth, symbol **L**

fig., figure, figurative, -ly, figure

figurant/ (Fr. m.), a member of the ballet chorus, *pl.* **-s**; *fem.* **-e,** *pl.* **-es.** *Also* (It. m. *or* f.) **-e,** *pl.* **-i**

figure-head (hyphen)

figures, term used by some printers for arabic numerals; BS 2961 recommends the term **numerals** (q.v.) for both arabic and roman. *See also* **authorities, decimal currency, lakh**

FIHE, Fellow of the Institute of Health Education

F.I.Inst., Fellow of the Imperial Institute

FIJ, *use* **FJI**

Fiji/, W. Pacific, indep. 1970; adj. **-an** only of Polynesians, otherwise **Fiji**

filagree, *use* **filigree**

Fildes (Sir Luke), 1844–1927, English painter

filemot, a yellowish-brown, 'dead leaf' colour, *not* filamort, filmot, phil-

filet (Fr. cook. m.), fillet

filfot, *use* **fy-**

filibeg (Sc.), kilt, *not* the many variants

filibuster, one who obstructs public business by a long speech, the speech, *not* fill-

filigree/, ornamental metallic lace-work, adj. **-d,** *not* fila-, file-; — **letter** (typ.), an initial with filigree background

filing cabinet (two words)

Filipin/as, Sp. for **Philippine Islands**; **-o(s),** native(s) of the islands

fille de/ chambre (Fr. f.), chamber-maid, lady's maid, *pl.* *filles* — —; — *joie*, a prostitute

fillet/, in Fr. m. *filet*; **-ed, -ing**

fillibeg, *use* **fili-**

filling-station (hyphen)

fillip/, a stimulus; **-ed, -ing**

fillipeen, *use* **philippina**

fillister, a kind of plane-tool (two *l*s)

filmot, *use* **filemot**

filmsetting (typ.), composition and correction on photographic film instead of in metal

filoselle, floss silk (not ital.)

fils (Fr. m.), son, as Dumas fils (not ital.)

filter/, device for separating solid matter from liquid; **-bed, -paper** (hyphens). *See also* **philtre**

filucca, *use* **felucca**

F.I.Mech.E., Fellow of the Institution of Mechanical Engineers

Fin., Finland, Finnish

finable, liable to a fine

finale, conclusion (not ital.)

finalize, *not* -ise

fin de siècle (Fr. f.), end of the (nineteenth) century, decadent

fine (mus.), end of the piece

Fine Gael, Irish political party

fine-paper edition, abbr. **FP**

finesse, subtlety (not ital.)

fine-tooth comb (one hyphen)

finger-mark (hyphen)

finger/print, -tip (one word)

finick/ing, -y, fastidious

finis, the end (not ital.)

Finistère, dép. France

Finisterre/, weather-forecast sea area; — **(Cape),** Spain

finnan haddock, smoked haddock (two words), *not* the many variants

Finnegans Wake, novel by James Joyce, 1939 (no apos.)

Finnish/, abbr. **Fin.;** — **language** (typ.), is set in ordinary roman characters; *see* **accents**

Finno-Karelia, a former Soviet Socialist Republic. *See* **USSR**

Fin. Sec., Financial Secretary

F.Inst.P., Fellow of the Institute of Physics

FIOB, Fellow of the Institute of Building

FIOP, Fellow of the Institute of Printing

fiord (Norw.), arm of the sea, *not now* fjord (not ital.)

fioritura (mus.), decoration of melody

FIPA, Fellow of the Institute of Practitioners in Advertising

FIQS, Fellow of the Institute of Quantity Surveyors

fir., firkin, -s

Firbank (Ronald), 1886–1926, English novelist

Firdausi, ?930–1020, Persian poet, *not* Ferdausi, Firdousi, Firdusi

fire/-alarm, -engine, -escape, -guard, -hydrant, abbr. **FH** (hyphens)

fire/arm(s), -man, -place, -side (one word)

fire/ brigade, — **insurance** (two words)

Firenze, It. for Florence

fire-plug, abbr. **FP**

fireproof (one word)

firkin/, small cask; *pl.* **-s;** abbr. **fir.**

firman/, an edict, *pl.* **-s**

firn, névé, granular snow not yet compressed into ice at head of glacier (not ital.)

first (adj.), abbr. **1st**

first/ aid (two words), — **class** (hyphen when attrib.)

first/-born, -fruit (hyphens)

firsthand (adj., one word), *but* **at first hand** (two words)

first proof (typ.), the first impression taken, and corrected by the 'copy'

first-rate (hyphen)

First World War, 1914–18 (caps.); *also* **World War I**

firth, an estuary, *not* fri-

FIS, Fellow of the Institute of Statisticians

FISA, Fellow of the Incorporated Secretaries Association

Fischer (Robert James), b. 1943, US chess master

Fischer-Dieskau (Dietrich), b. 1925, German baritone

fisgig, *use* **fiz-**

fish/ cake, — **finger** (two words)

fishing/-line, -rod (hyphens)

fissile, able to undergo fission, *not* fissionable

FIST, Fellow of the Institute of Science Technology

fist (typ.), the ☞

fisticuffs, *not* fisty-

fistul/a, opening of internal organ to the exterior or to another organ, *pl.* **-ae** (not ital.)

fit (arch.), section of a poem, *not* fytte

FitzGerald/, fam. name of Duke of Leinster; — **(Edward),** 1809–83, English poet and translator (one word), cap. G

Fitzgerald/ (Francis Scott Key), 1896-1940, US novelist; — **(George Francis),** 1851–1901, Irish physicist, devised the theory of — **contraction;** lower-case *g*

Fitzwilliam/ College, — **Museum,** Cambridge

fivefold (one word)

fixed star, sign ✿ or ✳

fizgig, flirtatious girl, small firework, *not* fis-, fizz-

fizz, a sound, *not* fiz

FJI, Fellow of the Institute of Journalists, *not* FIJ

fjord, *use* **fiord**

FL, (naval) Flag Lieutenant, Florida (off. postal abbr.)

Fl., Flanders, Flemish

fl., florin, fluid, (Neth. etc.) gulden

fl., *flores* (flowers), *floruit* (flourished)

f.l., *falsa lectio* (a false reading)

Fla., Florida (off. abbr.)

FLA, Fellow of the Library Association

flabbergast, to dumbfound, *not* flaba-, flaber-

flabell/um (eccl. and bot.), a fan; *pl.* **-a**

flag/-boat, -day (hyphens)

flageolet, (mus.) small flute, (bot.) kidney bean, *not* -elet (not ital.)

flag-pole (hyphen)

flag/ship, -staff (one word)

flak, anti-aircraft fire, *not* -ck

flambé (Fr. cook.), served in flames

flambeau, a torch; *pl.* **-s** (not ital.)

flamboyant, showy, (archit.) with flamelike lines (not ital.)

flamenco/, type of gypsy song or dance from Andalusia, *pl.* **-s** (not ital.)

flame-thrower (hyphen)

flamingo/, trop. bird; *pl.* **-s**

flammable, inflammable

Flamsteed (John), 1646–1719, first English Astronomer Royal (1675–1719), *not* -stead

Flanders, abbr. **Fl.**

flân/erie (Fr. f.), lounging; **-eur,** *fem.* **-euse,** an idler

flannelette, cotton imitation of flannel, *not* -llette

flannelled, *not* -eled

flash/back, -light, -point (one word)

flat/ (mus.), sign ♭; — **back** (bind.), *see* **square back;** — **impression** *or* — **pull** (typ.), *see* **rough pull**

flat-fish (hyphen)

flatworm (one word)

Flaubert (Gustave), 1821–80, French novelist

flautist, *not* flut-

flavour/, -ed, -ing, -less, *but* **flavorous**

flèche (Fr. f.), an arrow; **flèche** (archit., not ital.), a slender spire

flection, *use* **flexion**

fledgeling, *not* fledgling

Fleming, a native of Flanders

Fleming (Sir Alexander), 1881–1955, Scottish bacteriologist, discovered penicillin

fleur-de-lis, heraldic lily, *not* — - — -lys, *nor* flower-de-luce; *pl.* **fleurs-de-lis** (not ital.)

fleuret (bind.), flower-shaped ornament

fleuron (Fr. typ. m.), a type ornament

flexible, *not* -able

flexion, *not* -ction

flexitime, system of working variable hours, *not* flext-

flibbertigibbet, frivolous person

Fliegende Holländer (Der) (*The Flying Dutchman*), opera by Wagner, 1843

flier, *use* **flyer**

flintlock (one word)

floatation, *use* **flotation**

floccinaucinihilipilification, estimating as worthless (one word)

floccul/us (Lat.), a small tuft, (astr.) small solar cloud, (anat.) a small lobe in the cerebellum; *pl.* **-i**

flong (typ.), material used for making the mould for stereotyping

Flood (the), in Genesis (cap.)

flood/gate, -light (one word)

floozie, *not* -sie

flor., floruit (flourished)

flora/ (*collect. sing.*), the plants of a region or epoch; *pl.* **-s** (not ital.)

floreat (Lat.), may he, she, it flourish

Floren/ce, in It. **Firenze;** adj. **-tine**

flores (Lat.), flowers; abbr. *fl.*

florescen/ce, -t

floriat/e, -ed, florally decorated *not* -eate

Florida, off. abbr. **Fla.** or (postal) **FL**

florilegi/um, an anthology, *pl.* **-a**

florin, abbr. **fl.**

floruit (Lat.), flourished, abbr. *fl.,* or *flor.,* followed by date(s) (not ital. when used as noun)

flotation, *not* floatation

flotsam and jetsam (naut.), floating wreckage and goods thrown overboard (esp. when washed ashore)

flourished, abbr. *fl.* (*floruit*)

flow chart (two words)

flower-de-luce, *use* **fleur-de-lis**

flowers (typ.), type ornaments

flow rate (two words)

FLS, Fellow of the Linnean Society (off. spelling), *not* Linnae-

Flt. Lt., Flight Lieutenant

Flt. Sgt., Flight Sergeant

flu, influenza (no point)

fluffing (typ.), release of paper fluff or dust during printing

Flügel (Johann Gottfried), 1788–1855, German lexicographer

flugelhorn, kind of bugle

flugelman, *use* **fugle-**

fluid, abbr. **fl.**

fluidize, *not* -ise

fluky, *not* -ey

flummox, to confound, *not* -ix, -ux

flunkey/, liveried servant (now derog.), *pl.* **-s;** *not* -ky

fluoresce/, -nce, -nt, be made luminous, etc.

fluoridate, add fluoride to

fluorinate, introduce fluorine into

fluorine, *not* -in; symbol **F**

fluoroscope, instrument for X-ray examination, with fluorescent screen

fluorspar (chem.), calcium fluoride as mineral

flush (typ.), set to margin of column or page

Flushing, Netherlands, in Du. **Vlissingen**

flustr/a, a seaweed; *pl.* **-ae** (not ital.)

flutist, *use* **flautist**

fluty, flutelike, *not* -ey

fluxions (math.), *not* -ctions

fly/, light carriage, *pl.* **-s**

flyer, *not* flier

fly/leaf (typ.), a blank leaf at the beginning or end of a book, also blank leaf of a circular; **-sheet**, a two- or four-page tract (one word)

flyover, road or railway bridge (one word)

fly-past (noun, hyphen)

flywheel (one word)

FM, field magnet, Field Marshal, Foreign Mission, frequency modulation

Fm, fermium (no point)

fm., fathom

FMS, Fellow of the Medical Society

f.n., fn., footnote

FO, Field Officer, (naval) Flag Officer, (RAF) Flying Officer, Foreign Office

fo., folio

FOC, father of the chapel (*see* **chapel**)

focalize, *not* -ise

Foch (Ferdinand), 1851–1929, French general

fo'c'sle, *use* **forecastle**

focus/, *pl.* **-es**, (sci.) **foci**

focus/ed, -es, -ing, *not* -uss-

foehn, *use* **föhn**

foenugreek, *use* **fe-**

Foerster (Friedrich Wilhelm), 1869–1966, German philosopher and polit. writer

foet/id, -or, *use* **fe-**

foet/us, child or young animal developed in womb, *pl.* **-uses**; **-al, -ation, -icide**; *not* fet-

fog (meteor.), abbr. **f.**

foggy, obscured by fog

fog/y, one with antiquated notions, *not* -ey, -ie; *pl.* **-ies**

föhn, Alpine south wind, *not* foehn

foie (Fr. m.), liver

Fokine (Michel), 1880–1942, Russian-American choreographer

fol., folio, following

fold, as suffix forms one word (e.g. threefold)

fold-out, oversize page in book (hyphen)

foldstool, *use* **fald-**

foliaceous, leaflike, *not* -ious

foliate, (verb) number folios consecutively, (adj.) leaflike

Folies Bergère (ital.)

folio/ (typ.), a sheet of MS or copy; a page number; a book based on two leaves, four pages, to the sheet; *pl.* **-s**; as verb, **-ed**, *not* -'d; abbr. **fo., fol.**

foli/um (Lat.), a leaf; *pl.* **-a**

folk-dance (hyphen)

Folketing, Danish Parliament; *not* -thing

folklor/e, -ism, -ist, -istic (one word)

folk/-song, -tale (hyphens)

follic/le (bot., med.), small sac, *not* -cule; *but* **-ular, -ulated**

following, abbr. **f.** *or* **fol.**

fonda (Sp.), an inn

fondant, a sweetmeat (not ital.)

fondue (Fr. cook. f.), melted cheese, eggs, etc., *not* fondu (ital.)

fons et origo (Lat.), source and origin

font (US typ.), Eng. **fount**

Fontainebleau, dép. Seine-et--Marne, France

Fonteyn (Dame Margot), b. 1919, English prima ballerina

foodstuff (one word)

foolproof (one word)

foolscap, former size of paper, $17 \times 13\frac{1}{2}$ in.; — 4to, $8\frac{1}{2} \times 6\frac{3}{4}$ in.; — 8vo, $6\frac{3}{4} \times 4\frac{1}{4}$ in. (untrimmed). *See also* **book sizes**

foot, in metric system 30.48 centimetres; abbr. **ft.**, sign ′; not now in scientific use

foot-and-mouth disease (hyphens)

football (one word)

foot/-brake, -bridge (hyphens)

foot candle (two words)

foot/hills, -hold, -lights, -loose (one word)

footnotes, in copy should be written either at the bottom of the folio or on a separate sheet, with reference figures for identification; (typ.) all references in text to be by superior figures outside the punctuation mark or quote, except in math. setting (*see* **reference marks**). For a second set of cues *use* superior letters. At page make-up, leave apparent white between end of text and first line of notes

foot/path, -plate, -print (one word)

Foots Cray, London

Footscray, Vic., Australia

foot/sore, -step, -stool, -way, -wear (one word)

for., foreign, forestry

foram/en (anat., bot.), an orifice; *pl.* **-ina**

forasmuch (one word)

foray, a raid, *not* forr-

forbade, *see* forbid

for/bear, to abstain, past **-bore,** partic. **-borne.** *See also* **fore-bear**

Forbes-Robertson (Sir Johnston), 1853–1937, English actor

for/bid, past **-bade,** partic. **-bidden, -bidding**

force majeure (Fr. f.), circumstances beyond one's control

forcemeat, meat finely chopped and seasoned, *not* forced- (one word)

forceps, surgical pincers; *pl.* same

forcible, *not* -eable

Ford (Ford Madox), orig. surname **Hueffer,** 1873–1939, English writer

forearm (noun and verb, one word)

forebear, ancestor. *See also* **fore-bear**

forebod/e, -ing, *not* forb-

forecast

forecastle, *not* fo'c'sle

foreclose

foredge (bind.), the edge of a book opposite the binding, *not* fore-edge. *See also* **margins**

fore-edge (hyphen in non-technical uses)

fore-end, *not* forend

Forefathers' Day (US), anniversary of landing of the Pilgrims at Plymouth, Mass., 21 Dec. 1620, usu. celebrated on 2 Dec.

forefend, *use* forf-

fore/finger, -front (one word

foregather, *use* forg-

fore/go, to go before, past **-wen** partic. **-gone, -going.** *See also* **forgo**

forehead

foreign, abbr. **for.**

Foreign and Commonwealth Office, formed, 1968, from the two separate Offices; abbr. **FCC**

Foreign/ Mission, abbr. **FM ;** — **Office** (caps.), abbr. **FO,** combined, 1968, with the Commonwealth Office

forel, early vellum-like covering for books, *not* forr-

fore/leg, -lock, -mast, -play (one word)

fore/run, past **-ran,** partic. **-run; -runner** (one word)

fore/said, -see, -shadow, -shore, -shorten, -sight, -skin, -stall, *not* for- (one word

Forester (Cecil Scott), 1899– 1966, English novelist

foretell, *not* fort-, fortel

for ever, for always (two words); **forever,** continually (one word)

forewarn, *not* for- (one word)

foreword (of book), *not* forw-. *See also* **preliminary matter**

forfeit

forfend, avert, *not* fore-

forgather, *not* fore-

rget-me-not (bot.)

rgett/able, -ing

rgivable, *not* -eable

r/go, to abstain from, past **-went,** partic. **-gone, -going.** *See also* **forego**

rmalin, a germicide or pre-servative, *not* -ine

rmalize, *not* ise

rmat/ (typ.), the size (octavo, quarto, etc.) of a book; (loosely) ts general typ. style and ap-pearance; (comp.) a frequently-used set of typographical commands stored in the key-board (not ital.); as verb, past **-ted,** partic. **-ted. -ting**

rme (typ.), a body of type secured in the frame called a chase; *not* form

rmer, first of two. *See also* **latter**

rmul/a, *pl.* **-as,** (sci.) **-ae** (not ital.); abbr. **F.**

rray, *use* **foray**

rrel, *use* **forel**

rsaid, *use* **fore-**

ors Clavigera, by Ruskin, 1871–84

orster (**John**), 1812–76, English biographer (see also **Foster**); — (**Edward Morgan**), 1879–1970, English novelist

rsw/ear, -ore, -orn

rsythia, ornamental shrub

ort, cap. F when with name, as Fort Southwick, Tilbury Fort; abbr. **Ft.**

ort., fortification, fortified

orte/, person's strong point (not ital.); (mus.), strong and loud, abbr. *f*; — **-piano,** loud, then immediately soft (hyphen), abbr. *fp*; — **piano,** early form of pianoforte (one word)

ortell, *use* **fore-**

ortissimo/ (mus.), very loud; as noun, *pl.* **-s**; abbr. *ff*

ortississimo/ (mus.), as loud as possible; as noun, *pl.* **-s**; abbr. *fff*

rtiter in re (Lat.), bravely in action. *See also* **suaviter in modo**

FORTRAN, Formula Transla-tion, a computer compiler lan-guage

forty-eightmo (typ.), a book based on 48 leaves, 96 pages, to the sheet; abbr. **48mo**

forwarn, *use* **fore-**

forzando (mus.), forced; abbr. *fz* (no point). *See also* **sforzando**

foss/a (anat.), a cavity; *pl.* **-ae**

fosse, a ditch, *not* foss

fossilize, *not* -ise

Foster (**Birket**), 1825–99, Eng-lish painter; — (**John**), 1770–1843, English essayist (*see also* **Forster**); — (**Stephen Col-lins**), 1826–64, US song-writer; — (**Sir Harry Hylton-**), *see* **Hylton-Foster**

Fotheringhay Castle, Nor-thants, 1066–1604

Foucault (**John Bernard Léon**), 1819–68, French physi-cist

Foucquet, *use* **Fouq-**

Foulahs, *use* **Fulahs**

foulard, a fabric (not ital.)

foul/, adv. **-ly; -up** (hyphen as noun)

foul proof (typ.), a proof--reader's marked proof, as op-posed to the corrected (or **clean,** q.v.) proof which succeeds it

foundry, *not* -ery; — **proof,** final proof from forme which has been prepared for plating

fount (typ.), a complete set of type of one particular face and size; in US **font**

Fouqué (**Friedrich, Baron de la Motte**), 1777–1843, German poet and dramatist

Fouquet (**Jean**), 1416–80, French painter; — (**Nicolas**), 1615–80, French statesman; *not* Foucq-

four-colour process (typ.), printing in yellow, magenta, cyan, and black to give a com-plete colour reproduction

fourfold (one word)

Fourier

Franc

Fourier (François Marie Charles), 1772–1837, French socialist, hence **Fourier/ism, -ist, -ite**; — (**Jean Baptiste Joseph**), 1768–1830, French mathematician and physicist, hence **Fourier series**

four-poster, four-post bed

four/score, -some (one word)

four-stroke (hyphen)

Fourth of July, US Independence Day (caps.)

Fourth of June, George III's birthday, day of celebration at Eton

Fowler (Francis George), 1870–1918, English lexicographer; — (**Henry Watson**), 1858–1933, English lexicographer, author of *Modern English Usage* and joint-author, with brother F. G., of *The King's English*

Fox (Charles James), 1749–1806, English politician; — (**George**), 1624–91, English preacher, founder of Society of Friends (Quakers)

Foxe (John), 1516–87, English martyrologist

foxed paper, stained with yellowish-brown spots

fox/glove, -hole, -hound (one word)

fox/-hunt, -terrier (hyphens)

foxtrot (one word)

foyer, theatre lounge (not ital.)

FP, Fine Paper (the best edition of a work); fire-plug; (Sc.) former pupil(s)

f.p., freezing-point

fp (mus.), forte-piano, loud, then immediately soft

FPA, Family Planning Association, Foreign Press Association

F.Ph.S., Fellow of the Philosophical Society of England

FPS, Fellow of the Pharmaceutical Society of Great Britain

Fr, francium (no point)

Fr., Father, France, French, Friar, Friday; (Ger.) Frau (Mrs, wife); (It.) *Fratelli* (Brothers)

fr., fragment, franc, from, (Ger.) *frei* (free)

fra (It.), brother, friar (no point)

fracas, noisy quarrel; *pl. same* (not ital.)

fractionalize, *not* -ise

fractions (typ.), spell out simple fractions in textual matter, e.g. one-half, two-thirds, one and three-quarters; hyphenate compounded numeral in compound fractions such as nine thirty-seconds, forty-seven sixty-fourths; in statistical matter use one-piece fractions where available ($\frac{1}{2}$, $\frac{2}{3}$, $\frac{3}{4}$, etc.), but if these are not available use **split fractions,** i.e. those with dividing line attached to denominators and the numerator justified above it, e.g. nineteen hundredths, $\frac{19}{100}$; but to avoid difficult handwork complex non-displayed fractions are now often set in fount-size numerals with a solidus between, e.g. 19/100. For math. setting, *see Hart's Rules*, pp. 55–6

fractions, decimal, *see* **decimal fractions**

FRAD, Fellow of the Royal Academy of Dancing

fraenum, *not* fre-

F.R.Ae.S., Fellow of the Royal Aeronautical Society

F.R.Ag.Ss., Fellow of the Royal Agricultural Societies

FRAI, Fellow of the Royal Anthropological Institute

Fraktur (typ.), German name for German style of black letter, as 𝔉𝔯𝔞𝔨𝔱𝔲𝔯

FRAM, Fellow of the Royal Academy of Music

framable, *not* -eable

framework (one word)

franc (Fr. m.), coin (not ital.); abbr. **f.** *or* **fr.,** *pl.* **f.** *or* **frs.,** to be put *after* the figures, as 10 f. 50 c., or 10.50 fr.

française (à la) (Fr.), in the French style, *not* cap. F

France, abbr. **Fr.**

ranche-Comté, province, France

ranchise, the right to vote, *not* -ize

rancium, symbol Fr

ranck (César Auguste), 1822–90, Belgian-born French organist and composer

ranco (General Francisco), 1892–1975, Spanish head of state 1939–75

ranc-tireur (Fr. m.), an irregular sharpshooter; *pl.* ***francs-tireurs***

ranglais (Fr. m.), French regarded as including too many borrowings from English

rankenstein (in Mary Shelley's novel), the maker of the monster, *not* the monster

Frankfort, Ind. and Ky. (US)

Frankfurt/-on-Main, W. Germany; — **-on-Oder,** E. Germany; in Ger. **Frankfurt/ am Main,** — **an der Oder** (no hyphens)

rankfurter, seasoned smoked sausage

Frankfurter Allgemeine Zeitung, German newspaper, *not* Frankfor-, -für-

Franz Josef Land, Arctic Ocean (USSR)

frappant (Fr.), striking, affecting

frappé (Fr.), iced

FRAS, Fellow of the Royal Astronomical Society, ditto Asiatic Society

Fraser, family name of Barons Lovat, Saltoun, and Strathalmond. *See also* **Frazer**

Fraser River, Brit. Columbia, Canada

Fraser's Magazine

frat/e (It.), a friar; *pl.* *-i*

fraternize, *not* -ise

Frau/ (Ger. f.), Mrs, wife, *not* Fräu; abbr. **Fr.;** *pl.* **-en**

Fräulein (Ger. n.), Miss, unmarried lady; *pl. same*; abbr. **Frl.**

Fraunhofer/ (Joseph von), German physicist, discoverer of

— **lines** in sun's spectrum (no umlaut)

Frazer (Sir James George), 1854–1941, English anthropologist. *See also* **Fraser**

FRBS, Fellow of the Royal Botanic Society; Fellow of the Royal Society of British Sculptors

FRCGP, Fellow of the Royal College of General Practitioners

FRCM, Fellow of the Royal College of Music

FRCO, Fellow of the Royal College of Organists

FRCOG, Fellow of the Royal College of Obstetricians and Gynaecologists

FRCP, Fellow of the Royal College of Physicians, London

F.R.C.Path., Fellow of the Royal College of Pathologists

FRCS, Fellow of the Royal College of Surgeons, England

FRCVS, Fellow of the Royal College of Veterinary Surgeons, London

F.R.Econ.S., Fellow of the Royal Economic Society

Fred., Frederic, Frederick. When it is a full name, or a diminutive of familiarity, it takes no point

Fredericton, New Brunswick, Canada, *not* -ck-

free and easy (three words)

freeboard (one word)

free-born (hyphen)

freedman, an emancipated slave; **freeman,** one to whom the freedom of a city has been given

free/-for-all, -hand (adj.) (hyphens)

free/hold, -holder (one word)

free lance (hyphen when attrib. or as verb)

Freemantle, *use* Frem-

Freemason/, -ry (cap., one word)

free-range (adj., hyphen)

free/-thinker, -thought (hyphens)

free will, the power of self-determination (two words); **freewill** (adj., one word)

freeze, to convert to ice. *See also* **frieze**

freezing-point, abbr. **f.p.**

frei (Ger.), free; abbr. *fr.*

Freiberg, Saxony, Germany

Freiburg/ im Breisgau, Baden, Germany; Ger. abbr. — **i.B.**

Freiburg, Switzerland, *use* **Fribourg**

Freiherr, Ger. title; abbr. *Frhr.*

Freischütz (*Der*), opera by Weber, 1819

freize, *use* **frieze**

Fremantle, W. Australia, *not* Free-

French, abbr. **Fr.**; (typ.), alphabet as English; *see* **accents.** There are strict rules for capitalization, division of words, spacing, etc. The following must suffice here, but reference to *Hart's Rules,* pp. 88–102, is strongly recommended. Division of words is according to spoken syllables, rarely etymology: single consonant goes with following vowel, including consonant + *r* or ‑ *l*; take over *gn* (e.g. *sei-gneur*); doubled consonants may be divided. No caps. for adjectives of nationality, the first pers. pronoun, days of week, months, names of cardinal points, names indicating rank, as *anglais, je, lundi, mars, le nord, le duc.* **Accents: acute** (´) used only over *e*; when two *es* come together the first always has acute accent, as *née*; **grave** (`) used over *a, e, u*; **circumflex** (^) used over any vowel. **Cedilla c** (ç) used only before *a, o, u.* **Diaeresis** as in English; the digraph *œ* is not to be separated (e.g. *Œuvre, cœur*). Quotation marks are usu. guillemets (« »), though rules are more commonly used in conversational passages

french chalk (not cap.)

French groove (bind.), extra space between board and spine

Frenchified, Frenchlike (cap.)

french polish/, -er, french windows (two words, not cap.

frenetic, delirious, frantic, not ph-

frenum, *use* **frae-**

freq., frequent, -ly, -ative

frère (Fr. m.), brother, friar

FRES, Fellow of the Royal Entomological Society

fresco/, water-colour done on damp plaster; *pl.* **-s** (not ital.)

freshman, first-year man at university (one word)

freshwater (adj., one word); **fresh water** (noun, two words

Fresnel (Augustin Jean), 1788–1827, French physicist

fret/saw, -work (one word)

Freud/ (Sigmund), 1856–1939 Austrian neurologist and founder of psychoanalysis; adj. **-ian**

Freytag (Gustav), 1816–95, German writer

FRG, Federal Republic of Germany

FRGS, Fellow of the Royal Geographical Society

F.R.Hist.S., Fellow of the Royal Historical Society

Frhr. (Ger.), *Freiherr* (a title)

FRHS, Fellow of the Royal Horticultural Society

Fri., Friday

friar's balsam, tincture of benzoin, *not* -s'

FRIBA, Fellow of the Royal Institute of British Architects (*pl.* **FF-**)

Fribourg, Switzerland, *not* Frei-: in Ger. **Freiburg**

FRIC, Fellow of the Royal Institute of Chemistry, *now* **FRSC**

fricandeau/, braised and larded fillet of veal; *pl.* **-x** (not ital.)

fricassee/, a white stew (not ital., no accent), *pl.* **-s**

FRICS, Fellow of the Royal Institution of Chartered Surveyors

Friday, abbr. **F., Fr.,** *or* **Fri.**

Friedman (Milton), b. 1912, US economist

frier, one who fries, *use* **fryer**

Friesian, breed of cattle, *not* Fris-

frieze, cloth, (archit.) part below cornice, *not* frei-

FRIPHH, Fellow of the Royal Institute of Public Health and Hygiene

frippery, tawdry finery

Fris., Frisia (Friesland, in Netherlands), Frisian. *See also* **Friesian**

fris/ette, curls on forehead, *not* friz-; **-eur,** hairdresser; **-ure,** mode of hairdressing

frisson (Fr. m.), a shudder

frit/ (Fr. cook.), *fem.* **-e,** *pl.* **-(e)s,** fried

frith, estuary, *use* **firth**

frizz, to roughen, curl, *not* friz

Frl. (Ger.), Fräulein (Miss)

F.R.Med.Soc., Fellow of the Royal Medical Society

F.R.Met.S., Fellow of the Royal Meteorological Society

FRMS, Fellow of the Royal Microscopical Society

FRNS, Fellow of the Royal Numismatic Society

fro (no point)

Froebel (Friedrich), 1782–1852, German educationalist, founder of kindergarten system

frolic/, -ked, -king

fromage (Fr. m.), cheese

Fronde, French rebel party during minority of Louis XIV

frondeur, member of Fronde, political rebel

Frontignan, a muscat grape or wine (not ital.)

frontispiece (typ.), illustration facing title-page (one word)

frost/-bite, -bitten (hyphens)

Froude (James Anthony), 1818–94, English historian

Froufrou, comedy by Meilhac and Halévy

frou-frou, a rustling of dress (not ital.)

frowsty, musty, stuffy

frowzy, unkempt, slatternly

FRPS, Fellow of the Royal Photographic Society

FRS, Fellow of the Royal Society, in Lat. **SRS** (*Societatis Regiae Sodalis*)

frs., francs

FRSA, Fellow of the Royal Society of Arts

FRSC, Fellow of the Royal Society of Chemistry

FRSE, Fellow of the Royal Society of Edinburgh

FRSH, Fellow of the Royal Society of Health

FRSL, Fellow of the Royal Society of Literature

FRSM, Fellow of the Royal Society of Medicine

FRST, Fellow of the Royal Society of Teachers

frumenty, boiled wheat with milk, sugar, etc., *not* the many variants

frust/um (geom.), lower portion of intersected cone or pyramid; *pl.* **-a;** *not* -rum

FRVA, Fellow of the Rating and Valuation Association

fryer, one who fries, *not* frier

FSA, Fellow of the Society of Antiquaries

FSAA, Fellow of the Society of Incorporated Accountants and Auditors

FSE, Fellow of the Society of Engineers

FSIAD, Fellow of the Society of Industrial Artists and Designers

FSS, Fellow of the Royal Statistical Society

FSVA, Fellow of the Incorporated Society of Valuers and Auctioneers

Ft., fort

ft., feint (paper), flat, foot *or* feet, fortified

FTCD, Fellow of Trinity College, Dublin

FTCL, Fellow of Trinity College (of Music), London

FTI, Fellow of the Textile Institute

fuchsia (bot.) (not ital.)

fuc/us (Lat.), a seaweed; *pl.* **-i**

Fuehrer (Ger.), *use* **Führer**

fuelled, *not* -eled

fugleman, leader in mil. exercises, *not* flugel-, flugle-, fugal-, fugel-

fug/ue (mus., psych.) (not ital.); *adj.* **-al**

Führer, leader, title assumed by Hitler in Nazi Germany

Fujiyama, mt., Japan, *properly* **Fujisan,** *not* Fusi-

-ful, suffix denoting amounts, *pl.* **-fuls**

Fulahs, Sudanese, *not* Felláh, Fellani, Feulhs, Foulahs, Fulbe. *See also* **fellah**

fulcr/um, point of purchase for lever; *pl.* **-a** (not ital.)

fulfil/, -led, -ling, -ment

fulgor, splendour, *not* -gour

full-bound (bind.), completely cased in the same material (hyphen)

full/-length, not shortened, **-scale,** not reduced (hyphens)

fullness, *not* fulness

full out (typ.), set to margin of column or page

full point (typ.), the **full stop;** *see* **punctuation** V

fulmar, a petrel

fulness, *use* **full-**

fulsome, excessive and cloying, *not* full-

fumatory, a place for smoking. *See also* **fumi-**

fumigator, *not* -er

fumitory (bot.), a plant. *See also* **fuma-**

function (math.), abbr. **F** or **f**

fun-fair (hyphen)

fung/us, *pl.* **-i**; *adj.* **-ous** (not ital.)

funny/-bone, -face (hyphens)

fur., furlong

für (Ger.) for; abbr. *f.*

furbelow, a flounce, *not* -llow

furfur/, dandruff; *adj.* **-aceous**

furlong, eighth of mile; abbr. ▶ *or* **fur.**

furmenty, *use* **fru-**

Furness (Christopher, Baron), 1852–1912, shipping magnate; — **(Horace Howard),** 1833–1912, and his son, 1865–1930, US Shakespearian scholars

Furniss (Harry), 1854–1925, caricaturist

furniture (typ.), spacing material

Furnivall (Frederick James) 1825–1910, English philologist

furor (Lat.), rage (ital.)

furore (It.), enthusiastic admiration, uproar (not ital.)

furry, furlike

furth/er (adv., adj., and verb), **-est** (adv. and adj.), now the dominant spellings; use in all cases, *not* far-

Furtwängler (Adolf), 1853–1907, German archaeologist; — **(Wilhelm),** his son, 1886–1954, German conductor

fusable, *use* **fusi-**

fusain, porous coal, artists' fine charcoal crayon

fus/e, -ee, fuselage, *not* fuz-

fusible, *not* -able

fusil, musket, *not* -zil

fusilier, *not* -leer

fusillade, *not* -ilade

Fusiyama, mt., Japan, *use* **Fujiyama**

fut., future

futhorc, runic alphabet, *not* -ark, -ork

fuz/e, -ee, -elage, -il, *use* **fus-**

f.v., *folio verso* (on the back of the page)

FWA, Family Welfare Association

fylfot, the swastika, *not* fil-

fytte, *use* **fit**

Fyzabad, Uttar Pradesh (India) *not* Faizabad

fz (mus.), forzando

FZS, Fellow of the Zoological Society

G

G, gauss, the seventh in a series, (as prefix) giga-

G., Graduate, Grand, Gulf, (naval) gunnery

g, gram(s) (no point in scientific and technical work), (dyn.) local gravitational acceleration

g., guinea, -s, (Fr.) *gauche* (left), *gros/, -se* (big), (meteor.) gale

GA, General Assembly (Sc. Ch.); Georgia, US (off. postal abbr.)

Ga, gallium (no point)

Ga., Gallic; Georgia, US (off.)

gabardine, durable cloth

gabbey, *use* **gaby**

gabbro/ (geol.), an igneous rock; *pl.* **-s**

gaberdine, a loose cloak

Gabon/, W. Africa, indep. republic 1960; adj. **-ese**

Gaborone, Botswana

gaby, a simpleton, *not* -ey, gabbey, gawby

Gadarene swine

Gaddi, family of Florentine painters, 1259–1396

Gadhel/, a Gael of the Irish, Highland Scottish, or Manx branch; adj. **-ic**

Gaditanian, of Cadiz, SW Spain

gadolinium, symbol **Gd**

Gadshill, Kent, site of **Gad's Hill Place,** Charles Dickens's residence 1860–70; character in Shakespeare's *1 Henry IV* (one word)

Gaekwar, title of prince of Baroda, India, *not* Gaik-

Gaelic/, abbr. **Gael.;** — **alphabet,** same as English, but no *j, k, q, v, w, x, y, z. See also* **accents**

Gaeltacht, region of Ireland where vernacular language is Irish

gaga, senile, incapable (one word)

Gagarin (Yuri Alekseyevich), 1934–68, Russian cosmonaut, first to orbit earth, 1961

gage, a, *or* to, pledge. *See also* **gauge**

gaiety, *not* gay-

Gaikwar (title), *use* **Gaekwar**

gaillardia (bot.), a plant

gaily, *not* gayly

gairfish, *use* **garfish**

gairfowl, *use* **gare-fowl**

gairish, *use* **garish**

Gair Loch, Strathclyde, *use* **Gare**

Gairloch, Highland, *not* Gare-

Gaitskell (Hugh Todd Naylor), 1906–63, British statesman

Gaius, Roman praenomen, *not* Caius; abbr. **C.**

gal, unit of acceleration, with numbers **Gal** (no point); in SI units 1 cm/s²

Gal., Galatians

gal., gallon, -s

gala/, festive occasion; *pl.* **-s** (not ital.)

galaena, *use* **-lena**

galangale, *use* **gali-**

galantine, white meat served cold in jelly, *not* gall-

Galantuomo (Il Re), King Victor Emmanuel I of Italy

galanty show, a shadow pantomime, *not* -tee, gallantee, -ty

Galatea (Acis *or* **Pygmalion and)**

Galati, Romania, *not* -acz, -atch, -atz

Galatia, Asia Minor

Galatians, abbr. **Gal.**

galavant, *use* **galli-**

Galaxy (astr.), the Milky Way (system) (cap.)

gale (meteor.), wind moving at 40–70 miles per hour

galena, lead ore, *not* -aena

galera (Sp. typ.), a galley

galère (Fr. f.), galley (ship); *qu'allait-il faire dans cette — ?* (how did he get into this scrape?)

Galilean, of Galilee, or of Galileo

147

Galileo [Galilei], 1564–1642, Italian astronomer and mathematician; in Fr. *Galilée*, It. *Galilei*

galingale, a sedge with medicinal root, *not* gala-

galiot, a vessel, *use* **galliot**

galipot, a resin. *See also* **gall-**

galivant, *use* **galli-**

gall., gallon, -s

gallantine, *use* **galantine**

gallanty show, *use* **gala-**

gallaway, *see* **gallo-**

gall-bladder (hyphen)

Galle, town, Sri Lanka, *formerly* Point de Galle

gallery, in Fr. f. *galerie*

Galles (Fr. f. sing.), Wales; adj. *gallois*/, *fem.* -e

galley/ (typ.), a flat oblong tray for holding composed type; — **proofs**, those supplied in 'slips' about 18 in. long (i.e. impressions taken from type on galleys). *See also* **proof**

Gallic, of Gaul, French; abbr. Ga.

gallice (Lat.), in French

Gallic/ism, French idiom; **-ize**, to make Gallic or French, *not* -ise

Galli-Curci (Amelita), 1889–1963, Italian soprano

galligaskins, *pl.*, breeches, *not* -in

gallimaufry, a medley

Gallio, a typical sceptic (Acts 18: 17)

galliot, Dutch cargo-boat

Gallipoli, S. Italy, Turkey

gallipot, a small jar. *See also* **gali-**

gallium, symbol Ga

gallivant, to gad about, *not* gala-, gali-

gallon/, -s, abbr. **gal.** *or* **gall.**

galloon, a dress trimming

galloot, *use* **galoot**

gallop/, a horse's movement; **-ed, -er, -ing**. *See also* **galop**

gallopade, Hungarian dance, *not* galop-, gallopp-

Gallovidian, of Galloway

Galloway, SW Scotland

galloway, a horse, *also* breed o cattle, *not* galla-

gallows, treated as sing.

gallstone (one word)

Gallup/ (Dr George Horace b. 1901, founded American In stitute of Public Opinion, 1936 — **poll**, measure of public opi ion (two words, one cap.)

galoot, an awkward fellow, *not* gall-, geel-

galop, a dance. *See also* **gallop**

galore, in abundance

galosh/, an overshoe, *not* gol-, -oshe; **-ed**

galumph, to gallop triumphantly

galv., galvan/ic, -ism

Galvani (Luigi), 1737–98, discoverer of galvanism

galvanize, *not* -ise

Galway, W. Ireland

Galwegian, of Galloway

Gama (Vasco da), 1467–1524 Portuguese navigator, first round Africa to India

Gambia (The), W. Africa, indep. 1965

Gambier, Ohio, US

gambier, a gum, *not* -beer, -bir

gambit, chess opening in whic pawn or piece is risked to obtai advantageous position

gamboge, yellow pigment, *not* -booze

gambol/, to frisk; **-led, -ling**

game/bird, -keeper (one word)

gamin/, a street urchin, *fem.* -e (not ital.)

gamma/, unit of magnetic flux density; abbr. γ; in SI units $1nT$; — **globulin**, a protein; — **ray** (two words)

gammon, a cured ham, *not* gamon

gamy, having the flavour or scent of game left till high, *not* -ey

Gand, Fr. for **Ghent**, Belgium

Gandhi/ (Mohandas Karam chand) (Mahatma Gandhi),

1869–1948, Indian nationalist leader; — (**Mrs Indira**), b. 1917, Prime Minister of India 1966–77, 1980– , dau. of Pandit Nehru, not related to preceding

angli/on, knot on nerve; *pl.* **-a**

angue, rock or earth in which ore is found

angway (one word)

anister, a hard stone, *not* gann-

antlet, *use* **gaun-**

antry, platform to carry travelling crane, etc., *not* gau-

Ganymede (Gr. myth.), cup-bearer to the gods

aol/, -er, *not* jai-. *See also* **goal**

aol/bird, -break (one word)

aramond, a typeface

Garamont (Claude), *c.*1500–61, French type-designer, -cutter, and -founder

Garcilaso de la Vega, 1503–36, Spanish poet; —————('**the Inca'**), 1540–1616, Spanish historian

arçon (Fr. m.), bachelor, boy, waiter (ital.)

Garde nationale (Fr.), national guard (one cap.)

ardenia (bot.), an ornamental shrub (not ital.)

Gardens, abbr. **Gdns.**

are (Fr. f.), railway station

arefish, *use* **garfish**

are-fowl, the great auk, *not* gair-, gar- (hyphen)

Gare Loch, Strathclyde, *not* Gair —

Gareloch, Highland, *use* **Gair-**

arfish, similar to pike, *not* gair-, gare-

argantuan, enormous (not cap.)

argoyle (archit.), grotesque spout, *not* -ile, -oil

Garhwal, Uttar Pradesh, India

ari, *use* **gharry**

arish, gaudy, *not* gair-

arlic, but **garlicky**

Garmisch-Partenkirchen, W. Germany

arnet, red precious stone

arn/i (Fr.), *fem.* **-ie,** furnished

garrott/e, to throttle, *not* -ote, garotte; **-er**

Garter King-of-Arms, herald, *not* -at-

gas (US colloq.), gasoline

gaseous

Gaskell (Mrs Elizabeth Cleghorn), 1810–65, English novelist

gasoline, volatile liquid from petroleum, esp. (chiefly US and techn.) petrol, *not* -ene

Gaspé, peninsula, cape, and town (Quebec prov., Canada)

gas poisoning (two words)

Gast/haus (Ger. n.), an inn; *pl.* *-häuser* (cap.)

Gast/hof (Ger. m.), a hotel; *pl.* *-höfe* (cap.)

gastronom/e, a judge of good eating; **-ic, -y**

gastropod, any member of the Gastropoda, mollusc class (not ital.); *not* -ster-

gât/é (Fr.), *fem.* **-ée,** spoiled

gâteau/ (Fr. m.), cake, *pl.* **-s**

gatecrash/, -er (one word)

gate/fold, folded oversize page (one word)

gate/keeper, -post, -way (one word)

gather (bind.), to assemble the printed and folded sections in sequence

GATT, General Agreement on Tariffs and Trade

gauch/e, awkward; **-erie,** awkwardness (not ital.)

gauche (Fr.), left; abbr. *g.*

gaucho/, a mounted herdsman, native of the pampas, *not* gua-; *pl.* **-s**

gaud/y, annual entertainment, esp. college dinner; (adj.) showy; **-ily, -iness**

gaug/e, a measure; **-ing,** *not* guage. *See also* **gage**

Gauguin (Eugène Henri Paul), 1848–1903, French painter

Gaul, anc. France; adj. **Gallic,** (Fr.) *gaulois/,* *fem.* **-e**

Gaullist, follower of de Gaulle

Gauloise, propr. term for French cigarettes

gauntlet/, a long glove, *not* gant-; **-ed**

gauntr/ee, -y, *use* **gantry**

gaur, Indian ox, *not* gour

Gauss (Karl Friedrich), 1777–1855, German mathematician and physicist

gauss, unit of magnetic induction, abbr. **G;** *pl. same*

Gautama Buddha, founder of Buddhism; *see* **Buddha**

Gauthier-Villars, publishers, Paris

Gautier (Théophile), 1811–72, French writer

gauzy, *not* -ey

gavel, a president's hammer

gavial, Asian crocodile, *not* gharial

gavotte, a dance, music for it

gawby, *use* **gaby**

gay/ety, -ly, *use* **gai-**

Gay-Lussac (Joseph Louis), 1778–1850, French chemist and physicist

gaz., gazett/e, -ed, -eer

gazebo/, summer-house, belvedere; *pl.* **-s**

gazett/e, -ed, -eer, abbr. **gaz.**

gazpacho, cold vegetable soup

gazump, to raise price to would--be buyer after agreement but before completion

GB, Great Britain

GBE, Knight *or* Dame Grand Cross of the Order of the British Empire

GBH, grievous bodily harm

G.B.S. (George Bernard Shaw, 1856–1950), Irish-born British dramatist and critic

GC, George Cross; Golf Club

GCA, Ground Controlled Approach (radar)

GCB, Knight Grand Cross of the Order of the Bath

GCE, General Certificate of Education

GCF *or* **g.c.f.** (math.), greatest common factor

GCHQ, Government Communications Headquarters

GCI, Ground Controlled Interception (radar)

GCIE, Knight Grand Commander of the Order of the Indian Empire

GCLH, Grand Cross of the Legion of Honour

GCM, general court martial

GCM *or* **g.c.m.** (math.), greatest common measure

GCMG, Knight *or* Dame Grand Cross of the Order of St. Michael and St. George

GCSI, Knight Grand Commander of the Order of the Star of India

GCVO, Knight *or* Dame Grand Cross of the Royal Victorian Order

GD, Grand/ Duchess, — Duchy — Duke

Gd, gadolinium (no point)

Gdańsk, Poland, *formerly* **Danzig**

Gdns., Gardens

GDP, gross domestic product

GDR, German Democratic Republic

Gdsm., Guardsman

Ge, germanium (no point)

g.e. (binding), gilt edges

gearbox (one word)

gear-lever (hyphen)

gearwheel (one word)

geb. (Ger.), *geboren* (born), *geburden* (bound)

GEC, General Electric Company

gecko/, a house-lizard; *pl.* **-s**

gee/-ho, — **-up,** call to horses, *not* jee-

geeloot, *use* **gal-**

geezer (slang), an old man. *See also* **geyser**

gefuffle, *use* **ker-**

Geiger counter (in full, **Geiger-Müller counter**), instrument for detecting radioactivity

Geisenheimer, a white Rhine wine

geisha/, Japanese dancing-girl; *pl.* **-s**

Geissler (Heinrich), 1814–79

German physicist, inventor of
— vacuum tube

el (phys.), (form) semi-solid
colloidal solution

elatin/e, -ize, *not* -ise, **-ous**

eld (Ger. n.), money (cap.)

elderland, E. Netherlands, *not*
Guel-

elder rose, *use* gue-

elée (Fr. f.), frost, jelly

elée (Claude), *see* **Lorrain**

elert, faithful dog of Welsh
legend

**ellert (Christian Fürchte-
gott),** 1715–69, German poet

elsemium (bot.), *not* -inum

em (typ.), name for a former
size of type, about 4 pt.

emel, finger-ring, hinge, etc.,
not gemew, gimbal, gimmal,
gimmer

emm/a, a bud; *pl.* **-ae**

emütlich (Ger.), leisurely,
agreeab/le, comfortab/le, **-ly**

en., General, Genesis

en., gender, genera, general,
-ly, genitive, genus

endarme/ (Fr. m.), *pl.* **-s** (not
ital.)

endarmerie (Fr. f.), body of
soldiers used as police

ender, abbr. **gen.**

ên/e (Fr. f.), constraint; **-é,** *fem.*
-ée, constrained

ene/, the carrying unit of a cell,
the unit of heredity; *pl.* **-s**

enealog/y, family's pedigree;
adj. **-ical,** *not* -olog-

enera, *see* **genus**

eneral, abbr. **Gen.**

eneral Assembly (Sc. Ch.),
abbr. **GA**

eneral election (not caps.)

eneralia (Lat. pl.), general
principles

eneralissimo/, supreme commander, *pl.* **-s**

eneralize, *not* -ise

enerator, *not* -er

enesis, abbr. **Gen.**

enes/is, *pl.* **-es**

enet, kind of civet-cat. *See also*
jennet

Geneva, Switzerland, in Fr. **Ge-
nève,** Ger. **Genf,** It. **Ginevra;** adj. **Genevan; Genevese**
(sing. and pl.), native(s) of —,
in Fr. *genevois/,* *fem.* **-e(s)**

Geneviève (Sainte), patron of
Paris

Genghis Khan, 1162–1227,
Mongol conqueror of N. China
and Iran, *not* Jenghiz —

gen/ie, a spirit in Muslim myth.,
pl. **-ii.** *See also* **jinnee**

genit, *use* **jennet**

genitive, abbr. **gen.** *or* **genit.**

genius/, (person of) consum-
mate intellectual power; *pl.* **-es**
(**genii** is pl. of **genie**)

geni/us loci (Lat.), the pervad-
ing spirit of a place; *pl.* **-i loci**

Gennesaret (Sea of), *not* -eth

gennet, *use* **jennet**

Genoa, Italy; in Fr. **Gênes,** It.
Genova; adj. **Genoese,** *not*
-ovese

genocide, deliberate extermin-
ation of a race

genre (art), a painting of the
ordinary scenes of life (not ital.)

gens (Lat.), a clan; *pl.* **gentes**

Gens de Lettres (Société des),
French society of authors

Gensfleisch, *see* **Gutenberg**

Gent, Fl. and Ger. for **Ghent,**
Belgium

gent., gentle/man, -men

genteel/, fashionable, snobbishly
refined, adv. **-ly**

Gentele's green, a colour

gentil/ (Fr.), *fem.* **-le,** gentle,
kind

Gentile, anyone not a Jew
(cap.). *See also* **goy**

gentilhomme (Fr. m.), nobleman, gentleman; *pl.* **gentils-
hommes**

gentleman-at-arms (hyphens)

Gentleman's Magazine (*not*
-men's)

Gents (the), men's lavatory
(cap., no point)

genuflexion, a bending of the
knee, *not* -ction

genus, pl. **genera,** abbr. **gen.**
See also **botany, zoology**
Geo., George
geod., geode/sy (large-scale earth measurement), -tic
**Geoffroy Saint-Hilaire
(Étienne),** 1772–1844, French zoologist, *not* -frey (one hyphen only)
Geoffrey-Lloyd (Baron)
geog., geograph/er, -ical, -y
geographical qualifiers, forming everyday terms, usually have lower-case initials, as chinese white, indian ink, roman type. *But* Brussels sprouts, London pride. *See* **capitalization** and *Hart's Rules,* pp. 12–13
geol., geolog/ical, -ist, -y
geologize, *not* -ise
geology, names of formations to have caps., as Old Red Sandstone
geom., geome/ter, -trical, -try
Geordie, native of Tyneside
Georg, Ger. for George
George, abbr. **Geo.**
George-Brown (Baron)
Georgetown, Guyana
George Town, Cayman Is., Malaysia (two words)
Georgia, US; off. abbr. **Ga.** or (postal) **GA**
Georgia, a Soviet Socialist Republic. *See also* **USSR**
Georgian, of Georgia or the Georges
Georgium sidus (astr.), old name for Uranus
Ger., German, Germany
ger., gerund, -ial
geranium/ (bot.), pl. **-s** (not ital.)
Gerard (John), 1545–1612, English botanist, *not* -arde
gerbil, desert rodent, *not* jer-
gerfalcon, *not* gyr-, jer-
Géricault (Jean Louis André Théodore), 1791–1824, French painter
gerkin, *use* ghe-
German, abbr. **Ger.** (typ.), now usually set in roman (*Antiqua*)

type; *see* **accents.** The *Fraktur* alphabet (listed in *Hart's Rules,* p. 103) has no small caps. or italic, emphasis being shown by letter-spacing. For details of German printing-practice (note esp. the rules on capitalization, word-division, and ligatures) see *Hart's Rules,* pp. 102–10
german (cousin)
germane to, relevant
germanium, symbol **Ge**
Germanize, *not* -ise
Germany, abbr. **Ger.**
Gérôme (Jean Léon), 1824–1904, French painter
gerrymander, to manipulate unfairly, *not* je-
gerund/, -ial, abbr. **ger.**
Ges. (Ger.), *Gesellschaft* (a company or society) (cap.)
gesammelte Werke (Ger. pl.), collected works (one cap.)
Gesellschaft (Ger. f.), a company or society (cap.), abbr. ***Ges.***
gesso/, gypsum used in art, pl.. **-es**
gest. (Ger.), *gestorben,* deceased
Gestalt/ (Ger. f.), a shape, the whole as more than the sum of its parts; pl. ***-en;*** also used as adj. to describe philosophy or psychology based on this
Gestapo, German secret police (under Hitler)
Gesta Romanorum (Lat. pl.), medieval collection of anecdotes
gesticulator, one who moves his hands and arms in talking, *not* -er
gestorben (Ger.), deceased; abbr. ***gest.,*** sign †
gesundheit! (Ger.), your health (said also to one who sneezes)
get-at-able, accessible (hyphens)
gettable (two *t*s)
Gettysburg, Pennsylvania, scene of battle and of Lincoln's address, 1863
Geulinex (Arnold), 1625–69, Dutch philosopher

GeV, giga-electron-volt

Gewandhaus (Ger. n.), Cloth-workers' Hall, concert-hall in Leipzig

gewgaw, gaudy plaything (one word)

geyser, hot spring, water-heat⌃r (not ital.). *See also* **geezer**

GFS, Girls' Friendly Society

GG, Girl Guides, Governor-General, Grenadier Guards

Gg, gigagram

g. gr., a great gross, or 144 dozen

GGSM, Graduate of Guildhall School of Music

Ghadames, Libya

Ghana, *formerly* Gold Coast Colony, indep. 1957, republic 1960; adj. and noun **Ghanaian**

Ghandi, *use* **Gandhi**

gharial, *use* **gavial**

gharry, a vehicle used in India, *not* gari

ghat (Anglo-Ind.), mountain pass, steps to a river, *not* ghât, ghát, ghaut

Ghazi, Muslim fighter against non-Muslims

ghee, Ind. butter, *not* ghi

Gheel, Belgian commune long celebrated for its treatment of the mentally ill

Ghent, Belgium, in Fr. **Gand**, Fl. and Ger. **Gent**

gherao, Ind. & Pak. lock-in of employers

gherkin, a small cucumber, *not* ge-, gi, gu-

ghetto/, (hist.) Jews' quarter, slum area; *pl.* **-s** (not ital.)

ghi, *use* **ghee**

ghiaour, *use* **gi-**

Ghibelline, one of the Emperor's faction in medieval Italian states, opp. to **Guelph**; *not* -in, Gib-, Guib-

Ghiberti (Lorenzo), 1378–1455, Italian sculptor, painter, goldsmith

Ghirlandaio (Domenico), 1449–94, Italian painter

Ghizeh, *use* **Giza**

Ghonds, *use* **Gonds**

Ghoorkas, *use* **Gurkhas**

ghoul, an evil spirit, *not* -ool, -oule, -owl

GHQ, General Headquarters

Ghurkas, *use* **Gurkhas**

GI (US), Government Issue, (colloq.) serviceman

Giacometti (Alberto), 1901–66, Swiss sculptor, painter, poet

giallo antico (It.), a rich-yellow marble

Giant's Causeway, Co. Antrim, *not* -ts'

giaour, Turk. name for non-Muslim, *not* ghiaour, giaur

Giaour (The), poem by Byron, 1813

Gib., Gibraltar

gibber, to chatter, *not* j-

gibbet/, -ed, -ing

Gibbon (Edward), 1737–94, English historian

Gibbons (Grinling), 1648–1720, English carver); — (**Orlando**), 1583–1625, English composer

gibbo/us, hump-backed, doubly convex like the moon between half and full; **-sity**

gib/e, to sneer; **-er, -ing.** *See also* **gybe**

Gibeline, *use* **Ghibelline**

Gibraltar/, abbr. **Gib.**; adj. **-ian**

gibus, opera-hat (not ital.)

Gide (André Paul Guillaume), 1869–1951, French writer

Gideon, member of body distributing bibles

Gielgud (Sir Arthur John), b. 1904, English actor and producer

Giessen, university city, W. Germany

giga-, prefix meaning one thousand million (10^9), abbr. **G**

gigolo/, *pl.* **-s** (not ital.)

gigot/, (cook.) leg of mutton; — **sleeve** (two words); *not* j-

gigue, a lively dance

Gilbert (Sir William Schwenck), 1836–1911, English librettist

Gil Blas, picaresque satire by Le Sage, 1715

gild, an association, *use* **guild**

Gilead, mountainous district east of R. Jordan

Gill (Arthur Eric Rowton), 1882–1940, artist, sculptor, type-designer and typographer; designed Perpetua and Gill Sans founts

Gillette (King Camp), 1855–1932, US inventor of safety razor; — **(William),** 1857–1937, US actor and writer

Gillray (James), 1757–1815, English caricaturist

gillyflower, clove-scented flower, *not* jilli-

gilt, abbr. **gt.**

gimbal, *use* **gemel**

gimcrack, trumpery, *not* jim-

gimlet, a tool, *not* gimb-

gimm/al, -er, *use* **gemel**

gimmick, a contrivance, device

gimp, a trimming, a fishing-line, *not* gui-, gy-

Ginevra, It. for **Geneva**

ginger/ ale, — beer (two words)

ginglym/us (anat.), a hingelike joint such as the elbow, *pl.* **-i;** adj. **-oid**

ginkgo, oriental tree, *not* gingko

ginn, *see* **jinnee**

Gioconda (La), see Mona Lisa; opera by Ponchielli, 1876; play by d'Annunzio, 1898

Giorgione (Giorgio Barbar- elli da Castelfranco), 1475–1510, Italian painter

Giotto di Bondone, 1266–1337, Italian painter and archi- tect; adj. **Giottesque**

gipsy, *use* **gyp-**

girandole, a firework (not ital.)

girasol, a fire-opal, *not* -ole

Giraudoux (Jean), 1882–1944, French novelist and playwright

girkin, *use* **ghe-**

Girl Guide, *now* **Guide**

Giro, a Post Office system of money transfer, *not* Gy-

Girond/e, dép. SW France; **-ist,**

French moderate republican, 1791–3

Giscard d'Estaing (Valéry), b. 1926, Pres. of France 1974–81

gitan/o (Sp.), *fem.* **-a,** a gypsy

gite, a stopping-place; in Fr. m. *gîte*

Giulia, Giulio, Giuseppe, It. names, *not* Gui-

Giulini (Carlo Maria), b. 1914, Italian conductor

Giulio Romano, *c.* 1499–1546, Italian architect

giuoco piano (It.), quiet play (a chess opening)

giveable, *not* givable

Giza, site of pyramids, Egypt, *no* El-Ghiz-, -eh

Gk., Greek

glace (Fr. cook. f.), ice

glacé (Fr.), glazed

gladiol/us (bot.), *pl.* **-i** (not ital.)

Gladstone/ bag, — cap (two words)

Gladstonian, *not* -ean

Glagolitic, pertaining to former Slavonic alphabet

Glam., Glamorgan

glamorize, *not* -ise

glam/our, *but* **-orous**

Glas., Glasgow

glaserian fissure (anat.), *not* glass-

Glasse (Mrs Hannah), wrote *The Art of Cookery* in 1747

glassful, *pl.* **glassfuls**

glasshouse, greenhouse, (slang) mil. prison

Glaswegian, of Glasgow

Glauber/ (Johann Rudolf), 1604–68, German chemist; —**'s salt,** a cathartic (apos.)

glaucoma, an eye-disease

glaucous, greenish; (bot.) covered with a bloom

Glazunov (Alexander Kon- stantinovich), 1865–1936, Russian composer

GLC, Greater London Council

Gleichschaltung (Ger. f.), standardization of institutions

Glen, written separately when

referring not to a settlement
but to the glen itself, as Glen
Almond, Glen Coe

Glenalmond, Tayside

Glencoe, Highland

Glendower (Owen), Welsh
chieftain d. 1416; in Welsh
Owain Glyndwr

Glenealy, Co. Wicklow

Gleneely, Co. Donegal

Glengarry, a Scotch cap

Glenlivet, a whisky, *not* -at, -it
(one word)

Glenrothes, Fife, 'new town',
1948

glissade, a slide down a steep
slope (not ital.)

glissando (mus.), slurred, in a
gliding manner

glissé, sliding step in dance

glockenspiel/, orchestral percussion instrument, *pl.* **-s**

Gloria/ (liturgy), *pl.* **-s**

Gloria/ in Excelsis, — Patri,
hymns

Gloria Tibi, glory to Thee
(cap. *T*)

Glos., Gloucestershire

gloss, a superficial lustre, a marginal note

gloss., glossary, a collection of
glosses

Gloucestershire, abbr. **Glos.**

Gloucestr:, sig. of Bp. of
Gloucester (colon)

**glove/ box, — compartment,
— puppet** (two words)

glower, to gaze angrily, *not* glour

glow-worm (hyphen)

gloxinia/ (bot.); *pl.* **-s** (not ital.)

**Gluck (Christoph Willibald
von),** 1714–87, German composer, *not* Glü-

glu/e, -ed, -ey, -ing

glue-pot (hyphen)

glut/en, nitrogenous part of
wheat-flour, *not* -in; *but* **-inize,
-inous**

glycerine, *not* -in

Glyndwr, *see* **Glendower**

GM, George Medal, Grand
Master

gm., gram, *use* **g**

gm², *properly* **g/m²,** grams per
square metre, metric method for
measuring weight of paper

G-man (US colloq.), special
(police) agent of the FBI

G.m.b.H., GmbH (Ger.), *Gesellschaft mit beschränkter Haftung*
(limited liability company)

GMC, General Medical Council

GMT, Greenwich Mean Time

GMWU, General and Municipal
Workers' Union

gn., guinea

GNC, General Nursing Council

gneiss (geol.), a laminated rock
(not ital.)

gnocchi (It. pl.), small dump-
lings

gnos/is, knowledge of spiritual
mysteries; *pl.* **-es; -tic,** having
spiritual knowledge; **Gnosti-
cism,** an eclectic philosophy of
the redemption of the spirit from
matter through knowledge

GNP, gross national product

Gnr. (mil.), gunner

gns., guineas

gnu/, antelope; *pl.* **-s**

GO, general order, great (*or*
grand) organ

go/ (noun), *pl.* **-es**

goal, the objective at games, etc.
See also **gaol**

goalkeeper (one word)

goatee, chin-tuft like a goat's
beard

goat/herd, -skin, -sucker
(bird) (one word)

goat's-beard, plant *or* fungus
(hyphen)

goaty, goatlike

gobbledegook, pompous jargon, *not* -dyg-

Gobbo (Launcelot), in Shakespeare's *The Merchant of Venice*;
— (Old), his father

Gobelin tapestry, *not* -ins

gobemouche/, credulous per-
son, *pl.* **-s;** in Fr. m. sing. and pl.
gobe-mouches

Gobi, desert in Mongolia and E.
Turkistan

goby, a fish (not ital.)

GOC, General Officer Commanding

go-cart, miniature racing-car, *use* **-kart**

God-awful (hyphen, cap.)

godchild (one word)

god-daughter (hyphen)

godfather (one word)

God-fearing (hyphen, cap.)

godhead, divine nature, *but* **the Godhead** (cap.)

godless (one word)

Godley (**Arthur**), *see* **Kilbracken**

godlike (one word)

god/mother, -parent (one word)

God's acre, a burial ground (apos., one cap.)

god/send, -sent (one word)

godson (one word)

Godspeed (one word, cap.)

Godthaab, Greenland

godwit, a marsh bird

Goebbels (**Joseph Paul**), 1897–1945, German minister of propaganda, 1933–45, *not* Gö-

Goehr (**Alexander**), b. 1932, German composer; — (**Walter**), 1903–60, his father, German conductor

Goering (**Hermann**), 1893–1946, German field marshal, *not* Gö-

Goeth/e (**Johann Wolfgang von**), 1749–1832, German poet and playwright, *not* Gö-, Gœ-; adj. **-ian**

goffer, to crimp, iron used, *not* gauffer; animal, wood, *use* **gopher**

Goffs Oak, Herts. (no apos.)

goi (Heb.), a Gentile, *use* **goy**

Goidel, a member of Gadhelic peoples, a Gael

go-kart, miniature racing-car, *not* -cart (hyphen)

Golconda, a rich source of wealth (from a ruined city near Hyderabad)

gold, symbol **Au**

gold blocking (bind.), *see* **blocking**

gold-dust (hyphen)

Golders Green, NW London (no apos.)

gold-field (hyphen)

golf ball (two words), *but* **golf -ball typewriter** (comp.), use in strike-on (q.v.) composition (hyphen)

golf-club, the implement (hyphen)

golf club, the premises or association (two words)

Goliath, Philistine giant, type crane, *not* -iah

Gollancz (**Sir Israel**), 1864–1930, English writer; — (**Sir Victor**), 1893–1967, British publisher and writer

golliwog, *not* golly-

golosh, *use* **galosh**

GOM, Grand Old Man (esp. W. E. Gladstone)

Gomorr/ah (OT); **-ha** (NT), *but* **-ah** in NEB

Goncharov (**Ivan Alexandrovich**), 1812–91, Russian novelist

Goncourt (**Edmond Louis Antoine**), 1822–96, and his brother — (**Jules Alfred Huot**), 1830–70, French novelists; **Prix** — (named after Edmond —)

gondola/, *pl.* **-s** (not ital.)

Gonds, a tribe of cent. India, *not* Gh-

Gongo, tributary of the Zaïre

Góngora y Argote (**Luis de**), 1561–1627, Spanish poet

gongorism, Spanish form of euphuism

gonorrhoea, *not* -œa

Gonville, *see* **Caius**

good-afternoon (salutation) (hyphen)

goodbye (one word)

good/-day, -evening (salutations) (hyphens)

Good Friday (caps., two words)

good humour (two words); **good-humoured** (hyphen)

good-morning (salutation) (hyphen)

ood nature (two words);
good-natured (hyphen)
ood-night (salutation) (hyphen)
5ood/ Samaritan, — Templar (caps.)
roodwill, of a business, etc. (one word)
5oorkhas, use **Gur-**
,ooroo, use **guru**
'oose, *pl.* **geese,** (in tailoring) **gooscs**
,oose/-flesh, -skin (hyphens)
;oose step (two words)
'oosey, dim. of goose, *not* -sie, -sy
5oossens, 19th–20th c. Anglo--Belgian family of musicians
5OP, Grand Old Party (Republican party in US)
'opher, a kind of wood, burrowing animal. *See also* **goffer**
'ordian knot, tied by Gordius, cut by Alexander the Great
5ordonstoun School, Elgin, Grampian
;ordonstown, Grampian
'orgio/, non-gypsy; *pl.* **-s**
'orgonzola, a cheese
5orky/ (Maxim), 1868–1936, pseud. of A. M. Peshkov, Russian writer; —, USSR, *formerly* **Nijni Novgorod**
;ormand, use **gour-**
;ormandize, to eat greedily, *not* gour-, -ise
;ors/e, furze; adj. **-y**
;orsedd, meeting of Welsh bards and druids
5oschen (Viscount)
Göschen (Georg Joachim), 1752–1828, German publisher, grandfather of the first Viscount Goschen (1831–1907)
Goshen, a land of plenty
go-slow (noun, hyphen)
Gospodin (Russ.), Lord, Mr; *fem.* **Gospoja.** (In modern Russian *tovarishch* (comrade) is off. use)
Goss (Sir John), 1800–80, English organist and composer
Gosse (Sir Edmund Wil-

liam), 1849–1928, English writer; — **(Philip Henry),** 1810–88, his father, English naturalist
gossip/, -ed, -er, -ing, -y (one *p*)
Göteborg, Sweden, in Ger. **Gothenburg**
Gotham (wise men of), i.e. fools; (US colloq.), New York City
Gothic, architecture, etc. (cap.); abbr. **Goth.**
Gothic (typ.), loose name for (esp. bold) sanserif faces; former name for Old English faces
Götterdämmerung (Twilight of the Gods), last part of Wagner's *Ring des Nibelungen,* 1876
Göttingen University, W. Germany
gouache, a method of water--colour painting
Gouda, a Dutch cheese
gouge, a concave chisel
goujon (Fr. m.), gudgeon fish
gouk, use **gowk**
gour, Indian ox, use **gaur**
gourmand/, a glutton; **-ise,** indulgence in gluttony. *See also* **gormandize**
gourmet, an epicure
goût (Fr. m.), taste
Goutte d'or, a white burgundy wine
gov., govern/or, -ment
Government, meaning the State (cap.); abbr. **Govt.**
Governor-General, abbr. **Gov.-Gen.** *or* **GG**
Gowers (Sir Ernest), 1880–1966, English civil servant, author of *Plain Words,* reviser of Fowler's *Modern English Usage*
gowk, a fool, *not* gouk
goy/, Jewish name for Gentile; *pl.* **-im**
Goya y Lucientes (Francisco José de), 1746–1828, Spanish painter
GP, general practitioner, Graduate in Pharmacy, Grand Prix, Gloria Patri (glory be to the Father)

Gp., Group

g.p., great primer

Gp/Capt (RAF), Group Captain (no points)

GPDST, Girls' Public Day School Trust

GPO, General Post Office

GR, *Georgius,* or *Gulielmus, Rex* (King George, *or* William)

Gr., Grand, Greater, Grecian, Greece, Greek

gr., grain, -s; *for* gram *use* **g.**

Graafian follicle, in ovary

Graal, *use* **Grail**

Gracchus (Gaius Semproni-us), 153–121 BC, and his brother — **(Tiberius Sempronius),** 163–133 BC, the Gracchi, Roman reformers

grace-note (mus.) (hyphen)

Gracián (Baltasar), 1601–58, Spanish moralist

gradatim (Lat.), step by step

gradus ad Parnassum (Lat.), step(s) to Parnassus; Lat. or Gr. poetical dictionary; any series of graded exercises; in short **gradus** (not ital.)

Graeae (Gr. myth.), three sisters who guarded the abode of the Gorgons

Graec/ism, a Greek characteristic; **-ize, -ophil**

Graf (Ger.), a count; *fem.* **Gräfin** (cap.)

graffit/o, scribbling, usu. on wall; *pl.* **-i.** See also **sgraffito**

Grahame (Kenneth), 1859–1932, English author

Graian Alps, France and Italy

Grail (the Holy), in medieval legend, the platter used by Christ at the Last Supper, and by Joseph of Arimathea to catch Christ's blood, *not* Graal, Graile

grain, apothecaries', avoirdupois, or troy weight, all the same, being 0.0648 gram; abbr. **gr.**

Grainger (Percy Aldridge), 1882–1961, Australian-born US composer

gram, unit of mass; abbr. **g.** (*but*

no point in scientific and technical work); in SI units 0.001 k

gram., gramm/ar, -arian, -atica

gramaphone, *use* **gramo-**

graminivorous, feeding on grass, *not* gramen-

gramm/ar, -arian, -atical, abbr. **gram.**

grammar school (two words

gramme, *use* **gram**

Grammont, E. Flanders, Belgium

Gramont (Philibert, comte de), 1621–1707, French courtier, adventurer, and soldier, *nc* Gramm-

gramophone, *not* grama-, grammo-

Gram's stain (bacteriol.) (cap.

Granada, Spain

granadilla, one of the passion-flowers, *not* gren- (not ital.)

Gran Chaco (El), region in Bolivia, Paraguay, and Argentina, S. America

Grand, abbr. **G.** *or* **Gr.**

grandad, *not* grand-dad

grandam, grandmother, old woman, *not* -dame

grand-aunt (hyphen)

Grand Canyon, Arizona, US

grandchild (one word)

Grand Coulee Dam, Washington, US

grand-dad, *use* **grandad**

granddaughter (one word)

Grand/ Duchess, — Duchy, — Duke (two words, both caps if used as title); abbr. **GD**

grande/ dame (Fr.), dignified lady; **— passion,** violent love-affair; **— tenue,** *or* **— toilette,** full dress

grandeur naturelle (Fr. f.), life-size

grandfather (one word)

Grand Guignol, horrific drama (not ital.)

grand jury (not caps.)

grand mal (Fr. m.), serious form of epilepsy (ital.)

grandmaster, chess-player of highest class

Grand Master (caps.), abbr. **GM**

grand' messe (Fr. f.), high mass

Grand Monarque (*le*), Louis XIV

grand monde (*le*) (Fr.), the Court and nobility

grandmother (one word)

grand/-nephew, -niece (hyphen)

Grand Old Party (US), the Republican party; abbr. **GOP**

grandparent (one word)

Grand Prix (Fr.), international motor-racing event; *pl.* **Grands —** (not ital.); abbr. **GP**

Grand Rapids, town, Michigan, US

grand signior, one of high rank; caps. the Sultan of Turkey; in Fr. *grand seigneur*; It. *gran signore*; Sp. *gran señor*

grandson (one word), abbr. **g.s.**

grand-uncle (hyphen)

grangerize, to illustrate a book with cuttings etc. from other sources, to produce a book with blank leaves for these; *not* -ise (not cap.)

granny, *not* -ie

Grant Duff (Sir Mountstuart Elphinstone), 1829–1906, Scottish politician (no hyphen)

granter, one who grants

Granth, the Sikh scriptures, *not* Grunth

grantor (law), one who makes a grant

gran turismo (It.), touring-car

Granville-Barker (Harley), 1877–1946, English playwright, producer, and actor

grapefruit, *pl. same*

graphology, study of character from handwriting, *not* graphio-

gras (Fr.), *fem.* **grasse,** fat. *See also gros*

Grasmere, Cumbria

grass/ (typ.), casual work; **— hand,** one casually employed

Grass (Günter Wilhelm), b. 1927, German novelist

Grasse, dép. Alpes-Maritimes, France

grass/hopper, -land (one word)

gratia Dei (Lat.), by the grace of God

gratin, see au gratin

gratis, for nothing, free (not ital.)

Grattan (Henry), 1746–1820, Irish statesman and orator

Gratz, Austria, *use* **Graz**

Grätz, Czechoslovakia, *use* **Hradec**

Grätz, Poland, *use* **Grodzisk**

Graubünden, Ger. for **Grisons**

grauwacke, *use* **grey-**

gravamen/, chief ground of complaint; *pl.* **-s**

grave (mus.), slow, solemn

grave accent (`)

gravel/, cover with gravel, puzzle; **-led, -ling**

Graves, a Bordeaux white wine

Graves' disease, exophthalmic goitre (apos.)

grave/stone, -yard (one word)

gravitas (Lat.), solemn demeanour

gravure (typ.), intaglio printing process (from **photogravure**)

gray, colour, *use* **grey**

gray, unit of absorbed radiation dose; abbr. **Gy** (no point); in SI units 1 J/kg

Gray (Asa), 1810–88, US botanist; **— (Louis Harold),** 1905–65, English radiobiologist; **— (Thomas),** 1716–71, English poet. *See also* **Grey**

grayling, a fish, *not* grey-

Gray's Inn, London

graywacke, *use* **grey-**

Graz, cap. of Styria, Austria, *not* Gratz

grazier, one who pastures cattle, *not* -zer

GRCM, Graduate of the Royal College of Music

greasy, *not* -ey

Great Britain, abbr. **GB**

Greater Manchester, metropolitan county

great gross, 144 dozen; abbr. **g.gr.**

Great Power (caps. with hist. reference)

great primer (typ.), name for a former size of type, about 18 pt.; *pron.* primmer; abbr. **g.p.**

Greats, Oxford BA final examination for honours in Lit. Hum.

Grecian, almost entirely superseded by **Greek** except as adj. with ref. to archit. and facial outline, and as noun for a boy in the top form at Christ's Hospital, or a Greek-speaking Jew of the Dispersion; *also* — **bend,** — **knot,** — **slippers**; abbr. **Gr.**

Grec/ism, -ize, -ophil, *use* **Grae-**

Greco (El), 1541–1614, Spanish painter, born in Crete, real name **Domenico Theotocopuli** *or* **Kyriakos Theotokopoulos**

Greece, abbr. **Gr.**

greegree, Afr. fetish, *not* gri-gri

Greek (typ.): for details of alphabet, word-division, accentuation, numerals, etc. *see Hart's Rules*, pp. 110–15. *See also* **accents.** With a few exceptions for early texts, the same sorts are used in classical and modern work, but modern initial ρ does not take a breathing. In England a sloping fount (esp. Porson, q.v.) is preferred for classical work; in Greece and the continent generally, upright founts predominate in all setting, and in some (not scholarly) work accents and breathings are replaced by superscript dots. Abbr. **Gr.** or **Gk.**

Greek calends (at the), never; in Lat. *ad kalendas Graecas*

Greeley (Horace), 1811–72, US journalist and politician

Greely (Adolphus Washington), 1844–1935, US general and Arctic explorer

Green (John Richard), 1837–83, English historian

Greene (Sir Conyngham), 1854–1934, British diplomatist; — (Graham), b. 1904, English novelist; — (Plunket), 1865–1936, English singer; — (Robert), 1560–92, English dramatist and pamphleteer

green/gage, -grocer, -house (one word)

Greenland/, -er, -ic

green-room, room off-stage for actors (hyphen)

greensand, a sandstone; **Lower** and **Upper Greensand,** two strata of the Cretaceous system

Greenwich Mean Time (caps.), abbr. **GMT**

greetings card, *not* greeting — (two words)

gregale, the Mediterranean NE wind, *not* -cale, grigale (not cap.)

grège, colour between beige and grey, *not* grei-

gregory-powder, an aperient

Greifswald, E. German university town

Grenada, W. Indies, indep. 1974

grenadilla, *use* **gran-**

grenadine, a thin silk fabric; (cook.) dish of veal or poultry fillets; a syrup made from currants or pomegranates, in Fr. m *grenadin*

Grenadines, chain of islands between Grenada and St. Vincent, W. Indies

Gresham/'s Law, 'bad money drives out good', from **Sir Thomas** —, 1519–79, English financier

Grétry (André Erneste Modeste), 1741–1813, French composer

Greuze (Jean Baptiste), 1725–1805, French painter

Grévy (François Paul Jules), 1807–91, French President, 1879–87

grey, colour, *not* gray

Grey (Lady Jane), 1537–54, proclaimed Queen of Eng., 1553, deposed, beheaded. *See also* **Fallodon; Gray**

greybeard, old man, large jug, clematis (one word)

Greyfriars College, Oxford

grey-hen, *fem.* of blackcock (hyphen)

greyhound (one word)

greyling, *use* **gra-**

grey matter (two words)

greywacke (geol.), a sedimentary rock, *not* grau-, gray- (not ital.)

Grieg (Edvard Hagerup), *not* Edward, 1843–1907, Norwegian composer

Grieve, *see* **MacDiarmid**

griffin, a fabulous creature, *not* -on, gryphon

griffon, vulture, breed of dog

grigale, *use* **gre-**

gri-gri, *use* **greegree**

grill, to cook under grill, etc.

grille, a grating (not ital.)

grill/é, *fem.* **-ée** (Fr. cook.), broiled

grimalkin, a cat, spiteful old woman

Grimm (Jakob), 1785–1863, and his brother — **(Wilhelm),** 1786–1859, German philologists, collectors of fairy-tales; **Grimm's Law,** deals with consonantal changes in Germanic languages

Grimond (Joseph), b. 1913, British Liberal politician

Grimsetter Airport, Orkney

grimy, begrimed, *not* -ey

Grindelwald, Switz., *not* Grindle-

grindstone (one word)

gringo/ (Sp. Amer.), a foreigner, esp. English-speaking (derog.); *pl.* **-s** (not ital.)

grippe (la), Fr. for influenza

Griqua/, half-breed of Cape Dutch and Hottentot parents; **-land,** S. Africa

grisaille, a method of decorative painting (not ital.)

grisette (Fr. f.), a working girl

grisly, terrible, *not* grizz-

Gris-Nez (Cap), dép. Pas-de--Calais, France (hyphen)

Grisons, Swiss canton, in Ger. **Graubünden**

grissini (It. pl.), long sticks of crispbread

gristly, having or like gristle, *not* -ey

Grizel, a proverbial meek wife

grizzly/, grey-haired; — **bear** (two words)

gro., gross

Grodzisk, Poland, *not* Grätz

groin, the fold between belly and thigh; (archit.) line of intersection of two vaults. *See also* **groyne**

gros/ (Fr.), *fem.* **-se,** big. *See also* **gras**

grosbeak, the hawfinch

groschen (m.), Austrian and old German small coin; *pl.* same

grosgrain, corded silk fabric

gros point (Fr. m.), cross-stitch embroidery on canvas

gross (144) is *sing.* and *pl.*; abbr. **gro.**

Grosseteste (Robert), 1175–1253, Bp. of Lincoln

grosso modo (It.), approximately

Grote (George), 1794–1871, English historian

grotesque (typ.), nineteenth century sanserif typeface; abbr. **grot.**

grotesquerie, *not* -ery (not ital.)

Grotius (Hugo), 1583–1645, Dutch jurist and statesman (Latinized form of De Groot)

grotto/, *pl.* **-es**

ground/floor, — level (two words, hyphen as adj.)

ground/-plan, -rent (hyphens)

ground/sheet, -work (one word)

grovel/, -led, -ler, -ling

Grove's Dictionary of Music and Musicians, now The New Grove Dictionary of Music and Musicians

groyn/e, a breakwater; **-ing.** *See also* **groin**

GRSM, Graduate of the Royal Schools of Music (the Royal Academy and the Royal College)

GRT, gross registered tonnage

Grub Street, London, former haunt of literary hacks, *later* Milton Street, *now* demolished

grummet (naut.), a rope ring, *not* gro-

Grundy (**Mrs**), in Thos. Morton's *Speed the Plough*, 1798, type of conventional propriety

Grunth, the Sikh scriptures, *use* **Granth**

Gruyère (Fr. m.), a French Swiss cheese

Gruyères, Switzerland

Gryphius (**Sébastien**), 1491–1556, German-born French printer, *not* Gryptinus

gryphon, *use* **griffin**

grysbok (Afrik.), an antelope, *not* -buck

GS, General/ Secretary, — Service (mil.), — Staff; Grammar School

g.s., grandson

GSM, Guildhall School of Music

g.s.m., grams per square metre. *See* **gm²**

GSO, General Staff Officer

GSP (naval), Good Service Pension

G/-string, -suit (hyphens)

GT, Good Templar; Gran Turismo

gt., gilt, great, gutta

g.t. (bind.), gilt top

gtt., guttae

gu., gules

guacho, *use* **gaucho**

Guadalupe/, Spain; — **Hidalgo,** Mexico; — **Mts.,** Texas and New Mexico

Guadeloupe, W. Indies

guage, *use* **gauge**

guaiacum (bot.), trop. American tree; resin obtained from it

Guaira (**La**), Venezuela, *not* Guayra

Guam, Marianas Islands, *not* Guaham, Guajam

guana, a lizard, shortened from **iguana**

guano, fertilizer from sea-birds' excrement

guarant/ee, use in all senses of noun and verb, *not* -y; **-or,** one who gives a guarantee

guaranty, undertaking of liability, *use* **guarantee**

guards (bind.), strips of paper or linen to which the inner margins of odd leaves of plates etc. are pasted before sewing

guard-ship (hyphen)

Guareschi (**Giovanni**), 1908–68, Italian novelist

Guarneri, violin makers, *not* -nieri, -nerius

Guatemala, Cent. America

Guayaquil, Ecuador

Guayra, *use* **La Guaira**

guazzo (It.), same as **gouache**

Guedalla (**Philip**), 1889–1944 British writer

Guelderland, *use* **Geld-**

guelder rose, *not* ge-

Guelph, one of the Pope's faction in medieval Italian states, opp. to **Ghibelline,** *not* Guelf

Guenevere (*Defence of*), by W. Morris, 1858. *See also* **Guinevere**

Guernica, Spain; subject of painting by Picasso, 1937

Guernsey, a Channel Island, in Fr. **Guernesey**

guernsey, a heavy knitted woollen jersey

guerre (Fr. f.), war

guerrilla/, a person or body engaged in — **warfare,** irregular fighting by small bodies, *no* -eri-

guesswork (one word)

Gueux, league of Dutch nobles, 1565–6

Guevara de la Serna (**Ernesto, 'Che'**), 1928–69, Cuban revolutionary leader

Guglielmo, It. for **William**

Gui (Vittorio), 1885-1975, Italian conductor and composer

Guiana (British), *now* **Guyana; — (Dutch),** *now* **Surinam**

guichet, ticket-office window

Guide, member of girls' organization, *formerly* **Girl Guide** (cap.)

guidebook (one word)

Guido, It. for **Guy**

Guilbert (Yvette), 1869-1944, French *diseuse*

guild, an association, *not* gild

guilder, Anglicized form of **gulden**

Guildford, Surrey, *not* Guilf-

Guildford:, sig. of Bp. of Guildford (colon)

guild-hall, *not* gi- (hyphen); *but* **Guildhall,** London

Guilford (Earl of)

Guili/a, -o, It. names, *use* **Giu-**

guillemets (typ.), Fr. etc. quotation marks « »; see *Hart's Rules*

guillemot, a sea-bird

guilloche, archit. ornament (not ital.)

guillotine, a beheading apparatus, a paper-cutting machine, a device for terminating parliamentary debates

guimp, *use* **gimp**

Guinea/, *formerly* part of Fr. W. Africa, indep. 1958, adj. **-n;** **— -Bissau,** *formerly* Portuguese Guinea, indep. 1974; — **(New),** E. Indies

guinea/, 21/-, £1·05, *pl.* **-s**; abbr. **g., gn., gns.**

guinea-pig, a small S. American rodent; the subject of an experiment (hyphen)

Guinevere, wife of King Arthur. *See also* **Guenevere**

Guiranwala, Pakistan

Guisborough, Cleveland

Guiscard (Robert), 11th c., Norman leader

Guiseppe, It. name, *use* **Giu-**

Gujarat, area in India, *not* Guze-

Gujrat, town in Pakistan

gulden, Dutch silver florin; *also* med. German coin and Austrian unit of account

gules (her.), red, abbr. **gu.**

Gulf/, cap. when with name, as Gulf of Corinth, Persian Gulf; abbr. **G.; — Stream** (no hyphen)

Gulielmus, Lat. for **William**

gullible, easily cheated, *not* -able

Gulliver's Travels, by Swift, 1726

gully, a channel, *not* -ey

gum arabic (two words)

gumbo/, kind of soup, *pl.* **-s**

gum/boil, -boot (one word)

gum-shield (hyphen)

gum-tree (hyphen)

gun., gunnery

gun/boat, -fire (one word)

gun dog (two words)

Gungl (Josef), 1810-89, Hungarian composer

gun-lock (hyphen)

gunman, armed person (one word)

gunnel, *use* **gunwale**

gunnery (naval), abbr. **G.** *or* **gun.**

gunny, sacking

gun/point, -powder, -ship, -shot, -smith (one word)

Gunter's chain, a surveyor's, 66 ft. long

Gunther, character in the *Nibelungenlied* (no umlaut)

gunwale, upper edge of ship's side, *not* gunnel

Gurkhas, Nepalese soldiers in Indian or British service, *not* Ghoor-, Ghur-, Goor-

gurkin, *use* **gher-**

gurnard, a fish, *not* -net

guru, religious teacher, esp. Sikh, *not* gooroo

Gutenberg *or* **Gensfleisch (Johann),** *c.*1399-1468, German inventor (so generally assumed) of movable metal types

gutt/a, a drop, abbr. **gt.**; *pl.* **-ae,** abbr. **gtt.**

gutta-percha, hard rubber-like substance (hyphen, not ital.)

gutter (typ.), space between imposed pages of type allowing for two foredge margins; in a book, the inner space between two columns or pages

guttural, connected with the throat

Guyan/a, NE South America, *formerly* Brit. Guiana, indep. 1966; adj. **-ese**

Guyot (Yves), 1843–1928, French economist

Guy's Hospital, London (apos.)

Guzerat, India, *use* **Gujarat**

Gwalior, India

Gwent, Welsh county

GWR, Great Western Railway (prior to nationalization)

Gwyn (Nell), 1650–87, English actress, mistress of Charles II, *not* -nne

Gwynedd, Welsh county

Gwynn (Stephen), 1864–1950, English writer

Gy, gray (no point)

gybe (naut.), change course. *See also* **gibe**

gym/, colloq. for gymnasium (n point); — **-slip** (hyphen)

gymkhana, an athletic display a pony-club competition, *not* -kana (not ital.)

gymnasium/, *pl.* **-s** (not ital.); colloq. **gym**

gymnot/us, the electric eel; *pl.* **-i** (not ital.)

gymp, *use* **gimp**

gynaeceum (Gk. and Rom.), women's apartments in a house

gynoeci/um (bot.), female organ of flower, *pl.* **-a**

gynaecology, study of women's diseases; abbr. **gyn.**

gypsum, hydrated calcium sulphate

gypsy, *not* gipsy

gyrfalcon, *use* ger-

Gyro, *use* **Gi-**

gyro/, *pl.* **-s**

H

H, henry (unit of inductance), hydrogen, (pencils) hard (see also **HH, HHH**), (former film-censorship classification) horrific, (mus.) B natural in German system, the eighth in a series

H., harbour, hydrant

h, hour, -s (in scientific and technical work), (as prefix) hecto-

h., hardness, height, husband, (meteor.) hail

H. (Ger.), *Heft* (number, part)

h, Planck's constant, q.v.

h̷, Planck's constant divided by 2π

HA, Historical Association, Horse Artillery

Ha, hahnium (no point)

ha, hectare (no point in scientific and technical work)

h.a., *hoc anno* (in this year), *hujus anni* (this year's)

Haag (den), Dutch for **The Hague** (informal, cf. **'s-Gravenhage**)

Haakon, name of seven Norwegian kings

haar, a raw sea-mist (E. coast of England and Scotland)

Haarlem, Netherlands. *See also* **Harlem**

hab., habitat

Habakkuk (OT), abbr. **Hab.**

Habana, Sp. for **Havana**

habanera, a Cuban dance

habeas corpus, writ to produce a person before court; abbr. **hab. corp.**

Habeas Corpus Act (caps., not ital.)

habendum (law), part of deed defining estate or interest granted

Haberdashers' Aske's School, Elstree

habet (Lat.), he is hit. *See also* **hoc habet**

habile, skilful, able, *not* -ille (not ital.)

habitat, normal abode of animal or plant (not ital.); abbr. **hab.**

habitu/é (Fr.), *fem.* *-ée,* a frequenter (ital.)

Habsburg (House of), Austrian Imperial family, *not* Hap-

HAC, Honourable Artillery Company

Hachette et Cie, publishers, Paris and London

hachis (Fr. cook. m.), minced meat, hash

hachisch, *use* **hashish**

hachure, line used in map hill-shading (usu. in *pl.*)

hacienda (Sp.), large estate with mansion

Hackluyt, *use* **Hak-**

hackney/, to make trite; **-ed**

hack-saw (hyphen)

hac lege (Lat.), with this proviso

Hades (Gr. myth.), abode or god of the dead (cap.)

hadji, a Muslim who has made the pilgrimage (hadj) to Mecca (not ital.); **hajji** is strictly correct, but not in common use

hadn't, to be printed close up

Hadramaut, S. Arabia, *not* Hadhra-

hadst (no apostrophe)

haecceity (philos.), individual quality

Haeckel (Ernst Heinrich), 1834–1919, German philosopher, *not* Hä-

haema-, haemo-, prefix meaning blood

haemorrh/age, -oids

Haffner, serenade and symphony by Mozart

hafiz (Arab., Pers.), a Muslim who knows the Koran by heart

Hāfiz, d. 1388, Persian poet, real name **Shams ud-din Muhammad**

hafnium, symbol **Hf**

Hag., Haggai

hagberry, the bird-cherry, *not* hack-

Haggadah, legendary part of the Talmud, or part of Jewish Passover ritual, not Agadah, Hagada, -ah

Haggai (OT), abbr. **Hag.**

haggard, wild-looking, an untamed hawk

haggis, Sc. dish, *not* -ess, -ies

Hague (Cap de La), NW France

Hague (The), seat of government of Netherlands; in Dutch 's-Gravenhage or den Haag, in Fr. La Haye (caps.)

ha ha, laughter (two words)

ha-ha, a sunk fence, *not* aha, haw-haw (not ital.)

hahnium, symbol Ha

Haidarabad, *use* **Hyder-**

haik, an Arab garment, *not* -ck

haiku, Jap. verse-form; *pl.* same

hail, *see* **hale**

Haile Selassie, 1892–1975, Emperor of Ethiopia 1930–74

Haileybury College, Herts.

hail-fellow-well-met (adj.), friendly (three hyphens)

hail/stone, -storm (one word)

Hainault, London

Hainaut, province, Belgium

hair/bell, — -brain, *use* **hare-**

hair/breadth, -brush, -cut (one word)

hair-do/, coiffure (hyphen), *pl.* -s

hairdress/er, -ing (one word)

hair-line (hyphen)

hairlip, *use* **hare-**

hairpin (one word)

hair's breadth (two words)

hair shirt (two words)

hair-space (typ.), very thin space formerly used for letter-spacing (hyphen)

hair/-style, -stylist, -trigger (hyphens)

Haiti, W. Indies, *not* Hayti; *formerly* the island of Hispaniola, *now* its western part

hajji, *see* **hadji**

hakenkreuz (Ger. n.), the swastika

hakim, oriental medical man, oriental ruler, *not* -keem

Hakluyt (Richard), *c.*1553–1616, English historian and geographer; — **Society;** *not* Hack-

halal, meat prepared according to Muslim law

halberd/, combined spear and battleaxe, *not* -ert; **-ier**

Halcyone, *use* **Alcyone**

hale, to drag, *not* hail

Halevi (Judah), *c.*1085–1140, Spanish-Jewish philosopher and poet

Halévy (Élie), 1870–1937, French historian; — **(Jacques François Fromental Élie),** 1799–1862, his great-uncle, French composer; —**(Joseph)** 1827–1917, French traveller; — **(Ludovic),** 1834–1908, father of Élie, French playwright and novelist

half, *see* **fractions**

half/ a crown, — a dozen, — an hour, etc. (no hyphens), *but* **half/-crown, -dozen, -hour** (etc.)

half-and-half (hyphens)

half/-binding (bind.), when the spine and corners are bound in a different material from the sides (hyphen); — **-bound** (hyphen)

half/-breed, -caste, -hearted (hyphens)

half/ holiday, — moon, — pay (two words)

half/-mast, -past (hyphens)

halfpenny (one word)

halfpennyworth (one word), colloq. **ha'p'orth**

half/-price, -term, -time (hyphens)

half-title (typ.), the short title printed on the leaf before the full title. *Also called* **bastard title**

halftone (typ.), technique whereby the various tones are represented by varying sizes of minute dots (one word)

half/-way, -wit(ted), -year(ly) (hyphens)

Haliburton (Thomas Chandler), 1796–1865, Canadian--born British writer; pseud. Sam Slick

halibut, a fish, *not* hol-

halieutic, of fishing

hallabaloo, *use* **hulla-**

Hallé/ (Sir Charles), 1819–95, German-born Mancunian pianist and conductor, founded — **Orchestra** 1857

Halle an der Saale, E. Germany; abbr. **Halle a/S.**

hallelujah, *see* **alleluia; 'Hallelujah Chorus',** in Handel's *Messiah*

Halles (Les), Paris, the central market until 1968

Halley (Edmond, *not* -und), 1656–1742, English astronomer

halliard, *use* **halyard**

Halliwell-Phillipps (James Orchard), 1820–89, English Shakespearian scholar

hallmark, mark used at Goldsmiths' Hall and by Govt. assay officials for marking standard of gold and silver

hallo, *not* he-, hu-

Hallow/e'en, 31 Oct.; **-mas** (one word), 1 Nov., All Saints' Day

hall/ux, the great toe; *pl.* **-uces**

halm, stalk or stem, *use* **haulm**

halo/, ring of light round moon, head, etc.; *pl.* **-es**

Hals (Frans), 1580 (or 84)–1666, Dutch painter

halva, sweetmeat of sesame flour and honey, *not* -ah

halyard (naut.), a rope for raising sail, *not* halli-, hauly-

hamadryad/, a wood-nymph, a serpent, a baboon; *pl.* -s

Hambleden, Bucks.; — **(Viscount)**

Hambledon, Hants, Surrey

Hambleton, Lancs., Leics., N. Yorks.

Hambros Bank, Ltd. (no apos.)

Hamburg/, city in W. Germany, a fowl, a grape; **-er** (not cap.), cake of chopped steak usu. in a roll

'Hamelin (Pied Piper of)', by R. Browning, 1842

Hamitic, group of Afr. languages

hammam, Turkish bath, *not* hummum, -aum

Hammarskjöld (Dag Hjalmar Agne Carl), 1905–61, Sec.-Gen. of UN 1953–61

Hammergafferstein (Hans), pseud. of Henry W. Longfellow

Hammerklavier (Ger.), former name of pianoforte (opp. harpsichord); (quot. marks, not ital.) name given to a piano sonata by Beethoven

Hampden (John), 1594–1643, English patriot and statesman

Hampshire, abbr. **Hants**

Hampton Court, London

hamster, household pet

ham/string, past and partic. correctly **-stringed,** but **-strung** favoured by usage

Hamtramck, Mich., US

hamza, *see* **accents**

hand/bag, -bell, -bill, -book, -brake (one word)

Hand/buch (Ger. n.), manual (cap.), *pl.* *-bücher*

h. & c., hot and cold (water)

handcuffs (one word)

Handel (George Frideric), 1685–1759, German composer resident in England; in Ger. **Händel.** *See also* **Handl**

Handels/blatt (Ger. n.), trade journal (cap.), *pl.* *-blätter*

handful/, *not* -ll; *pl.* **-s**

handicap/, -per, -ping

handiwork, *not* handy-

H. & J. (comp.), hyphenation and justification

handkerchief/, *pl.* **-s,** abbr. **hdkf.**

Handl (Jacob), 1550–91, Austrian composer, *not* Hä-. *See also* **Handel**

hand/list, -made, -maid(en) (one word)

hand-out (noun, hyphen)

hand-picked (hyphen)

handrail (one word)

Handschrift (Ger. f.), MS; abbr. **Hs.** (cap.)

handsel/, earnest-money; **-ling,** *not* hans-

handset, combined mouthpiece and ear-piece (one word)

hand-setting (typ.), manual composition from previously--cast metal sorts assembled in a fount-case (hyphen)

hand/shake, -writing (one word)

handyman (one word)

handywork, *use* **handi-**

hangar/, a shed; **-age**

hangdog, shamefaced (one word)

hanged, past tense or partic., used of capital punishment; in other senses use **hung**

hanger, one who, or that which, hangs; *also* a sword, a wood

hanging/ paragraph *or* — **indent** (typ.), short paragraph or listed items (as in bibliographies) set with second and following lines indented under first line (as here)

hangover, after-effects of drinking (one word)

Hanover, in Ger. **Hannover**

Hans/ (Du., Ger.), Jack; — **Niemand,** 'Mr Nobody'

Hanse/, medieval German league; adj. **-atic**

hansel, *use* **hand-**

hansom, a two-wheeled cab

Hants, Hampshire (no point)

hapax legomenon/on (Gr.), a word found once only; *pl.* **-a**

haphazard (one word)

ha'p'orth (colloq.), a halfpennyworth

happi, loose Japanese coat

happy-go-lucky (hyphens)

happy hunting-ground (one hyphen)

Hapsburg, *use* **Hab-**

hara-kiri (Jap.), suicide, *not hari-kari,* hurry-curry

haram, *use* **-em**

harangu/e, to address like an orator; **-ed**

harass, *not* harr-

harbour (US **-or**), abbr. **H.**

hard (pencils), abbr. **H**

hard/-a-lee, -a-port, -a-starboard, -a-weather (two hyphens each)

hard/back (book), **-board** (one word)

hard-boiled egg (one hyphen)

hard copy (comp.), printed or typed record of keyboard input (two words). *See also* **soft copy**

Hardecanute, accepted variant of **Harthacnut,** 1019–42, king of Denmark and England

hard hit, severely affected (two words)

Hardie (James Keir), 1856–1915, British socialist

hardi/hood, -ness, *not* hardy-

Harding (Baron, of Petherton)

Hardinge (Viscount); — **(Baron, of Penshurst)**

hardline (adj.), unyielding (one word)

hardness (mineral.), abbr. **h.**

hards, coarse flax, *not* hur-

hard shoulder, on motorway (two words)

Hardt Mountains, Bavarian Palatinate. *See also* **Harz** —

Hardwicke (Earl of)

hardwood (one word), *but* **hard-wooded**

hard-working (hyphen)

harebell (bot.), *not* hair- (one word)

hare/-brain, -brained, *not* hair- (hyphens)

Harefoot, sobriquet of Harold I

harelip, *not* hair- (one word)

harem (Arab.), the women's

part of a house, its occupants, *not* -am, -eem, -im (not ital.)

harem-scarem, *use* **harum- -scarum**

Hargreaves (James), 1720– 78, English weaver, inventor of the spinning-jenny

haricot (Fr. cook. m.), a ragout usu. of mutton and beans; — *de mouton,* Irish stew; —*s blancs,* kidney beans; —*s d'Espagne,* scarlet runners; —*s verts,* French beans

haridan, *use* **harridan**

harier, *use* **harrier**

harijan, Indian Untouchable

hari-kari, use hara-kiri

harim, *use* **harem**

Haringey, Greater London borough, 1965, *not* -ay. *See also* **Harringay**

Harington (Sir John), 1561– 1612, English poet and pamphleteer. *See also* **Harri-**

hark, listen, *but* (arch.) **hearken**

harl, a fibre, *not* -le

Harlech, Gwynedd, *not* -ck

Harleian, of Harley

Harlem, New York. *See also* **Haarlem**

Harlesden, London

Harleston, Devon, Norfolk, Suffolk

Harlestone, Northants

Harlow (Jean), 1911–37, US film-actress

Harlow, Essex, 'new town', 1947

Harlowe (Clarissa), by Richardson, 1748

harmattan, W. African land- -wind

Harmen, *see* **Arminius**

harmonize, *not* ise

Harper's Bazaar, US magazine of fashion, founded 1867; also the British version, founded 1929, *now* **Harper's Queen**

Harpers Ferry, West Virginia, US (no apos.)

Harper's Magazine, US monthly magazine, founded 1850

harquebus/, a portable gun, *not* -ss, arquebus; **-ier**

Harraden (Beatrice), 1864– 1936, English writer

harrass, *use* **harass**

harridan, a haggard old woman, *not* hari-

harrier, one who harries, a hound for hunting hares, a hawk, a cross-country runner, *not* harier

Harringay, N. London. *See also* **Haringey**

Harrington (Earl of). *See also* **Hari-**

Harrogate, N. Yorks., *not* Harrow-

Harrovian, member of Harrow School

Hart (Horace), Printer to the University of Oxford 1883– 1915

hartal, closing of Indian shops as protest or gesture

Harte (Francis Bret), 1836– 1902, US novelist and story- -writer

hartebeest, S. Afr. antelope, *not* hartb-, -bees

Hartford, Conn., US

Harthacnut, more correct form of **Hardecanute**

Hartlepool, Cleveland

hartshorn, ammonia (one word)

Hart's Rules for Compositors and Readers, 38th edition, completely revised, 1978

Hartz Mountains, *use* **Harz** —. *See also* **Hardt**

harum-scarum, reckless (hyphen), *not* harem-scarem

Harun-al-Rashid, 763–809, a caliph of Baghdad, hero of the *Arabian Nights, not* -oun, -ar-, -sch- (two hyphens)

Harvard system of references, *see* **authorities**

Harvard University, abbr. **Harv.** or **HU,** at Cambridge, Mass., US

Harv/ey (William), 1578– 1657, English physician,

Harvey (*cont.*)
discovered blood circulation;
-eian, *not* -eyan
Harz Mountains, cent. Germany, *not* Hartz —. *See also*
Hardt
has-been/, person or thing that
is no longer of use, *pl.* **-s** (hyphen)
Hašek (Jaroslav), 1884–1923,
Czech satirist
Hashemites, Arab princely
family
hashish, hemp smoked or
chewed as drug, *not* hach-,
hasch-, -eesh, -isch
Hasid/, member of mystical Jewish sect; adj. **-ic,** *not*
Hass-, Ch-
Haslemere, Surrey. *See also*
Haz-
hasn't, to be printed close up
Hasse (Johann Adolph),
1699–1783, German composer
Hassler (Hans Leo), 1564–
1612, German composer
hatable, *not* -eable
hatband (one word)
hatch/back (car with door at
rear), **-way** (one word)
hatha yoga, system of exercises
in yoga
hat-pin (hyphen)
Hatshepsut, Queen of Egypt,
*c.*1500 BC
hatti/, short form of — *-humayun or* — *-sherif,* Turkish edict
made irrevocable by Sultan's
mark
hat trick (two words)
hauler, one who *or* that which
hauls
haulier, a man employed in
hauling something, esp. coal in
a mine; a firm or person engaged
in road transport
haulm (bot.), a stalk or stem, *not*
halm
haulyard, *use* hal-
Hauptmann (Gerhart), 1862–
1946, German poet and dramatist; — **(Moritz),** 1792–1868,
German composer

Hausa, of Cent. Sudan, *not* -ssa,
Housa
hausfrau, Germanic housewife
(ital., not cap.)
**Haussmann (Georges
Eugène, baron),** 1809–91,
Paris architect
haussmannize, to open out and
rebuild
haut/bois, -boy (mus.), *use* oboe
haute/ bourgeoisie (Fr. f.), upper middle class; — *couture,*
elegance inspired by the high-fashion houses; — *cuisine,*
high-class cooking; — *école,*
advanced horsemanship
Haute/-Garonne, dép. SW
France; -**Loire,** dép. cent.
France; -**Marne, -Saône,** déps.
E. France; -**Savoie,** dép. SE
France (hyphens)
Hautes-Alpes, dép. SE France
(hyphen)
Hautes-Pyrénées, dép. SW
France (hyphen)
hauteur, haughty demeanour
(not ital.)
Haute-Vienne, dép. cent.
France (hyphen). *See also*
Vienne
haute volée (Fr. f.), the upper
ten
haut-goût (Fr. m.), high flavour
haut monde (Fr. m.), fashionable society
Haut-Rhin, dép. E. France (hyphen)
Hauts-de-Seine, dép. Paris region, France (two hyphens)
Havana, Cuba, *not* -ah, -annah;
in Sp. **Habana**
Havas, a French news agency
haven't, to be printed close up
haver (Sc.), to talk nonsense (*not*
to vacillate)
Haverfordwest, Dyfed
Havergal (Frances Ridley),
1836–79, English hymn-writer
haversack, bag for carrying
food, *not* -sac
havoc/ (noun); as verb, **-ked,
-king**
Havre, France, *use* Le Havre

Hawai/i, one of the Hawaiian Islands, formerly Sandwich Islands; these islands as a group; state of US comprising most of them, off. abbr. (postal) **HI**; adj. **-ian**

haw-haw, a sunk fence, *use* **ha- -ha**

Haw-Haw (Lord), nickname of William Joyce, American-born German propagandist in Second World War, executed 1946

hawk's-bill, a turtle

hawk-eyed (hyphen)

hawse (naut.), part of ship's bows

hawthorn (one word)

Hawthorne (Julian), 1846– 1934, US writer; — **(Nathaniel),** 1804–64, his father, US novelist

hay, a country dance, *not* hey (but *Shepherd's Hey*, by P. A. Grainger)

Haydn (Johann Michael), 1737–1806, and his brother — **(Franz Joseph),** 1732–1809, Austrian composers; — **(Joseph),** d. 1856, compiled *Dictionary of Dates*

Haydon (Benjamin Robert), 1786–1846, English painter

hay fever (two words)

hay/field, -maker, -rick, -stack, -wire (one word)

Hayti, *use* **Haiti**

hazel/-hen, -nut (hyphens)

Hazlemere, Bucks. *See also* **Hazlitt (William),** 1778–1830, English essayist; — **(William Carew),** 1834–1913, English bibliographer

hazy, indistinct, *not* -ey

HB (pencils), hard and black

Hb, haemoglobin (no point)

HBM, Her, *or* His, Britannic Majesty('s)

H-bomb, hydrogen bomb (no point, hyphen)

HC, habitual criminal, Heralds' College, High Church, Holy Communion, Home Counties, (Fr.) *hors concours* (not competing), House of Commons, House of Correction

h.c., *honoris causa*

HCF *or* **h.c.f.** (math.), highest common factor; **HCF,** Honorary Chaplain to the Forces

hd., head

hdkf., handkerchief

hdqrs., headquarters

Hdt., Herodotus

HE, His Eminence, — *or* Her Excellency; high explosive

He, helium (no point)

head (typ.), the blank space at the top of a page. *See also* **margins**

headache (one word)

headachy, *not* -ey

head/band, band worn round head, (bind.), strengthening band of multicoloured silk etc. sewn or stuck to head (and sometimes tail) of back of book; **-board** (one word)

head-dress (hyphen)

headgear (one word)

headings (typ.), should be graded to show relative importance, and used consistently throughout

head/lamp, -land (one word)

Headless Cross, Hereford & Worc.

headlight (one word)

headlines *or* **running titles** (typ.), at head of pages; various combinations can be adopted, including book, part, section, or chapter title, subhead, or summary of page content, shortened if necessary to keep it to a single line; in bookwork chapter title generally most suitable on both recto and verso, dividing long titles across opening; pagination, q.v., may be included in outer corners.

headman, a chief (one word)

head/master, -mistress, as general term (one word)

Head/ Master, — Mistress, official title at certain schools (two words); abbr. **HM**

head-note, summary at head of chapter or page, (mus.) a tone produced in the head register (hyphen)

head-on (hyphen)

head/phone(s), -piece (one word)

headquarters, used as sing. of the place, pl. of the occupants (one word); abbr. **HQ** *or* **hdqrs.**

head-rest, support for the head (hyphen)

headroom (one word)

head-sail, one before the fore-mast (hyphen)

headscarf (one word)

head sea, waves from forward direction (two words)

headship, position of chief (one word)

headsman, executioner (one word)

headstock, bearings in machine (one word)

head/stone, -strong, -way (one word)

head wind (two words)

headword (typ.), emphasized word opening a paragraph or entry, etc. (as here) (one word)

head-work (hyphen)

heal, *see* **hele**

health/ centre, — certificate, — food (two words)

healthful, *not* -ull

health/ officer, — visitor (two words)

Heap, *see* **Heep**

hear, hear!, exclamation of agreement, *not* here, here

hearken, *see* **hark**

Hearn (Lafcadio), 1850–1904, US author

heart/ache, -beat (one word)

heart-break/, -er, -ing, heart--broken (hyphens)

heartburn (one word)

heart-disease (hyphen)

heart failure (two words)

heartfelt (one word)

hearth/rug, -stone (one word)

heart/-rending, -searching (hyphens)

heart's-ease, a pansy (apos., hyphen)

heart/sick, -sore (one word)

heart/-strings, -throb, -whole (hyphens)

heathenize, *not* -ise

heather mixture, fabric of mixed hues (two words)

heat-resistant (hyphen)

heave ho! (two words)

Heaven, cap. when equivalent to the Deity; l.c. when a place, as 'heaven is our home'

heaven/-born, -sent (hyphens

Heaviside layer, layer of the atmosphere that reflects radio waves, *not* Heavy-

heavy/-duty, -handed (hyphens)

heavy water (two words)

heavyweight (one word)

Heb., Hebrew(s), Epistle to the Hebrews (NT)

hebdomad/, a group of seven, a week, *not* -ade; adj. -**al**

Hebraize, to make Hebrew, *not* -ise, -aicize

Hebrew (typ.), 22 letters (all consonants) and many strokes and points (representing vowel sounds), etc. It is read from right to left: hence if any passage is divided, the right-hand words must go in the first line, and the left-hand in the second. Abbr. **Heb.**

Hebrews, abbr. **Heb.**

Hebridean, *not* -ian

Hecat/e, a Greek goddess; adj. -**aean**

hecatomb, sacrifice of many (literally 100) oxen

heckelphone (mus.), baritone oboe

Heckmondwike, W. Yorks.

Hecla, mt., Western Isles, *not* Hek-. *See also* **Hekla**

hectare, 10,000 sq. metres; abbr. **ha** (no point in scientific and technical work)

hecto-, prefix meaning 100;

abbr. **h**; **hectogram** (**hg**), 100
grams; **hectolitre** (**hl**), 100
litres; **hectometre** (**hm**), 100
metres (abbrs. no point in scientific and technical work)

Iedda Gabler, play by Ibsen,
1890

hedgehog/, adj. **-gy**

hedgerow (one word)

hedgrah, use **hegira**

hee-haw, a, or to, bray, not he-
eel, see **hele**

Ieep (**Uriah**), in Dickens's
David Copperfield, not Heap

Heer (Ger. n.), army (cap.)

Ieft (Ger. n.), number, part,
abbr. *H.*

Iegel/ (**Georg**, not -e, **Wilhelm
Friedrich**), 1770–1831, German philosopher; adj. **-ian**

hegemon/**y**, leadership, esp. political; adj. **-ic**

hegira, the flight of Muhammad
from Mecca to Medina, 16 July
AD 622, from which the Muslim
era is reckoned (Arab. *hijrah*);
not *hedgrah*, heijira (not ital.).
The calendar follows a 355-day
lunar cycle, and it is difficult to
equate AD and AH dates

Heidegger (**Martin**), 1889–
1976, German philosopher

Heidelberg, W. German university town

Heidsieck, a champagne

Heifetz (**Jascha**), b. 1901, Russian-born US violinist

heighday, use **hey-**

heigh-ho, an audible sigh, not
hey-

height to paper (typ.), overall
height of type, usu. 2.33 cm.

heijira, use **hegira**

heil (Ger.), hail!

Heiland (*der*) (Ger.), the Saviour (cap.)

heilig (Ger.), holy; abbr. *hl.*

Heilige Schrift (Ger. f.), Holy
Scripture (caps.); abbr. *Hl.S.*

Heimskringla, 'the round
world', a history of Norse kings
by Snorri Sturluson, 1178–1241

Heimweh (Ger. n.), homesickness

Heine/ (**Heinrich**, *but signed*
Henri), 1797–1856, German
poet; adj. **-sque**

Heinemann (**William**), **Ltd.**,
publishers

Heinrich, Ger. for **Henry**

Heinz, Ger. for **Harry**

heir apparent, legal heir,
whoever may subsequently be
born (two words); abbr. **heir
app.**

heirloom (one word)

heir presumptive, legal heir if
no nearer relative should be
born (two words); abbr. **heir
pres.**

Heisenberg (**Werner Karl**),
b. 1901, German physicist, associated with principle of uncertainty

Hejaz, 'the boundary', not Hi-.
See also **Saudi Arabia**

hejira, use **hegira**

Hekla, volcano, Iceland, not
Hec-. See also **Hecla**

Hel (Norse myth.), originally,
the abode of the dead; later, the
abode of the damned, opp. **Valhalla**; later still, the goddess of
the dead (*also* **Hela**)

HeLa, strain of human cells (one
word, two caps.)

Heldentenor (Ger. m.), 'hero-
-tenor', with robust operatic
voice

hele, to set (plants) in the ground,
not heal, heel

Helensburgh, Strathclyde, not
-borough

Helicon (Gr. myth.), mountain
range in Boeotia, home of the
Muses, site of fountains Aganippe and Hippocrene

Heliogabalus, use **Elag-**

heliogravure, a gravure process
(not ital.)

heliport, landing-place for helicopters (one word)

helium, symbol **He**

hel/**ix**, a spiral curve like the
thread of a screw; *pl.* **-ices**

hell (not cap.)

hell (Ger.), clear, bright

hell-bent (hyphen)

Helle (Gr. myth.), sister of Phrixus, fell from the golden ram into strait afterwards named **Hellespont** (the Dardanelles)

Hellen (Gr. myth.), son of Deucalion and Pyrrha, progenitor of the Greek people

Hellen/e, a Greek; *pl.* **-es**; *adj.* **-ic** (cap.)

Hellenist, one skilled in Greek; one who has adopted Greek ways, esp. a Jew of the Dispersion

Hellenistic, of the Greek language and culture of the period after Alexander the Great

Hellenize, to make Greek, *not* -ise

hell/-fire, -hole (hyphens)

hello, *use* **hallo**

helmet/, -ed (one *t*)

Helmholtz (**Hermann Ludwig Ferdinand von**), 1821–94, German scientist

helmsman, one who steers (one word)

Héloïse, 1101–64, and Abélard, *not* El-, -sa

helpmate, *not* -meet (one word)

Helsingør, Danish for **Elsinore**

Helsinki, Finland, in Swed. **Helsingfors**

helter-skelter (hyphen)

Helvellyn, mountain, Cumbria

Hely-Hutchinson, Irish name, fam. name of Earl of Donoughmore (hyphen); — - — (**Victor**), 1901–47, British composer

hema-, prefix, *use* **haema-**

Hemel Hempstead, Herts. (two words)

Hemingway (**Ernest Miller**), 1899–1961, US novelist

hemistich (prosody), half a line

hemo-, prefix, *use* **haemo-**

hem-stitch (hyphen)

hence/forth, -forward (one word)

hendiadys, one idea expressed by two words joined by a conjunction, as 'try and come'

henna/, oriental shrub, leaves used for dyeing nails and hair; **-ed**

henpeck/, -ed (one word)

Henri, Fr. for **Henry**

henr/y, international unit of inductance, *pl.* **-ies**; abbr. **H**

Henry, abbr. **Hy**

Henry (**O.**), pseud. of William Sydney Porter (1862–1910), US author

Henschel (**Sir George**), 1850–1934, German-born English musician

Henslow (**John Stevens**), 1796–1861, English botanist and geologist

Henslowe (**Philip**), d. 1616, English theatrical manager, wrote 'Diary'

Hephaestus, *not* Hephaistos, except in Gr. contexts

Hepplewhite, eighteenth-century style of furniture, from **George** —, d. 1786, *not* Heppel-

her., heraldry

her., *heres* (heir)

Herakles, *see* **Hercules**; *not* -cles

heraldry, abbr. **her.**

Heralds' College, abbr. **HC** (apos.); **College of Arms**

Hérault, dép. France

Herausgeber (Ger. m.), editor

herbaceous, *not* -ious

herbari/um, a collection of dried plants; *pl.* **-a** (not ital.)

Herbart (**Johann Friedrich**), 1776–1841, German philosopher and educationist

Hercegovina, Yugoslavia, *not* Herz-

Herculaneum, Roman town overwhelmed by Vesuvius, AD 79, *not* -ium

herculean, strong, difficult, not cap. unless referring to Hercules, as 'Herculean labours'

Hercules, *not* Herakles, except in Gr. contexts

erd-book (hyphen)

ere/about(s), -after, -by, -in, -of, -on, -out (one word)

Iereford:, sig. of Bp. of Hereford (colon)

Iereford & Worcester, county of England

ere, here!, use **hear, hear!**

ere/s (Lat.), heir; *pl.* **-des,** abbr. **her.**

ere/to, -tofore, -under, -upon, -with (one word)

Iergesheimer (Joseph), 1880–1954, US novelist

Ieriot-Watt University, Edinburgh, 1966

ieritrix, an heiress, *not* -tress, here-

Ier Majesty('s) (caps.), abbr. **HM**

iermeneutics, the science of interpretation, treated as sing.

iernia/, a rupture; *pl.* **-s**

iero/ (**a,** *not* an); *pl.* **-es;** adj. **-ic**

Ierodotus, *c.*484–*c.*420 BC, Greek historian, abbr. **Hdt.** or **Herod.**

ieroin, a drug, *not* -ine

ieroine, heroic woman, chief female character in story, etc.

ieroize, to make a hero of, *not* -ise

ierpes, a skin disease; **herpes zoster,** shingles

ierpetolog/y, study of reptiles; **-ist,** *not* er-

Ierr/ (Ger. m.), Mr, Sir, *pl.* **-en**

Ierr (der) (Ger.), the Lord (cap.)

ierr (Dan., Norw., Swed.), Mr, abbr. **hr.**

Ierrenvolk (Ger. n.), master race, (in Nazi ideology) the German people

Herrgott (Ger.), Lord God

ierring-bone, a stitch (hyphen)

Ierrnhuter, one of the sect of the Moravians (not ital.)

Ier Royal Highness (caps.), abbr. **HRH**

iers (no apostrophe)

Ierschel (Caroline Lucre- tia), 1750–1848, German-born English astronomer; — (**Sir John Frederick William**), 1792–1871, English astronomer, son of — (**Sir William,** orig. **Friedrich Wilhelm**), 1738–1822, German-born English astronomer

Herschell (Baron)

Herstmonceux, E. Sussex, site of Royal Observatory, *not* Hurst-

hersute, use **hir-**

Hertford/, -shire

Herts., Hertfordshire

Hertz (Heinrich Rudolf), 1857–94, German physicist, discoverer of **Hertzian waves,** used in radio-communication

hertz, SI unit of frequency, *pl.* same; abbr. **Hz**

Hertzog (James Barry Munnik), 1866–1942, S. African statesman, *not* Herzog

Herz (Ger. n.), heart (cap.)

Herzegovina, use **Herce-**

Herzog/ (Ger. m.), duke, *pl.* **-e; -in** (f.), duchess; **-tum** (n.), duchy, *not* -thum (caps.)

Hesperis, a genus of plants

Hesperus, evening star

Hesse, German state, in Ger. **Hessen; Hessian,** inhabitant of Hesse, in Ger. **Hess/e,** *f.* **-in**

Hesse (Hermann), 1877–1962, German writer

hessian, a coarse cloth

het (Du. n.), the; abbr. **'t,** as van 't Hoff

hetaer/a (Gr.), a courtesan, *pl.* **-ae; -ism,** *not* -tair-, -tar-

heterogeneous, dissimilar, *not* -nous

heteroousian (theol.), believing Father and Son to be of unlike substance, *not* heterou-. *See also* **homoiousian, homoousian**

Hetton-le-Hole, Tyne & Wear (hyphens)

heu (Lat.), alas!

heureka, use **eu-**

heuristic, inciting to find out (an educational method). *See also* **eureka**

hex-, Gr. prefix for six; in Lat. **sex-**

hey, a dance, *use* **hay**

heyday, prosperity, *not* heigh- (one word)

heyduck, Hungarian of an ennobled mil. class (Hung. *hajdú, pl.* **hajdúk**)

Heyerdahl (Thor), b. 1914, Norwegian ethnologist (*Kon- -Tiki* expedition, 1947)

hey-ho, *use* **heigh-ho**

hey presto, conjuror's exclamation (two words)

HF, (pencils) hard firm, high frequency, Home Fleet, — Forces

Hf, hafnium (no point)

hf., half

h.f., high frequency

HFRA, Honorary Fellow of the Royal Academy

HG, His, *or* Her, Grace, High German, Home Guard, Horse Guards

Hg, *hydrargyrum* (mercury) (no point)

hg, hectogram, -s (no point in scientific and technical work)

HGV, heavy goods vehicle

HH, His, *or* Her, Highness; His Holiness (the Pope)

HH (pencils), double hard; *also* **2H**

hh., hands (measure of horse's height)

hhd., hogshead, -s

HHH (pencils), treble hard; *also* **3H**

HI, Hawaii (off. postal abbr.), Hawaiian Islands, *hic iacet* (here lies)

hiatus/, *pl.* **-es** (not ital.)

Hiawatha, American-Indian hero of Longfellow's poem

hibernate, to spend the winter in torpor or in a warm climate, *not* hy-

hic (Lat.), this, here

hiccup/, *not* -cough, -kup; **-ed, -ing** (one *p*)

hic et ubique (Lat.), here and everywhere

Hichens (Robert Smythe), 1864–1950, English novelist

hic iacet (*or* **jacet**) (Lat.), here lies; —— *sepult/us,* fem. **-a,** here lies buried; abbr. **HIS** *or* **HJS**

hidalgo/, Spanish gentleman b birth; *pl.* **-s**

hide-and-seek, a game (hyphens)

hidebound, narrow-minded (one word)

hieing, *not* hy-

hier (Ger.), here; *Hier sprich man Deutsch,* German spoke here

hieratic, priestly, esp. of ancien Egyptian sacred writing

hieroglyph/, stylized figure representing word; **-ic(s), -ist, -ize** (*not* -ise)

hifalutin, *use* high-

hi-fi, high-fidelity (no points)

higgledy-piggledy, haphazard, in confusion

highbrow (one word)

high chair (two words)

High Church (hyphen when attrib.), abbr. **HC; High- -Churchman** (hyphen, caps.)

high-class (adj., hyphen)

high-falutin, bombast(ic), *not* -en, -n', -ng, hifalutin

high-fidelity (radio), reproducing sound faithfully; colloq. **hi-fi**

high/-flown, -flyer, -flying (hyphens)

high frequency (hyphen when attrib.)

highjacker, *use* **hijacker**

high jump (two words)

Highland, region of Scotland

highlight (one word)

high pressure, (hyphen when attrib.); abbr. **HP** *or* **h.p.**

high priest, abbr. **HP** (two words)

high-rise (adj.), having many storeys (hyphen)

high/ road; — seas (the), outside territorial waters (two words)

high-water mark (one hyphen), *abbr.* **HWM**

highwayman (one word)

IHH, His, *or* Her, Imperial Highness

hijack/, seize control of aircraft, etc.; steal goods in transit; **-er, -ing**

Hijaz, *use* **Hejaz**

hijra/, -h, *use* **hegira**

Hil., Hilary

Hilary, a session of the High Court of Justice, a term of Oxford and Dublin universities; from St. Hilary of Poitiers, d. *c.*367, festival 13 Jan.

hill, when with name to have cap., as Box Hill

Hillary (Sir Edmund), b. 1919, NZ apiarist and mountaineer, climbed Everest 1953

hill fort (hyphen)

Hillingdon, London

Hillington, Norfolk, Strathclyde

hill/side, -top (one word)

HIM, His, *or* Her, Imperial Majesty

Himachal Pradesh, Indian state

Himalaya/ *or* **the —s,** India and Tibet; adj. **-n**

hinc (Lat.), hence; **— *illae lacrimae*,** hence those tears

Hinckley, Leics., *not* Hink-

Hind., Hindu, -stan, -stani

Hindi, a language of N. India, or a literary Hindustani, *not* -dee (not ital.)

hind leg (two words)

hindmost (one word)

Hindoo, *use* **Hindu**

hindquarters (one word)

hindrance, *not* -erance

hindsight, wisdom after the event (one word)

Hindu/, *not* -doo, abbr. **Hind.** (not ital.); **-ism, -ize**

Hindu Kush, mountains, Afghanistan

Hindustan, abbr. **Hind.,** *not* Indostan

Hindustani, the Indian lingua franca; abbr. **Hind.**

hing/e, -ed, -ing

Hinkley Point, Som., site of nuclear power station

Hinshelwood (Sir Cyril Norman), 1897–1967, British chemist

hinterland, the 'back country' (one word)

hip/-bath, -bone, -flask, -joint, -pocket (hyphens)

hippie, person of unconventional behaviour, dress, etc., *not* -y

hippo/, hippopotamus, *pl.* **-s**

Hippocrat/es, *fl.* 400 BC, Greek 'Father of Medicine'; **-ic**

Hippocrene, fountain of the Muses on Mt. Helicon

hippogriff, a fabulous monster, *not* -gryph (not ital.)

hippopotamus/, *pl.* **-es**

hirdy-girdy (Sc.), in disorder. *See also* **hurdy-gurdy**

hireable, obtainable for hire, *not* -rable

hire-car (hyphen)

hire-purchase (hyphen), abbr. **HP** *or* **h.p.**

hirly-birly, *use* **hurly-burly**

Hiroshima, Japan

hirsute, hairy, *not* her-

HIS, *hic iacet sepult/us, -a* (here lies buried)

His/ Eminence, abbr. **HE; — Excellency, HE; — Majesty('s), HM** (caps., not ital.)

Hispanicize, to render Spanish, *not* -ise

His Royal Highness (caps.), abbr. **HRH**

hist., histor/ian, -ic, -ical

historical eras and events, *see* **capitalization**

Hitchens (Sydney Ivon), b. 1893, English painter

Hitchin, Herts., *not* -en

HJ, *hic jacet* (here lies)

HJS, *hic jacet sepult/us, -a* (here lies buried)

HK, Hong Kong; House of Keys, Isle of Man

HL, House of Lords

hl, hectolitre(s) (no point in scientific and technical work)

hl. (Ger.), *heilig* (holy)

Hl.S. (Ger.), *Heilige Schrift* (Holy Scripture)

HM, Head Master, — Mistress; Her, *or* His, Majesty('s); Home Mission

hm, hectometre(s) (no point in scientific and technical work)

h.m., *hoc mense* (in this month), *huius mensis* (this month's)

HMC, Her, *or* His, Majesty's Customs; Headmasters' Conference; Royal Commission on Historical Manuscripts

HMG, Her, *or* His, Majesty's Government

HMI, Her, *or* His, Majesty's Inspector

HMP, *hoc monumentum posuit* (erected this monument)

HMS, Her, *or* His, Majesty's Service, *or* Ship

HMSO, Her, *or* His, Majesty's Stationery Office

HMV, His Master's Voice

HNC, HND, Higher National Certificate, — — Diploma

HO, Home Office, (Ger.) *Handels-organisation* (state shop in GDR)

Ho, holmium (no point)

ho., house

Hoangho, *use* **Hwang-Ho**

hoard, store of money or possessions, to save these. *See also* **horde**

hoar-frost (hyphen)

hoarhound, *use* **hore-**

Hobbema (Meindert), 1638–1709, Dutch painter

Hobbes (John Oliver), pseud. of **Pearle Mary Teresa Craigie,** 1867–1906, English writer; — **(Thomas),** 1588–1679, English philosopher

hobbledehoy, raw youth, *not* the many variants

hobby-horse (hyphen)

hob-nob (hyphen)

hobo/ (US), a tramp; *pl.* **-es**

Hoboken, NJ, US

hoboy (mus.), *use* **oboe**

hoc/ age (Lat.), attend!; — *anno,* in this year, abbr. *h.a.*; — *genus omne,* all of this kind; — *habet,* he has a hit (of gladiators)

Hoccleve (Thomas), ?1370–?1450, English poet, *not* O-

hochepot (Fr. cook. m.), hotch potch, stew, ragout

Ho Chi Minh/, 1892–1969, Vietnamese politician; — **City,** Vietnam, *formerly* **Saigon**

Hocking (Joseph), 1855–193 and his brother, — **(Silas Kitto),** 1850–1935, British novelists

hoc/ loco (Lat.), in this place; — *mense,* in this month, abb. *h.m.*; — *monumentum posuit,* erected this monument, abbr. **HMP;** — *sensu,* in this sense, abbr. **h.s.**; — *tempore* at this time, abbr. **h.t.**; — *titulo* in, *or* under, this title, abbr. **h.t**

hocus/, to hoax, drug; **-ed, -in**

hocus-pocus, jugglery, deception (hyphen)

Hodder & Stoughton, Ltd., publishers

hodgepodge, a medley (one word). *See also* **hotchpotch**

hodie (Lat.), today

hodmandod, a snail

hodograph (math.), a curve

hodometer, *use* **od-**

Hoe, Plymouth

Hoe, US printers: **Robert,** 1784-1833, his son **Richard March,** 1812–86, and grandson **Robert,** 1839–1909

hoeing, *not* hoing

Hofer (Andreas), 1767–1810, Tyrolese patriot

Hoffmann (August Heinrich), 1798–1874, German writer ('Hoffmann von Fallersleben'; *Deutschland über Alles,* 1841); — **(Daniel),** 1576–160

German theologian; — (**Ernst Theodor Amadeus Wilhelm**), 1776–1822, German writer and composer, source of Offenbach's *Tales of Hoffmann*; — (**Friedrich**), 1660–1742, German chemist. *See also* **Hofmann**

Hoffnung (Ger. f.), hope (cap.)

Hoffnung (**Gerard**), 1925–59, English humorous artist and musician

Hofmann (**August Wilhelm von**), 1818–92, German chemist; — (**Josef Casimir**), 1876–1957, Polish pianist and composer; — (**Johann Christian Conrad von**), 1810–77, German theologian. *See also* **Hoffmann**

Hofmannsthal (**Hugo von**), 1874–1929, Austrian poet, librettist of some operas by Richard Strauss

hoggin, gravel mixture

Hogmanay (Sc.), the last day of the year

hogshead/, -s, abbr. **hhd.**

Hogue (**La**), dép. Manche, France

Hohenzollern (**House of**), Prussian Imperial family

hoiden, *use* **hoy-**

hoing, *use* **hoe-**

hoi polloi (*not* the — —) (Gr.), the masses, *not* oi —

hokey-pokey, ice-cream, *not* hoky-poky (hyphen)

hokum, stage business used for cheap effect

Holbein (**Hans**), German painters: **'the Elder'**, 1465–1524; **'the Younger'**, 1497–1543

Hölderlin (**Johann Christian Friedrich**), 1770–1843, German poet

hold-up (noun), delay, robbery (hyphen)

holey, having holes

Holi, Hindu religious festival

holibut, *use* **halibut**

holiday-maker (hyphen)

holily, in holy manner

Holinshed (**Raphael**), d. 1580, English chronicler, *not* -ings-, -head

Holland, the country, *use* **Netherlands; but North —, South —** (provinces), **Parts of** — (former division of Lincolnshire)

holland, a linen (not cap.)

hollandais/ (Fr.), *fem.* **-e,** Dutch (not cap.)

hollandaise, a creamy sauce (not ital.)

Hollands, Dutch gin

hollow (bind.), paper reinforcement of back, sometimes also of spine, of book

hollyhock, a plant

Hollywood, Los Angeles, Calif., US

Holman-Hunt (**William**), 1827–1910, English painter (hyphen)

Holmes (**Oliver Wendell**), 1809–94, US author, and his son, 1841–1935, US lawyer

Holmes-McDougall, Ltd., publishers

holmium, symbol **Ho**

Holm Patrick (**Baron**)

holocaust, whole burnt offering, wholesale sacrifice or destruction (cap. with hist. reference to 1939–45)

Holofernes, Assyrian general (-ph- in NEB); a pedantic teacher (Shakespeare: *Love's Labour's Lost*)

holograph, a document wholly in the handwriting of the person from whom it proceeds

holus-bolus, all at once (hyphen)

Holy Communion (caps.), abbr. **HC**

Holy Cross Day (three words, caps.), Holy Rood Day, feast of the Exaltation, 14 Sept.

Holy/ Family, — Ghost, — Land (two words, caps.)

Holyoake (**George Jacob**), 1817–1906, English writer and

Holyoake (*cont.*)
agitator; — (**Rt. Hon. Sir Keith Jacka**), b. 1904, Prime Minister of New Zealand, 1960–72

holy of holies, inner chamber of the Jewish tabernacle (not caps.)

Holy Roman Empire (caps.), abbr. **HRE**

Holy Rood Day, *see* **Holy Cross Day**

Holyroodhouse, Palace of, Edinburgh

Holy Saturday, day before Easter Sunday (caps.)

Holy Spirit, as Deity (caps.)

holystone (naut.), (to scour decks with) soft sandstone (one word)

Holy Thursday, Ascension Day in English Church; but Thursday in Holy Week, or Maundy Thursday, in Roman Church

Holy Week, the week before Easter (two words, caps.)

Holywood, Co. Down, N. Ireland

Hom., Homer

homage, public acknowledgement of allegiance, in Fr. **hommage** (m.), q.v.

homard (Fr. m.), lobster

hombre (Sp.), man

Homburg, soft felt hat

Home/, surname, *pron.* Hume; — **of the Hirsel** (**Baron**)

home/-brewed, -coming (hyphens)

Home Counties (**the**), Essex, Herts., Kent, Surrey (sometimes includes Berks., Bucks., Sussex); abbr. **HC**

home/-grown (hyphen)

Home Guard (two words, caps.), (member of) British citizen army formed in 1940, abbr. **HG** (no points). *See also* **LDV**

homeland (one word)

home-made (hyphen)

homeopathy, *use* homoeo-

Homer, *c.* 9th c. BC, Greek poet, abbr. **Hom.**

Home Rule (two words, caps.) abbr. **HR**

homesick/, -ness (one word)

homespun (one word)

home/ straight, — town (two words)

homework (one word)

homey, *use* homy

homing, *not* -eing

homin/id, member of family Hominidae; **-oid,** manlike animal

Hommage/ d'auteur, — de l'auteur (Fr.), with the author's compliments; — *d'éditeur,* — *de l'éditeur,* ditto publisher's, *not* editor's

homme/ d'affaires (Fr.), business man, not — *des* —; — *de bien,* a respectable man; — *de cour,* a courtier; — *de lettres* author; — *de paille,* man of straw; — *d'épée,* a mil. man; — *de robe,* lawyer; — *d'esprit,* man of wit; — *d'état* statesman; — *de tête,* man of resource; — *du monde,* man of fashion

Homo (zool.), genus of man

hom/o (Lat.), human being; *pl.* -ines

homoeopath/y, treatment of diseases by minute doses of drugs that excite similar symptoms; **-ic, -ist** (*not* homeo-)

homogene/ous, consisting of parts all of the same kind; noun -ity

homogenize, to make (milk) homogeneous, *not* -ise

homogenous, having common descent

homoiothermic, warm-blooded

homoiousian (theol.), believing Father and Son to be of like substance (not ital.). *See also* **heteroousian, homoousian**

homologize, be *or* make homologous, *not* -ise

homonym, a word of same form but different sense, *not* -me

homoousian (theol.), believing

Father and Son to be of the same substance, *not* homou- (not ital.). *See also* **heteroousian, homoiousian**

1omophones, words spelt differently but pronounced alike

Iomo sapiens (zool.), the species modern man

Ioms, Syria

Iomy, homelike, *not* -ey

Ion., Honourable (son or daughter of a peer; MP), Honorary

Ionble., Honourable (former Indian title)

Iondura/s, Cent. America; adj. **-n**

Ionegger (Arthur), 1892–1955, French-born Swiss composer

1oney-bee (hyphen)

1oneycomb (one word)

1oneydew, a sticky substance, a melon, a tobacco (one word)

1oneyed, sweet, *not* -ied

1oney/moon, -suckle (one word)

Iong Kong (S. coast of China) (no hyphen), abbr. **HK**

Ioni soit qui mal y pense, shamed be he who thinks evil of it (motto of the Order of the Garter)

1onorand, one to be honoured

1onorarium/, voluntary fee for professional services; *pl.* **-s** (not ital.)

1onorary/, (of office, etc.) bestowed as an honour, unpaid; — **secretary,** abbr. **Hon. Sec.**

1onorific, (utterance) expressing honour or respect

1onoris/ causa, or — *gratia* (Lat.), for the sake of honour

1onourable, abbr. for son or daughter of a peer, or for MP, **Hon.;** for Indian title, **Honble.**

Hons., Honours

Hon. Sec., honorary secretary

1oodwink (one word)

1oof/, *usual pl.* **-s,** *not* -ves

Hooghly, India, *not* Hugli

Hook (Theodore Edward), 1788–1841, English humorist. *See also* **Hooke**

hookah, oriental pipe, *not* the many variants

hook and eye (no hyphen)

Hooke (Robert), 1635–1703, English physicist. *See also* **Hook**

Hooker (Sir William Jackson), 1785–1865, and his son — **(Sir Joseph Dalton),** 1817–1911, English botanists

hooping cough, *use* **wh-**

hoopoe, S. European bird

hoor/ah, -ay, *use* **hurr/ah, -ay**

Hoover, propr. term for vacuum cleaner (cap.)

Hopi/, American Indian, *pl.* **-s,** *not* -ki, -qui

Hopkins (Gerard Manley), 1844–89, English poet; — **(Johns,** *not* John), 1795–1873, US financier; — **(Johns)** University, Baltimore, Md., US (no apos.)

hop-o'-my-thumb, a dwarf (hyphens)

hop-picker (hyphen)

Hoppner (John), 1758–1810, English painter

hopscotch, children's game (one word)

hor., horizon, -tal

hor/a (Lat.), hour, *pl.* **-ae;** *horae/ canonicae,* hours for prayer; — *subsecivae,* leisure hours

Horace (in Lat., **Quintus Horatius Flaccus),** 65–8 BC, Roman poet, abbr. **Hor.**

Horatius Cocles, who kept the Sublician Bridge

horde, troop of nomads, crowd of people. *See also* **hoard**

horehound, a plant (juice used for coughs), *not* hoar-

horizon/, -tal, abbr. **hor.**

horn (English), one of the oboes, *use* **cor anglais;** — **(French),** circular coiled horn, usu. with valves

hornblende, a mineral, *not* -d

hornpipe, a dance, music for it (one word)

horology, abbr. horol.

horresco referens (Lat.), I shudder to mention it

hors/ (Fr.), beyond, out of; — *concours,* not for competition (*not* de); — *de combat,* disabled; — *de la loi,* outlaw; — *de pair,* without an equal; **hors-d'œuvre,** appetizer

Horse Artillery (caps.), abbr. **HA**

horseback (one word)

horse/-box, -chestnut, *not* -chesnut; **-flesh, -fly** (hyphens)

Horse Guards, abbr. **HG** (two words, caps.)

horse/hair, -play (one word)

horsepower (abbr. **hp**), unit of power (7.46×10^2 watts)

horse-rac/e, -ing (hyphen), *but* **Horserace** (one word) **Totalisator Board,** and **Horserace Betting Levy Board**

horse-radish (hyphen)

horseshoe (one word) *but* **horse-shoeing** (hyphen)

Horseshoe Fall, Niagara; **Horse Shoe Falls,** Guyana

horse-tail, Turkish standard; a plant (hyphen)

horsewhip (one word)

'Horst Wessel Song', song of the German Nazi party, named from the writer of the words

horsy, horselike, *not* -ey

hort., horticulture

hortus siccus, collection of dried plants

Hos., Hosea

Hosanna/, a shout of praise, 'save, we pray', *not* -ah (not ital.); — **Sunday,** Palm Sunday

hospital/, abbr. **hosp.**; **-ize,** *not* -ise

hospitaller, one of a charitable brotherhood (US **-aler**)

hotbed (one word)

hotchpot (law), commixture of property to secure equable distribution (one word)

hotchpotch (cook.) (one word) *See also* **hodgepodge**

hôte (Fr. m.), innkeeper, host; *also* guest

hotel, *see* a; — **(name of),** to b roman quoted to avoid ambiguity, as 'The Farmhouse'

Hôtel des Invalides, Paris, founded 1670 as hospital for disabled soldiers; contains Napoleon's tomb (caps.)

hôtel/ de ville (Fr. m.), town hall; —*-Dieu* (*l*'), chief hospital of a town (one cap.); — *garni* — *meublé,* furnished lodging

hotelier, hotel-keeper (not ital no accent)

hot/foot, -head, -house, -plat (one word)

hot-pot (cook.) (hyphen)

hot metal (typ.), justified mechanical composition in purpose-cast metal sorts or slug by casting-machine driven by punched paper roll from keyboard

houdah, *use* how-

Houdan, breed of fowls

Houdin (Jean Eugène Robert), 1805–71, French conjuro

Houdini (Harry), real name Ehrich Weiss, 1874–1926, US escapologist

Houdon (Jean-Antoine), 1741–1828, French sculptor

Houghton-le-Spring, Tyne & Wear (hyphens)

hour/, -s, abbr. **hr., hrs.,** but **h** (no point, same in pl.) in scientific and technical work; **-glass** (hyphen)

houri/, a nymph of the Muslim paradise; *pl.* **-s**

Housa, *use* Hausa

house, number of, in a street, ha no comma after, as 6 Fleet Street; abbr. **ho.**

House (the), the Stock Exchange, Christ Church (Oxford), the House of Commons, (*formerly*) the workhouse

house-agent (hyphen)

house arrest (two words)

ouseboat (one word)

ousebote (law), tenant's right to wood to repair house (one word)

ousebreak/er, -ing (one word)

ouse/-flag, distinguishing flag of a shipping company; **-fly** (hyphens)

ouse/holder, -keeper, -keeping, -maid (one word)

ouse party (two words)

ouse/-physician, abbr. **HP;** **-surgeon,** abbr. **HS** (hyphens)

ouse style (typ.), the custom of a printing establishment as to both the literal aspect (use of capitals, spellings, abbreviations, italics, word-division, etc.) and the general layout and design

ouse/wife, -work (one word)

Housman (Alfred Edward), 1859–1936, English scholar and poet; — **(Laurence),** 1865–1959, his brother, English writer and artist

Houssaye (Arsène), 1815–96, French novelist and poet; — **(Henri),** 1848–1911, his son, French historian

Houyhnhnm, one of a race of horses with noble human characteristics (Swift's *Gulliver's Travels*), *pron.* hwinnim; contrasted with **Yahoo,** q.v.

Howards End, by E. M. Forster, 1910 (no apos.)

owbeit (one word)

owdah, elephant-seat, *not* -a, houda, -ah, -ar

ow-do-you-do, *or* how-d'ye--do,** awkward situation (hyphens)

owitzer, a cannon

oyden, tomboy, not hoi-

IP, high pressure (*also* **h.p.**), high priest, hire-purchase (*also* **h.p.**), house-physician, Houses of Parliament

ıp, horsepower (no point)

IQ, Headquarters

HR, Home Rule, House of Representatives

hr. (Dan., Norw., Swed.), *herr* (Mr)

hr., hrs., hour, -s (non-technical use)

Hradec/, Czechoslovakia, *not* Grätz; — **Králové,** *not* Königgrätz

Hrdlicka (Aleš), 1869–1943, Bohemian-born US anthropologist

HRE, Holy Roman Empire

HRH, Her, *or* His, Royal Highness

HRIP (Lat.), *hic requiescit in pace* (here rests in peace)

hrsg. (Ger.), *herausgegeben* (edited)

HS, *hic sepult/us, -a* (here buried), house-surgeon; symbol for sesterce, Roman coin equivalent to $\frac{1}{4}$ of a denarius

Hs. (Ger.), *Handschrift* (manuscript)

h.s. (Lat.), *hoc sensu* (in this sense)

HSC, Higher School Certificate, *now* GCE, A level

HSE, *hic sepult/us, -a est* (here lies buried)

HSH, His, *or* Her, Serene Highness

HSS, *Historicae Societatis Socius* (Fellow of the Historical Society)

ht., height

h.t., *hoc tempore* (at this time), *hoc titulo* (in, *or* under, this title); (elec.) high tension

ht. wkt. (cricket), hit wicket

HU, Harvard University

Huanghe, *see* **Hwang-Ho**

hubble-bubble, oriental pipe (hyphen, not ital.)

Huberman (Bronislaw), 1882–1947, Polish violinist

hubris/ (Gr. drama), presumptuous pride that invites disaster, *not* hy- (not ital.); adj. **-tic**

huckaback, rough-surfaced linen fabric, *not* hugga-

Hucknall, Notts., where Byron is buried

Hudson Bay, N. America, *but* Hudson's **Bay Company**

Hueffer, later Ford (**F. M.**), *q.v.*

huggaback, *use* **huck-**

hugger-mugger, secret(ly) (hyphen)

Hughenden, Bucks., *not* -don

Hugli, *use* **Hooghly**

Hugo (**Victor Marie**), 1802–85, French poet, novelist, and playwright

Huguenot (hist.), a French protestant, *not* -onot

huissier (Fr. m.), bailiff, door--keeper

huîtres (Fr. f. pl.), oysters

huius/ anni (Lat.), of this year, abbr. *h.a.;* — *mensis,* of this month, abbr. *h.m.*

hullabaloo, uproar, *not* the many variants

Hullah (**John Pyke**), 1812–84, English musician

hullo, *use* **hallo**

Hulsean Lectures, Cambridge

Humanae Vitae (Lat.), of human life, name of encyclical on contraception, by Pope Paul VI, 1968

Humaniora (Lat.), the humanities; abbr. *Hum. See also* **Lit. Hum.**

humanize, *not* -ise

humankind (one word)

Humberside, county of England

humble-bee (hyphen)

humble pie (**to eat**), *not* umble (two words)

Humboldt (**Friedrich Heinrich Alexander, Baron von**), 1769–1859, German naturalist; — (**Karl Wilhelm, Baron von**), 1767–1835, his brother, German philologist, statesman, and poet

humdrum, commonplace (one word)

Humean, of David Hume, 1711–76, Scottish historian and philosopher, *not* -ian

humer/us, upper-arm bone, *pl* -i (not ital.)

humming-bird (hyphen)

hummum, Turkish bath; *use* **hammam**

humoresque, a musical caprice, *not* humour-

humor/ist, -ize, *not* -ise

humorous/, -ly, -ness

humoursome (capricious)

humous, *see* **humus**

humpback(ed) (one word)

Humperdinck (**Engelbert**), 1854–1921, German composer

Humphrey, Duke of Glouces-ter, 1391–1447, son of Henry IV of England, Protector durin minority of Henry VI

Humphrey's Clock (*Master*) by Dickens, 1840

Humphry (*not* -ey) *Clinker,* b Smollett, 1771

Humpty-Dumpty (hyphen), short, squat person (from character in nursery rhyme)

hum/us, vegetable mould; adj. -ous

hunchback(ed) (one word)

hundert (Ger.), hundred

hundred/, symbol **C; -weight** -s (one word), abbr. **cwt**

hundred-and-first etc. (hyphens)

hundred-per-cent (adj.), entir

Hundred Years War, betwee England and France, 1337–1453 (caps., no apos.)

hung, *see* **hanged**

Hung., Hungar/y, -ian

Hungary, abbr. **Hung.;** in Fr. **Hongrie,** in Ger. **Ungarn,** in Hung. (Magyar) **Magyar-ország**

Hunstanton, Norfolk

Hunter's Quay, Strathclyde (apos.)

Huntingdon, Cambs.

Huntingdonshire, former county of England, abbr. **Hunts.**

Huntington, Hereford & Worc., Shropshire, Staffs., N.

Yorks.; — **Library,** San Marino, Calif., US

Huntley, Glos., Staffs.

Huntly/, Grampian; — **(Marquess of),** not -ey

Hunyadi János, Hungarian mineral water

Hunyady (János), 1387–1456, Hungarian general of Romanian descent, not -adi

hurds, use **hards**

hurdy-gurdy (mus.), not hirdy-girdy, q.v.

hurly-burly, commotion, not hi-bi-

hurrah or **hurray,** not hoo-

hurry-curry, use **hara-kiri**

hurry-scurry, pell-mell, not -sk-

Hurstmonceux, E. Sussex, use **Herst-**

Hurstpierpoint, W. Sussex (one word)

husband, abbr. **h.**

Husbands Bosworth, Leics. (no apos.)

Huss/ (John), 1369–1415, Bohemian religious reformer, in Ger. Hus **(Johann); -ite**

hussy, pert girl, not -zzy

Huygens (Christian), 1629–95, Dutch mathematician and astronomer, not -ghens

Huysmans (Joris Karl), 1848–1907, French novelist

huzzy, use **hussy**

h.w. (cricket), hit wicket

h/w, herewith

Hwang-Ho (river in China), not Hoangho; in Pinyin **Huanghe**

HWM, high-water mark

Hy, Henry (man's name) (no point)

Hyacinthe (Père), Charles Jean Marie Loyson, 1827–1912, French priest

hyaena, use **hyena**

hybernate, use **hi-**

hybrid (bot.), symbol × (see **botany**)

hybridize etc., not -ise

hybris, use **hu-**

hyd., hydrostatics

Hyderabad, Deccan, India; Sind, Pakistan; not Haidar-, Hydar-

hydrangea, a shrub, not -ia

hydro/, colloq. for hydropathic establishment, hydroelectric plant (no point), pl. **-s**

hydro/carbon, -dynamics, -electric (one word)

hydrogen, symbol **H**

hydro/lysis, decomposition by water; **-lyse,** not -lyze

hydrophobia, aversion to water, rabies

hydroplane, kind of motor boat (one word)

hydrostatics, abbr. **hyd.**

hydrotherapy (one word)

hyena, not hyae-

Hyères, dép. Var, France

Hygeia, (Gr.) goddess of health, not -gea, -giea, -gieia (the older Gr. spelling); adj. **hygeian,** healthy

hygien/e, science of health; adj. **-ic**

hying, use **hic-**

Hyksos, rulers of Egypt, 17th–16th c. B.C.

Hylton-Foster (Sir Harry), 1905–65, Speaker of the House of Commons 1959–65

hymen/ (anat.), virginal membrane (not ital.); adj. **-al**

Hymen (Gr. and Rom. myth.), god of marriage; adj. **hymeneal**

hymn-book (hyphen)

hyp., hypothesis, hypothetical

hyperaem/ia, excess of blood; adj. **-ic;** not -haemia

hyperbol/a, a curve; adj. **-ic**

hyperbol/e, exaggeration; adj. **-ical**

hypercritical, excessively critical

hypermarket, in theory larger than a supermarket (one word)

hyphens, see **punctuation** XIII

hypnotize, not -ise

hypochondria, morbid anxiety about health, not -condria

hypocri/sy, -te, -tical

hypotenuse (geom.), *not* hypoth-

hypothecate, to mortgage

hypothermia, lack of warmth in body

hypothes/is, provisional explanation; *pl.* **-es**; abbr. **hyp.**

hypothesize, to form a hypothesis, *not* -ise, -tize

hysterectomy (surg.), removal of the womb

hysteresis (phys.), a lagging ⟨ variation in effect behind variation in cause

hysteron proteron (Gr.), the reverse of natural order, the ca⟩ before the horse

hysterotomy (surg.), cutting into the womb

Hz, hertz, SI unit of frequency (no point)

I

I, iodine, (roman numeral) one (no point), the ninth in a series

I., Island, -s, Isle, -s, *imperator* (emperor), *imperatrix* (empress), (med. prescriptions) one, (math.) square root of minus one

i., id (that)

I or i (ital.), symbol for electric current

ι (Gr.), iota (no dot)

IA, Indian Army, infected area, Iowa (off. postal abbr.)

Ia., Iowa (off. abbr.)

IAAF, International Amateur Athletic Federation

IAAM, Incorporated Association of Assistant Masters, usu. called **AMA**

IAEA, International Atomic Energy Agency

IAHM, Incorporated Association of Headmasters. *See also* **AHMI**

IAM, Institute of Advanced Motorists

iamb/us, foot of two syllables (˘ ‐); *pl.* **-uses;** adj. **-ic; -ics,** iambic verse

Iaşi, Romania, *not* Jassy

IATA, International Air Transport Association

IBA, Independent Broadcasting Authority

Ibáñez (Vicente Blasco), 1867–1928, Spanish novelist

Iberian, pertaining to Spanish peninsula

ibex/, mountain-goat; *pl.* **-es**

ibidem (Lat.), in the same place, abbr. **ib.** *or* **ibid.** (not ital., not cap. except to begin sentence or note). *See also* **idem**

ibis/, wading bird; *pl.* **-es**

Ibiza, Balearic Is., *not* Iv-

Iblis, *use* E-

IBM, International Business Machines

Ibo, people and language of Nigeria, *not* Igbo

IBRD, International Bank for Reconstruction and Development (UN)

Ibsen (Henrik), 1828–1906, Norwegian writer

IC (comp.), integrated circuit

i/c, in charge, in command

ICA, Institute of Contemporary Arts

ICAN, International Commission for Air Navigation

ICAO, International Civil Aviation Organization (UN)

Icarian, of Icarus, *not* -ean

ICBM, inter-continental ballistic missile

ICE, Institution of Civil Engineers; internal-combustion engine

Ice., Iceland, -ic

iceberg (one word)

ice/-cap, -cream (hyphens)

Icelandic (typ.), roman alphabet now used; *see* **accents**

ice-pick (hyphen)

ICFTU, International Confederation of Free Trade Unions

ich (Ger.), I (cap. only at beginning of sentence)

Ich dien (Ger.), I serve (motto of the Princes of Wales)

I.Chem.E., Institution of Chemical Engineers

i ching, Chin. fortune-telling game

ichneumon, fly *and* mongoose

ichthyology, study of fishes; abbr. **ichth.**

ichthyosaur/us, extinct marine animal; *pl.* **-i**

Ichthys (Gr. 'fish', early Christian symbol), initial letters of *Iesous Christos Theou Uios Soter* (Jesus Christ, Son of God, Saviour)

ICI, Imperial Chemical Industries

187

Ici on parle français, French spoken here (l.c. *f*)

Icknield Street *or* **Way**

ICN, *in Christi nomine* (in Christ's name)

icon/, an illustration or portrait, esp. religious; *pl.* **-s; -ic;** *not* ik-, eik- (not ital.)

icon., iconograph/y, -ic

ICS, Indian Civil Service

ICU, International Code Use (signals)

ID (mil.), Intelligence Department, (off. postal abbr.) Idaho

I'd (I had, *or* I would), to be printed close up

id (in Freudian psychology), not cap.

id., *idem* (the same)

i.d., inner diameter

id (Lat.), that; abbr. **i.**

IDA, International Development Association (UN)

Idaho, off. postal abbr. **ID**

IDB (S. Afr.), illicit diamond buy/er, -ing

IDC, Imperial Defence College, now **RCDS**

idea'd, having ideas, *not* -aed

idealize, *not* -ise

idealogical etc., *use* **ideo-**

idée/ fixe (Fr. f.), fixed idea;— *reçue,* accepted opinion

idem/ (Lat.), the same, *or* as mentioned before; abbr. **id.** (not ital., not cap. except to begin sentence or note), *pl. same,* used generally to avoid repetition of author's name in footnotes or bibliographical matter (*see also* **ibidem**); *— quod,* the same as, abbr. *i.q.*

identif/y, -ication, *not* ind-

Identikit (picture), composite drawing of face (propr. term). *See also* **photofit**

ideo/gram, -graph, a character symbolizing the idea of a thing without expressing its name

ideologue, a visionary

ideolog/y, study of ideas, a way of thinking; **-ical, -ist;** *not* ideal-

ides (pl.), in Roman calendar

the fifteenth day of March, May, July, October, the thirteenth o other months

id est (Lat.), that is; abbr. **i.e.** (not ital., l.c., comma before)

id genus omne (Lat.), all of tha kind

idiolect, linguistic system of on person, *not* ideo-

idiosyncra/sy, a peculiarity o temperament, *not* -cy; **-tic**

idl/ing, -y, *not* -eing, -ey

IDN, *in Dei nomine* (in God's name)

idolater, *not* -or

idolize, *not* -ise

idol/um, a false mental image; *pl.* **-a**

Idumaea, *not* -mea

idyll, a work of art depicting innocence or rusticity, *not* -yl

-ie-, *not* -ei-, in: achieve, adieu, aggrieve, Aries, befriend, belie believe, besiege, bier, bombard ier, brief, brigadier, cashier, cavalier, chandelier, chief, chi fonier, fief, field, fiend, fierce, friend, frieze, gaselier, grena- dier, grief, griev/e, -ance, -ous, handkerchief, hygienic, lief, liege, mien, mischief, mischie- vous, niece, piece, -meal, pier, pierce, priest, relie/f, -ve, re- prieve, retrieve, review, rilieve shield, shriek, siege, sieve, spe- cies, thief, thieve, tier, tierce, tiercel, wield, yield. *See also* -ei-

IE, (Order of the) Indian Empire Indo-European

i.e., *id est* (that is) (l.c., not ital., comma before)

IEA, International Energy Agency

IEE, Institution of Electrical En gineers

IEEE, Institute of Electrical and Electronics Engineers (US)

Ieper, Fl. for **Ypres**

IERE, Institution of Electronic and Radio Engineers

Iesu (Lat.), Jesus (vocative)

IF, intermediate frequency

IFC, International Finance Corporation (UN)

IG, Indo-Germanic, (mil.) Inspector-General

I. Gas E., Institution of Gas Engineers

Igbo, *use* **Ibo**

igloo/, Eskimo snow-hut; *pl.* **-s**

ign., *ignotus* (unknown)

igneous, of *or* like *or* produced by fire, *not* -ious

ignis fatuus (Lat.), will-o'-the--wisp; *pl.* **ignes fatui**

ignitable, *not* -ible

ignoramus/, an ignorant person; *pl.* **-es**

ignoratio elenchi (Lat.), refuting a proposition differing from that one professes to be refuting

ignotum per ignotius (Lat.), the unknown by means of the more unknown, not *ignotus*

ignotus (Lat.), unknown; abbr. **ign.**

IGO, intergovernmental organization

IGY, International Geophysical Year

i.h., *iacet hic* (here lies)

IHC, same as **IHS,** C being a form of Gr. cap. S

ihm (Ger.), to him

Ihnen (Ger.), to you (cap.); to them (not cap.)

i.h.p., indicated horsepower

Ihr (Ger.), your (cap.)

ihr (Ger.), to her, her, their, (fam.) you (not cap.)

IHS, abbr. of Gr. *Iesous* (H being Gr. cap. long E); later interpreted as Lat. *Iesus Hominum Salvator* (Jesus Men's Saviour); *In Hoc Signo* (*vinces*), in this sign (thou shalt conquer); *In Hac* (*Cruce*) *Salus*, in this (cross) is salvation

IHVE, Institution of Heating and Ventilation Engineers

II (roman numeral), two (no point)

III (roman numeral), three (no point)

IIII, (roman numeral), four,

sometimes appears on clock--faced, otherwise *use* **IV**

IJ, water area, Netherlands, *not* Ij, Y

ij (med.), two; (typ.) in Dutch now used instead of **y**

i.J. (Ger.), *im Jahre* (in the year)

i.J.d.W. (Ger.), *im Jahre der Welt* (in the year of the world)

IJmuiden, town, **-ssel,** river, **-sselmeer,** *formerly* Zuider Zee, Netherlands, *not* Ij-, Y-

ikon, *use* **icon**

IL, Illinois (off. postal abbr.)

il (Fr.), he *or* it; (It., m. sing.), the

ILEA, Inner London Education Authority

île (Fr. f.), island

Île de France (Fr. region and prov.)

ileum (anat.), part of intestine

ilex/, the holm-oak; *pl.* **-es**

Iliad, of Homer (ital., not quoted)

Ilium (Lat.), Troy

ilium (anat.), part of pelvis

ilk (**of that**), of the same name as ancestral estate; (colloq.) of that kind

Ill., Illinois (off. abbr.)

ill., *illustrissimus* (most distinguished)

I'll, I shall *or* I will, to be printed close up

ill-advised/, -ly (hyphens)

Ille-et-Vilaine, dép. France

illegalize, *not* -ise

illegitimize, *not* -atize, -ise

ill/-fated, -gotten (hyphens)

ill/ health, — humour (two words)

illimitable, limitless

Illinois, off. abbr. **Ill.** or (postal) **IL**

ill nature (two words); **ill--natured** (hyphen)

ill/-treat, -use (hyphens)

illuminati (pl.), enlightened people, cap. with reference to particular movements (not ital.)

illus/ive, -ory, of more apparent than real value. *See also* **elusive**

illustrat/ed, -ion, abbr. **illus.**

illustrator, *not* -er

ill will (two words)

il n'y a pas de quoi (Fr.), don't mention it

ILO, International Labour Organization, ditto Office (UN)

ILP, Independent Labour Party

'Il Penseroso', poem by Milton, 1632

ILS, instrument landing system

ILTF, International Lawn Tennis Federation. *But see* **tennis**

IM, intramuscular

I'm, to be printed close up

imag/o, winged insect, *pl.* **-ines**; idealized mental picture, *pl.* **-s**

imam, Muslim priest, *not* -âm, -aum

I.Mar.E., Institute of Marine Engineers

imbed, *use* **embed**

imbroglio/, a tangle; *pl.* **-s**; *not* em- (not ital.)

imbrue, to stain, dye, *not* em-

imbue, to saturate, inspire (with feelings), *not* em-

IMCO, Inter-governmental Maritime Consultative Organization (UN)

I.Mech.E., Institution of Mechanical Engineers

IMF, International Monetary Fund (UN)

imfe, *use* **imphee**

I.Min.E., Institution of Mining Engineers

im Jahre (Ger.), in the year; abbr. *i.J.*

IMM, Institution of Mining and Metallurgy

immanent, inherent. *See also* **imminent**

immesh, *use* **en-**

imminent, impending. *See also* **immanent**

immobiliz/e, *not* -ise; **-ation**

immortalize, *not* -ise

immortelle, an everlasting flower (not ital.)

Immortels (*Les*), the members of the French Academy

immov/able, -ability, -ableness, -ably

immunize, to render immune, *not* -ise

immutable, unchangeable

imp., imperative, imperfect, imperial, impersonal, import/ed, -er, impression, imprimatur

imp., (Fr.) *imprimeur* (printer); (Lat.) *imperator* (emperor), *imperatrix* (empress)

impa/nel, -nnel, *use* **empanel**

impassable, that cannot be passed

impasse, a deadlock (not ital.)

impassible, that cannot feel

impassion/, to stir emotionally **-ed,** *not* em-

impasto/, the thick laying-on of colour, *pl.* **-s**

impayable (Fr.), invaluable, priceless

impedance (elec.), hindrance to alternating current

impedimenta (pl.), (mil.) baggage

impel/, -led, -ler, *not* -lor, **-ling**

imperative, (mood) expressing command; abbr. **imp.**

imperat/or (Lat.), *fem.* **-rix,** absolute ruler; abbr. **I.** *or* **imp.**

imperf., imperfect, (stamps) imperforate

imperfect, abbr. **imp.** *or* **imperf.**

imperfection (typ.), copy of book with printing or binding faults; good sheet needed to complete a binding order or replace faulty one

imperial, abbr. **imp.**

imperial, former size of paper, 22 × 30 in.; — 4to, 15 × 11 in.; — 8vo, 11 × 7½ in. (untrimmed). *See also* **book sizes**

imperil/, -led, -ling

imperium/ (Lat.), absolute power; — *in imperio,* an empire within an empire

impermeable, that cannot be passed through

impersonal, abbr. **imp.**

impetus/, momentum, incentive; *pl.* **-es**

mphee, a sugar-cane, *not* -fe, -phie (not ital.)

mpi (S. Afr.), a Zulu regiment, an armed band

mping/e, to make an impact; **-ing**

mplacable, *not* -ible

mport/ed, -er, abbr. **imp.**

mpose (typ.), to arrange pages of type in a 'forme' so that they will read consecutively when the printed sheet is folded

mpostor, *not* -er

mpracticable, incapable of being accomplished. *See also* **unpractical**

mpregn/able, that cannot be taken by force; **-atable,** that can be impregnated

mpresa (It.), an undertaking

mpresario/, manager of operatic or other cultural undertakings (one *s*, not ital.); *pl.* **-s**

mpressa (It.), an imprint

mpression (typ.), product from one cycle of a printing machine; all the copies of book etc. printed at one press-run from the same type, plates, etc., abbr. **imp.,** in trade practice 1/i, 2/i, etc. = 1st, 2nd, etc. imp.; indentation in paper by a printing surface; pressure exerted between printing and impression surfaces

mpressionable, *not* -ible

mprimatur, official licence to print, sanction; abbr. **imp.** (not ital.)

mprimatura (It.), coloured glaze

mprim/er (Fr.), to print; **-erie** (f.), printing office; **-eur** (m.), printer

mprimi potest (Lat.), formula giving imprimatur (ital.)

mprimis (Lat.), in the first place, not *in primis* (ital.)

mprint, any paper printed for circulation in the United Kingdom must show the name and address of the printer, who must for six months keep one copy

with the name and address of his customer; name and address of printer and publisher must appear on *all* parliamentary and municipal election work; matter bearing no message (e.g. business stationery) is exempt from these requirements. — (**publisher's**), the name of the publisher, place of publication, and date, usu. printed at the foot of the title-page. *See also* **preliminary matter**

impromptu/, (adv. and adj.) extempore, (noun) piano-piece, *pl.* **-s** (not ital.)

improvable, *not* -eable, -ible

improvis/e, to extemporize, *not* -ize; **-ator,** one who speaks or plays music extempore; **-er,** in general, one who improvises

improvvisat/ore (It.), improvisator; *pl.* **-ori,** *fem.* **-rice**

I.Mun.E., Institution of Municipal Engineers

IN, Indiana (off. postal abbr.)

In, indium (no point)

in., inch, -es, (q.v.)

in (Ger.), in, into

in- (Fr.), prefix in stating book sizes, as *in-8°*, octavo

in/ absentia (Lat.), while absent; — *abstracto,* in the abstract; — *aeternum,* for ever

inadmissible, *not* -able

inadverten/t, (of people) negligent, (of actions) unintentional; **-ce**

inadvisable, (of thing or course of action) not recommended. *See also* **unadvisable**

inamorat/o, *fem.* **-a,** *pl.* **-i,** a lover (not ital.); in It. *innamorat/o, -a; not* en-

in articulo mortis (Lat.), at the moment of death

inasmuch (one word)

Inauguration Day (of US President), 20 Jan.; before 1937, 4 Mar.

in banco (Lat.), as a full court of judges

Inbegriff (Ger. m.), epitome, embodiment (cap.)

in-built (hyphen)

INC, *in nomine Christi* (in Christ's name)

Inc. (US), Incorporated

Inca, one of the royal race of Peru, *not* Y-

incage, use **en-**

in camera, not in open court

in/capsulate, -case, use **en-**

in/ cathedra (Lat.), in the chair of office; — *cautelam,* for a warning

inch/, in metric system 25.4 millimetres (not now in scientific use); *pl.* **-es,** abbr. **in.,** sign ″

inchase, use **en-**

Inchcape Rock or **Bell Rock,** North Sea (two words)

incidentally, *not* -tly

incipit/ (Lat.), (here) begins; *-ur,* it is begun

incise, to cut into, engrave, *not* -ize

incl., including, inclusive

inclose, etc., *see* **en-**

includible, *not* -able

inclu/ding, -sive, abbr. **incl.**

'In Coena Domini', former annual papal bull against heretics

Incogniti, cricket club

incognito/ (noun, adj., and adv.), (person) with name concealed or disguised, the pretended identity; *pl.* **-s;** abbr. **incog.** (not ital.)

income/ group, — tax (two words)

in/ commendam (Lat.), temporarily holding a vacant benefice; — *concreto,* in the concrete

incommunicado, without means of communication (not ital.); in Sp. *incomunicado*

inconnu/ (Fr.), *fem.* **-e,** unknown

incorrigible, *not* -eable

increas/ed, -ing, abbr. **incr.**

incredible, *not* -able

incroach, use **en-**

incrust, *see* **en-**

incubous (bot.), having the upper leaf-margin overlapping the leaf above

incub/us, person or thing that oppresses like a nightmare; *pl.* **-uses** (*preferred*) or **-i**

incumber etc., *use* **en-**

incunabul/a, sing. **-um,** the earliest examples of any art; (bibliog.) books printed before 1501

incur/, -red, -ring, -rable

incurable, that cannot be cured

in curia (Lat.), in open court

incu/s (anat.), a bone in the ear *pl.* **-des**

Ind, poetical for India (no point)

Ind., India, -n; off. abbr. for Indiana, US

ind., independen/ce, -t, index, indication, industrial

IND, *in nomine Dei* (in God's name)

indeclinable (gram.), having no inflexions; abbr. **indecl.**

indefatigabl/e, tireless; **-y**

indefeasible, that cannot be forfeited, *not* -able

indefensible, that cannot be defended, *not* -able

indefinite, abbr. **indef.**

indelible, that cannot be blotted out, *not* -able, -eble

indemni/fy, to protect against harm or loss; **-ty;** *not* en-

indent (typ.), to begin a line, or lines, with a blank space, as here; *not* en-

indentation (**hanging** or **reverse**) (typ.), *see* **hanging paragraph**

indentif/y, -ication, use **id-**

indenture, sealed agreement, esp. one binding apprentice to master, *not* en-

independ/ence, -ent, abbr. **ind.** or **indep.**

Independence Day, US, 4 July

independency, a country that has attained independence

Independent Order of Odd Fellows, *not* Oddfellows, abbr. **IOOF**

in deposito (Lat.), in deposit
indescribable, not -eable, -ible
index/, *pl.* **-es,** abbr. **ind.** For
general instructions for setting
indexes see *Hart's Rules,* pp.
20–3
ind/ex (sci. and math.), *pl.* **-ices**
Index/ Expurgatorius' (Lat.),
index of the passages to be
expunged; '— — **Librorum/ Ex-**
purgandorum' (RCC), a list
of books which might be read
only in expurgated editions;
'— — **Prohibitorum'** (RCC),
a list of books which the Church
forbade to be read
India/, -n, abbr. **Ind.**
India/man, a large ship in the
Indian trade; *pl.* **-men**
Indiana (US), off. abbr. **Ind.** or
(postal) **IN**
Indianapolis, US
indian/ corn, — ink, *not* india
— (not cap.)
Indian summer, period of
warm weather in late autumn;
tranquil and productive late
period of life
India paper, thin book paper,
not -ian (no hyphen); **Oxford**
— —, very thin, strong bible
paper made only for the OUP
indiarubber (one word)
indication, abbr. **ind.**
indicative (gram.), abbr. **indic.**
ind/ices (sci. and math.), *not*
-exes; *sing.* **-ex**
indicia (*pl.*), signs, identifying
marks (not ital.)
indict/, to accuse by legal process; **-er,** *not* -or
indiction (later Rom. Emp.),
cycle of years for administrative
and dating purposes
indigestible, *not* -able
indiscreet, injudicious
indiscrete, not divided into distinct parts
indispensable, *not* -ible
indite, to put into written
words
indium, symbol **In**
individualize, *not* -ise

individu/um (Lat.), the indivis-
ible; *pl.* **-a**
Indo-China, unofficial collective name for the countries of
the SE peninsula of Asia
Indo-European, abbr. **IE** *or*
Indo-Eur.
Indo-German/, -ic, abbr. **IG** *or*
Indo-Ger.
Indonesia, indep. 1950, *formerly*
Dutch East Indies, with W. New
Guinea added 1962; abbr. **In-**
don.
indoor/, -s (one word)
indorse etc., *use* **en-**
Indostan, *use* **Hindu-**
indraught, a drawing in, *not*
-aft
indubitabl/e, *not* -ible; **-y**
induction, *not* -xion
indu/e, -re, *use* **en-**
industrial, abbr. **ind.**
inédit/, *fem.* **-e** (Fr.), unpublished
inedita (Lat.), unpublished compositions
ineducable, incapable of being
educated, *not* -atable, -ible
ineffaceable
ineligible, *not* -able
inept, out of place
inertia (not ital.), *but* **vis iner-**
tiae
in/ esse (Lat.), actually exist-
ing; **— excelsis,** in the highest
(degree); **— extenso,** in full;
— extremis, at the point of
death
in f., in fine (finally)
inf., infantry, inferior, infinitive
inf. (Lat.), *infra* (below)
infallib/le, *not* -able; **ilist,** a
believer in the Pope's infallibil-
ity, *not* -blist
Infanta, dau. of King and Queen
of Spain; **Infante,** younger son
of ditto
infantry, abbr. **inf.**
infantryman (one word)
infected area, abbr. **IA**
infer/, to draw a conclusion;
-red, -ring
infer/able, *not* -rr-; **-ence**
inferior, abbr. **inf.**

inferiors (typ.), small characters set at lower (right) side of ordinary characters

inferno/, a scene of horror, *pl.* **-s**; (cap. and ital.) the first part of Dante's *Divina Commedia*, describing his journey through Hell

in fieri (Lat.), in course of completion

infighting, boxing at close quarters, internal conflict (one word)

infill/, to fill in; **-ing**

infin., infinitive

in fine, finally; abbr. **in f.** (not ital.)

infinitive, abbr. **inf.** *or* **infin.**

infinity (math.), symbol ∞

in flagrante delicto (Lat.), in the very act of committing the offence

inflatable, *not* -eable

inflater, one who or that which inflates, *not* -or

inflexible

inflexion, modulation of voice, gram. termination

infold etc., use **en-**

inforce etc., use **en-**

in forma pauperis (Lat.), as a pauper

infra/ (Lat.), below, abbr. **inf.**; — *dignitatem,* undignified, abbr. *infra dig.*

infra-red (hyphen)

infrastructure (one word)

infula/, each of the ribbons of a bishop's mitre, *pl.* **-e**

infuser, one who or that which steeps something in a liquid, *not* -or

infusible, that cannot be melted

in futuro (Lat.), in, *or* for, the future

Inge (**William Ralph, Dean**), 1860–1954, English divine

Ingelow (**Jean**), 1820–97, English poet

in genere (Lat.), in kind

ingenious, inventive

ingénue (Fr. f.), an artless girl

ingenuity, inventiveness

ingenuous/, free from guile; noun **-ness**

ingle-nook, chimney corner (hyphen)

Ingoldsby Legends, by R. H. Barham, 1840

ingraft, ingrain (verb), use **en-**

ingrain/ (adj.), dyed in the yarn **-ed** (adj., less specific), deeply rooted, inveterate. *See also* **en-grain**

Ingres (**Jean Auguste Dominique**), 1780–1867, French painter (no accent)

in/gross, -gulf, use **en-**

in-group, small exclusive group of people (hyphen)

inhabitant, abbr. **inhab.**

in hac parte (Lat.), on this part

in/ hoc (Lat.), in this respect; — *hoc salus,* safety in this; — *infinitum,* for ever

in-house (adj., hyphen)

INI, *in nomine Iesu* (in the name of Jesus)

Inishfail, poet. for Ireland, *not* Inn-

init., initio (Lat.), in the beginning

initial/, -led, -ling

initial letter (typ.), large letter used at beginning of chapter

Initial Teaching Alphabet (caps.), abbr. **i t a** (spaced l.c., no points)

injuri/a (Lat.), a wrong; *pl.* **-a**

ink-blot test (one hyphen)

Inkerman, Crimea, *not* -ann

ink/pot, -stand (one word)

ink-well (hyphen)

in-laws, relatives by marriage (hyphen)

in/ limine (Lat.), at the outset, abbr. *in lim.*; — *loco,* in place of; — *loco citato,* in the place cited; — *loco parentis,* in the position of parent; — *medias res,* into the midst of affairs; — *medio,* in the middle; — *medio tutissimus ibis,* the middle course is safest; — *memoriam,* to the memory (of)

innamorat/o, *fem.* **-a,** mod. It. spelling of **inam-**

inner (typ.), the side of a sheet containing the second page

Innes (Cosmo), 1798–1874, Scottish historian; — **(James Dickson),** 1887–1914, British painter; — **(Michael),** pseud. of John Innes Mackintosh Stewart, b. 1906, British novelist; — **(Thomas),** 1662–1744, Scottish historian

Inness (George), 1825–94, US painter

innings (US **inning**), portion of game played by one side; *pl. same*

Innisfail, *use* **Inishfail**

Inniskilling Dragoon Guards. *See also* **Enniskillen**

innkeeper (one word)

Innocents' Day, 28 Dec. (caps.)

innoculate, *use* **ino-**

in/ nomine (Lat.), in the name (of a person)

inn (name of), *see* **hotel**

Innsbruck, Austria

Inns of Court (the), Inner Temple, Middle Temple, Lincoln's Inn, Gray's Inn

in/ nubibus (Lat.), in the clouds; — **nuce,** in a nutshell

innuendo/, an injurious insinuation, *pl.* **-es,** *not* inu- (not ital.)

Innuit, a Canadian Eskimo. *See also* **Yuit**

inoculate, to inject an immunizing serum into, *not* en-, inn-

in/ pace (Lat.), in peace; — **pari materia,** in an analogous case; — **partibus infidelium,** in the regions of unbelievers, abbr. **i.p.i.** *or* **in partibus**; cf. **Bishop**

in-patient (hyphen)

in petto (It.), secretly

in/ pontificalibus, in pontifical vestments; — **posse,** potentially; — **potentia,** potentially; — **primis,** *use* **imprimis**; — **principio,** in the beginning, abbr. **in pr.**; — **propria persona,** in his, *or* her, own person; — **puris naturalibus,** naked

input/ (verb), **-ting,** past and partic. **input**

inquire, to undertake formal investigation. *See also* **enquire**

inquiry, an off. investigation. *See also* **enquiry**

inquorate, lacking a quorum

in/ re (Lat.), in the matter of; — **rem** (law), relating to a matter; — **rerum natura,** in the nature of things

INRI, *Iesus Nazarenus, Rex Iudaeorum* (Jesus of Nazareth, King of the Jews)

in/road, -rush (one word)

ins., insurance

in saecula saeculorum (Lat.), for ever and ever

insconce, *use* **en-**

in se (Lat.), in itself, in themselves

Insecta (zool.), (Lat. pl.) insects, *not* -ae

insert (bind.), bookmark, advertisement, etc., slipped loose inside bound pages

inset (bind.), a folded section (min. 4 pages) placed inside (usu. at centre) of another section, so that sewing passes through both; a small map etc. printed within the borders of a larger one

inshallah (Arab.), (if) God (is) willing

insign/ia, badges of office, is pl.; sing. **-e** (not ital.)

insisten/ce, -t, *not* -ance, -ant

in situ (Lat.), in position

insnare, *use* **en-**

in so far (three words)

insomuch (one word)

insoucian/ce, lack of concern; **-t,** unconcerned (not ital.)

Inspector, abbr. **Insp.**

Inspector-General, abbr. **IG** *or* **Insp.-Gen.**

INST, *in nomine Sanctae Trinitatis* (in the name of the Holy Trinity)

inst., instant, institut/e, -es, -ion

Inst. Act., Institute of Actuaries

install/, -ation (two *l*s)

instalment (one *l*)

instant, of this month, abbr. **inst.**

instantaneous, occurring or done instantly

instanter, at once (jocular) (not ital.)

in statu/ pupillari (Lat.), in a condition of pupillage; —— *quo* (*ante, prius*; *nunc*), in the same state (as formerly; as now)

Inst./D. Institute of Directors; —F., ditto Fuel

instil/, to inculcate gradually; **-led, -ling**

Institut de France, the association of five French Academies

Institut Français du Royaume-Uni, the French Institute, London

Institute, *see* separate entries beginning **I.** *or* **Inst.** (Institute of), **FI** (Fellow of the Institute of), **RI** (Royal Institute of)

Institution, *see* separate entries beginning **I.** (Institution of), **FI** (Fellow of the Institution of), **RI** (Royal Institution of),

institutionalize, *not* -ise, -ionize

institutor, *not* -er

Inst./P., Institute of Physics; —R., ditto Refrigeration

instruct/or, *not* -er; *fem.* **-ress**

instrument/, -al, abbr. **instr.**

insurance, when effected against a risk, but **assurance** (q.v.) of life; **insur/e, -er** in all insurance senses; abbr. **ins.**

inswathe, *use* **en-**

int., interest, interior, interjection, internal, international, interpreter

intaglio/, incised design, *pl.* **-s** (not ital.); *see also* **cameo**; (typ.), a printing process based on an etched or incised plate

intailed, *use* **en-**

intangible, *not* -able

intarsia, mosaic woodwork

integration (sign of, in math.), ∫

intelligentsia (collect. noun), the intellectual part of population, *not* -zia

Intelsat, International Telecommunications Satellite Consortium

in tenebris (Lat.), in darkness, in doubt

inter/, to bury; **-red, -ring**

inter/ (Lat.), between; — *alia,* among other things; — *alios,* among other persons

inter., intermediate

intercom, internal telephone system (no point)

inter-continental (hyphen)

interest, abbr. **int.**; in Fr. m. *intérêt*

interface (comp.), boundary or link between two computer units (one word)

interim, (noun) the mean while; (adj.) temporary (not ital.)

interior, abbr. **int.**

interjection, abbr. **int.** *or* **interj.**

interlea/f, an extra leaf, usu. blank, inserted between the regular leaves of a book for notes or to protect illustration; caption pasted on inner margin of plate; *pl.* **-ves; -ve,** to insert such a leaf

interlinear matter (typ.), small type between lines of larger

intermarr/iage, -y (one word)

intermedi/ate, abbr. **inter.**; **-um,** an intervening agent, *pl.* **-a** (not ital.)

intermezz/o, short piece of entertainment, *pl.* **-i**; (cap., ital.) opera by Richard Strauss, 1924

intermit/, to stop for a time; **-ted, -tent, -ting**

internal/, abbr. **int.**; — **-combustion engine** (one hyphen)

international, abbr. **int.** *or* **internat.**

International (**the first**), association of working classes of all countries (Marxist, 1862–73); — (**the second**) (French,

1889–   ); — (**the third**) (Russian communist, 1919–43), in Fr. **Internationale,** also called the Comintern; — (**the fourth**) (Trotskyist, 1938–   )

Internationale (The), French socialistic hymn

internationalize, *not* -ise

inter nos (Lat.), between ourselves

internuncio/, papal ambassador, *pl.* **-s** (not ital.)

interoceanic (one word)

interpellate (Parl.), interrupt to demand explanation

interplanetary (one word)

Interpol, the International Criminal Police Commission

interpolate, make insertions

interpret/, -ed, -er, abbr. **int., -ing; -ative,** *not* -pretive

interregn/um, period between one ruler and another; *pl.* **-ums**

interrelat/ed, -ionship (one word)

interrog., interrog/ation, -ative, -atively

in terrorem (Lat.), as a warning

interrupter, *not* -or

inter/ se (Lat.), among, *or* between, themselves; — *vivos,* from one living person to another

Intertype, propr. term for a composing machine that casts lines of type, and for (computer--driven) photosetting systems developed by firm of this name

inter-war, of period between two wars (hyphen)

in testimonium (Lat.), in witness

inthron/e, -ization, *use* en-

intitule, etc., *not* en-, *but* **entitle**

intonaco (It.), plaster surface for fresco painting, *not* -ico

in toto (Lat.), entirely

intra/ (Lat.), within; — *muros,* privately

intra/muscular, abbr. **IM; -venous,** abbr. **IV**

intrans., intransitive

intransigent (adj.), uncompromising; in Fr. *intransigeant* (m. and adj.) (no accent)

intrap, *use* **en-**

intra-uterine, within the womb (hyphen)

intravenous, within a vein (one word)

intra vires (Lat.), within one's powers

in-tray (hyphen)

intreat, *use* **en-**

intrench, *use* **en-**

intrigant/, *fem.* **-e,** intriguer

introduction, abbr. **introd.** *See also* **preliminary matter**

introvert, *not* intra-

intrust, *use* **en-**

intussusception (physiol.), taking in of foreign matter by living organism, withdrawal of one portion of tube into another (double *s*)

intwine, intwist, *use* **en-**

inuendo, *use* **inn-**

inure, to accustom, (law) to take effect, *not* en-

in/ usu (Lat.), in use; — *utero,* in the womb; — *utroque iure,* under both laws (canon and civil)

in vacuo (Lat.), in empty space

Invalides (Hôtel des), Paris, *see* **Hôtel**

invenit (Lat.), designed this; abbr. *inv.*

invented, abbr. **inv.**

inventor, *not* -er; abbr. **inv.**

Inveraray, Strathclyde, *not* -ry

inverness, man's sleeveless coat with removable cape (not cap.)

Inverness-shire, former county of Scotland (hyphen)

Invertebrata (coll. noun), all animals other than vertebrates

inverted commas, *see* **quotation marks**

inverter, *not* -or

Inverurie, Grampian

investor, *not* -er

inv., invenit (designed this)

inv., invent/ed, -or, invoice

in vino veritas (Lat.), a drunken man speaks the truth

invita Minerva (Lat.), uninspiredly

in vitro (Lat.), in the test-tube

in vivo (Lat.), in the living organism

invoice, abbr. **inv.**

involucre (anat., bot.), a covering, envelope (not ital.)

inweave, *not* en-

inwrap, inwreathe, *use* en-

I/O (comp.), input/output

io (Gr., Lat.), exclamation of triumph (ital.)

iodine, symbol **I**

IOF, Independent Order of Foresters

IOGT, International Order of Good Templars

I.o.M., Isle of Man (also **IOM**)

Ion., Ionic

Ion/ian, (hist.) of Ionia, (mus.) the mode; **-ic** (of) dialect and archit. order

ionize, to convert into ions, *not* -ise

IOOF, Independent Order of Odd Fellows

IOP, Institute of Painters in Oil Colours

IOR, Independent Order of Rechabites

iota/, the Gr. *i* (*ι*, no dot); — **adscript,** printed after lower-case *α, η, ω* : *αι, ηι, ωι*; — **subscript,** printed beneath: *ᾳ, ῃ, ῳ.* See also *Hart's Rules*, p. 111

IOU, I owe you

IOW, Isle of Wight (*also* **I.o.W., IW**)

Iowa, off. abbr. **Ia.** or (postal) **IA**

IP, input primary

IPA, International Phonetic Alphabet, ditto Association

IPCS, Institution of Professional Civil Servants

IPD (Sc. law), *in praesentia Dominorum* (in the presence of the Lords [of Session])

ipecacuanha (bot., med.), a purgative root

i.p.i., *in partibus infidelium* (in the regions of unbelievers)

IPI, International Press Institute

IPM, Institute of Personnel Management

I.Prod.E., Institution of Production Engineers

i.p.s., inches per second

ipse dixit (Lat.), he himself has said it

ipsissima verba (Lat.), the very words

ipso facto (Lat.), by the fact itself

IQ, intelligence quotient

i.q., *idem quod* (the same as)

Iquique, Chile

IR, infra-red, Inland Revenue

Ir, iridium (no point)

Ir., Irish

IRA, Irish Republican Army

irade (Turk.), a written decree signed by the Sultan himself

Iran/, *formerly* Persia; **-ian, -ic,** abbr. **Iran.**

Iraq/, *formerly* Turkish Mesopotamia, *not* Irak; adj. **-i**

Irawadi, *use* **Irrawaddy**

IRBM, intermediate range ballistic missile

IRC, International Red Cross

Ireland, abbr. **Ire.**

Irena, Ireland personified

irenicon, *use* ei-

Irian Jaya, Indon. for W. New Guinea

iridescen/ce, play of rainbow colours; **-t,** *not* irr-

iridium, symbol **Ir**

iris/ (anat., bot.), *pl.* **-es**

Irish, abbr. **Ir.**

Irishism, *not* Iricism

Irkutsk, E. Siberia, *not* Irkoo-, Irkou-

IRO, International Refugee Organization, Inland Revenue Office

iron, symbol **Fe** (*ferrum*)

ironclad (one word)

Iron Curtain (two words, caps.)

iron-mould (hyphen)

iron ration (two words)

ronside/, sobriquet of Edmund II; — (**William Edmund, Baron**), 1880–1959, British soldier; **-s,** Cromwell's troops in the Civil War

ron/ware, -work (one word)

roquois, American Indians

rrawaddy, river in Burma, *not* Irawadi

rreconcilable, *not* -eable, -iable

rredentist, one who (re)claims regions for his country on grounds of language, esp. (*cap.*) Italian

rrefragable, unanswerable, *not* -ible

rreg., irregular, -ly

rrepairable, of things broken

rreparable, of losses

rreplaceable, *not* -cable

rresistibl/e, -y, *not* -abl/e, -y

rridescen/ce, -t, *use* **iri-**

rtysh, river, cent. Asia

rvine, Strathclyde, 'new town', 1966

S, input secondary, Irish Society

s., islands, isles

sa., Isaiah

saian, of the prophet Isaiah, *not* Isaiahian

SBN, International Standard Book Number

SC, Imperial Service College (Windsor), now combined with Haileybury

ise, *see* **-ize**

sère, river and dép., SE France

seult, Tristram's lady-love, *not* the many variants (*but* **Isolde** in Wagner's opera)

sfahan, Iran, *not* Isp-

SI, International Statistical Institute, Iron and Steel Institute

sl., island, -s, isle, -s

slam, lit. 'surrender (to God)', the Muslim religion. *See also* **Muslim**

sland/, -s, abbr. **I.** *or* **isl.**; when with name to have cap., as Cape Verde Islands, Isle of Man. In Fr. f. *île*

sland, Ice. for Iceland

Island (Ger. and Dan. n.), Iceland; *Isländer* (Ger. m.), an Icelander

Islay, Strathclyde, *not* Isla

isle/, -s, abbr. **I.** *or* **isl.**

Isle of/ Man, abbr. **I.o.M.** *or* **IOM**; — — **Wight,** abbr. **IOW, I.o.W.,** *or* **IW**

Isleworth, London

ism, the suffix used as generic noun

ISM, Imperial Service Medal; Incorporated Society of Musicians

Ismaili, (member) of a Muslim sect

isn't, is not, to be printed close up

ISO, (Companion of the) Imperial Service Order, International Organization for Standardization

isobar (meteor.), *not* -are (not ital.)

Isocrates, 436–338 BC, Athenian orator

isola (It.), island

Iso/ld, -lde, -lt, -lte, -ulde, *see* **Iseult**

isosceles (geom.), of a triangle, having two sides equal

isotop/e, a form of an element differing from other forms in the mass of its atoms; adj. **-ic**

isotron, a device for separating isotopes by accelerating ions

isotrop/ic (phys.), having same properties in all directions; noun **-y**

Ispahan, *use* **Isfahan**

Israel/i (noun and adj.), (a citizen) of the modern state of Israel, est. 1948; **-ite** (noun), one of the Jewish people, adjs. **-itic, -itish**

ISSN, International Standard Serial Number

issue (**second**) (bibliog.), state resulting from sheets of same **edition** (q.v.) being bound up with new title-page and additional matter, or in a different order. The original state of that

issue (*cont.*)
edition is then called the **first issue**. *See also* **impression**
Istanbul, Turkey, *formerly* Constantinople
isth., isthmus
Isthmian games, one of the four principal Panhellenic festivals
I.Struct.E., Institution of Structural Engineers
It., Italian, Italy
ITA, Independent Television Authority, *now* **IBA**
i t a (spaced l.c., no points), Initial Teaching Alphabet
ital., italic
Italian (typ.), same alphabet as Eng., omitting *k, w, x, y* (and *j* except in names). Some words take a grave accent (or acute, if the author prefers it) on the final vowel: with even full caps. a final apostrophe may be used. For word-division and spacing, see *Hart's Rules,* pp. 116–17. Abbr. **It.**
italic (typ.), *a style of type, as this,* indicated in copy by a single underline. Use italic in English for foreign words and phrases not naturalized; for words or letters mentioned by name, as the letter *a*; for titles of books, very long poems, plays, films, operas, works of art, newspapers, and periodicals; and as a method of emphasizing or distinguishing. There is no italic in German Fraktur type, Greek, or Bernard Shaw's plays, letters being interspaced instead. Abbr. **ital.** *See also* **botany, quotation marks, zoology,** the many words and phrases in this dictionary where a preference for italic or roman is indicated, and a fuller treatment in *Hart's Rules,* pp. 23–8
italice (Lat.), in Italian
italicize, to print in italic type, *not* -ise

italienne (*à l'*) (Fr.), in Italian style
Italiot, of the ancient Greek colonies in S. Italy, *not* -ote
Italy, abbr. **It.**; in Fr. f. *Italie*; in Ger. n. *Italien*; in It. *Italia*
item, a separate thing
item (Lat.), also, likewise
itemize, to give item by item, *not* -ise
itin., itinerary
ITO, International Trade Organization
its, poss. pronoun (no apos.)
it's, it is; (typ.) no space before apos. *See also* **punctuation** VIII. 2
ITU, International Telecommunication Union (UN)
ITV, Independent Television
IU, international unit
IUCD, intra-uterine contraceptive device
IV, intravenous, (roman numeral) four (no point); *see also* **IIII**
Ivanovich (**Ivan**), nickname for a Russian, as in Eng. John Bull
I've, I have, to be printed close up
Iveagh (**Earl of**)
ivied, clothed with ivy, *not* ivyed
Iviza, *use* **Ibiza**
Ivory Coast, W. Africa, indep. 1960
ivy, *pl.* **ivies**
Ivybridge, Devon (one word)
IW, Isle of Wight
IWES, Institution of Water Engineers and Scientists
Iwo Jima, island in N. Pacific, taken by US Marines, 1945
IWW, Industrial Workers of the World (US)
IX (roman numeral), nine (no point)
Izaak (*not* Isaac) **Walton,** q.v.
-ize, use in preference to -ise as verbal ending where both spellings are in use. Use **-ise** in advertise, advise, apprise, arise chastise, circumcise, comprise,

ompromise, demise, despise,
evise, disfranchise, disguise,
mprise, enfranchise, enterprise, excise, exercise, franchise,
mprovise, incise, merchandise,
premise (verb), prise (open),

promise, reprise, supervise, surmise, surprise, televise. Many
examples are given in this book.
See also *Hart's Rules*, pp. 85–6

I Zingari, cricket club

İzmir, Turk. for **Smyrna**

J

J (phys.), joule, Joule's mechanical equivalent of heat; it is not used in the numeration of series

J., jack (at cards), judge, *judex* (judge), (after judge's name) Justice, (Ger.) *Jahr* (year)

j (med. prescriptions), one; (math.) square root of minus one

JA, Judge Advocate

Jabalpur, Madhya Pradesh, India, *not now* Jubbulpore

jabot, ornamental frill on woman's bodice *or* (hist.) man's shirt-front

Jac., *Jacobus* (James)

jacana, small tropical wading bird, *not* jaç-, jass-

jacconet, *use* **jaco-**

jackanapes (noun, sing.), a pert fellow (one word)

Jack and Gill, correct, but popularly **Jill**

jackaroo (Austral. slang), a newcomer (usu. from England), *not* -eroo

jackass, a male ass, a stupid person

jackboot (one word)

jacket/, -ed, -ing

jack-in-the-box (hyphens)

Jack Ketch, the hangman, *not* — Ca-, — Ki-

jack of all trades (four words)

Jack Russell, *see* **Russell**

Jackson (**Andrew**), 1767–1845, US President 1829–37; — (**Thomas Jonathan, 'Stonewall'**), 1824–63, Confederate general at the first battle of Bull Run, 1861

Jacob/ean, of the reign of James I of England; of St. James the Less. *See also* **-ian**

Jacobi (**Karl Gustav Jakob**), 1804–51, German mathematician, hence (in math.) **Jacobian.** *See also* **-ean**

Jacobin, a French Dominican monk; a member of a French revolutionary society which met, 1793–9, in a former Jacobin convent; **jacobin,** a hooded pigeon; *not* -ine

Jacobite, an adherent of James II of England after his abdication in 1688, or of his line

Jacobus, Lat. for **James;** abbr. **Jac.**

jacobus, a gold coin of James I

jaconet, a medium cotton cloth, *not* jacc-

Jacquard (**Joseph Marie**), 1752–1834, French weaver; hence — **loom**

Jacquerie, the French peasant revolt of 1358

Jacques Bonhomme, good fellow James, popular name for French peasant

jacta est alea (Lat.), the die is cast

jactation, boasting

jactitation (law), false pretence of marriage; (path.) restless tossing of the body

j'adoube (Fr.), 'I adjust', said by a chess-player touching, but not moving, a man

Jaeger, propr. term for a woollen clothing material from which vegetable fibres are excluded (cap.)

jaeger, *see* **Jäger**

JAG, Judge Advocate General

Jäger (Ger. m.), huntsman, rifleman; Anglicized as **jaeger,** *not* yager

jag/gernaut, -anath, *use* **juggernaut**

jaghire, Indian land tenure, *not* -gheer, -geer, -gir

Jahr/ (Ger. n.), year, *pl.* **-e,** abbr. **J.** (*not* Jä-); **-buch** (n.), year-book, abbr. **Jb.; -esbericht** (m.), annual report; **-gang** (m.), year's issues etc., abbr. **Jg.** (caps.)

jährlich (Ger. adj.), annual

Jahveh, *use* **Yahveh,** q.v.

ai alai (Sp.), game like pelota

ail/, -er, *use* **gaol/, -er**

Jaipur, Rajasthan, India. *See also* **Jeypore**

Jakarta, Indonesia, *not* Djakarta

Jakob, Ger. for **Jacob, James**

Jakutsk, Siberia, *use* **Yakutsk**

Jalalabad, Afghanistan and Uttar Pradesh, India, *not* Jela-

Jalandhar, *use* **Jullundur**

jalopy, a dilapidated motor car, *not* -ppy

jalousie, external window shutter

jam/, to pack tightly; **-med, -ming.** *See also* **jamb**

Jamaica, indep. 1962; abbr. **Jam.**

jamb, a side post, as of a door. *See also* **jam**

jambon/ (Fr. cook. m.), a ham; **-neau,** small ham

jamboree, rally of Scouts

James, abbr. **Jas.**

Jamesone (George), 1588–1644, Scottish painter

Jameson/ Raid, S. Africa, 1895–6, from **Sir Leander Starr —,** 1853–1917, S. Afr. statesman

James's Day (St.), 25 July (caps., apos.)

Jamieson (John), 1759–1838, Scottish lexicographer. *See also* **James-**

jam satis (Lat.), enough by this time

Jan., January

Janáček (Leoš), 1854–1928, Czech composer

Janeite, an ardent admirer of Jane Austen's writings

Jane's year-books on aircraft, ships, etc. (apos.)

janizary, Turkish soldier, *not* -issary

Jan Mayen, Arctic island (Norwegian)

Jansen/ (Cornelius), 1585–1638, RC Bishop of Ypres; **-ist,**

one who believes, with Jansen, that the natural human will is perverse

Janssen (Cornelius), 1590–1665, Dutch painter; — **(Johannes),** 1829–91, German historian; — **(Pierre Jules César),** 1824–1907, French astronomer

Janssens (Abraham), 1569–1631, Dutch painter

January, abbr. **Jan.**

Janus, Roman god of doors

janvier (Fr. m.), January; abbr. *janv.* (not cap.)

Jap/ (colloq., usu. derog.), a Japanese; *pl.* **-s** (no point)

Jap., Japan/, -ese

Japan, abbr. **Jap.;** native name **Nippon**

japan/, to lacquer with a hard varnish; **-ned, -ner, -ning**

Japanese/, (typ.) printed in characters numbering several thousand, alternating with signs for some fifty syllables (each of which has two different signs), which are phonetically transcribed into the Latin alphabet according to various systems, of which the Hepburn system is the best known; abbr. **Jap.;** — **paper,** handmade (usu. in Japan). Used for proofs of etchings and engravings

Japheth, third son of Noah, *not* -et

Jaques, in Shakespeare's *As You Like It*

jar/, -red, -ring

Jardin/ d'Acclimatation, — des Plantes, Paris, bot. and zool. gardens

jardinière (Fr. f.), ornamental flower-pot

jargonelle, a pear, *not* -el (not ital.)

jarl, old Norse or Danish chieftain, *not* y-

Järnefelt (Edvard Armas), 1869–1958, Finnish composer

Jaroslav, Russ., *use* **Yaroslavl;** Pol., *use* **Jarosław**

jarrah, Australian mahogany gum-tree

jarvey/, hackney-coachman, driver of Irish car, *not* -vie, -vy; *pl.* **-s**

Jarvie (Bailie Nicol), in Scott's *Rob Roy*

Jas., James

jasmine (bot.), *not* -in, jessamine, -in

jaspé, mottled, veined, of materials and floor-coverings (not ital.)

jassana, *use* **jacana**

Jassy, Romania, *use* **Iaşi**

Jaunpur, Uttar Pradesh, India

Jaurès (Jean Léon), 1859–1914, French socialist

Jav., Javanese

javelle water, a bleach or disinfectant, *not* -el, -elle's (not cap.)

jaw-bone (hyphen)

jawohl (Ger.), yes (one word)

jay-walk/, -er (hyphens)

Jb. (Ger.), *Jahrbuch* (year-book)

JC, Jesus Christ, Julius Caesar, Justice Clerk

J.-C. (Fr.), *Jésus-Christ* (hyphen)

JCR, Junior Common Room, (Camb.) — Combination Room

JD, Junior Deacon, — Dean, *Jurum Doctor* (Doctor of Laws)

je (Fr.), I (cap. only at beginning of sentence)

Jeaffreson (John Cordy), 1831–1901, English historical writer

Jeanne d'Arc (Fr.), Joan of Arc

Jean Paul, pseudonym of J. P. F. Richter, q.v.

Jedda, Saudi Arabia, *not* Jidda, -ah

jee/-ho, -up, *use* **gee-**

Jeejeebhoy (Sir Jamsetjee), 1783–1859, Indian philanthropist

Jefferies (Richard), 1848–87, English naturalist

Jefferson (Thomas), 1743–1826, third President of US

Jefferys (Thomas), *fl.* 1732–71, English cartographer

Jeffrey (Francis, Lord), 1773–

1850, Scottish jurist and literary critic

Jeffreys (George, Baron), 1648–89, the infamous judge

jehad, *use* **jihad**

Jehlam, *use* **Jhelum**

Jehovah, traditional form of **Yahveh,** q.v.

jejune, meagre, insipid (not ital.)

Jekyll (Dr.) and Mr. Hyde (Strange Case of), by R. L. Stevenson, 1886

Jelalabad, Jellalabad, *use* **Jala-**

jellaba, Arab's hooded cloak, *not* dj-

jellify, to convert into jelly, *not* -yfy

jelly baby (two words)

jellyfish (one word)

Jemappes (battle of), Belgium, French victory over Austrians, 1792, *not* Jemm-

jemimas, elastic-sided boots (not cap.)

je ne sais/ quoi (Fr.), an indescribable something ; — — — *trop,* I don't exactly know

Jenghis Khan, *use* **G-**

Jenisesi, *use* **Yeniseisk**; **Jen-issei,** *use* **Yenisei**

Jenkins's Ear, incident which precipitated war with Spain, 1739

jennet, a Spanish horse, *not* genit, gennet, -tt. *See also* **genet**

jeopardize, to endanger, *not* -ise

Jephthah, judge of Israel: *Jephtha,* oratorio about him by Handel, 1752

jequirity, Indian shrub with ornamental and medicinal seeds, *not* -erity

Jer., Jeremiah

jerbil, *use* **ger-**

jeremiad, a lamentation, *not* -de

Jerez, Spain, *not* Xeres

jerfalcon, *use* **ger-**

jeroboam, winebottle containing 10–12 quarts

Jérôme Bonaparte, 1784–1860, brother of Napoleon I

jerry/-builder, builder of

unsubstantial houses (hyphen);
-building, -built

errymander, *use* **ge-**

ersey, a knitted fabric, a garment of same (not cap.)

ervaulx Abbey, N. Yorks.

es., Jesus

espersen (Jens Otto Harry),
1860–1943, Danish philologist,
not **-son**

essamin/, -e, *use* **jasmine**

esse, father of David (OT)

esuits (Order of), *Societas Jesu*
(Society of Jesus); abbr. **SJ**

esus, abbr. **Jes.;** in voc. (arch.)
Jesu

et-black (hyphen)

et/ engine, — lag (two words)
et-propelled (hyphen)
et propulsion (two words)

etsam (naut.), goods thrown
overboard, *not* **-som, -some, -son**

ettison (naut.), the act of throwing goods overboard

eu/ (Fr. m.), game, *pl.* **-x;** — ***de
mots,*** play upon words; — ***de
paume,*** real tennis(-court);
Musée du Jeu de Paume
(caps.), Paris, exhib. of impressionist paintings; — ***d'esprit,*** a
witty trifle

eune fille (Fr. f.), a girl

eune/ premier (Fr.), a stage
lover; *fem.* — ***première***

eunesse dorée (Fr. f.), gilded
youth

Jewel (John), 1522–71, English
bishop

ewel/, -led, -ler, -lery

Jew's harp, small lyre-shaped
instrument played against the
teeth

eypore, Orissa, India. *See also*
Jaipur

g. (Ger.), *Jahrgang* (year's issues)
etc.)

helum, Pakistan, *not* Jehlam

hind, India, *use* **Jind**

HS, *use* **IHS**

ib, of horse, *also* a sail, *not* jibb

ibbah, Muslim man's long coat,
not dj-

ibber, to chatter, *use* **g-**

jibe, *see* **gibe, gybe**

Jibuti, *formerly* French Territory
of Afars and Issas, *not* Djibouti

JIC, Joint Industrial Council

Jidda, *use* **Jedda**

jiffy, a short time, *not* -ey

jigger, W. Indian parasite, *use*
chigoe

jigot, *use* **gigot**

jigsaw/ (one word); — **puzzle**
(two words)

jihad, a holy war (Muslim), *not*
jc-

jillaroo (Austral. slang), a female station-hand, *not* -eroo

jilliflower, *use* **gilly-**

jimcrack, *use* **g-**

Jímenez (Juan Ramón),
1881–1958, Spanish poet

Jind, town and former state,
India, *not* Jh-

Jingis Khan, *use* **Ge-**

jingo/, in declaration, **by —!;** a
fanatical patriot, *pl.* **-es; -ism,
-istic**

jinnee, a spirit in Muslim
myth., *not* djinn, ginn; *pl.* **jinn,**
often used as sing. *See also*
genie

jinricksha, *use* **rickshaw**

jiu-jitsu, *use* **ju-jitsu**

JJ., Justices

Jno., John, but use only in exact
reprints of documents etc.

jnr., junior, *use* **jun.**

JO, *Journal Officiel*

joannes, *use* **johan-**

jobbing work (typ.), minor
pieces of printing

job lot (two words)

jobmaster, livery-stable keeper
(one word)

'Jock o' Hazeldean', mainly by
Sir W. Scott

jock-strap (hyphen)

jodel, *use* y-

Jodhpur, Rajasthan, India

jodhpurs, long riding-breeches

johannes, a gold coin of John V
of Portugal, *not* joa-

Johannesburg, S. Africa

Johannine, of the apostle John,
not Johannean

Johannisberg, Hesse, W. Germany

Johannisberger, Rhine wine, *not* -berg (not ital.)

John, abbr. **J.** *See also* **Jno.**

John Dory, a fish, *not* — -ey; *also* **dory**

Johnian, (member) of St. John's College, Cambridge

johnny, fellow, man (not cap.)

Johnny Crapaud, nickname for a Frenchman

John o' Groat's House, Highland

Johns Hopkins University, Baltimore, Md., US, *not* John — — (no apos.)

Johnson (Lyndon Baines), 1908–73, President of US 1963–9; — **(Samuel),** 1709–84, English prose-writer and lexicographer. *See also* **Jonson**

Johnsonese, a stilted style, *not* Jon-

John the Baptist (caps.)

Johore, *see* **Malaya (Federation of)**

joie de vivre (Fr.), joy of living (no hyphens)

joint/ capital, — **stock,** etc. (hyphen when attrib.)

Jokjakarta, Indonesia

jole, *use* **jowl**

jolie laide (Fr. f.), fascinatingly ugly woman

Jon., Jonathan

Jonah (Book of), not to be abbr.

jonquil, a narcissus (not ital.)

Jonson (Benjamin—'Ben', 1572–1637, English playwright. *See also* **Johnson**

Joppa, anc. name of Jaffa, Israel

Jordaens (Jakob), 1593–1678, Dutch painter

Jordan/, a united kingdom from 1950; adj. **-ian**

Jorrocks' Jaunts, by R. S. Surtees, 1831–4

Jos., Joseph

Josephine (the Empress), Marie Josèphe-Rose Tascher de la Pagerie, widow of Visc. Beau-

harnais, m. Napoleon I 1796, divorced 1809

Josh., Joshua (OT)

jostl/e, to push; **-er, -ing**

jot/, -ted, -ting

joule (phys.), SI unit of energy, abbr. **J**

jour., journal, journey

jour/ (Fr. m.), day, abbr. *jr.*; — *de fête,* a festival; — *de l'an,* New Year's Day; — *des morts,* All Souls' Day, 2 Nov.; — *gras,* flesh-day; — *maigre,* fish-day

journal, abbr. **jour.**

journ/al (Fr. m.), *pl.* **-aux,** newspaper; *journal intime,* a private diary

Journal Officiel, the French and EEC equivalents of *London Gazette,* abbr. *JO*

journey, abbr. **jour.**

journeyman (typ.), a compositor, printer, or bookbinder who has completed his apprenticeship

joust, knightly combat, *not* just

jowl, the jaw, *not* jole

JP, Justice of the Peace

JR, *Jacobus Rex* (King James)

jr., junior, *use* **jun.**

jr. (Fr.), *jour* (day)

jt., joint

JTC, Junior Training Corps (in schools)

Juan, Sp. for **John**

Jubbulpore, *use now* **Jabalpur**

Jubilate Deo, Ps. 100 as canticle

jud., judicial

Judaean, of Judaea, *not* Judean

Judaeo-, Jew(ish), *not* -deo-

Judaize, to make Jewish, *not* -ise

Judg., Judges (OT)

Judge/, abbr. **J.**; — **Advocate/** (no hyphen), abbr. **JA;** — — **General** (no hyphen), abbr. **JAG** (caps.)

judgement, *but* **judgment** in legal works

Judges (OT), abbr. **Judg.**

judicial, abbr. **jud.**

Judith (Apocr.), not to be abbr.

judo/, modern form of ju-jitsu; **-ist**

ug/, -ged, -ging, -ful

uge/ de paix (Fr. m.), justice of the peace; — d'instruction, examining magistrate

ugendstil, Ger. name of art nouveau (ital., cap.)

uggernaut, (cap.) incarnation of Vishnu, carried on huge wheeled vehicle under which devotees were said to sacrifice themselves; (not cap.) large heavy vehicle, not ja-

ugoslavia, use Yugoslavia

ugular vein (in the neck)

uillet (Fr. m.), July (not cap.)

uin (Fr. m.), June (not cap.)

u-jitsu, Japanese wrestling, not jiu-. See also judo

uke-box (hyphen)

ulep, a medicated drink; (US) drink of spirits, sugar, ice, and mint, not -ap, -eb

ulian/, 'the Apostate' Roman emperor, 331–63; — Alps, Italy–Yugoslavia

ulien (Saint-), a claret (hyphen)

ulienne (Fr. f.), a clear soup containing vegetables in strips, also a pear

uliet cap (two words, not cap.)

ullundur, India, not Jalandhar

uly, not to be abbr.

umble sale (two words)

umbo/, big person or thing; pl. -s; — jet (two words)

umelles (Fr. f. pl.), opera--glasses

un., Junius

un., junior

unction, abbr. junc.

une, not to be abbr.

uneau, Alas., US

ung (Carl Gustav), 1875–1961, Swiss psychiatrist

unior, abbr. jun., not jnr., jr., junr.; J. Smith, jun., Esq., not Esq., jun.

unker (Ger. m.), a young squire or noble (cap.)

unket/, -ed, -ing

unk-shop (hyphen)

unr., junior, use jun.

junta, political clique, a council in Spain or Italy

Junto, the Whig chiefs in reigns of William and Anne

jupe (Fr. dress. f.), jupon (m.), a skirt or petticoat

Jurassic (adj.), a geol. period (cap., not ital.)

jure/ divino (Lat.), by divine right; — humano, by human law

jurisp., jurisprudence

Juris utriusque Doctor (Lat.), Doctor of both civil and canon law; abbr. JUD

jury-box (hyphen)

jury/man, -woman (one word)

jus (Fr. cook. m.), gravy

jus (Lat.), law; — canonicum, canon law; — civile, civil —; — divinum, divine —; — gentium, law of nations; — gladii, the right of the sword

jusjurandum (Lat.), an oath; pl. jurajuranda

jus/ mariti (Lat.), right of husband to wife's property; — naturae, law of nature; — primae noctis, droit de seigneur; — relictae, right of the widow

Jussieu (Adrien de), 1797–1853, and his father, — (Antoine Laurent de), 1748–1836, French botanists

Just., Justinian

just, a knightly combat, use joust

juste milieu (Fr. m.), the golden mean

Justice, a judge; abbr. J., pl. JJ.

Justice Clerk (Lord), second highest Scottish judge (caps.); abbr. JC

Justice General (Lord), highest Scottish judge (caps.)

Justice of the Peace, abbr. JP

justiciar/, -y, a judge, not -er, -itiar (not ital.)

Justiciary (High Court of), supreme Scottish criminal court

justify (typ.), to put equal spaces between the words in a line of type; to adjust types of differing

justify (*cont.*)
(or the same) founts etc. so that
they range
Justinian, 483–565, Roman
Emperor of the East, codified
Roman law; abbr. **Just.**
just/itiar, *use* **-iciar**
jut/, **-ted, -ting**
Juvenal (in Lat., **Decimus**

Junius Juvenalis), *c.*60–140,
Roman poet, abbr. **Juv.**
juvenescen/ce, -t, (process
of) passing from infancy to
youth
juvenilia (pl.), works produce
in one's youth (not ital.)
j'y suis, j'y reste (Fr.), here I
am, here I stay

K

, *kalium* (potassium), Kelvin
temperature scale), kelvin
unit of temperature), (chess)
ing (no point); (comp.) unit of
ore-memory size, = 1,024
often taken as 1,000) words;
he tenth in a series
., (assaying) carat, king(s),
King('s), Köchel (Mozart the-
natic catalogue no.)
(as prefix), kilo-
(meteor.), cumulus
(phys.), Boltzmann constant
2, the second highest mountain
n the world, in Karakoram
Mts.

aaba, the most sacred shrine at
Mecca, in Arab. *al-Ka'bah*; not
Caaba
aan, *use* khan
abbala, *use* c-
abul, Afghan., *not* C-
adi, *use* cadi
affir, S. African Bantu, *not*
Caffre
affraria, S. Africa
afir, native of Kafiristan, cent.
Asia, *not* Caffre
afka/ (Franz), 1883–1924,
Austrian poet and novelist; adj.
-esque
aftan, kagoule, *use* c-
ail, *see* kale
aiman, *use* cay-
ainozoic, *use* C-
aisar-i-Hind, (Anglo-Ind.)
(the Caesar of India), former
Indian title of English monarch;
not Q-, -er-; — **medal,** abbr.
K.i.H. *or* KIH
aiser (Ger. m.), emperor (cap.,
not ital.)
akemono (Jap.), a wall-
-picture
al., kalendae
ale/, the cabbage genus;
—-yard, a cabbage garden,
kail-yaird only in strict Scots
aleidoscop/e, -ic, (an optical

toy) producing a changing variety of colours and shapes
kalendae (Lat.), the calends
(first day of the month), not *c*-;
abbr. *kal.*
kalend/ar, -er, -s, *use* c-
Kalevala, national epic of Finland, *not* -wala
kali/f, -ph, *use* caliph
Kalimantan, Indon. name for
part of Borneo
Kaliningrad, USSR, *formerly*
Königsberg
kalium (Lat.), potassium; symbol **K**
kalmia (bot.), American evergreen shrub, *not* c-
Kalmuck, member or language
of a Mongolian people, *not* -muk,
C-
kamarband, *use* **cummerbund**
Kamboja, *use* **Cambodia**
Kamchatka, E. Siberia, *not*
Kams-, -mtchatka, -mtschatka
Kamerad (Ger.), comrade
(word used by soldier surren-
dering)
Kamerun, *use* **Cameroon**
kamikaze, suicide pilot (not
ital.)
Kampuchea, *see* **Cambodia**
kamsin, *use* **kh-**
Kan., Kansas (off. abbr.)
kanaka, a South Sea Islander;
esp. an indentured labourer, *not*
canaker
Kanar/a, district of W. India, *not*
C-; adj. **-ese**
kanaster, a tobacco, *use* c-
Kanchenjunga, *use* **Kinchin-**
Kandahar, Afghan., *not* C-
Kandy, Sri Lanka
Kannada, Kanarese language
Kanpur, Uttar Pradesh, India,
correct modern form of **Cawnpore**
Kans/as, off. abbr. **Kan.** or
(postal) **KS;** adj. **-an**

Kant (Immanuel), 1724–1804, German philosopher

KANU, Kenya African National Union

kaolin, a fine white clay

Kap. (Ger.), *Kapitel* (chapter)

kapellmeister, director of orchestra or choir (not cap., not ital.)

Kapurthala, E. Punjab, India

kaput (slang), broken, ruined (not ital.), from Ger. *kaputt*

Karachi, Pakistan, *not* Kurrachee

Karafuto, *use* **Sakhalin**

Karaite, a member of Jewish sect which interprets scriptures literally

Karajan (Herbert von), b. 1908, Austrian conductor

Karakoram Mts., Kashmir, *not* -um

karakul, (fur of) Asian sheep, *not* c-, -cul

karat, *use* **carat**

karate, the Japanese art of unarmed combat; **karateka,** an expert in this

kari, *use* **karri**

Karl-Marx-Stadt, E. Germany, *formerly* **Chemnitz** (hyphens)

Karlovingian, *use* **Carolingian**

Karlsbad, Czechoslovakia, *use* trad. Eng. **Carlsbad** or mod. Czech **Karlovy Vary**

Karls/krona (Sweden), **-ruhe** (W. Germany), etc., *not* C-

karma (Budd.), destiny (not ital.)

Karnatic, *use* C-

Kärnten, Ger. for **Carinthia,** Austria

karoo, high pastoral plateau in S. Africa, *not* karr-

Karpathian Mts., *use* C-

karri, Australian blue gum-tree, *not* kari. *See also* **kauri**

karroo, *use* **karoo**

Kartoum, *use* **Khartoum**

kasbah, fortress of Arab quarter, *not* c-

Kashmir, NW India, *not* Cashmer(e)

Kassel, Germany, *not* C-

Kathay, *use* C-

Katherine, *see* C-

kathism/a, section of Greek Psalter, *pl.* **-ata,** not *ca-*

Kathmandu, cap. of Nepal, *r* Katm-

kathode, *use* c-

kation, *use* c-

Kattegat (Dan.), **Kattegatt** (Swed.), the strait between Denmark and Sweden, *not* C-

Kauffmann (Angelica), 174 1807, English painter

Kaufman (George Simon), 1889–1961, US playwright

Kaunas, Lithuania, *not* Kovno

kauri, N. Zealand coniferous tree, etc., *not* the many varian (not ital.). *See also* **cowrie, karri**

Kavafis, *use* **Cavafy**

kavass, Turkish armed attendant, *not* the many variants

kayak/, Eskimo canoe, *not* the many variants; **-er, -ing**

Kaye-Smith (Sheila), 1889– 1955, English novelist (hyphen

Kazakhstan, a Soviet Socialis Republic; adj. **Kazakh.** *See* **USSR**

Kazan, USSR, *not* Kas-

Kazantzakis (Nikos), 1885– 1957, Greek writer

KB (chess), king's bishop (no points), King's Bench, Knight Bachelor, Knight of the Order of the Bath

KBD, King's Bench Division

KBE, Knight Commander of th Order of the British Empire

KBP (chess), king's bishop's pawn (no points)

KC, King's College, — Counsel

kc, kilocycle. *See also* **kc/s**

KCB, Knight Commander of th Order of the Bath

KCIE, Knight Commander of the Order of the Indian Empir

KCL, King's College, London

KCMG, Knight Commander o

the Order of St. Michael and St. George

kc/s, *better* **kHz,** q.v.

kčs, koruna

KCSI, Knight Commander of the Order of the Star of India

KCVO, Knight Commander of the Royal Victorian Order

KD, knocked down

KE, kinetic energy

Keats House, Hampstead (no apos.)

kebab, cubes of meat roasted on a skewer, *not* the many variants

keblah, *use* **kiblah**

Kedah, *see* **Malaya (Federation of)**

kedgeree (cook.), a dish of rice, fish, etc., *not* the many variants (not ital.)

Kedron, *see* **Kidron**

Keele, Staffs.; university 1962

keelhaul, naval punishment (one word)

keelson, *use* **kelson**

keepsake (one word)

keep standing (typ.), keep type or film stored in page form after first printing for possible reprint; abbr. **KS**

keeshond, Dutch breed of dog

keffiyeh, Bedouin Arab head-dress, *not* the many variants

Kelantan, *see* **Malaya (Federation of)**

kell/eck, -ick (naut.), *use* **killick**

Kellner/ (Ger.), *fem.* **-in,** waiter (cap.)

Kellogg Pact, Paris, 1928, fifteen leading nations renounced war

Kelly's Directories

Kelmscott Press, 1891–8, founded by W. Morris

kelpie, a water-spirit, *not* -y

kelson (naut.), inboard keel, *not* keel-

Kelt/, -ic, -icism, *use* **C-**

kelter, *use* **ki-**

kelvin, sci. unit of temperature;

symbol **K** (no point). *See also* **Thomson**

Kempis (Thomas à), 1379–1471, German monk, reputed author of *Imitatio Christi*

kendo, Japanese art of fencing with bamboo swords

Kenia, *use* **-ya**

Kenmare, Kerry

Kenmore, Highland, Tayside

kennel/, -led, -ling

Kenney (James), 1780–1849, Irish playwright

kentle, *use* **quintal**

Kents Bank, Cumbria (no apos.)

Kentucky, off. abbr. **Ky.** or (postal) **KY**

Kenwigs (Morleena), in Dickens's *Nicholas Nickleby*

Kenya/, E. Africa, indep. 1963, *not* -ia; adj. **-n**

Kenyatta (Jomo), 1893–1978, Kenyan statesman

kephalic, *use* **c-**

kepi (no accent, not ital.), Anglicized form of Fr. *képi,* mil. cap

Kepler (Johann), 1571–1630, German astronomer, *not* Kepp-

keramic etc., *use* **c-**

kerb/, -stone (one word). *See also* **curb**

Kerch, Crimea, *not* Kertch

kerfuffle, disorder, fuss, *not* cur-, ge-, kur-

Kerguelen Island, Indian Ocean

Kerkyra, mod. Gr. name of **Corfu**

kermis (Du.), a fair or entertainment, *not* -ess, *kirmess*

kerned (typ.), said of a sort which has any part of the face projecting beyond the body

kernel/, seed within a hard shell; **-led, -ly**

kerosene (US, Canada, Austral., and NZ), paraffin, *not* -ine

kerseymere, *not* cassimere. *See also* **cashmere**

Kertch, *use* **Kerch**

ketch (naut.), a two-masted vessel

Ketch (**Jack**), the hangman, *not* Ca-, Ki-

kettledrum (one word)

Keuper (geol.), uppermost division of the Trias

keV, kilo-electron-volt

key, a wharf, *use* **quay**

keyboard (mus., etc.) (one word); (typ.), *see* **hot metal**; (comp.) input of terminal, VDU, etc.

key-bugle (hyphen)

key/hole, -note (one word)

Keynes (**John Maynard, Baron**), 1883–1946, British economist

key-ring (hyphen)

Keys (**House of**), Isle of Man (caps.); abbr. **HK**

keystone (one word)

Key West, Florida, US (two words)

keyword (one word)

KG, Knight of the Order of the Garter, (Ger.) *Kommanditgesellschaft* (limited partnership)

kg., kilogram(s) (*but* no point in scientific and technical work)

KGB, USSR 'Committee of State Security', 1953– . See also **MGB**

Kgs. (**1, 2**), Kings, First, Second Book of (OT)

Khachaturyan (**Aram Il-yich**), 1903–78, Armenian composer

Khaibar Oasis, Saudi Arabia

Khaiber Pass, *use* **Khyber**

khaki

khal/eefate, -ifat, *use* **caliphate**

khalif/, -a, *use* **caliph**

khamsin, Egyptian hot wind, *not* ka-, -seen

khan, Eastern inn in a town or village, *not* kaan. See also **caravanserai**

kharaj (Turk.), tax on Christians, not *-ach, -age, caratch*

Khartoum, cap. of Sudan, *not* Ka-, -tum

Khayyám (**Omar**), ?1050–1123, Persian poet

khediv/e, a viceroy of Egypt between 1867 and 1914, cap. : title; **-a,** his wife; **-ate,** his office adj. **-al**

khidmutgar (Ind.), a male waiter (*not* the many variants)

Khmer/, language of SE Asia; — **Republic,** name of Cambodia 1970–5; — **Rouge,** Cambodian communists

Khrushchev (**Nikita**), 1894–1970, Soviet politician

Khyber Pass, NW Frontier Province, Pakistan, *not* Khaibe

kHz, kilohertz, 1,000 hertz (no point, cap. H)

kiak, *use* **kayak**

kibbutz/, Israeli communal agricultural etc. settlement, *pl.* **-im; -nik,** a member of it

kibitz/, to look on at cards, to b meddlesome; **-er,** one who doe this, *not* -bb-

kiblah, the point to which Muslims turn in prayer, not *ke-, qib*

kickback, recoil, payment for favour (one word)

kick-off (noun, foot.) (hyphen

kickshaw, a trifle (one word)

kiddie (colloq.), young child, *n* -y

kidmutgar, use **khid-**

kidnap/, -ped, -per, -ping, *no* -aped etc.

kidney bean (two words)

Kidron (*or* **Ce-**), Palestine, **Ke** in NEB; Arabic, **Wadi en Na**

Kieff, Ukraine, *use* **Kiev**

Kierkegaard (**Søren Aaby**), 1813–55, Danish philosopher

Kiev, cap. of Ukraine, *not* -eff

K.i.H. (*or* **KIH**), Kaisar-i-Hind medal

kil., kilderkin, liquid measure o 18 gallons

Kilauea, volcano in Hawaii, *n* -aua

Kilbracken (**Baron**) (Sir Arthur Godley, of the India Office), 1847–1932

Kilimanjaro, mt., E. Africa (one word, divide Kilima-njaro)

ill (typ.), to cancel a line etc.; to distribute composed type

Killala, Mayo, *also* Bp. of

Killaloe, Clare, *also* Bp. of

Killaloo, Co. Londonderry

Killea, Co. Donegal

Killeagh, Co. Cork

Killen, Highland

Killianwala, *use* **Chilianwala**

illick, a stone used as anchor, *not* -ock, kcllcck, -ick

Killiecrankie, Tayside

Killin, Central

Killylea, Co. Armagh

Killyleagh, Co. Down

Kilmansegg (Miss)', by T. Hood, 1828, *not* -eg

kilo-, prefix meaning 1,000, abbr. **k**

kilo/, shortened form of kilogram or kilometre (no point); *pl.* **-s**

kilocycle, abbr. **kc,** *but use* **kHz** for **kc/s**

kilogram, SI unit of mass, *not* -me; abbr. **kg.** (*but* no point in scientific and technical work)

kilometre/, 1,000 metres; abbr. **km.** (*but* no point in scientific and technical work); **-age,** number of kilometres

kilowatt, 1,000 watts; abbr. **kW**

kilter (US and dial.), good condition, *not* ke-

Kimeridgian (geol., *not* -mm-), an Upper Jurassic clay found at Kimmeridge, Dorset (cap.)

kimono/ (Jap.), kind of gown, *pl.* **-s**

Kimric, Welsh, *use* **Cym-**

kinaesthesia, the sense of muscular effort, *not* kine-

Kincardine(shire), former county of Scotland

Kincardine O'Neil, Grampian (three caps.)

Kinchinjunga, mt. in Himalayas, *not* Kanchen-

kindergarten/, a school for young children (not ital., not cap.); **-er,** its teacher

kinesthesia, *use* **kinae-**

King/, abbr. **K.;** (chess, no point); (typ.) print as Edward VII, *or* the Seventh, *not* the VII; — **Charles's spaniel** (two caps.)

kingd., kingdom

kingmaker (one word)

King-of-Arms, *not* —-at—- (hyphens)

kingpin (one word)

King's, abbr. **K.**

Kings, First, Second Book of (OT), abbr. **1 Kgs., 2 Kgs.**

Kingsale (Baron). *See also* **Kinsale**

King's Bench, abbr. **KB** (apos.)

Kingsbridge, Devon, Som. (one word)

King's/ College, abbr. **KC;** — **Counsel, KC;** — **County,** Ireland, *now* **Offaly;** — **Cross,** London and Sydney (apos.)

king-size(d) (hyphen)

King's/ Langley, Herts.; — **Lynn,** Norfolk (apos.)

Kings Norton, Leics., —— **(Baron)** (no apos.)

King's Printer, *see* **printer**

Kingsteignton, Devon (one word)

Kingston, Jamaica, New York, Ontario

Kingstone, Hereford & Worc., Som., Staffs., S. Yorks.

Kingston/ upon Hull, Humberside; — **upon Thames,** London (no hyphens)

Kingstown, Dublin, off. Dun Laoghaire

Kingswinford, W. Midlands (one word)

Kington, Hereford & Worc.

Kinloss (Baroness)

Kinnoull (Earl of)

Kinross (Baron)

Kinross-shire, former county of Scotland (hyphen)

Kinsale, Co. Cork. *See also* **King-sale**

Kinshasa, cap. of Zaïre; *formerly* Léopoldville

kintle, *use* **quintal**

Kintyre, Strathclyde, *not* Cantire

Kioto, Japan, *use* **Kyoto**

Kirchhoff (Gustav Robert), 1824–87, German physicist

Kirghizia, a Soviet Socialist Republic; adj. **Kirghiz.** *See also* **USSR**

Kiribati, W. Pacific, *formerly* Gilbert Is.; adj. *same*

Kirkby in Ashfield, Notts. (three words)

Kirkbymoorside, N. Yorks. (one word)

Kirkcaldy, Fife

Kirkcudbright, Dumfries & Galloway

Kirkpatrick (Ralph), b. 1911, US harpsichordist, compiler of catalogue of Domenico Scarlatti's keyboard works

kirmess, use kermis

kirschwasser, a cherry liqueur, *not* kirschen-

Kishinev, cap. of Moldavia, USSR

kist, use cist

kitbag (one word)

kit-cat, a portrait 36 × 28 in., like those painted of members of the Kit-cat Club in the reign of James II

Kitch (Jack), *use* **Ke-**

kitchen garden/, -er, -ing (two words)

kitchen-maid (hyphen)

kitmutgar, use khid-

kitsch, worthless pretentious art, *not* -itch

Kit's Coty House, Aylesford, Kent, a dolmen, *not* — Coity —, — Cotty —

Kitzbühel, Austria

Kk., prefix to enumeration of D. Scarlatti's works (*see* **Kirkpatrick**)

KKK, Ku-Klux-Klan

KKt, (chess) king's knight; **KKtP,** king's knight's pawn (no points)

Klang (Ger. m.), quality of musical sound

kleistogam/ic, -ous, *use* c-

Klemperer (Otto), 1885– 1973, German conductor

klepsydra, *use* c-

kleptomania, irresistible tendency to theft, *not* c-

klieg light, arc light used in making motion pictures

klinometer, *use* c-

Klischograph (typ.), propr. term for electronic photo-engraving machine

KLM, Royal Dutch Air Lines (Koninklijke Luchtvaart Maatschappij NV)

Klondike, Yukon, Canada, *not* -yke

Klopstock (Friedrich Gottlieb), 1724–1803, German poet

km., kilometre(s) (*but* no point in scientific and technical work)

K.Mess., King's Messenger

KN (chess), king's knight (no points)

kn., knot, -s

kneecap, the patella (one word)

kneel/, -ed (*or* **knelt**), **-ing**

Knesset, the Israeli Parliament

knick-knack (hyphen), *not* nicknack

knight, abbr. **K., Knt., Kt.,** (chess) **Kt** *or* **N** (no point). *See also* **KB, KBE, KCB, KCIE, KCMG, KCSI, KCVO, KG, KP, KT**

Knightbridge, Cambridge professorship, *not* Knights-

knight errant, pl. **knights errant** (two words), *but* **knight-errantry** (hyphen)

Knightsbridge, London

Knights/ Hospitallers, charitable military brotherhood (otherwise — **of St. John,** — **of Rhodes,** — **of Malta**), maintaining hospital for pilgrims in Jerusalem from middle of eleventh century

knit/, -ted, -ting

knitting/-machine, -needle (hyphens)

knobby, knob-shaped

knobkerrie (S. Africa), short stick with knobbed head, *not*

kerry, -kiri; in Afrik. **knop-
kierie**

nock/-about (noun); **-down**
(adj.); **-knee(d)**; **-out** (hyphens)

nole, Kent, Som.

nopkierie, *use* **knobkerrie**

nossos, Crete, *not* Cnossus

not/, -ted, -ting; as measure of
speed, abbr. **kn.**

now-how (noun, hyphen)

nowl, W. Yorks.

nowle, Avon, Devon, Shrop-
hire, Som., W. Midlands

nowledgeable, *not* -dgable

NP (chess), king's knight's
pawn (no points)

nur/, a knot; — **and spell,** a
game, *not* -rr

nurl, a small projection, *not*
url

nut, *use* **Cnut**; *see also* **Canute**

nutsford, Ches.

O *or* **k.o.,** kick-off (foot.),
nock-out (boxing)

oala, Australian arboreal mar-
upial

obenhavn, Dan. for **Copenhagen**

oblenz, W. Germany, *not* C-,
ce

och (Ludwig), 1881–1974,
German-born, English domi-
iled musician and recorder of
bird-song; — **(Robert),** 1843–
1910, German bacteriologist

öchel (Ludwig von), 1800–
77, Austrian scientist and
musicologist, cataloguer of
Mozart's works; abbr. **K.**

ock (Charles Paul de),
1794–1871, French novelist

odaikanal Observatory,
Madras, India

odály (Zoltán), 1882–1967,
Hungarian composer

oedoe, *use* **koodoo**

oestler (Arthur), 1905–83,
Hungarian-born British novel-
st

oh-i-noor diamond, *not* -núr,
nûr (hyphens)

ohlrabi, a turnip-rooted vege-

table of the cabbage family (one
word)

koine, lingua franca; **Koine,**
that of Greeks *c.* 300 BC–
AD 500

Kokoschka (Oskar), 1886–
1980, Austrian-born painter

kola, *use* **cola**

kolkhoz, collective farm in
USSR

Köln, Ger. for **Cologne**

Komintern, *use* **C-**

komitaji, *use* **comitadji**

Komsomol, Soviet youth organization, *not* **C-**

kongo, *use* **congou**

Königgrätz, Czechoslovakia,
use **Hradec Králové**

Königsberg, *now* **Kaliningrad,** USSR (*formerly* E. Prussia)

Konrad, *see* **Conrad**

kontaki/on, liturgical hymn in
Eastern Church, *pl.* **-a,** *not c-*

Konzert/meister (Ger. m.),
leader of orchestra; **-stück** (n.),
concert-piece

koodoo, *use* **kudu**

kookaburra, the laughing jackass, an Australian bird

Koords, *use* **Kurds**

Kootenai, Idaho & Mont., US

Kootenay, BC, Canada

kopek, *use* **copeck**

kopje (S. Afr.), a small hill, in
Afrik. **koppie**

kopro-, *use* **copro-**

Koran, Muslim sacred book,
from Arab. **Qur'ān,** *not* Coran,
Q'ran

Korea/, indep. 1948, now divided into North and South
Korea; adj. **-n,** *not* C-

Kortrijk, Fl. for **Courtrai**

koruna, Czechoslovak unit of
currency, abbr. **kčs**

Kos, *use* **Cos**

KOSB, King's Own Scottish Bor-
derers

Kosciusko (Thaddeus),
1746–1817, Polish patriot; in
Pol. **Kościuszko (Tadeusz
Andrzej Bonawentura)**

kosher, (of food), prepared according to the Jewish law, *not* coshar, -er, koscher

kotow, *use* **kowtow**

Kotzebue (August Friedrich Ferdinand von), 1761–1819, German playwright

koumiss, preparation from mare's milk, *not* ku-

kourie, *use* **kauri**

Koussevitsky (Serge Alexandrovich), 1874–1951, Russian conductor

Kovno, Lithuania, *use* **Kaunas**

kowrie, *use* **kauri**

kowtow/, Chinese form of submission, to act obsequiously, *not* kot-; **-ed, -ing**

KP (chess), king's pawn (no points), Knight of the Order of St. Patrick

k.p.h., kilometres per hour

KR (chess), king's rook (no points), King's Regiment, — Regulations

Kr, krypton (no point)

kr., kreuzer, krona, króna, krone

kraal (Afrik.), enclosure; native village

Krafft-Ebing (Richard von, Baron), 1840–1902, German psychiatrist

Krakatoa (Mount), Straits of Sunda, *not* -au

Kraków, Polish for **Cracow**

krans (Afrik.), cliff, *not* krantz

Krapotkine (Prince), *use* **Kropotkin**

kreese, *use* **kris**

Krefeld, W. Germany, *not* C-

Kreisler (Fritz), 1875–1962, Austrian-born US violinist

Kremer (Gerhard), *see* **Mercator**

Kremlin (the), Moscow citadel, used for the Russian government

kremlin, any Russian citadel (not cap.)

kreosote, *use* **creosote**

Kreutzer (Rodolphe), 1766–1831, German-French violinist; **'Kreutzer' Sonata** (by Beet-

hoven, 1803); *The Kreutzer Sonata* (by Tolstoy, 1889)

kreuzer, Austrian and old German copper coin; abbr. **kr.**

kriegspiel, the war-game, in Ger. n. *Kriegsspiel*

krill, tiny plankton

kris, Malay dagger, *not* creese, kreese

kromesky, Russian dish of minced chicken fried in bacon, in Fr. m. *cromesquis*

krone/, silver coin, Austrian (*pl.* **-n**), Danish (*pl.* **-r**), Norwegian (*pl.* **-r**), Swedish (**kron/a**, *pl.* **-or**), and Icelandic (**krón/a**, *pl.* **-ur**), all abbr. **kr.**; Czechoslovak **koruna,** abbr. **kčs** (*sing.* and *pl.*)

Kronos (Gr. myth.), *not* C-, -us

Kroo, W. African Negro, *not* -o, -u (not ital.)

Kropotkin (Prince Peter Alexeivich), 1842–1921, Russian geographer, revolutionist, and author, *not* Kra-, -ine

KRP (chess), king's rook's pawn (no points)

Kru, *use* **Kroo**

Kruger (Stephanus Johannes Paulus), 1825–1904, S. Afr. statesman

krugerrand, S. Afr. gold coin (one word, not cap.)

krummhorn, *not* crumhorn

krypton, chem. element, *not* c- symbol **Kr**

KS, Kansas (off. postal abbr.); King's Scholar; (typ.) keep standing

Kshatriya, the warrior caste of the Hindus (cap., not ital.)

K.St.J., Knight of the Order of St. John of Jerusalem

KT, Knight/ of the Order of the Thistle, — Templar

Kt (chess), knight (no point)

Kt., Knight (Bachelor)

Kt. Bach., Knight Bachelor

κτλ. (Gr.), *kai ta loipa* (and the rest *or* etc.)

Ku, kurchatovium (no point)

Kuala Lumpur, cap. of Malaysia (and of Selangor)

Kublai Khan, 1214–94, first Mongol Emperor of China

Kubla Khan', poem by Coleridge, 1816

Kuch Behar, *use* **Cooch** —

kudos (Gr.), renown (not ital.)

kudu, a S. Afr. antelope, *not* koodoo; in Afrik. **koedoe**

Kufic, of a form of Arabic writing, *not* C-

Ku-Klux-Klan, US secret society (not ital.); abbr. **KKK**

ukri (Ind.), a curved knife, *not* the many variants

kulak, Russian peasant proprietor

Kultur (Ger. f.), culture

Kulturkampf (Ger. m.), the war of culture (between Bismarck's government and the Catholic Church, *c.*1872–7)

Kumasi, Ghana, *not* Coo-, -assie

koumis(s), *use* **koumiss**

kümmel, a liqueur

kummerbund, *use* **cummerbund**

kumquat, orange-like citrus fruit in S. China and Malaysia; small citrus fruit or tree in Australia, *not* c-

kung fu, Chinese karate

Kunstlied (Ger. n.), art-song

Kuomintang, the Chinese nationalist people's party, founded 1912

kupfernickel (mineral.) (one *f*, not cap.)

kupfferite (mineral.) (two *f*s)

kurchatovium, symbol **Ku**

Kurds, of Kurdistan, *not* Koo-

kurfuffle, *use* **ker-**

Kurile Is., NW Pacific

Kurrachee, *use* **Karachi**

Kursaal (Ger. m.), a hall at a spa or resort (ital., cap.)

Kursiv, Kursivschrift (Ger. typ. m., f.), italic type

Kutch, India, *not* C-

kutch, *use* c-

Kutchuk-Kainardji, treaty, 1774

Kuwait/, Persian Gulf; adj. **-i**

Kuyp, *use* **C-**

kV, kilovolt (no point)

kVA, kilovolt-ampere (no point)

kvass (Russ.), rye-beer, *not* quass

kW, kilowatt (no point)

kWh, kilowatt-hour

KWIC (indexing), keyword in context

KWOC (indexing), keyword out of context

KY, Kentucky (off. postal abbr.)

Ky., Kentucky (off. abbr.)

Kyd (Thomas), 1558–94, English playwright

Kymric, Welsh, *use* **C-**

Kyoto, Japan, *not* Ki-

Kyrie eleison (eccl.), 'Lord, have mercy'; abbr. **Kyrie,** *not* — eleëson

Kyrle (John), 1637–1724, English philanthropist, 'the Man of Ross'

217

L

L, fifty, learner (motor vehicle), the eleventh in a series

L., Lady, Lake, Latin, (theat.) left (from actor's point of view), (Lat.) *liber* (book), Liberal, licentiate, (biol.) Linnaeus, lir/a, -e, (Fr.) *livre* (book, pound), Loch, (Lat.) *locus* (place), London, Lough, prefix to enumeration of D. Scarlatti's works (*see* **Longo**)

l. *or* **£,** pound, the form £ to be used and placed *before* figures, as £50. If *l.* is used, it must be placed *after*, as 50*l.* **£E,** Egyptian pound (100 piastres); **£m.,** (one) million pounds. *See also* **decimal currency**

l. (*but* no point in scientific and technical work), litre(s). When there may be confusion, the word should be spelt out in full

l., leaf, league, line, (Ger.) *lies* (read), (meteor.) lightning, line, link. *See also* **l.** *or* **£**

L (elec.), symbol for inductance

l (mech.), symbol for length

Ł, ł, Polish letter

LA, law agent, Legislative Assembly, Library Association, Local Authority, Los Angeles, Louisiana (off. postal abbr.)

La, lanthanum (no point)

La., Lane, Louisiana (off. abbr.)

la (mus.), *use* **lah**

laager, encampment, esp. in circle of wagons; in Afrik. *laer*. *See also* **lager**

Lab., Labour (party), Labrador

lab (no point), laboratory

label/, -led, -ling

labi/um, (anat.) a lip or liplike structure; *pl.* **-a**

Labor Day (US), first Monday in Sept.

lab/our, -orious

Labour Day (UK and many other countries), 1 May

Labourers Act (caps., no apos.)

labourite, supporter of Labour Party (not cap.)

labour-market (hyphen)

Labrador, abbr. **Lab.**

labr/um (Lat.), lip of a jug, etc; *pl.* **-a**

La Bruyère (Jean de), 1645–96, French writer and moralist

labyrinth/, structure with many confusing passages; adjs. **-ian, -ic, -ine**

LAC, Leading Aircraftman, Licentiate of the Apothecaries' Company, London Athletic Club

lac, a resin. *See also* **lakh**

Laccadive Islands, off Madras west coast

lace-up (adj. and noun), (shoe) having laces (hyphen)

lace/wing, -wood (one word)

lace-work (hyphen)

Lachaise (Père), Paris cemetery (two words, caps.)

lâche (Fr.), lax, cowardly (ital.)

laches (law), negligence or unreasonable delay (not ital.)

Lachesis (Gr. myth.), the Fate that spins the thread of life

lachrym/al, of tears; **-ation, -atory, -ose;** *not* lacry-; **lacri** is a correct form, now usual in scientific use

lackadaisical, listless

lacker, *use* **lacquer**

lackey, a footman, *not* -quey

Lackland, sobriquet of King John

Laconian, Spartan

laconic, concise (not cap.)

lacquer, a varnish, *not* lacker

lacquey, *use* **lackey**

lacrimal etc., *see* **lachry-**

La Crosse, Wis., US

lacrosse, a ball game

Lacryma Christi, a sweet red or white wine from Mt. Vesuvius area

lacrymal etc., *use* **lachry-**

acun/a, a missing portion, esp. in ancient MS; *pl.* **-as** *or* **-ae** (not ital.)

ACW, Leading Aircraftwoman

acy, lacelike, *not* -ey

addie, young fellow, *not* -y

adies (the), women's lavatory (cap., no apos.)

Ladies'/ Gallery, House of Commons; — **Mile,** Hyde Park, London

adies' man (apos., two words)

Ladikia, *use* **Latakia**

Ladin, Engadine dialect

Ladino/, Spanish dialect, non-Indian in Cent. Amer., *pl.* **-s**

Ladismith, Cape Province. *See also* **Ladysmith**

Lady, abbr. **L.**

Lady (Our) (caps.)

adybird (one word)

Lady Day/, 25 Mar. (two words, caps.); — — **in Harvest,** 15 Aug.

adyf/y, -ied, *not* -if-

ady-in-waiting (hyphens)

adylike (one word)

Lady Margaret Hall, Oxford, abbr. **LMH**

ady's maid, *pl.* **ladies' maids** (no hyphen)

Ladysmith, Natal. *See also* **Ladismith**

ady-smock, cuckoo flower (hyphen)

ady's slipper (apos.)

aemergeier, *use* **lammergeyer**

aer (Afrik.), *use* **laager**

aesa majestas (Lat.), *lèse-majesté*

Laetare Sunday, fourth in Lent

aevo-, prefix meaning left, *not* le-

Lafayette/ (Marie Joseph, marquis de), 1757–1834, French general, aided Americans in Revolution (one word); — **College,** Easton, Pa.

Laffitte (Jacques), 1767–1844, French statesman

Lafite (Château-), a claret (hyphen)

Lafitte (Jean), 1780–*c.*1826, buccaneer of unknown origin

La Follette (Robert Marion), 1855–1925, US politician

La Fontaine (Jean de), 1621–95, French writer

LAFTA, Latin-American Free Trade Association

lager, light beer. *See also* **laager**

Lagerkvist (Pär), 1891–1974, Swedish novelist

Lagerlöf (Selma Ottilia Lovisa), 1858–1940, Swedish novelist

Lagrange (Joseph Louis, comte), 1736–1813, French mathematician

Lagting, Upper House of Norwegian Parliament

La Guaira, Venezuela, *not* -yra

LaGuardia/ (Fiorello Henry), 1882–1947, Mayor of New York (one word, two caps.); — **Airport,** New York, US

LAH, Licentiate of Apothecaries' Hall, Dublin

lah (mus.), *not* la

La Hague (Cape), NW France

La Haye, Fr. for The Hague

La Hogue, in Fr. **La Hougue,** a roadstead, NW France

Lahore, Pakistan, *not* -or

Laibach, Yugoslavia, Ger. for **Ljubljana**

laicize, to secularize, *not* -ise

laid paper, that which when held to the light shows close-set parallel lines

Laïs, a Greek courtesan and beauty of fourth century BC

laissez/-aller (Fr.), absence of restraint; — *-faire,* letting people do as they think best, let well alone!; — *-passer* (m.), pass, permit (for persons and things); not *laisser-*

Lake, cap. when with name, as Bala Lake, Lake Superior; abbr. **L.**

Lakeland, the Lake District in Cumbria (one word)

lakh/ (Anglo-Ind.), 100,000, *not*

lakh (*cont.*)

-ck, -c (not ital.); — **of rupees,** pointing above one lakh is with a comma after the number of lakhs: thus 25,87,000 is 25 lakhs 87 thousand rupees. *See also* **crore**

Lalitpur, Uttar Pradesh, India, *not* Lalat-

Lalla Rookh, by Moore, 1817

'L'Allegro', by Milton, 1632

Lam., Lamentations (OT)

lama, Buddhist priest (cap. when part of title). *See also* **llama**

Lamarck/ (Jean Baptiste Pierre Antoine de Monet, chevalier de), 1744–1829, French naturalist; adj. **-ian**

Lamarque (comte Maximilien), 1770–1832, French general

Lamartine (Alphonse Marie Louis de), 1790–1869, French poet and statesman

lamasery, a monastery of lamas

Lambaréné, Gabon, site of Albert Schweitzer's hospital

lambaste, to beat, criticize, *not* -bast

lambda, Gr. letter (*Λ, λ*); (phys.) symbol for wavelength

lamb's fry (cook.) (apos., two words)

lambskin (one word)

lamb's-wool (apos., hyphen)

LAMDA, London Academy of Music and Dramatic Art

lamé (adj. and noun), (material) with inwoven gold or silver thread (not ital.)

lamell/a, a thin plate; *pl.* **-ae**

Lamentations (Book of), abbr. **Lam.**

lamin/a, a thin plate; *pl.* **-ae**

Lammas, 1 Aug.

lammergeyer, the bearded vulture, *not* lae-, le-, -geier (not ital.)

lampblack (one word)

lamp-holder (hyphen)

lamp/lighter, -post, -shade (one word)

Lancashire, abbr. **Lancs.**

Lancaster (County, *or* **County palatine, of),** formal name of Lancashire

Lance-Bombardier, abbr. **L.Bdr.** *or* **L/Bdr.**

Lance-Corporal, abbr. **L/Cpl**

lancewood, a tough W. Indian wood (one word)

Lancing College, W. Sussex

Lancs., Lancashire

Land (Ger. n.), province of W. Germany or Austria; *pl.* *Länder* (ital., cap.)

landau/, *pl.* **-s; -let,** a carriage (not ital.)

landdrost (S. Afr.), a district magistrate (not ital.)

landfall (one word)

landgrav/e, a German count; *fem.* **-ine**

landgraviate, a landgrave's territory, *not* -vate

landholder (one word)

ländler, Austrian dance

land-locked (hyphen)

land/lubber, -mark (one word)

land-mine (hyphen)

L. & N.W.R., London and North-Western Railway, became part of **LMSR** until nationalization

Landor (Walter Savage), 1775–1864, English poet and prose writer

landowner (one word)

landscape (typ.), book, page, or illustration, of which the width is greater than the depth. *See also* **portrait**

Land's End, Cornwall

landslide (one word)

Landsmål, Norwegian peasant dialect given literary form *c.*1850, and now called **Nynorsk,** new Norwegian

lands/man, *pl.* **-men**

L. & S.W.R., London and South-Western Railway, became part of **SR** until nationalization

land-tax (hyphen)

L. & Y.R., Lancashire and Yorkshire Railway, became part of **LMSR** until nationalization

Lane

Lane, abbr. **La.**

Lang (Andrew), 1844–1912, Scottish man of letters; — **(Cosmo Gordon),** 1864–1945, Abp. of Canterbury

lang., language

Langeberg Mountains, S. Africa; in Afrik. **Langeberge**

langouste (Fr. f.), a spiny lobster

langsam (Ger. mus.), slowly

lang syne, long ago (two words)

language, abbr. **lang.**

langue *de chat* (Fr. f.), finger-shaped biscuit or chocolate; — *d'oc,* medieval French spoken south of the Loire; — *d'oïl,* ditto north of the Loire

Languedoc, former French province, between R. Loire and Pyrenees

languor/, lassitude; **-ous**

langur, Indian monkey

laniard, *use* **lany-**

Lankester (Sir Edwin Ray), 1847–1929, English zoologist

lanolin, fat in sheep's wool, *not* -inc

Lansdown, Avon, Glos.; — **(battle of),** 1643

Lansdowne (Marquess of)

Lansing, Mich., US

lantern-slide (hyphen)

lanthanum, symbol **La**

lanyard, short rope attached to something, *not* lani-

Lao, people and language of SE Asia

Laocoön, Trojan priest, subject of a famous sculpture

Laodicean, lukewarm (of feelings)

Laois, county of Ireland, formerly **Leix**

Laos, independent 1949, *formerly* the kingdoms of Luang Prabang and Vientiane and the principality of Champassac (Fr. Indo-China); adj. **Laotian**

Lao-tzu, *c.*605–530 BC, founder of Taoism (hyphen)

Lap., Lapland. *See also* **Lapp**

La Paz, Bolivia

lap-dog (hyphen)

larynx

lapel/, the lap-over of a coat, *not* -elle, lappelle; **-led**

lapis lazuli, a rich blue stone or its colour (two words, not ital.)

Laplace (Pierre Simon, marquis de), 1749–1827, French astronomer

Lapland, abbr. **Lap.**

Lapp, a native of Lapland, (adj. and noun) Lappish

lappelle, *use* **lapel**

lapsus/ (Lat.), a slip; — *calami,* ditto of the pen; — *linguae,* ditto tongue; — *memoriae,* ditto memory; *pl.* same

Laputa, flying island in Swift's *Gulliver's Travels*; **-n,** visionary, absurd

lar/, Roman household god; *pl.* **-es; lares and penates,** one's home

lardon, bacon for larding, *not* -oon

La Reyne/ le veult, — — *s'avisera,* forms of *Le Roy* etc. when queen is reigning

large crown (metric), *see* **book sizes**

large-paper, special copies of a book, with large margins, etc.; also termed *édition de luxe;* abbr. **LP**

largess, a free gift, *not* -esse (not ital.)

larghetto (mus.), fairly slow

largo (mus.), slow, broad

lariat, rope for tethering animals, *not* -iette, larriet (not ital.)

larkspur (bot.) (one word)

La Rochefoucauld (François, duc de), 1613–80, French writer

La Rochelle, dép. Charente-Maritime, France

Larousse/ (Pierre Athanase), 1818–75, French lexicographer; —, French publishers of reference books

larrikin, Australian street rowdy, *not* lari-

larv/a, *pl.* **-ae** (not ital.)

laryn/x (anat.), *pl.* **-ges; -geal, -gitis**

lasagne (It. pl.), pasta in wide ribbons

Lasalle (Antoine Chevalier Charles Louis, comte de), 1775–1809, French general

La Salle (Antoine de), *c.*1400–60, French soldier and poet; — **(St. Jean Baptiste de)**, 1651–1719, French founder of the order of Christian Brothers; — **(René Robert Cavelier)**, 1643–87, French explorer

Lascar, E. Indian sailor (not ital.)

Las Casas (Bartolomé de), 1474–1566, Spanish missionary, 'the apostle of the Indies'

Las Cases (Emmanuel Augustin Dieudonné, comte de), 1766–1842, friend of Napoleon I on St. Helena

Lascaux, SW France, site of palaeolithic cave-art

laser, device using *l*ight *a*mplification by *s*timulated *e*mission of *r*adiation (acronym). *See also* **maser**

lashkar, a body of Indian irregular troops (not ital.)

Las Meninas, by Velázquez, *not* -iñas

La Spezia, NW Italy, *not* -zzia

Lassa, Tibet, *use* **Lhasa**; —, Nigeria, source of — **fever**

Lassalle (Ferdinand), 1825–64, German Socialist (one word). *See also* **Lasa-**

Lassell (William), 1799–1880, English astronomer

lassie, young girl, *not* -y

lasso/, as noun, *pl.* **-s**; as verb, **-ed, -es, -ing**

Lassus (Orlandus), 1532–94, Belgian-born composer

Last Supper (the) (caps.)

Lat., Latin; **lat.,** latitude

Latakia, Syria, *not* Ladi-, -ieh, -yah

latchkey (one word)

lateish, *use* **latish**

La Tène ware (anc. pottery)

Lateran/ (St. John), church in Rome; — **Council,** one of five held there

latex/, rubber fluid, *pl.* **-es**

lath, a thin strip of wood

lathe, machine for turning

lathi, heavy stick carried by Indian police

Latin, abbr. **L.** *or* **Lat.**; (typ.) alphabet same as English without *w.* Accents, e.g. *â,* and ligatures, e.g. *æ,* falling into disuse; most scholars use *i* for *j,* many use lower-case *u* for *v* and cap. V for U. For use of caps. and word-division, see *Hart's Rules,* p. 113

Latin Cross, †

latine (Lat.), in Latin

Latinity, the quality of one's Latin

Latinize, to make Latin, *not* -ise

latish, fairly late, *not* lateish

latitude, abbr. **lat.**

Latour (Château-), a claret (hyphen)

La Trobe, university, Melbourne

Latrobe, Pa., US, and Tas., Austral.

latten, metal like brass (not ital.)

latter, second of two. *See also* **former**

Latter-day Saints, Mormons' name for themselves

Latvia, a Soviet Socialist Republic, *see* **USSR**; adj. **Latvian** *or* **Lettish**

Latymer Upper School, Hammersmith, London

laudator temporis acti (Lat.), a praiser of past times

laughing/-gas, -stock (hyphens)

launderette, *not* -drette

laura/ (hist.), a group of hermits' cells; *pl.* **-s**

laurel/, -led

Laurence (Friar), in Shakespeare's *Romeo and Juliet.* *See also* **Law-**

Laurentian Mts. and (geol.) rocks, near the St. Lawrence River, Canada

laurustinus, an evergreen, *not* laures-, lauris-

Laus Deo (Lat.), Praise (be) to God, abbr. **LD**

lav, colloq. for lavatory (no point)

Lavoisier (Antoine Laurent), 1743–94, French chemist (one word)

law (typ.), practically no punctuation used in legal documents. Copy must be followed. Spell out all figures; **law/ agent,** abbr. **LA; — -binding, — calf, — sheep,** binding in smooth pale brown calfskin or sheepskin, formerly much used for law books

lawcourt (one word)

Law Courts (the) (caps., no hyphen)

lawgiver (one word)

lawn tennis, *see* **tennis**

Lawrence (David Herbert), 1885–1930, English novelist; — **(Sir Thomas),** 1769–1830, English painter; — **(Thomas Edward),** 1888–1935, English archaeologist and soldier, 'Lawrence of Arabia', (from 1927) Aircraftman Shaw; — **(St.),** Canadian river. *See also* **Lau-**

lawrencium, symbol **Lw**

Laws, abbr. **LL**

law sheep, *see* **law**

law-stationer (hyphen)

lawsuit (one word)

Laxness (Haldór Kiljan), b. 1902, Icelandic writer

lay, untilled land, *use* **lea**

Layamon, *fl.* 1200, English priest

layabout (one word)

lay brother (etc.) (two words)

lay-by/, short 'siding' on main road, *pl.* **-s** (hyphen)

layette, complete outfit for a baby (not ital.)

layout (noun, one word)

lazaretto/, a place for quarantine (not ital.); *pl.* **-s**

lazy/-bones, -tongs (hyphen)

lazzaron/e, one of a low class at Naples, *not* lazar-; *pl.* **-i**

lb., pound, pounds (weight) (not now in scientific use)

l.b. (cricket), leg-bye

L.B. & S.C.R., London, Brighton, and South Coast Railway, became part of **SR** until nationalization

L.Bdr., L/Bdr., Lance-Bombardier

lbw. (cricket), leg before wicket, *not* l.b.w.

LC, (theat.) left centre, Legislative Council, (US) Library of Congress, Lord Chamberlain, — Chancellor, letter of credit

l.c., *loco citato* (in the place cited), (typ.) lower case, that is *not* caps. (without points as a proof-correction mark)

LCB, Lord Chief Baron

LCC, London County Council, -lor. See **GLC**

L.Ch., Licentiate in Surgery

LCJ, Lord Chief Justice

LCM *or* **l.c.m.** (math.), least common multiple; **LCM,** London College of Music

LCP, Licentiate of the College of Preceptors

L/Cpl., Lance-Corporal

LD, Lady Day, *Laus Deo* (praise be to God), Low Dutch

Ld., Lord

Ldg., (naval) Leading

L.d'H., Légion d'honneur

Ldp., Lordship

LDS, Licentiate in Dental Surgery

LDV, Local Defence Volunteers (later **Home Guard**)

£E, Egyptian pound(s)

LEA, Local Education Authority

lea, untilled land, *not* lay, lee. See *also* **ley**

lead (typ.), a thin strip of metal less than type high, used to separate lines of type

lead (chem.), symbol **Pb**

leaded/ matter, — type, having the lines separated by leads

leader, *see* **leading article**

leaders (typ.), dots used singly or in groups to guide the eye across the page

leading (typ.), the action of placing leads between lines of type; the leads so placed

leading article, an article appearing in a newspaper as the expression of editorial opinion on any subject; also called **leader**

lead pencil (two words)

leaf (typ.), single piece of paper, two pages back to back; abbr. **l.,** *pl.* **ll.**

leaflet (typ.), minor piece of printing, usu. 2, 4, or 6 pp.

leaf-mould (hyphen)

league, abbr. **l.**

Leamington Spa, War. *See also* **Lem-, Lym-**

lean/, -ed, *or* **-t**

lean-to (hyphen)

leap/, -ed, *or* **-t**

leap-frog/, a game (hyphen); also as verb, **-ged, -ging**

leap year (two words)

learn/, -ed, *or* **-t**

leasehold/, -er (one word)

leatherette, a strong machine- -glazed paper, coloured and embossed to resemble binding- -leather

leaves, abbr. **ll.**

Leban/on, independent 1943; adj. **-ese**

Lebensraum (Ger. m.), territory for natural expansion

Lebesgue/ (Henri Léon), 1875–1941, French mathematician; — theory of integration (math.)

Lebewohl! (Ger. n.), farewell!

Lecocq (Alexandre Charles), 1832–1918, French composer

Leconte de Lisle (Charles Marie René), 1818–94, French poet

Le Corbusier, pseud. of Charles Édouard Jeanneret, 1887–1965, Swiss architect

Lecouvreur (Adrienne), 1692–1730, French actress; ital.

when name of play (1849) by E Scribe and E. Legouvé; *Adriana Lecouvreur,* opera by Ciléa, 1902

lect., lecture

lectern, church reading-desk, *no* -urn

lecture/ship, post of lecturer, but **-rship** at Oxford University

LED, light-emitting diode

lederhosen (Ger. f. pl.), leather shorts

ledger, book of accounts; — - -tackle, in fishing. *See also* **leger line**

lee, untilled land, *use* **lea**

Leeuwenhoek (Anton van), 1632–1723, Dutch microscopist

leeway (one word)

Le Fanu (Joseph Sheridan), 1814–73, Irish writer

left/ (theat., from actor's point of view), abbr. **L.;** — **centre,** abbr. **LC**

left/-hand, -handed, adjs. (hyphens)

left wing (two words), *but* **left- -winger** (hyphen)

Leg., legislat/ive, -ure

leg., legal, *legit* (he, *or* she, reads), *legunt* (they read, pres. tense)

legalize, *not* **-ise**

Le Gallienne (Richard), 1866–1947, English author and journalist; — **(Eva),** his daughter, b. 1899, US actress

legato (mus.), smooth

leg-bye (cricket), abbr. **l.b.**

legenda (Lat.), things to be read

Léger (Alexis St-Léger), b. 1887, French poet, pseud. St. John Perse

legerdemain, sleight of hand (not ital.)

leger line (mus.), *not* led-

leges (Lat.), laws; abbr. **ll.**

leggiero (mus.), light, swift, delicate

Legh, family name of Baron Newton. *See also* **Leigh**

Leghorn, former English name for Italian port **Livorno; leg-**

horn, a straw plait, a breed of domestic fowl

Légion d'honneur (la), French order of merit

legionnaire, a member of a legion, a legionary (not ital.)

legislat/ive, -ure, abbr. **Leg.**

Legislative Assembly, abbr. **LA** (no points). *See also* **Assemblies**

legit (Lat.), he, *or* she, reads; abbr. **leg.**

legitim/ate (verb), (law) to render (a child) legitimate; elsewhere *use* **-ize**, *not* -atize, -ise

leg/-pull, -rest, -room (hyphens)

légumes (Fr. m. pl.), table vegetables

legunt (Lat.), they read (pres. tense); abbr. **leg.**

Le Havre, dép. Seine-Maritime, France, *not* Havre

Lehigh University, Bethlehem, Pa. (one word)

lei, Polynesian garland. *See also* **leu**

Leibniz (Gottfried Wilhelm, Baron von), 1646–1716, German mathematician and philosopher, *not* -itz

Leicestershire, abbr. **Leics.**

Leicester:, sig. of Bp. of Leicester (colon)

Leiden, Netherlands. *See also* **Leyden**

Leigh (Baron). *See also* **Legh**

Leighton Buzzard, Beds. (two words)

Leipzig, E. Germany, abbr. **Lpz.** *or* **Leip.**

leishmaniasis, disease caused by parasite

leitmoti/v, -f (mus.), theme associated with person, situation, or sentiment, *not* -ive (one word, not ital.)

Leix, county of Ireland, *formerly* Queen's County, *now* **Laois**

le juste milieu (Fr.), the golden mean. *See also* **milieu**

L.E.L., pseud. of Letitia Elizabeth Landon, 1802–38

Lely (Sir Peter), 1618–80, Dutch-English painter

Lemberg, Ger. for **Lvov**

Lemington, Glos., Tyne & Wear. *See also* **Lea-, Ly-**

lemm/a, a title or theme, proposition taken for granted, *pl.* **-as**; heading of an annotation, *pl.* **-ata** (not ital.)

lemmergeyer, *use* la-

Lemprière (John), 1765–1824, English lexicographer

Le Nain (Antoine), 1588–1648, — **(Louis),** 1593–1648, and — **(Mathieu),** 1607–67, brothers, French painters

Lenclos (Ninon de), 1616–1706, a French beauty

lending library (two words)

length, abbr. **l.,** (mech.) symbol *l*

Lenin, assumed name of **Vladimir Ilich Ulyanov,** 1870–1924, Russian political leader

Leningrad, name for Petrograd, q.v., since 1924

lenis, Greek smooth breathing (')

Lennoxtown, Strathclyde (one word)

Lenox Library, New York

lens/, *not* lense; *pl.* **-es**

Lent, from Ash Wednesday to Easter (cap.)

lento (mus.), slow

Leonardo da Vinci, 1452–1519, Italian painter, sculptor, engineer

Leoncavallo (Ruggiero), 1857–1919, Italian composer

Leonid/, a meteor; *pl.* **-s** (not ital.)

Léopoldville, Zaïre, *now* **Kinshasa**

Lepidoptera (zool.), butterflies and moths

LEPRA, British Leprosy Relief Association

leprechaun, Irish sprite, *not* lepra-, -awn

Le Queux (William Tufnell), 1864–1927, English novelist and traveller

Lermontov (Mikhail Yure-vich), 1814–41, Russian poet

Le Roy/ le veult, the royal assent to Bills in Parliament; — — *s'avisera,* ditto dissent

lès or lez (Fr. topog.), near (with names of towns)

Le Sage (Alain René), 1668–1747, French novelist and dramatist

Lesbian, (inhab.) of Lesbos

lesbian, a homosexual woman (not cap.)

lèse-majesté (Fr. f.), treason (ital.) ; Anglicized as **lese-majesty** (not ital.)

L. ès/ L., *Licencié ès lettres,* Licentiate in (*or* of) Letters; — — **S.,** *Licencié ès sciences,* ditto Science

Lesotho, S. Africa, indep. 1966, *formerly* Basutoland

Lethe/, a river in Hades; adj. **-an**

let-in notes (typ.), those set inside the text-area (usu. in a smaller-size type), as distinct from side-notes

Letraset, propr. term for system of lettering using transfers (cap.)

letter/, -ed, -ing

letterhead (typ.), printed (heading on) stationery (one word)

letterpress (typ.), printing from raised type and/or blocks; text as opposed to illustrations (one word)

letterset (typ.), printing by letterpress on to a blanket and then offsetting on to the paper (one word)

letters of distinction, as FRS, LL D, etc., are usually put in large caps. Even s. caps. often improve general effect

letterspacing (typ.), shown in copy by short vertical dashes between letters (one word)

letters patent, formal writing conferring patent or privilege (two words)

Lettish, *see* **Latvia**

lettre de/ cachet (Fr. f.), warrant for imprisonment, bearing

the royal seal; — — *créance,* — — *crédit,* letter of credit; — — *marque,* letter of marque q.v.; *pl.* **lettres de —**

lettuce, *not* -ice

Letzeburgesch, German dialect of Luxemburg

leu, Romanian monetary unit; *pl.* **lei**

leucotomy, a brain operation, *not* -k-

leukaemia, excess of white cor puscles in blood, *not* -c-, -ch-, -kemia

Leuven, Fl. for **Louvain**

Leeuwenhoek, *use* **Leeuwenhoek**

Lev., Leviticus

lev/, *pl.* **-a,** Bulgarian monetary unit

levant morocco (binding), a superior quality with prominen grain

levee, an assembly (no accent, not ital.) ; US river embankmen

level/, -led, -ler, -ling

lever de/ rideau (Fr. m.), open ing piece at theatre, 'curtain-raiser'; — — *séance,* closing of a meeting

leviable, that may be levied

Leviathan, book by Hobbes, 1651

leviathan, sea-monster

Lévi-Strauss (Claude), b. 1908, French anthropologist

Leviticus, abbr. **Lev.**

Levkosia, Gr. for **Nicosia**

levo-, prefix, *use* **laevo-**

Lewes, E. Sussex

Lewes (Charles Lee), 1740–1803, English actor; — **(George Henry),** 1817–78, English philosopher and critic. *See also* **Lewis**

Lewis, Western Isles

Lewis (Sir George Cornewall, *not* Cornw-), 1806–63, English statesman and man of letters; — **(Cecil Day),** 1904–72, Poet Laureate 1968–72 ; — **(Clive Staples),** 1898–1963, English writer; — **(Harry**

Sinclair), 1885–1951, US novelist; — (**Isaac Newton**), 1858–1931, US soldier, inventor of — **gun**; — (**Matthew Gregory, 'Monk'**), 1775–1818, English writer of romances. *See also* **Lewes, Wyndham Lewis**

lex (Lat.), law; *pl. leges*

lexicog., lexicograph/er, -y, -ical

lexicon/, dictionary, esp. of Greek, Hebrew, Arabic; *pl.* **-s**; abbr. **lex.** (not ital.)

lex/ *loci* (Lat.), local custom; — *non scripta*, unwritten law; — *scripta*, statute law; — *talionis*, 'an eye for an eye'; — *terrae*, the law of the land

ley, untilled land, *use* **lea**, but **ley farming**, grass-growing

Leyden/, Netherlands, *use* **Leiden**, but — **jar** (elec.)

Leys School, Camb.

lez, *see* **lès**

LF *or* **l.f.**, low frequency

Lfg. (Ger.), *Lieferung* (q.v.)

LG, (gunpowder, leather, wheat) large grain, Life Guards, Low German

L.Ger., Low German

L.Gr., Late Greek, Low Greek

LGSM, Licentiate of the Guildhall School of Music

LGU, Ladies' Golf Union

l.h., left hand

LHA, Lord High Admiral

Lhasa, Tibet, *not* -ssa, Lassa

LHC, Lord High Chancellor

LHD, *Literarum Humaniorum Doctor* (literally doctor of the more humane letters)

LHT, Lord High Treasurer

LI, Light Infantry, Long Island (US)

Li, lithium (no point)

Liadov (**Anatol Konstantinovich**), 1855–1914, Russian composer

liaise (colloq.), to form a liaison, *not* -ase

liaison, illicit amour; joining of words; (mil.) connection, touch (now also in general use)

liana, tropical climbing plant, *not* -ne

Liapunov (**Sergei Mikhailovich**), 1859–1924, Russian composer

Lib, colloq. abbr. of (Men's, Women's, etc.) Liberation (cap., no point)

lib./, librarian, library; — **cat.**, library catalogue

lib. (Lat.), *liber* (a book)

libel/, **-led, -ler, -ling, -lous**

liber (Lat.), a book; abbr. **L.** *or* **lib.**

Liberal/, abbr. **L.**; — **Unionist** (caps., no hyphen)

liberalize, *not* -ise

Liberia, republic, W. Africa

libid/**o**, sex drive, *pl.* **-os**; adj. **-inal**

libr/**a**, pound, *pl.* **-ae**; abbr. **£, l., lb.** *See also* **l.** *or* **£**

librair/**e** (Fr. m.), bookseller; **-ie** (f.), bookseller's shop

librar/**ian, -y**; abbr. **lib.**

librett/**o** (It.), text of an opera, etc.; *pl.* **-os, -i** (not ital.); **-ist**, writer of this

libris (**ex-**), *see* **ex-libris**

libr/**o** (It.), a book; *pl.* **-i**

Libya, N. Africa, indep. 1951, *formerly* Tripolitania, Cyrenaica, and Fezzan (Italian)

Libyan, of anc. N. Africa or mod. Libya

licence (noun), a permit; (US **-se**)

licens/**e** (verb), to authorize; **-ee, -er, -ing**

licensed victualler

licentiate, abbr. **L.**

licet (Lat.), legal; it is allowed

lichee, *use* **litchi**

lichen, epiphytic veg. growth

Lichfield, Staffs. *See also* **Litchfield**

Lichfield:, sig. of Bp. of Lichfield (colon)

lich-gate, roofed gateway of churchyard, *not* lych-

lichi, *use* **litchi**

lickerish, desirous, greedy. *See also* **liquorice**

Lick Observatory, California

licorice, use **liquor-**
Liddell (Henry George),
1818–98, English lexico-
grapher, father of the original
of *Alice in Wonderland*
lido/, open-air swimming-pool,
pl. **-s**
Lie (Jonas Lauritz Edemil),
1833–1908, Norwegian novelist; — **(Trygve Halvdan),**
1896–1968, Sec.-Gen. of UN
1946–53
Liebfraumilch, a hock; in Ger.
Liebfrauenmilch
Liebig/ (Justus, Baron von),
1803–73, German chemist; —,
a beef extract first prepared by
him (cap.)
Liechtenstein, principality on
Upper Rhine
Lied/ (Ger. n.), a song; *pl.* **-er**
Lieder ohne Worte, songs without words
Lieferung (Ger. f.), a part of a
work published in instalments,
abbr. *Lfg.*
Liège, Belgium, *formerly* Lié-; in
Fl. **Luik; Liégeois/,** *fem.* **-e,** an
inhabitant of Liège.
lieu (in), in place (of) (not ital.)
Lieutenant/, abbr. **Lt.**
or **Lieut.; — -Colonel,** abbr.
Lt.- *or* **Lieut.-Col.; —-**
-Commander, abbr. **Lt.-** *or*
Lieut.-Com.; — -General,
abbr. **Lt.-** *or* **Lieut.-Gen.; — -**
-Governor, abbr. **Lt.-** *or*
Lieut.-Gov. (hyphens)
life/belt, -boat (one word)
life cycle (two words)
life/-guard, -jacket (hyphens)
Life Guards, regiment of household cavalry (two words); **Life-**
guardsman (hyphen)
life insurance is the general
term. *See also* **assurance**
life/like, -line (one word)
lifelong, lasting for life (one
word). *See also* **livelong**
life-size(d) (adj., hyphen)
lifetime (one word)
ligature (typ.), two or more
letters joined together and form-

ing one character or type, as
Æ, œ, ffi. See also **diphthongs,**
and *Hart's Rules,* pp. 62–3
light (verb), past **lit** (preferred)
partic. **lit** when predic. (the fire
is lit), **lighted** when attrib. (a
lighted match)
light bulb (two words)
lightening, making less heavy
lighthouse (one word)
Light Infantry, abbr. **LI** *or* **Lt.**
Inf.
lighting-up time (one hyphen)
light meter (two words)
lightning (meteor.), abbr. **l.**
light-o'-love, fickle woman (hyphens, apos.)
lightship (one word)
lightweight (one word)
light-year (astr.), unit of distance (hyphen)
ligneous, of wood, woody
-like. In formations intended as
nonce-words, or not generally
current, the hyphen should be
used. It may be omitted when
the first element is of one syllable, but nouns in *-l* always
require it, e.g. eel-like
likeable, *not* lika-
likelihood, *not* -lyhood
Lilienthal (Otto), 1849–96,
German aviator
Liliput, liliputian, use **Lilli-,**
lilli-
lillibullero, a seventeenth-cen-
tury song refrain, *not* the many
variants (not ital.)
Lilliput, country of the pygmies
in Swift's *Gulliver's Travels*; **lilli-**
putian, diminutive (not cap.)
lily of the valley (no hyphens)
limbo/, the borderland of Hell,
a place of oblivion; *pl.* **-s** (not
ital.)
limbus/ fatuorum (Lat.), a
fool's paradise; — *infantum,*
limbo of unbaptized children;
— *patrum,* limbo of pre-Christian good men
lime/-juice, -kiln (hyphens)
lime/light, -stone (one
word)

'imes (Lat.), Roman frontier (ital.)

Limey (derog.), an English immigrant in former British colonies; (US slang), a British sailor, an Englishman. *See also* **limy**

Limited, abbr. **Ltd.**

limy, limelike, sticky. *See also* **Limey**

lin, line/al, -ar

linable, able to be covered on the inside, *not* linea-

linage, the number of lines, payment by the line. *See also* **lineage**

linament, *use* **linea-** *or* **lini-**

linchpin, *not* ly- (one word)

Linch's law, *use* **lynch** —

Lincoln:, sig. of Bp. of Lincoln (colon)

Lincoln Center, New York, a nexus of theatres, opera-house, concert-hall, etc.

Lincolnshire, abbr. **Lincs.**

Lindbergh (Charles Augustus), 1902–74, US aviator

Lindley (John), 1799–1865, English botanist

Lindsay (Earl of), family name Lindesay-Bethune; — **(Sir Coutts),** 1844–1913, English artist; — **(Sir David),** 1490–1555, Scottish poet; — **(Nicholas Vachel),** 1879–1931, US poet

Lindsey (Earl of), family name Bertie

line, abbr. **l.**; *pl.* **ll.**

lineage, ancestry. *See also* **linage**

line/al, -ar, abbr. **lin.**

lineament, a feature, *not* lina-. *See also* **liniment**

Linear/ A, earlier of two forms of anc. Cretan writing; — **B,** later form, found also on Gr. mainland (no hyphens)

line block (typ.), letterpress (q.v.) block for lines and solids

linen-draper (hyphen)

linenfold, carved scroll (one word)

lingerie, linen articles collec-

tively, women's underwear (not ital.)

lingo/ (colloq.), a (foreign) language, *pl.* **-s**

lingua franca, an international language, esp. a mixture of It., Fr., Gr., and Span. used in the Levant (not ital.)

liniment, an embrocation, *not* lina-. *See also* **lineament**

lining numerals (typ.), those that align at top and bottom, 1234567890

lining paper (binding), that inside the cover

link, 7.92 in.; one-hundredth of a chain; abbr. **l.**

Linn/aean, abbr. **Linn.**; *but* **-ean Society,** London (off. spelling); abbr. **LS**

Linnaeus (Carolus), 1707–78, Swedish naturalist; in Sw. **Carl von Linné;** abbr. **L.** *or* **Linn.**

Linnhe (Loch), Highland

Linotype, firm *formerly* producing hot-metal composing machine casting each line of type as a continuous unit (abbr. **Lino**), *now* various computer--driven filmsetting systems not on the slug (q.v.) principle

linsey-woolsey, a thin coarse fabric of linen and wool; gibberish; *not* linsy-, -wolsey

Linson (bind.), strong paper used in place of bookcloth (propr. term)

lintel/, -led

liny, full of lines, *not* -ey

Lion (Gulf of the) *or* **Lion Gulf,** off Mediterranean Fr. coast; in Fr. **golfe du Lion,** *not* Lions, Lyon, Lyons

Lionardo da Vinci, *use* **Le-lionize,** *not* -ise

Lippi (Fra Filippo *or* **Lippo),** *c.*1406–69, and — **(Filippino),** *c.*1457–1504, his son, Italian painters

lip/salve, -stick (one word)

lip-service (hyphen)

liq., liquid, liquor

lique/faction, -factive, -fiable, -fy, *not* liqui-

liqueur/, a strong alcoholic liquor, sweetened and flavoured (not ital.); — **glass** (hyphen)

liquid, abbr. **liq.**

liquidambar, a genus of balsam-bearing trees, *not* -er (one word, not ital.)

liquor/, abbr. **liq.; — on draught,** *not* draft

liquorice, *not* licor-

liquorish, desirous, greedy, *use* **licker-**

lir/a, unit of Italian currency; *pl.* -e; abbr. **L.**

lissom, supple, *not* -e

Liszt (Abbé Franz), 1811–86, Hungarian pianist and composer

lit., literal, -ly, literary, literature, litre, little

Litchfield, Hants, *also* Conn., US; — **(Earl of).** *See also* **Lich-**

litchi, Chinese fruit(-tree), also grown in Bengal, *not* the many variants (not ital.)

Lit.D., *Literarum Doctor* (Doctor of Letters). *See also* **D.Lit., D.Litt., Litt.D.**

lite pendente (Lat.), during the trial

liter/al, -ally, -ary, -ature; abbr. **lit.**

literal (typ.), printing of wrong sort, of turn (q.v.), or of sort in battered state or wrong fount

literalize, etc., to render literal, *not* -ise

litera (*or* **littera**) **scripta manet** (Lat.), the written word remains

literat/i, the learned as a class; *sing.* **-us,** *not* litt-

literatim (Lat.), letter for letter

Lith., Lithuanian

lithium, symbol **Li**

lithography (typ.), planographic (q.v.) printing process from smooth plate, originally a stone; abbr. **litho** (no point)

Lithuania, a Soviet Socialist Republic. See *also* **USSR**

Lit. Hum., *Literae Humaniores,* Faculty (Classics and Philosophy) at Oxford

litigious, fond of going to law

litmus-paper (hyphen)

Litolff (Henri Charles), 1818–91, French pianist, composer, music-publisher

litotes (rhet.), a form of meiosis (q.v.), esp. one using ironical negative, e.g. no mean feat

litre/, abbr. **l.** (*but* no point in scientific and technical work) *or* **lit.,** one-thousandth of a cubic metre, 1.76 pint; **-age,** number of litres

Litt.B., *Literarum Baccalaureus* (Bachelor of Letters)

Litt.D., *Literarum Doctor* (Doctor of Letters, Camb. and TCD). *See also* **D.Lit., D.Litt., Lit.D.**

littera, *see* litera

littérateur, a literary man (not ital.)

litterati, *use* lite-

Little Englander (hist., two words, caps.)

Littlehampton, W. Sussex (one word)

Littleton, family name of Baron Hatherton; — **(Sir Thomas),** ?1407–81, English jurist. *See also* **Lyttelton**

littoral, a region by a coast

Littré (Maximilien Paul Émile), 1801–81, French lexicographer

liturg., liturg/ies, -ical, -y

liv. (Fr.), *livre* (m. book, f. pound)

liveable, *not* liva-

livelong, intensive and emotional form of **long.** *See also* **lifelong**

Liver/politan, -pudlian, (inhab.) of Liverpool

Liverpool:, sig. of Bp. of Liverpool (colon)

livestock (one word)

Livingston, Lothian, 'new town', 1962

Livingstone (David), 1813–73

Scottish explorer and missionary

Livorno, Italian seaport, *formerly* in Eng. **Leghorn**

livraison (Fr. f.), a part of a work published in instalments

livre (Fr. m.), book; (f.) pound; abbr. **L.** *or* **liv.**

Livy (in Lat., **Titus Livius**), 59 BC–AD 17, Roman historian

L J, Lord Justice; **L JJ,** Lords Justices (thin space)

Ljubljana, Yugoslavia; in Ger. **Laibach**

LL, late, law, *or* Low, Latin, Lord-Lieutenant, -s

ll, in Spanish and Welsh, a separate letter, not to be divided

ll., leaves, lines, (Lat.) *leges* (laws)

ll (words ending in), followed by -ful *or* -ly, usu. omit one *l*

llama, S. American ruminant. *See also* **lama**

Llandeilo, Dyfed; — **Group** (geol.) (caps.)

Llandrindod Wells, Powys

Llanelli, Dyfed (*not now* -y)

llano/, S. American plain, *pl.* **-s**

Llantwit Major, S. Glamorgan, in Welsh **Llanilltud Fawr**

LL B, *Legum Baccalaureus* (Bachelor of Laws) (no points, thin space)

LL D, *Legum Doctor* (Doctor of Laws) (no points, thin space)

LL M, *Legum Magister* (Master of Laws) (no points, thin space)

Lloyd (**Norddeutscher**), the North-German Lloyd Steamship Co. (two words); abbr. **NDL**

Lloyd's/, the association of underwriters, *not* -s'; — **marks,** in — **Register of Shipping,** in order of merit, for wooden ships, **A1, A1** (in red), **Æ,** and **E;** for iron or steel, **100 A1, 90 A1, 80 A1.** *See also* **Loyd**

Lloyds Bank, Ltd. (no apos.)

LM, Licentiate in Midwifery; (mus.) long metre

£m., (one) million pounds

l.M. (Ger.), *laufenden Monats* (of the current month)

lm, lumen(s) (no point)

LMBC, Lady Margaret Boat Club (St. John's College), Camb.

LMD, long metre double

LMH, Lady Margaret Hall (Oxford)

LMR, London Midland Region

LMS, London Missionary Society

LMS(R), London, Midland, & Scottish Railway (prior to nationalization)

LMSSA, Licentiate in Medicine and Surgery, Society of Apothecaries

ln, natural logarithm (no point)

LNER, London and North Eastern Railway (prior to nationalization)

LO, Liaison Officer

loadstar, *use* **lodestar**

loadstone, magnetic oxide of iron, *not* lode- (one word)

load-water-line (hyphens); abbr. **LWL**

Loanda, Angola, *not* St. Paul de —, *but use now* **Luanda**

loath, averse, *not* loth

loathe, to hate

loathsome, *not* loth-

LOB, Location of Offices Bureau

Lobachevski (**Nikolai Ivanovich**), 1793–1856, Russian mathematician

'Lobgesang', Mendelssohn's 'Hymn of Praise', 1840

lobscouse, a sailor's dish

lobworm (one word)

locale, scene of operations (erron. form, but well established)

localize, *not* -ise

loc. cit., *loco citato* (in the place cited) (not ital.)

loch, Scottish lake, *not* -ck; capital when with name, abbr. **L.**

Lochalsh, Highland

lochan, small Scottish lake

Loch Awe, Strathclyde

Lochearnhead, Central (one word)

Lochgilphead, Strathclyde

Loch Leven, Highland, Tayside

Lochnagar, mountain in Grampian (one word)

Lock/e (John), 1632–1704, English philosopher; adj. **-ian,** *not* **-ean**

lockjaw (one word)

lock-out, employers' refusal of work; *pl.* **lock-outs** (hyphen)

locksmith (one word)

lock-up (noun *or* adj., hyphen); — (typ.), exertion of pressure to hold various elements in the forme together

loco/, locomotive, *pl.* **-s** (no point)

loco/ (Lat.), in the place; — *citato,* ditto cited, abbr. **l.c.** *or* **loc. cit.** (not ital.); — *laudato,* ditto cited with approval, abbr. **loc. laud.;** — *sigilli,* ditto of the seal, abbr. **LS;** — *supra citato,* ditto cited above, abbr. **l.s.c.** (not ital.)

Locofoco (US), extreme section of Democratic Party, 1835

locum/-tenency (hyphen); — **tenens,** a substitute, *pl.* — **tenentes** (not ital.)

loc/us (Lat.), a written passage, a curve, *pl.* **-i;** *locus/ citatus,* the passage quoted; — *classicus,* an authoritative passage from a standard book, *pl.* **loci classici;** — *communis,* a commonplace; — *delicti,* the place of a crime; — *in quo,* the place in which; — *poenitentiae,* a place of repentance; — *sigilli,* the place of the seal, abbr. **LS;** — *standi,* recognized position, (law) right to appear

LOD, Little Oxford Dictionary

lodestar, star steered by, *not* load-

lodestone, *use* **load-**

lodg/e, -eable, -ement, -ing

lodging-house (hyphen)

Łódź, Poland

Loeb (James), 1867–1933, US banker, founder of the — Library of classical authors

loess (geol.), deposit of fine yellowish loam in certain valleys, *not* loëss, löss

L. of C., line of communication

Lofoten Islands, Norway, *not* -den, -ffoden

log, logarithm (no point)

log., logic

logan-stone, a rocking-stone, *not* loggan-, logging-

logarithm, abbr. **log;** — **(natural),** to base *e,* abbr. **ln**

log-book (hyphen)

loge (Fr. f.), theatre stall (ital.)

loggia/ (It.), a gallery; *pl.* **-s**

logic symbols: and, $\wedge$; belongs to, $\in$; does not belong to, $\notin$; is equivalent to, $\leftrightarrow$; there exists, $\exists$; for all, $\forall$; implies, $\rightarrow$; i included in, $\subset$; is not included in, $\not\subset$; includes, $\supset$; does not include, $\not\supset$; intersection, $\cap$; or $\vee$; union, $\cup$

logi/on, a saying of Christ not i the Gospels; *pl.* **-a** (not ital.)

logo/, (colloq.), a device used as printed badge etc. of organization, *pl.* **-s** (no point)

logotype (typ.), several letters, or a word, cast as a single sort

log-roll/, -er, -ing, (to give, on who gives) mutual aid among politicians or reviewers (hyphen)

logwood (one word)

Lohengrin, German hero; (ital.) opera by Wagner, 1850

Loir, Loire, Loiret, French rivers

lollipop/, a sweetmeat, *not* lolly-; — **man** (colloq.), traffic-controller

Lombroso (Cesare), 1836–1909, Italian criminologist

Lomé, Togo

Londin:, sig. of Bp. of London (colon)

London/, abbr. **L.** *or* **Lond.;** — **Apprentice,** hamlet in Cornwall; — **County Council/, -lor,** abbr. **LCC** (no points). *See also* **GLC**

ong., longitude

ong-bodied (typ. adj.), denoting metal sorts whose body is cast a point or more longer than the size of letter upon them normally takes, producing the effect of leading, q.v.

ongbow (one word)

ong-distance (adj., hyphen)

onge, *use* lu-

ong/eval, long-lived, *not* -aeval; **-evity**

ongfellow (Henry Wadsworth, *not* Words-), 1807–82, US poet

onghand (one word)

ong Island, US; abbr. **LI**

ongitude, abbr. **long.**

ong jump (two words)

ongleat House, Wilts.

ong/ letter (typ.), ā, ē, etc.; — **mark,** that placed over the long letter; the macron

ongman Group Ltd., publishers

ong Mynd (The), Shropshire; **Longmynd Group** (geol.) (caps.)

ongo (Alessandro), 1864–1945, compiler of edition of Domenico Scarlatti's keyboard works

'ongo intervallo (Lat.), at a long interval

ong page (typ.), one having a line or lines more than its companion pages

ong primer (typ.), name for a former size of type, about 10 pt.; *pron.* primmer

Longridge, Lancs.

long s, former letter f, italic ſ

Longshanks, sobriquet of Edward I

Longships, islands and lighthouse off Cornwall

long-shore (adj., hyphen), *not* 'long-; *but* **longshoreman** (one word)

long ton, 2,240 lb., abbr. **l.t.**

Longton, Lancs., Staffs.

Longtown, Cumbria, Hereford & Worc.

longueur, tedious passage in book, play, film, usu. *pl.* (ital.)

long vowel (typ.), ā, ē, etc.

long/ways, -wise (one word)

looking-glass (hyphen)

look-out/, *pl.* **-s** (hyphen)

loophole (one word)

loosestrife (bot.) (one word)

Lope de Vega, *see* **Vega Carpio**

loping, with long strides, *not* lope-

lop-sided (hyphen)

loquitur (Lat.), he, *or* she, speaks; abbr. *loq.*

Lorca (Federico García), 1899–1936, Spanish poet

Lord (with name), may be substituted for **Marquess, Earl,** or **Viscount,** and is more often used than **Baron**

Lord, abbr. **Ld.**

Lord/ Chamberlain, — Chancellor, abbr. **LC;** — **Chief Baron,** abbr. **LCB;** — **Chief Justice,** abbr. **LCJ;** —**Justice,** abbr. **LJ,** *pl.* **L JJ** (thin space); — **-Lieutenant,** *pl.* — **-s** (hyphen), abbr. **LL;** — **Mayor** (two words, caps.); — **of hosts, — of lords,** as Deity (one cap.); — **President of the Council,** abbr. **LPC;** — **Privy Seal,** abbr. **LPS;** — **Provost,** abbr. **LP**

Lord's Cricket Ground, London (apos.)

Lord's Day (caps.)

Lordship, abbr. **Ldp.**

Lord's/ Prayer, — Supper, — Table (caps.)

Lorelei, siren of the Rhine, *not* -ey, Lurlei

Lorenzo, *see* **Lourenço**

Loreto, Colombia, Italy, Mexico, Peru

Loretto, Anglicized version of the It. **Loreto;** *also* Scottish school

lorgnette (Fr. f.), opera-glass, or pair of eye-glasses with long handle

loris, small lemur, *not* lori, lory

Lorrain (Claude) (*né* **Gelée**), 1600–82, French painter, *not* -aine

Lorraine, *see* **Alsace-Lorraine**

lory, one of the parrots. *See also* **loris**

losable, *not* -eable

Los Angeles, Calif., US

löss (geol.), *use* **loess**

Lot-et-Garonne, dép. France

loth/, -some, *use* **loath**(-)

Lothario/, 'the gay', in Nicholas Rowe's *The Fair Penitent*, 1703; *pl.* **-s**

Lothian, region of Scotland

Loti (Pierre), pseud. of Julien Viaud, 1850–1923, French writer

lotus/, Egyptian and Asian water-lily; *pl.* **-es**

lotus-eater, *not* lotos-; but 'The Lotos-Eaters' by Tennyson

louche (Fr.), disreputable (ital.)

loudspeaker (one word)

lough, Irish lake, capital when with name, abbr. **L.**

Louisiana, off. abbr. **La.** or (postal) **LA,** *not* Lou.

Louis-Philippe, 1773–1850, French king

Louis/-Quatorze, reign 1643–1715; — **-Quinze,** 1715–74; — **-Seize,** 1774–93; — **-Treize,** 1610–43 (Louis XIV, XV, XVI, XIII), art styles (hyphens)

loung/e, -ing

lounge/ bar, — suit (two words)

lour, to frown, *not* lower

Lourenço Marques, Mozambique, *now* Maputo; *not* Lorenzo —, -ez

Louvain, Belgium, in Fl. **Leuven**

louver, shutter, ventilator, *not* -re

lovable, *not* -eable

love/-affair, -bird, -hate (relationship), **-letter** (hyphens)

love/lock, -lorn, -sick (one word)

Love's Labour's Lost, by Shakespeare, 1598 (two apos.)

love/-song, -story (hyphens)

Low Church (hyphen when attrib.) ; **Low-Churchman** (hyphen, caps.)

low-down, (adj.) mean, (noun, colloq.) the basic facts (hyphen)

lower, to frown, *use* **lour**

Lower California (caps.)

lower case (typ.), the case containing the small letters, hence the small letters a–z; abbr. **l.c.** (no points as proof-correction mark)

Low (Sampson), Marston, & Co., publishers

Low Sunday, first after Easter (caps.)

low water (two words)

low-water mark (one hyphen) ; abbr. **LWM**

Loyd (Sam), 1841–1911, US chess-player. *See also* **Lloyd/, -s, -'s**

LP, large paper, long-playing record, Lord Provost, low pressure (also **l.p.**), (paper) large post

LPC, Lord President of the Council

LPG, liquefied petroleum gas

L-plate, learner-driver's sign (hyphen)

LPO, London Philharmonic Orchestra

L'pool, Liverpool

LPS, Lord Privy Seal

Lpz. (Ger.), Leipzig

lr., lower

LRAD, Licentiate of the Royal Academy of Dancing

LRAM, Licentiate of the Royal Academy of Music

LRCP, Licentiate of the Royal College of Physicians

LRCS, Licentiate of the Royal College of Surgeons

LRCVS, Licentiate of the Royal College of Veterinary Surgeons

LRSC, Licentiate of the Royal Society of Chemistry

LS, Leading Seaman, Linnean Society, *loco* or *locus sigilli,* (in) the place of the seal

l.s., left side

LSA, Licentiate of the Society of Apothecaries

l.s.c., *loco supra citato* (in the place cited above)

LSD, the hallucinatory drug *d*-lysergic acid diethylamide, *librae, solidi, denarii* (pounds, shillings, and pence, also **£.s.d.**), Lightermen, Stevedores, and Dockers

LSE, London School of Economics and Political Science

LSI (comp.), large-scale integration

LSO, London Symphony Orchestra

L. (&) S.W.R., London and South Western Railway (became part of **SR** until nationalization)

Lt., Lieutenant

l.t., long ton, low tension

LTA, Lawn Tennis Association, London Teachers' Association

L.T. & S.R., London, Tilbury, and Southend Railway (became part of **LMS** until nationalization)

LTCL, Licentiate of Trinity College of Music, London

Lt. Col., Lieutenant Colonel

Lt.-Com., Lieutenant-Commander

Ltd., Limited

LTE, London Transport Executive

Lt.-Gen., Lieutenant-General

Lt.-Gov., Lieutenant-Governor

L.Th., Licentiate in Theology (Durham)

Lt. Inf., Light Infantry

LTM, Licentiate in Tropical Medicine

Lu, lutetium (no point)

Luanda, Angola, *not now* Loanda

Luang Prabang, Laos

Lübeck, W. Germany, *not* Lue-

Lubumbashi, Zaïre, formerly **Élisabethville**

Lucan, of St. Luke, *not* Luk-; —

(in Lat., **Marcus Annaeus Lucanus**), 39–65, Roman poet

lucarne, a dormer window

Lucerne, Switzerland, in Ger. **Luzern**

lucerne, a plant, *not* -ern

Lucian, 2nd c., Greek writer

Lucknow, Uttar Pradesh, India

lucus a non lucendo (Lat.), *approx.* inconsequent or illogical

LUE (theat.), left upper entrance

Luebeck, *use* **Lübeck**

Luftwaffe, German air force

Luggnagg, island in Swift's *Gulliver's Travels*

lug/hole, -sail, -worm (one word)

Luik, Fl. for **Liège**

Luke (St.) (NT), not to be abbreviated; adj. **Lucan,** *not* Luk-

lukewarm/, -ness, tepid(ity) (one word)

Lulsgate Airport, Bristol

lumbar (anat.), of the loins

lumber, to move clumsily; unused furniture, etc.

lum/en (anat.), a cavity, *pl.* -ina; **lumen/** (phys.), the unit of luminous flux, *pl.* -s, abbr. **lm**

lumpenproletariat, the lowest elements of the proletariat, term orig. used by Karl Marx (one word, not cap.)

lunation, time from one new moon to next, about 29½ days

lunge, long rope for exercising horses, *not* lo-

lunging, *not* lungeing

lupin, a garden plant, *not* -e

lupine (adj.), of a wolf

Lurlei, *use* **Lorelei**

Lusiads (The), by Camoens, 1572

Lusitania, Portugal

lustr/um (Lat.), a five-year period; *pl.* -a

lusus naturae (Lat.), a freak of nature; *pl.* same

lutenist, a lute-player, *not* -anist

lutetium, *not* -tec-, symbol **Lu**

Luth., Lutheran

Lutine bell, bell rung at Lloyd's, once to announce the loss, twice for the arrival, of a vessel over-due

Lutosławski (Witold), b. 1913, Polish composer

lux (phys.), unit of illuminance, abbr. **lx**

Luxembourg, Gardens and Palace, Paris

Luxemburg, prov. of Belgium; *also* **Grand Duchy of** —; in Fr. **-ourg**

Luxemburg (Rosa), 1870–1919, German socialist

Luzern, Ger. for **Lucerne**

LV, luncheon voucher

Lvov, USSR, in Ger. **Lemberg,** in Pol. **Lwów**

Lw, lawrencium (no point)

LWL, load-water-line

LWM, low-water mark

Lwów, *see* **Lvov**

lx, lux (no point)

LXX, the Septuagint, seventy

Lya/dov, -punov, *use* **Lia-**

lycée (Fr. m.), higher secondary school

Lyceum (the), Aristotle's school of philosophy; **lyceum/,** a college of literary studies, *pl.* -s

lychee, *use* **litchi**

lych-gate, *use* **lich-**

lyddite, an explosive (two *d*s)

lying-in, childbed (hyphen)

lyke-wake, night-watch over dead body

Lyly (John), ?1554–1606, English author. *See also* **Euphues**

Lymington, Hants. *See also* **Lea-, Le-**

lynch law, *not* -'s, Linch's (two words)

lynchpin, *use* **li-**

lynx/, animal of cat-tribe; *pl.* **-es**

Lyon King-of-Arms, chief Scottish herald

Lyonnais (Crédit), French banking corporation

Lyons, in Fr. **Lyon.** *See also* **Lion**

lysin (chem.), lysing substance

lysine (chem.), an amino acid

Lyte (Henry Francis), 1793–1847, English hymn-writer

Lytham St. Annes, Lancs. (no apos.)

Lyttelton, family name of Viscounts Chandos and Cobham (*see also* **Littleton**); — **(Humphrey Richard Adeane),** b. 1921, English jazz musician

Lytton, *see* **Bulwer-Lytton**

M

M, 1,000; (chem.) molar; motor-way; (as prefix) mega-; the twelfth in a series

M., Majesty, Marqu/ess, -is, Member, metronome, middle, militia, (in Peerage) minor, Monday, (Fr.) *main* (hand), *mille* (a thousand), monsieur, (It.) *mano* (hand), *mezz/o, -a* (half), (Lat.) *magister* (master), *medicinae* (of medicine)

M', *see* **Mac**

m, metre(s) (in scientific and technical work), (as prefix) milli-

m., male, married, masculine, merid/ian, -ional, metre, -s, mile, -s, mill, million (q.v.), minute, -s, (meteor.) mist, month, -s, moon, (Lat.) *meridies* (noon)

m, (chem.) *meta*-, (mech.) mass, (Fr.) *mois* (month)

ℳ, minim (drop)

MA, *Magister Artium* (Master of Arts), Massachusetts (off. postal abbr.), Military Academy

mA (elec.), milliampere (no point)

m/a (book-keeping), my account

ma (Italian), but

ma'am, *see* **madam**

Maartens (Maarten), pseud. of Joost Marius van der Poorten-Schwartz, 1858–1915, Dutch novelist, wrote in English

Maas, *see* **Meuse**

Maastricht, Netherlands, *not* Maes-

Mabinogion, collection of ancient Welsh romances

Mac (the prefix), spelling depends upon custom of the one bearing the name, and this must be followed, as: MacDonald, Macdonald, McDonald, Mᶜ-Donald, M'Donald (the turned comma is usual here, not the apos.). In alphabetical arrange-

ment it should, however spelt, be treated as Mac

mac, colloq. for mackintosh, *not* mack (no point)

macadamize, to cover with layers of small broken stone, each rolled in, *not* -ise; from **John Loudon McAdam,** 1756–1836, British engineer

Macao, E. Asia, *not* -au

macao, a parrot, *use* **macaw**

macaron/i (*not* macc-), long tubes of pasta; an eighteenth--century dandy, *pl.* **-ies; -ic,** in burlesque verse, Latinized modern and modernized Latin

MacArthur (Douglas), 1880–1964, US general

Macaulay (Dame Rose), 1881–1958, British novelist; — **(Thomas Babington, Baron),** 1800–59, British writer

macaw, a parrot, West Indian palm, *not* macao, maccaw

Macc. (1, 2), Maccabees, First, Second Book of

Maccabean, of the Maccabees, *not* -bæan

McCarthy (Eugene Joseph), b. 1916, US politician; — **(Joseph Raymond),** 1909–57, US politician; — **(Mary),** b. 1912, US writer

MacCarthy (Sir Desmond), 1878–1952, British writer

Macchiavelli, *use* **Mach-**

McCormack (John), 1884–1945, Irish-born US singer

McCoy (the real), the genuine article

MacCunn (Hamish), 1868–1916, Scottish composer

MacDiarmid (Hugh), 1892–1978, Scottish poet, pseud. of **Christopher Murray Grieve**

Macdonald (Alexandre), 1765–1840, Duke of Taranto, French marshal; — **(Flora),**

237

Macdonald (*cont.*)
1722–90, Jacobite heroine; —
(**George**), 1824–1905, Scottish
novelist and poet

**MacDonald (Rt. Hon. James
Ramsay),** 1866–1938, British
Prime Minister 1924, 1929–35

MacDowell (Edward Alexander), 1861–1908, US composer

Maced., Macedonian

macédoine, mixed fruit or
vegetables

McGill University, Montreal,
Canada

Macgillicuddy's Reeks,
mountains in Kerry

McGonagall (William), 1830–
1902, Scottish poet noted for his
bad verse

**McGraw-Hill Publishing
Co., Ltd.**

MacGregor, family name of
Rob Roy

ma chère (Fr. f.), my dear. *See
also* **mon cher**

machete, Central American
knife, *not* matchet

Machiavell/i (Niccolò), 1469–
1527, Florentine statesman,
playwright, writer on political
opportunism, *not* Macch-; adj.
-ian (cap. only in lit. or
hist. usage); noun, **-ism,** *not*
-ianism

machicolation (archit.), openings between supporting corbels

machina (Lat.), a machine. *See
also* **Deus ex** —

machinab/le, -ility, *not* machine-

machine, abbr. **M/C.**

machine revise, final proof
before actual printing

mach number, ratio of speed
of a body to speed of sound in
surrounding atmosphere (not
cap.), from **Ernst Mach,** 1836–
1916, Austrian physicist

Machpelah, burial place of
Abraham, *not* Macp-

Machu Picchu, Peru, site of
antiquities

**Mackenzie (Sir Edward
Montague Compton),** 1883–
1972, British novelist

McKinley (William), 1843–
1901, US President 1896–1901

mackintosh, a waterproof, colloq. **mac** (no point); patented
by **Charles Macintosh,**
1766–1843, Scottish chemist; *not*
maci-

mackle, printing blemish

macle, twin crystal, spot in a
mineral

MacLehose & Sons, printers,
Glasgow

MacLeish (Archibald), b.
1892, US poet

Macleod (Fiona), pseud. of
William Sharp, 1856–1905,
Scottish poet and novelist

McLuhan (Herbert Marshall), b. 1911, Canadian sociologist

MacMahon (Marie Edmé Patrice Maurice de), 1808–93,
Duke of Magenta, French marshal, President 1873–9

McMaster University, Hamilton, Ont., Canada

Macmillan (Maurice Harold), b. 1894, British statesman

Macmillan/ Publishers, Ltd.,
London; — **Publishing Company, Inc.,** New York

McNaghten Rules, on insanity
as defence in criminal trial; *not*
named from **MacNaghtan
(Edward),** 1830–1913, English
judge, but from the murderer of
Sir Robert Peel's private secretary, 1843

MacNeice (Louis), 1907–63,
Anglo-Irish poet, critic, traveller

Macon, Ga., US

Mâcon, dép. Saône-et-Loire,
France; *also* a burgundy, *not*
-çon

Macpelah, *use* Mach-

Macquarie University, Sydney

macramé lace, a trimming of
knotted thread, *not* -mi

Macready (William Charles), 1793–1873, English actor

macrocosm, the great world. *See also* **microcosm**

macron, a mark (¯) to indicate a long vowel or syllable

macroscopic, visible to the naked eye

macrurous (zool.), long-tailed, *not* macrourous

M'Taggart (John M'Taggart Ellis), 1866–1925, British philosopher

macul/a (Lat.), a spot; *pl.* **-ae**

Madagascar, off SE coast of Africa, independent 1960 as the **Malagasy Republic**

madam, *pl.* **-s;** colloq. abbr. **'m, ma'am** (cap. for the correct form of address to the Queen), **marm, m'm, mum**

Madame, abbr. **Mme,** *not* Mdme; *pl.* **Mesdames,** abbr. **Mmes** (no point after abbrs.)

Mädchen (Ger. n.), girl; *pl.* same (cap.)

Madeira, island, wine, cake

Mademoiselle, abbr. **Mlle,** *not* Mdlle; *pl.* **Mesdemoiselles,** abbr. **Mlles,** *not* Mdlles (no point after abbrs.)

Madhya Pradesh (India), formerly Central Provinces; abbr. **MP**

madonna, (representation of) the Virgin Mary (cap. as name)

madonna (It.), my lady, madam (in 3rd person only, not cap.)

madrasah (Ind.), a school or college (*not* the many variants)

Mad. Univ., Madison University, US; Madras ——

Maecenas/, a patron of the arts, *not* Me–; *pl.* **-es**

maelstrom, whirlpool, *not* mahl-, mal-

Maelzel (Johann Nepomuk), 1772–1838, German musician. *See also* **metronome**

maenad/, a female follower of Bacchus, *not* me–; *pl.* **-s**

maestoso (mus.), majestic, stately

Maestricht, Netherlands, *use* **Maas-**

maestr/o (mus.), master, composer, conductor; *pl.* **-i** (not ital.)

Maeterlinck (Maurice), 1862–1949, Belgian poet, playwright, essayist, philosopher

Mae West, inflatable life-jacket named after film actress (two words, caps.)

MAFF, Ministry of Agriculture, Fisheries, and Food

Mafia, Sicilian secret society, extending to US. *See also* **Camorra**

mafios/o, member of Mafia, *pl.* **-i** (not cap.)

ma foi! (Fr.), upon my word!

Mag., Magyar (Hungarian)

mag., magazine, magnetism

Maga, colloq. for *Blackwood's Magazine*

magazine, abbr. **mag.**

magazines (titles of) (typ.), when cited, to be in italic

magdalen, a repentant prostitute, a home for such; *but* **Mary Magdalene**

Magdalen College, Oxford

Magdalene College, Cambridge

Magdalenian (archaeol.), late palaeolithic culture or period

maggot, a grub. *See also* **magot**

magic/ (verb), **-ked, -king**

magilp, *use* **meg-**

magister (Lat. m.), master, abbr. **M.; —** *artium,* Master of Arts, abbr. **MA; —** *chirurgiae,* ditto Surgery, **M.Ch.**

magistrand, an arts student ready for graduation, esp. now at Aberdeen Univ.; at St. Andrews a fourth-year student

magma/, a mass; *pl.* **-ta,** or **-s** in gen. use

Magna Charta (1215), *not* —— Ca-

magnalium, alloy of magnesium and aluminium

magnesium, symbol **Mg**

magnetism, abbr. **mag.**

magnetize, *not* -ise

magneto/, type of electric generator; *pl.* **-s**

magnification sign, × (followed by a numeral)

magnifico/, Venetian grandee, *pl.* **-es** (not ital.)

magnif/y, -ied, -ying

magnifying glass (two words)

magnum, a reputed two-quart wine bottle

magnum bonum, a large, good variety, esp. of plums or potatoes; *pl.* **magnum bonums** (two words, not ital.)

magn/um op/us (Lat.), an author's chief work; *pl.* **-a -era.** *See also* **opus magnum**

magot, ape, *also* Chinese or Japanese figure. *See also* **maggot**

Magritte (René), 1898–1967, Belgian painter

mag/us, a wise man; *pl.* **-i;** *but* **the Magi** (cap.)

Magyar, dominant race in Hungary; a Hungarian; the Finno--Ugric speech of Hungary; abbr. **Mag.**

magyar, type of blouse or of its sleeves

Mahabharata, Indian epic (accent on third syllable)

Mahame/dan, -tan, *use* **Muhammadan,** q.v.

mahara/ja, Indian prince, *not* -jah (cap. as title); **-nee,** wife of maharaja, *not* -ni

maharishi, Hindu sage

mahatma, in esoteric Buddhism one possessing supernatural powers; (with cap.) title prefixed to exalted persons, esp. **Gandhi**

mahaut, *use* **mahout**

Mahican, (member of) an American Indian tribe between the Hudson River and Narragansett Bay. *See* **Mohegan** *and* **Mohican**

mah-jong, Chinese game played with tiles, *not* -ngg (hyphen)

Mahler (Gustav), 1860–1911, Austrian composer

mahlstick, painter's hand-rest, *use* **maul-**

mahlstrom, *use* **mael-**

Mahom/et, the trad. English spelling, *not* -ed; *but use* **Muhammad.** *See also* **Islam, Muhammadan, Muslim**

mahout, elephant-driver, *not* -aut

Mahratta, *use* **Maratha**

Mahratti, *use* **Marathi**

mahseer, large Indian freshwater fish, *not* the many variants

mahwa, Indian tree, *not* the many variants

mai (Fr. m.), May (not cap.)

Maia (Gr. myth.), mother of Hermes

maidan (Ind., Pers.), a plain, an esplanade, *not* -aun

maiden/hair, -head (one word)

maiden name (two words)

maieutic, of Socratic method of educing latent ideas

maigre day (RCC), one when no flesh is eaten

Maillol (Aristide), 1861–1944, French sculptor

maillot (Fr. m.), dancer's tights, woman's bathing-suit

main/ (Fr. f.), a hand, *also* a quire, abbr. **M.;** — *droite*, right hand, abbr. **MD;** — *gauche*, left hand, **MG**

Maine, US, off. abbr. **Me.** or (postal) **ME**

Maine-et-Loire, dép. France (hyphens)

main line (hyphen as adj.)

main/sail, -spring, -stay (one word)

Mainz am Rhein, W. Germany (no hyphens), *not* Mayence

maiolica, *use* **maj-**

maisonette, a small house, a flat; Anglicized form of Fr. f. *maisonnette*

maison garnie (Fr. f.), furnished house

Maisur, *see* **Mysore**

aître/, title of French advocate, abbr. **M^e^; — *d'hôtel*** (Fr. m.), house steward; *à la* — —, plainly prepared with parsley (no hyphen)

aîtresse (Fr. f.), mistress

aiuscol/a (It. typ.), capital letter; *-etto,* small capital letter

1aj., Major

naj., majority

Majesté (*Sa*) (Fr. f.), His, or Her, Majesty, not *Son* —

Majesty, abbr. **M.**

1ajlis, *see* **Assemblies**

najolica, an Italian glazed or enamelled earthenware, *not* maiol-; in It. *maiolica*

Major/, abbr. **Maj.; — -General** (caps. as title, hyphen), abbr. **Maj.-Gen.**

Majorca, Balearic Is.; in Sp. **Mallorca**

najor-domo/, a house steward, *pl.* **-s** (hyphen)

najority, abbr. **maj.**

najuscule, a capital, or upper-case, letter

nake-believe (hyphen)

nakeready (typ.), preparation of forme or plate, fitting of new offset blankets, on printing machine (one word)

nakeshift (one word)

nake-up (typ.), arrangement of matter into pages

nakeweight (one word)

1al., Malachi, Malayan

Malacca, *see* **Malaya (Federation of)**

nalacology, study of molluscs; abbr. **malac.**

naladroit, clumsy (not ital.)

nala/*fide* (Lat.), treacherously; **— *fides*,** bad faith; **— *in se*** (Lat.), acts which are intrinsically wrong

Málaga, Spain (accent); **Malaga,** wine from there

Malagasy, native or language of Madagascar

Malagasy Republic, *see* **Madagascar**

malagueña, dance from Málaga

malaise (Fr. m.), discomfort, uneasiness

malamute, Eskimo dog, *not* male-

Malaprop (Mrs), in *The Rivals,* by Sheridan, 1775

malapropos, inopportunely; in Fr. *mal à propos*

Malawi/, Cent. Africa, independent state within the Commonwealth, 1964, republic 1966; *formerly* Nyasaland; adj. **-an; — (Lake),** *formerly* Lake Nyasa

Malay, language and people

Malaya (Federation of): till 1957 consisted of nine states (Johore, Kedah, Kelantan, Negri Sembilan, Pahang, Perak, Perlis, Selangor, Trengganu) and the two British settlements of Penang and Malacca; adj. **Malayan,** abbr. **Mal.**

Malayalam, language of Malabar coast, SW India

Malaysia/, since 1957, independent state within the Commonwealth; consists of Malaya, Sabah, and Sarawak; adj. **-n**

malcontent, *not* male-

Małcużyński (Witold), b. 1914, Polish pianist

mal de mer (Fr. m.), seasickness

Malden, Surrey

mal de tête (Fr. m.), headache

Maldiv/e Islands, republic, SW of Sri Lanka; adj. **-ian**

Maldon, Essex

mal du pays (Fr. m.), homesickness

male, abbr. **m.** (bot., zool., sociol.), sign ♂

maleable, *use* mall-

malecontent, *use* malc-

malee, *use* mallee

malemute, *use* malamute

malentendu (Fr. m.), misunderstanding (one word)

Malesherbes (Chrétien Guillaume de Lamoignon de),

Malesherbes (*cont.*)
1721–94, French statesman. *See also* **Malherbe**
malgré (Fr.), in spite of
Malherbe (François de), 1555–1628, French writer. *See also* **Malesherbes**
mali, *use* **mallee**
Mali/, republic in NW Africa, 1960, *formerly* French Sudan; adj. **-an**
Malines, Belgium, *not* Mechlin, q.v.
Mallarmé (Stéphane), 1842–98, French poet
malleable, *not* male-
mallee, Australian eucalyptus, Indian gardener, *not* malee, mali
Mallorca, Sp. for **Majorca**
Mallow, Co. Cork
Malmaison, near Paris (one word)
Malmesbury, Wilts.
malmsey, a sweet wine, *not* -sie, -esie, -asye
Malone (Edmond, *not* -und), 1741–1812, British lit. critic
Malplaquet (battle of), 1709
malpractice, misbehaviour, *not* -se
malstrom, *use* **mael-**
Maltese cross, ✠. *See also* **cross, crux**
Malthus (Thomas Robert), 1766–1834, writer on population
Malton, N. Yorks.
mama, *use* **mamma**
mameluco, in Brazil, offspring of a White and an Indian
Mameluke (hist.), member of mil. body that ruled Egypt 1254–1811, *not* the many variants
mamill/a, nipple; *pl.* **-ae**; adj. **-ary**
mamma, child's name for mother, *not* mama
mamm/a, breast; *pl.* **-ae**; adj. **-ary**
mammee, tropical American tree, *not* the many variants

Mammon, wealth personified (cap.)
Man., Manila; Manitoba
man., manual
Man (Isle of), see **Isle of Ma**
man about town (three words
manacle, fetter, *not* -icle
Manacles, rocks off Cornish coast
manage/able, -ment
manakin, trop. American birc *See also* **manikin, mannequi minikin**
mañana (Sp.), tomorrow
Manassas, Virginia, scene of American Civil War battles
Manasseh (tribe of)
Manasses (Prayer of), Apocr., abbr. **Pr. Man.**
man-at-arms, *pl.* **men-** — - - (hyphens)
manatee, the sea-cow
Manaus, Brazil
manche (Fr. m.), a handle; (f.) a sleeve
Manche (La) (Fr.), the Englis Channel
Manche, dép. France
Manchester, abbr. **Manch.** *or* **M/C.**
Manchester:, sig. of Bp. of Manchester (colon)
man-child, *pl.* **men-children** (hyphens)
Manchu (noun), an inhabitant or the language of Manchuria (adj.), of the land, people, or language of Manchuria; *not* -choo, -chow
Manchukuo, an empire of NE Asia, 1932–45, formed by the Japanese out of Manchuria, Chinese Jehol, and part of Inne Mongolia
Manchuria, region in E. Asia, named (1643) from an invadin Mongolian people; *see* previou entry; returned to China, 1945
Mancunian, (inhabitant) of Manchester
Mandalay, Burma, *not* Mande-
mandamus, writ issued from a

higher court to a lower, *not* -emus (not ital.)

mandarin, western name for a Chinese official; Chinese language; a small Chinese orange; *not* -ine

mandatary (law, hist.), one to whom a mandate is given

mandatory (adj.), commanding, compulsory

M and B (*also* **M and B 693**), sulphonamide drug (from initials of manufacturers May and Baker)

Mandelay, mandemus, *use* -da-

mandioc, *use* **manioc**

mandolin, stringed instrument, *not* -ine

mandrel, a spindle, *not* -il

mandrill, a baboon

manège (Fr. m.), horsemanship, riding-school. *See also* **ménage**

Manet (**Edouard**), 1832–83, French painter

man/et (Lat., theat.), he, *or* she, remains; *pl.* *-ent*

manganese, symbol **Mn**

mangel-wurzel, a large beet, *not* mangle-, mangold-

mango/, Indian fruit, *not* -oe; *pl.* -es

mangold-wurzel, *use* **mangel- - -**

mangosteen, tropical fruit, *not* -an, -ine

mangy, having mange, *not* -gey

manhaden, *use* **men-**

manhandle (one word)

manhattan, a cocktail (not cap.)

Manhattan Island, New York

Manhattan Project, the production of the first atomic bombs in US

Manheim, W. Germany, *use* **Mann-**

man-hour/, work of one man per hour, *pl.* -s (hyphen)

man-hunt (hyphen)

Manichaean, pertaining to the **Manichees,** religious followers

of the Persian **Mani,** *c.*216–76; *not* -chean

manicle, a fetter, *use* -acle

manifesto/, declaration of policy; *pl.* -s

manikin, a dwarf, anatomical model, *not* mana-, manni-. *See also* **manakin, mannequin, minikin**

Manila, Philippine Islands, abbr. **Man.**

manila, hemp, paper, *not* -illa

manilla, African bracelet

manioc, the cassava plant, *not* the many variants

maniple, subdivision of Roman legion, vestment

manipulator, *not* -er

Manipur, NE India, *not* Munnepoor

Manitoba, Canada; abbr. **Man.**

man-made (hyphen)

mannequin, dressmaker's (live) model. *See also* **manakin, manikin**

Mannheim, W. Germany, *not* Manh-

mannikin, *use* **mani-**

mannish, manlike (usu. derog. of woman)

Mannlicher rifle

mano/ (It. f.), a hand, abbr. **M.**; — *destra*, right hand, abbr. **MD**; — *sinistra*, left hand, abbr. **MS**

manœuvrab/le, -ility (US **maneuverab/le, -ility**)

manœuvr/e, -ed, -ing (US **maneuver/, -ed, -ing**)

man-of-war, armed ship; *pl.* **men- — - —** (hyphens)

man-of-war's-man (apos., hyphens)

manqu/é (Fr.), *fem.* *-ée,* unsuccessful

Mansard (**François**), 1598–1666, French architect

mansard roof, lower part steeper than upper (not cap.)

Mansfield (**Katherine**), 1888–1923, NZ-born British writer of short stories, pseud. of Kathleen

Mansfield (*cont.*)
Beauchamp, later Murry. *See*
Murry

mansuetae naturae (law), adj.,
tame (of animals), lit. of a tame
nature

mantel, shelf above fireplace

mantelet, short cloak; movable
screen to protect gunners, *not*
mantlet

mantelpiece, *not* mantle-

mantilla (Sp.), a short mantle;
a veil covering head and shoul-
ders

mantis, orthopterous insect, *pl.
same*

mantle, cloak

mantlet, *use* **mantelet**

mantrap (one word)

Mantua, Italy, in It. **Mantova**

manual, organ keyboard; abbr.
man.

manufactur/e, -er, abbr. **mfr.;
-ed, mfd.; -ers, -es, mfrs.;
-ing, mfg.**

manum/it, to set (slave) free;
-itted, -itting; noun **-ission**

manu propria (Lat.), with one's
own hand

manus (Lat. f.), a hand; *pl. same*

manuscript, abbr. **MS** (**a,** *not*
an), *pl.* **MSS** (*but* spell out except
in bibliog. enumeration). Also
called **copy.** It should be cleanly
typed double-spaced on one side
of paper (which should be of the
same size throughout). One inch
blank margin on left-hand side.
Caps. I, J, S, T, and l.c. e, i, l, m,
n, t, u, if handwritten, to be
clear. Italics, bold, small capitals, etc. are indicated by under-
lining as in proof correction,
q.v. Copy must be folioed in
consecutive order throughout:
not each chapter separately.
Never place corrections or ad-
ditions on *back* of a leaf, but put
in an extra leaf and mark it, say,
23 A, B, or C; cue the alteration
clearly into the original text. If
a leaf is deleted, say 24, mark
previous one 23–24. All matter

not to be printed (e.g. folio
numbers, directions to the
printer) to be ringed round in
ink. Adhesive tape should be
used as little as possible, and
pins and staples should not be
used. Copy for extracts and
footnotes should be incorporated in sequence with the text.
See also **footnotes** and BS 526
part 1

manuscrit (Fr. m.), MS; *abbr.*
ms., *pl.* **mss.**

Manuskript (Ger. n.), MS; *als*
printer's copy

Manutius/ (**Aldus**), 1450–
1515, in It. **Aldo Manuzio;**
—— 'the younger', 1547–97;
— (**Paulus**), 1512–72; Italian
printers

Manx/, of the Isle of Man;
**-man, -men, -woman,
-women** (one word)

Manzanilla, a dry sherry

Mao/ism, doctrine of Mao Zedong; **-ist**

Maori/, member or language o
native race of New Zealand; *p*
-s

Mao/ Zedong (— **Tse-tung**),
1893–1976, Chairman of
Chinese Communist Party
1954–76

Maquis, the resistance movement in France in Second Worl
War; from Corsican *maquis,*
scrubland

mar/, -red, -ring

Mar., March

mar., maritime

Mar (**Earl of**), family name: o
Mar. *See also* **Mar and Kellie**

marabou/ feather, — **stork,**
not -bout, -bu

marabout, N. African monk or
hermit, *not* -but

Maracaibo, Venezuela, *not* -ybo

maranatha (Syriac), 'our Lord
cometh'

Mar and Kellie (**Earl of**),
family name Erskine. *See also*
Mar

maraschino, a liqueur

Marat (Jean Paul), 1743–93, French revolutionary

Maratha, an Indian people, *not* Mahratta

Marathi, Indian language, *not* Mahratti

Marazion, Cornwall

marbling (bind.), staining end-papers or book edges to resemble marble

marbly, marble-like, *not* -ley

narbré (Fr.), marbled; *also* marbled edges of books

Marcan, of St. Mark, *not* Mark-

narcato (mus.), emphasized

March (month of), abbr. **Mar.**

narch., marchioness

Märchen (Ger. n.), a fairy-tale; *pl. same*

narchese (It.), marquis; fem. *-a*, marchioness

narchioness, abbr. **march.** (cap. as title)

Marcobrunner, a hock

Marconi (Guglielmo, Marchese), 1874–1937, inventor of radio telegraphy; **marconigram,** message sent by his system

Marcuse (Herbert), 1898–1979, German-born US philosopher

mardi gras (Fr. m.), Shrove Tuesday

maréchal (Fr. m.), Field Marshal; his wife, *-e*

Maréchal Niel, a rose

mare (Lat.), sea, (astr.) lunar plain, *pl.* **maria; — clausum,** a sea under one country's jurisdiction; — **liberum,** a sea open to all countries

mare's-tail (bot.), marsh plant (apos., hyphen)

marg., margin, -al

margarine, butter-substitute, colloq. **marge**

margarite, a mineral or rock-formation

Margaux (Château-), a claret (hyphen)

marge, colloq. for **margarine,** *not* marg

marge (Fr. f.), margin

margin/, -al, abbr. **marg.**

marginalia (pl.), marginal notes (not ital.)

margins (typ.), the four are called back (at binding), head (top), foredge (opposite the binding), and tail (foot of page). An acceptable ratio for the size of margins, in order as above, is $1 : 1\frac{1}{2} : 2 : 2\frac{1}{2}$

Margoliouth (David Samuel), 1858 1940, British orientalist

Margrethe, b. 1940, Queen of Denmark 1972–

marguerite, ox-eye daisy

mariage de convenance (Fr. m.), marriage of convenience, *not marr-*

Mariamne, name of two wives of Herod the Great

Marianne, woman symbolizing France

Mariánské Lázně, Czechoslovakia, in Ger. **Marienbad**

Marie/ de' Medici, 1573–1642, wife of Henry IV of France (de', *not* de); in Fr. — **de Médicis**

marijuana (Sp.), hemp, *not* -huana

marin/ade (Fr. cook. f.), a pickle; **-ate,** to steep in it

Mariolatry (derog.), worship of the Virgin Mary, *not* Mary-

marionette, a puppet; in Fr. f. *marionnette*

maritime, abbr. **mar.**

marivaudage (Fr. m.), daintily affected style, from **Marivaux (Pierre Carlet de Chamblain de),** 1688–1763, French playwright and novelist

Mark (St.) (NT), not to be abbreviated; adj. **Marcan,** *not* Mark-

mark/, German coin, *pl.* in English contexts **-s;** the currency of the Federal Republic of Germany is based on the **Deutsche Mark,** q.v.

mark-down (noun), a reduction in price (hyphen)

market/, abbr. **mkt.**; **-ed, -ing**
market-day (hyphen)
Market/ **Drayton,** Shropshire;
  — **Harborough,** Leics.; —
  Weighton, Humberside (no
  hyphens)
market garden (two words)
**Marketors (Worshipful
  Company of)**
market overt (law), open market
market-place (hyphen)
market/ **town,** — **value** (two
  words)
marks of correction, *see*
  proof-correction marks
marks of reference (typ.),
  * † ‡ § ‖ ¶. *See also* **footnotes**
  and **reference marks**
mark-up (noun), an increase in
  price; (typ.), final copy-
  -preparation in printer's office
  (hyphen)
Marlburian, (member of
  Marlborough College
marlinspike, for separating
  strands of rope in splicing, *not*
  -ine-, -ing- (one word)
Marlow, Bucks., Hereford &
  Worc.
Marlowe (Christopher),
  1564–93, English playwright
Marmara (Sea of), *not* -ora
marmoset, a monkey
Marocco, *use* **Morocco**
**Marprelate Controversy
  (the),** war of pamphlets, 1588–
  9, between puritan 'Martin
  Marprelate' and defenders of
  the Established Church
marq., marquis
marque de fabrique (Fr. f.),
  trade mark
marque/, make (not type) of
  car; — **(letters of),** those
  authorizing reprisals at sea
marquee, large tent
Marquesas Islands, S. Pacific
marquetry, inlaid work, *not*
  -terie
mar/**quis, -quess,** abbr. **M.,** *or*
  marq.; *fem.* **marchioness,**
  abbr. **march.;** caps. as titles; in

Fr. *marquis*/, *fem.* **-e** (abbr.
  M^{ise})
Marrakesh, Morocco, *not*
  Mara-, -kech
marriage, in Fr. m. *mariage*
marriageable, *not* -gable
married, abbr. **m.**
marron (Fr. m.), chestnut;
  marron glacé, sugared chestnut (not ital.)
marrowfat, a pea
'Marseillaise (La)', French
  national song
Marseilles, S. France; in Fr.
  Marseille
marshal/ (noun and verb), *not*
  -ll; **-led, -ler, -ling**
Marshall/ **(Gen. George Cat
  lett),** 1880–1959, US soldier
  and statesman, initiator of —
  Plan for W. Eur. recovery
  1947
Marshall Islands, W. Pacific
Marsham (Viscount), eldest
  son of Earl of Romney. *See also*
  Masham
martellato (mus.), played with
  heavy strokes
Martello/, small circular fort,
  pl. **-s** (cap.)
marten, a weasel. *See also* **martin**
Martha's Vineyard, an island
  Mass., US
Martial (in Lat., **Marcus Valerius Martialis),** *c.*40–*c.*102,
  Roman poet, abbr. **Mart.**
martial, warlike
Martian, of the planet Mars
martin, a bird. *See also* **marten**
Martineau (Harriet), 1802–
  76, English writer
martingale, strap from horse's
  noseband to girth, *not* -gal
Martini, vermouth (propr.
  term)
Martinique, W. Indies
Martinmas, 11 Nov.
martyrize, etc., *not* -ise
marvel/, **-led, -ling, -lous**
Marvell (Andrew), 1621–78,
  English poet and satirist
Marx/ **(Karl),** 1818–83, Ger-

man socialist, author of *Das Kapital*; **-ian**; **-ism**; *not* -ianism

Mary, USSR, formerly **Merv**

Maryland, US, off. abbr. **Md.** or (postal) **MD**

Marylebone, London; — (**St.**), parliamentary constituency

Marymass, 25 Mar.

Maryolatry, *use* **Mari-*mas** (Lat.), a male; *pl.* **mares**

Masaryk (Thomas Garrigue), 1850–1937, first president of Czechoslovakia; — (**Jan Garrigue**), his son, 1886–1948, Czechoslovak politician

masc., masculine

Mascagni (Pietro), 1863–1945, Italian composer

masculine, abbr. **m.** *or* **masc.**

Masefield (John), 1878–1967, Poet Laureate 1930–67

maser, microwave amplification by stimulated emission of radiation (acronym). *See also* **laser**

mashallah! (Arab., Pers., Turk.), an exclamation of wonder

Masham (Baron), family name Lister; — **of Ilton (Baroness)**, family name Cunliffe-Lister. *See also* **Marsham**

mashie, a golf-club, *not* -y

Mashona/, -land, Zimbabwe, *not* Mashu-

masjid (Arab.), a mosque, not *mes-*, *mus-*

mask, cover for the face, etc. *See also* **masque**

masochism, a sexual perversion in which person delights in being cruelly treated; from **Leopold von Sacher-Masoch**, 1835–95, Austrian novelist

Mason/, -ic, -ry (Freemasonry) (cap.)

Mason-Dixon line, US, separated slave-owning South from free North; surveyed 1763–7 by Englishmen Charles Mason and Jeremiah Dixon (hyphen)

Masor/ah, Heb. tradition, *not* the many variants; **-etic Text**, abbr. **MT**

masque, entertainment. *See also* **mask**

mass (mech.), symbol *m*

Massachusetts, off. abbr. **Mass.** or (postal) **MA**

massé, stroke at billiards

Masséna (André), 1758–1817, French marshal

Massenet (Jules Émile Frédéric), 1842–1912, French composer

Massereene (Viscount)

masseu/r, *fem.* **-se** (not ital.)

massif (geol.), mountain mass (not ital.)

Massora, *use* **Masorah**

mastaba/ (archaeol.), anc. Egyptian tomb, *pl.* **-s**

Master, abbr. **M.**; **Master of/ Arts, MA**; — — **Commerce, M.Com.**; — — **Dental Surgery, MDS**; — — **Education, M.Ed.**; — — **Laws, LL M**; — — **Letters, M.Litt.**; — — **Music, M.Mus.** or (Camb.) **Mus.M.**; — — **Philosophy, M.Phil.**; — — **Science, M.Sc.** or **MS** (US); — — **Surgery, M.Ch., M.Chir., MS**, or **Ch.M.**; — — **Theology, M.Th.**

master-at-arms (naut.), first--class petty officer (hyphens)

master mariner, captain of a merchant vessel (two words)

master-mind (noun, adj., and verb, hyphen)

Master of, title of the eldest son of certain Scottish peers, e.g. The Master of Lovat

Master of the Rolls, abbr. **MR**

masterpiece (one word)

master printer, the head of a printing establishment

Master Printers Association, abbr. **MPA**

master/-stroke, -switch, -work (hyphens)

mast-head (hyphen)

mastic, a gum-resin, *not* -ich, -ick

mat, dull, *use* **matt**

Matabele/, *pl. same*; **-land,** Zimbabwe (one word)

matador, bullfighter, *not* -ore

match/board, -box (one word)

matchet, *use* **machete**

matchmaker (one word)

match point (two words)

matchwood (one word)

maté, Paraguay tea, *not* mate (not ital.)

matelot, a shade of blue, (slang) a sailor (not ital.)

matelot (Fr. m.), a sailor

matelote (Fr. cook. f.), a rich fish stew

mat/er (Lat.), mother; *pl. -res*

materialize, *not* -ise

materia medica, science of drugs (not ital.)

matériel (mil.), everything except personnel (ital.)

matey, sociable, *not* -ty

math., mathemat/ics, -ical, -ician

mathematics (typ.), reference to footnotes in math. works to be marks of reference (†, ‡, etc. *but not* *), and not superior figures, as these may be mistaken for indices. When letters are required for formulae, caps. and l.c. (*not* s. caps.) are usual. A formula, if detached from the text, is generally set in the middle of the line; and if it has to be carried on to the next, the break is made at an *equals, minus,* or *plus* sign, which is carried over. Abbr. **math.** *See also* **BSI, fractions, numerals**

Mathew (Lord Justice), 1830–1909

Mathews (Charles), 1776–1835, English actor; — **(Charles James),** 1803–78, his son, English actor and playwright; — **(Elkin),** publisher; — **(Shailer),** 1863–1941, US educator and theologian. The usual spelling of the name is **Matthews**

matin (Fr. m.), morning

matinée, afternoon entertainment; *matinée musicale,* ditto with music

matins, sometimes in Prayer Book **mattins**

matriculator, *not* -er

matr/ix (typ.), individual die for casting type; *pl.* **-ices**

matronymic, *use* **metro-**

matt, (of colour) dull, without lustre, *not* mat

Matt., St. Matthew's Gospel

Mattei (Tito), 1841–1914, Italian composer

matter (typ.), MS or copy to be printed, type that has been set

Matthews, the usual spelling, *but see* **Mathews**

mattins, *see* **matins**

Mau, Uttar Pradesh, India. *See also* **Mhow**

Maugham (William Somerset), 1874–1965, English novelist and playwright

Maugrabin (Hayraddin), in Scott's *Quentin Durward*

Maulmain, Burma, *use* **Moulmein**

maulstick, painter's hand-rest, *not* mahl-

Mau Mau, a Kikuyu secret society in Kenya, rebelled 1952 (two words)

Maundy Thursday, day before Good Friday, *not* Maunday (two words)

Maupassant (Guy de), 1850–93, French writer

Mauresque, *use* **Mor-**

Mauretania, ancient region of NW Africa

Mauretania, name of two successive Cunard liners

Maurist, member of the reformed Benedictine congregation of St. Maur (Fr.)

Mauritania, NW Africa, indep. 1960

Mauriti/us, former British colony, French-speaking, in Indian Ocean; indep. 1968; adj. **-an**

mausoleum/, a magnificent tomb; *pl.* **-s**

mauvaise honte (Fr. f.), shyness

mauvais/ goût (Fr. m.), bad taste; — *pas,* a difficulty; — *quart d'heure,* bad quarter of an hour, a short unpleasant time; — *sujet,* a ne'er-do-well; — *ton,* bad style

maverick, an unbranded animal, masterless person, rover

Max., Maximilian I or II, Holy Roman emperors

max., maxim, maximum

maxill/a, the jaw; *pl.* -**ae**

maximize, *not* -ise

maxim/um, the greatest; *pl.* -**a;** abbr. **max.** (not ital.)

maxwell, unit of magnetic flux; abbr. **Mx**

Maxwell (James Clerk), 1831–79, Scottish physicist, *not* Clerk-Maxwell

Maxwell-Lyte (Sir Henry Churchill), 1848–1940, English hist. writer

May (month of), not to be abbreviated

may (tree) (not cap.)

māyā (Hindu philos.), illusion

Maya/, one of an Indian people of Cent. America and S. Mexico; *pl.* -**s;** adj. -**n**

maybe, perhaps (one word)

may-bug (hyphen)

May Day, 1 May (caps., two words)

mayday, international radio distress signal (one word, not cap.)

Mayence on the Rhine, *use* **Mainz am Rhein**

Mayfair, London (one word), *but* **May Fair Hotel,** London

may/flower, -fly (one word)

mayhem, (legal and US) maiming

mayn't, to be printed close up

mayonnaise (Fr. f.), a salad-dressing (not ital.)

maypole (one word)

May queen (two words)

mayst (no apos.)

mazagran (Fr. m.), black coffee served in a glass

Mazarin Bible, 42-line, first book printed from movable type, by Gutenberg and Fust, *c.*1450. **Cardinal Mazarin,** 1602–61, had twenty-five copies

mazarine, deep blue

Mazeppa (Ivan), 1644–1709, Cossack chief, hero of Byron's poem *Mazeppa,* 1819

mazurka, a Polish dance, *not* mazou-

Mazzini (Giuseppe), 1805–72, Italian patriot

MB, *Medicinae Baccalaureus* (Bachelor of Medicine)

mb *or* **mbar,** millibar

Mbabane, Swaziland

MBCS, Member of the British Computer Society

MBE, Member of the Order of the British Empire

MBIM, Member of the British Institute of Management

MC, Master of Ceremonies, — of Surgery, (US) Member of Congress, — of Council, Military Cross

M/C., Manchester, machine

Mc, *see* **Mac**

MCC, Marylebone Cricket Club

M.Ch., M.Chir., *Magister Chirurgiae* (Master of Surgery)

M.Com., Master of Commerce

MCP, Member of the College of Preceptors; male chauvinist pig

MCR, Middle Common Room

MCS, Military College of Science

Mc/s, *better* **MHz,** q.v.

MCSP, Member of the Chartered Society of Physiotherapy

MD, managing director, Maryland, US (off. postal abbr.), (Lat.) *Medicinae Doctor* (Doctor of Medicine), mentally deficient, Middle Dutch, (It. mus.) *mano destra* (right hand), (Fr. mus.) *main droite* (right hand)

Md, mendelevium (no point)

Md., Maryland, US (off. abbr.)

Mdlle, *use* **Mlle** (Mademoiselle)

Mdme, *use* **Mme** (Madame)

MDS, Master of Dental Surgery

MDu., Middle Dutch

ME, Maine, US (off. postal abbr.), Marine Engineer, Mechanical Engineer, Middle English, Military Engineer, Mining Engineer, Most Excellent

Me., Maine, US (off. abbr.)

M^e (Fr.), *maître* (title of French advocate)

me (mus.), *not* mi

me (**it is**), **it is I,** both used in speech, but former should not be printed, except as a colloquialism

mea culpa (Lat.), by my fault

mealie/ (S. Afr.), ear or grain of maize; *pl.* **-s**

mealy-mouthed, hypocritical (hyphen)

mean time, that based on mean sun (two words)

meantime (adv.), *but in the* mean time

meanwhile (adv.), *but in the* mean while

measur/e, -able, -ing, abbr. **meas.**; (typ.) the width to which type is set, usu. stated in 12 pt. (pica) ems; **-ements,** *see* **numerals**

measuring-jug (hyphen)

meatball (one word)

Mebyon Kernow (Corn.), 'sons of Cornwall', movement for Cornish independence

MEC, Member of Executive Council

Mecca, cap. of Hejaz, and one of the federal capitals of Saudi Arabia, *not* Mekka, -ah, -eh. See also **Riyadh**

Mecenas, *use* **Mae-**

mech., mechan/ics, -ical

mechanics is *singular* as name of the science

Mechlin lace etc., *but see* **Malines**

Mecklenburgh Square, London, WC

Mecklenburg/-Schwerin, —-**Strelitz,** E. Germany

Mec Vannin (Manx), 'sons of Man', movement for Manx independence

M.Ed., Master of Education

med., medical, medicine, medieval, medium

medal/, -led, -lion, -list

Médecin malgré lui (*Le*), by Molière, 1666

media (**the**), conventional name for channels of information, is *plural*

mediaeval, *use* **medieval**

medic, slang for doctor or medical student (no point)

medical, abbr. **med.**; — **signs,** ʒ drachm, ♏ minim, **M** *misce* (mix), ℥ ounce, O pint, R/ *recipe,* ℈ scruple

Medic/i, Florentine ruling family, 15th–18th cc.; adj. **-ean.** *See also* **Marie de' Medici**

medicinae (Lat.), of medicine; abbr. **M.**

medicine, abbr. **med.**

medico/, colloq. for doctor or medical student, *pl.* **-s**

medieval/, abbr. **med.**; **-ism, -ist, -ize,** etc., *not* mediae-

Medina, river, IW

Medina, Arabia, in Arab. **al--Madinah**

Medit., Mediterranean

meditatio fugae (Lat., Sc. law), contemplation of flight

medi/um, *pl.* **-a,** in spiritualism **-ums** (not ital.); abbr. **med.** *See also* **media**

medium/, former size of paper, 18 × 23 in.; — **4to,** 11½ × 9 in., — **8vo,** 9 × 5¼ in. (untrimmed). *See also* **book sizes**

med. jur., medical jurisprudence

Médoc (Fr. m.), a claret

Meerut, Uttar Pradesh, India, *not* Merath, Mirat

mega-, prefix meaning a million, or very large, abbr. **M,** e.g. **megawatt, MW**

Megaera (myth.), one of the Furies

megalomania, the delusion of grandeur

megalopolis, a great city

Megara, city and port, anc. Greece

megavolt (elec.), a million volts; abbr. **MV**

megawatt, a million watts; abbr. **MW**

megilp, an artist's medium, vehicle for oil-colours, *not* mag-, -ilph

megohm (elec.), a million ohms; abbr. **MΩ**

megrim/, -s, headache, 'the blues'

mehrab, *use* mi-

Meilhac (Henri), 1831–97, French playwright

mein Herr (Ger.), a form of address, as Sir; *pl.* **meine Herren;** *fem.* **meine Dame,** *pl.* **meine Damen**

meiosis/is (rhet.), diminution, understatement (*see also* **litotes**); (biol.) cell-division before fertilization, *pl.* **-es**

Meiringen, Switzerland, *not* Mey-

Meirionnydd, district of Gwynedd

Meissen porcelain, from E. German town

Meissonier (Jean Louis Ernest), 1815–91, French painter

Meissonnier (Juste Aurèle), 1693–1750, French goldsmith, architect, furniture designer

Meistersinger von Nürnberg (Die), opera by Wagner, 1867

me judice (Lat.), in my opinion

Mekk/a, -ah, -eh, *use* Mecca

Melanchthon (Philip), Graecized form of Philipp Schwarzerd, 1497–1560, Luther's associate

Melanesia, the islands between the Equator and the Tropic of Capricorn, and between New Guinea and Fiji (an archbishopric, not a political unit)

mélange (Fr. m.), a mixture

Melchior (Lauritz), 1890–1973, Danish tenor

Melchi/zedek, king of Salem, OT; NT, **-sedec,** but **-zedek** in NEB

mêlée, a fray (not ital.)

Meliboean (poetry), alternating, *not* -aean, -bean

melodrama, romantic and sensational drama

Melos, Gr. island. *See also* **Milo**

Melpomene (Gr.), the muse of tragedy

melt, spleen, *use* **milt**

melting/-point, -pot (hyphens)

Melton Mowbray, Leics. (two words)

mem., memento, memorial

Member, abbr. **M.**

Member of Parliament (caps.); abbr. **MP,** *pl.* **MPs**

memento/, a souvenir, *pl.* **-es,** abbr. **mem.** (not ital.)

memento mori (Lat.), remember that you must die

Meml/ook, -uk, *use* **Mameluke**

memo/, memorandum, *pl.* **-s** (no point)

mémoire (Fr. m.), bill, report, treatise; (f.) memory

memorabilia (Lat.), noteworthy things, is *pl.*

memorand/um, a written note, *pl.* **-ums;** *pl.* **-a,** things to be noted. *See also* **memo**

memorial, abbr. **mem.**

Memorial Day (US), = Decoration Day

memorialize, to petition by a memorial, *not* -ise

memoria technica (Lat.), mnemonics

memorize, to commit to memory, *not* -ise

memsahib, *see* sahib

menad, *use* mae-

ménage (Fr. m.), a household. *See also* **manège**

menagerie, (place for) a collection of wild animals, *not* -ery

menarche, onset of menstruation

Mencken (Henry Louis), 1880–1956, US author

mendacity, falsehood

Mendel/ (**Gregor Johann**), 1822–84, Austrian botanist; **-ian, -ism**

mendelevium, symbol **Md**

Mendeleyev (**Dmitri Ivanovich**), 1834–1907, Russian chemist

Mendelssohn-Bartholdy (**Jacob Ludwig Felix**), 1809–47, German composer

mendicity, begging

meneer (Afrik.), Mr, sir

menhaden, a N. American fish of herring family, *not* man-

meningitis, inflammation of the meninges

meno (It. mus.), less

Menorca, Sp. for **Minorca**

mensa (Lat.), a table; *a mensa et toro,* from bed and board (a kind of divorce), not *thoro*

Menshevik, member of a minority group of the Russian Social Democratic Party before 1917. *See* **Bolshevik**

Men's Lib (caps., no point)

mens/ rea (Lat.), criminal intent; — *sana in corpore sano,* a sound mind in a sound body

menstru/um (Lat.), a solvent; *pl.* **-a**

mensur., mensuration

menswear (one word, no apos.)

menthe (Fr. cook. f.), mint, *not* mi-

Menton, Fr. Riviera, often given It. spelling **Mentone**

menu/ (Fr. m.), bill of fare, *pl.* **-s**

Menuhin (**Yehudi**), b. 1916, US violinist

MEP, Member of the European Parliament

m.e.p., mean effective pressure

mephistophelean, malicious, cynical, *not* -ian

mer., meridian, meridional

Merano, Italian Tyrol

Merath, *use* **Meerut**

Mercator's projection, a method of map-making, from Latinized form of surname of **Gerhard Kremer,** 1512–94,

Flemish-born German cartographer

Mercedes, German make of car; **Mercédès** (Fr.), **Mercedes** (Sp.), girl's name

mercerize, treat cotton under tension, *not* -ise

merchandise (noun *and* verb), *not* -ize

Merchant/ Company Schools, Edin.; — **Taylors Company; but —Taylors' School** (apos.)

merci (Fr. m.), thanks; no, thank you; (f.) mercy

mercury, symbol **Hg**

Meredith (**George**), 1828–1909, English novelist and poet — (**Owen**), *see* **Bulwer--Lytton** (**Edward Robert**)

meretricious, befitting a prostitute, showy

meridian, a great circle passing through the poles of a sphere and any given place; esp. that which cuts the observer's horizon at due north and south, and which the sun crosses at (local) noon; abbr. **m.** *or* **mer.**; (fig.) the highest point

meridies (Lat.), noon; abbr. **m.**

meridional, of a meridian, abbr. **m.** *or* **mer.**; (in the northern hemisphere) southern of S. Europe; (noun) an inhabitant of S. Europe, esp. of S. France

Mérimée (**Prosper**), 1803–70, French novelist and historian

merino/, sheep, *pl.* **-s**

meritocracy, government by the best people

meritorious, deserving reward or praise

merle, the common blackbird, *not* merl

merlin, falcon

merlon, solid part of embattled parapet

Merovingian, (king) of a Frankish dynasty in Gaul and Germany, *c.*500–700, *not* -j-

merry andrew, buffoon (two words, not cap.)

merry-go-round (hyphens)

merrythought, the wishbone (one word)

Merseyside, metropolitan county

Merthyr Tydfil, Mid Glam. (two words), *not* -vil

Merv, USSR, *now* **Mary**

mesa (Western US), flat-topped mountain

mésalliance (Fr. f.), marriage with a social inferior; *use* **misalliance**

Mesdemoiselles, abbr. **Mlles**, *not* Mdlles

mesjid, use **mas-**

mesmerize, to hypnotize, fascinate, *not* -ise

Mesolongi, *use* **Missolonghi**

Mesozoic (geol.) (cap.)

mesquite, N. American tree, *not* -it

Messerschmitt (Willy), 1898–1978, German aircraft-designer, *not* -dt

Messiaen (Olivier Eugène Prosper Charles), b. 1908, French organist and composer

Messiah, Messianic; *Messiah,* oratorio by Handel, 1742, not *The Messiah*

Messieurs (Fr.), abbr. **MM.**; *sing.* **Monsieur**

Messrs, *sing.* **Mr** (no point)

nestiz/o, one of Spanish and American-Indian blood, *not* -ino; *pl.* **-os**; *fem.* **-a,** *fem. pl.* **-as**

net., metallurg/y, -ical, -ist, metronome

meta- (chem. prefix) (ital.)

netal/, -led, -ling, -lize, *not* -ise

metallurg/y, -ical, -ist, abbr. **met.** or **metall.**

metal rule (typ.), a type-high strip of metal for printing a line on paper: en rule, em rule, two-em rule, and longer

metamorphos/e, to transform, *not* -ise, -ize; noun **-is,** *pl.* **-es**

metaph., metaphys/ics, -ical,

-ically, -ician, metaphor, -ical, -ically

meteor., meteorology

meter, measuring device

Meth., Methodist

methodize, *not* -ise

métier (Fr. m.), a handicraft

metonymy, naming of thing by its attribute; **metonym,** word thus used; abbr. **meton.**

metre/, SI unit of length, 39.37 in.; abbr. **m.** (*but* no point in scientific and technical work); **-age,** number of metres, *not* -rage

Métro (Le) (no point), abbr. of *Le Métropolitain,* the Paris underground railway

metrology, science of weights and measures; abbr. **metrol.**

metronome (mus.), instrument for fixing tempos; abbr. **M.** *or* **met.; Maelzel's —, MM**

metronymic, name taken from female ancestor, *not* ma-

metropol/is, *pl.* **-ises; -itan;** abbr. **metrop.**

meubl/é (Fr.), *fem.* **-ée,** furnished

meum/ (Lat.), mine; **— and tuum,** mine and thine, not *et tuum*

Meuse, river of France, Belgium, and Netherlands; in Du. **Maas**

MeV, mega-electron-volt; **meV,** milli-electron-volt

mews, a street of stables converted into garages or houses; *pl. same*

Mex., Mexic/o, -an

Meynell (Alice), 1847–1922, English poet; **— (Sir Francis),** 1891–1975, English typographer; **— (Wilfrid),** 1852–1948, English writer

Meyringen, *use* **Mei-**

MEZ (Ger.), *Mitteleuropäische Zeit,* time of the Middle European zone, one hour in advance of GMT

mezereon (bot.), a spring-flowering shrub, *not* -eum

mezzanine, a low storey between two others

mezza voce (It. mus.), not with full strength of sound, abbr. *m.v.*

mezzo/, short for mezzo--soprano, *pl.* **-s**

mezz/o (It. mus.), *fem.* **-a,** half, medium; abbr. **M.**

mezzo-rilievo (It.), half-relief

mezzo-soprano (mus., hyphen)

mezzotint (typ.), intaglio illustration process from a roughened plate

MF, (paper) machine finish; medium frequency; (typ.) modern face

mf (mus.), mezzo forte (fairly loud)

mfd., manufactured

mfg., manufacturing

MFH, Master of Foxhounds

M.F.Hom., Member of the Faculty of Homoeopathy

M.Fr., Middle French

mfr., manufactur/e, -er

mfrs., manufactur/ers, -es

m.ft., *mistura fiat* (let a mixture be made)

MG (Fr. mus.), *main gauche* (left hand); (paper) machine glazed; machine-gun; Morris Garages

Mg, magnesium (no point)

mg, milligram (no point in scientific and technical work)

MGB, USSR 'Ministry of State Security', 1946–53. *See also* **NKGB**

M.Glam., Mid Glamorgan

MGM, Metro-Goldwyn-Mayer

M.Gr., Middle Greek

Mgr., Manager, Monsignor, Monseigneur, *pl.* **Mgrs.**

MH, Master of Hounds (usu. beagles or harriers)

MHG, Middle High German

MHK, Member of the House of Keys (I.o.M.)

mho (elec.), unit of conductance. *See also* **ohm**

M.Hon., Most Honourable

Mhow, Madhya Pradesh, India. *See also* **Mau**

MHR, Member of the House of Representatives

MHz, megahertz, a million hertz (no point, two caps.)

MI, Michigan (off. postal abbr.) Military Intelligence, Mounted Infantry

M.I.5, Secret Service Division; **M.I.6,** Espionage Department

MIAE, Member of the Institute of Automobile Engineers

M.I.Ae.E., Member of the Institute of Aeronautical Engineers

miasm/a (Gr.), noxious emanation; *pl.* **-ata**

M.I.Biol., Member of the Institute of Biology

Mic., Micah (OT)

micaceous, pertaining to mica *not* -ious

Micawber/ (**Wilkins**), in Dickens's *David Copperfield*; **-ish,** unpractical but hopeful; **-ism**

MICE, Member of the Institution of Civil Engineers

Mich., Michaelmas; Michigan (off. abbr.)

Michelangelo Buonarroti, 1475–1564, Italian sculptor, painter, architect, poet (two words, not three)

Michelson/ (**Albert Abraham**), 1852–1931, German-American physicist; — **-Morley** experiment

M.I.Chem.E., Member of the Institution of Chemical Engineers

Michener (**James Albert**), b. 1907, US novelist

Michigan, off. abbr. **Mich.** or (postal) **MI**

mickey (to take the — out of), to make fun of, *not* micky

Mickey Finn (US), a drugged drink

Mickiewicz (**Adam**), 1798–1855, Polish poet

Micmac, a tribe of N. American Indians (one word)

micro-, prefix meaning very

254

small, or one-millionth, symbol **μ**, e.g. **microhm, μΩ**; **micro-metre** (preferred to **micron**), **μm**; **microvolt, μV**

microchip (one word)

microcosm, the little world. *See also* **macrocosm**

microfiche, a sheet of film, suitable for filing, containing microphotographs of pages of book etc.; *pl. same*

microfilm, film bearing microphotographs

microgram, one-millionth of a gram; abbr. **μg**

micrography, description or delineation of microscopic objects

microgroove, of gramophone record (one word)

micrometer, instrument for measuring minute distances or angles. *See also* **micrometre,** *under* **micro-**

micromillimetre *or* **milli-micron,** abbr. **mμ,** *use* **nanometre,** abbr. **nm**

micron, symbol **μ,** *but prefer* **micrometre,** q.v. *under* **micro-**

Micronesia, islands north of Melanesia

micro-organism (hyphen)

microphotograph, a photograph reduced to microscopic size; do *not* use for **photo-micrograph,** q.v.

microprocessor (one word)

micros., microscop/y, -ist

microsecond, one-millionth of a second; abbr. **μs**

microwave oven (two words)

mid., middle

midbrain (one word)

Mid Calder, Lothian (caps., no hyphen)

midday (one word)

middle, abbr. **M.** *or* **mid.**

middle age (two words), *but* **middle-aged** (hyphen)

Middle Ages (the), roughly from the fall of the Western Roman Empire to the Renais-

sance and the Reformation (two words, caps.)

middle class (hyphen as adj.)

Middle English, abbr. **ME**

middleman (one word)

Middlesbrough, Cleveland, *not* -borough

Middlesex, former county of England, abbr. **Middx.**

Middleton (Baron), family name Willoughby; — **(Thomas),** ?1570–1627, English playwright. *See also* **Midleton, Mydd-**

Middle West, *see* **Midwest**

Middx., Middlesex

middy, loose blouse, (colloq.) midshipman

midfield (noun and adj., one word)

Mid Glamorgan, county of Wales

midi, dress etc. of medium length

midinette, Parisian shop-girl (not ital.)

Midland Bank, Ltd.

Mid Lent, fourth Sunday in Lent (caps., no hyphen)

Midleton (Viscount), family name Brodrick. *See also* **Midd-, Mydd-**

Midlothian, former county of Scotland (one word)

mid/night, -rib, -ships (one word)

Midsomer Norton, Avon

midsummer (one word)

Midsummer Day, 24 June (two words, caps.)

midway (one word)

Midwest, the North-Central states of US (one word). *Also* **Middle West** (two words)

midwi/fe, *pl.* **-ves; -fery**

midwinter (one word)

MIEE, Member of the Institution of Electrical Engineers

Miehle (typ.), Chicago firm of press-builders, now also manufacturing in England

Miers (Sir Henry Alexander), 1858–1942, English

Miers (*cont.*)
mineralogist. *See also* **Myers, Myres**

Mies van der Rohe (**Ludwig**), 1886–1969, US architect

mightest, *not* mightst

mightn't, to be printed close up

mignonette, a fragrant plant; in Fr. f. *mignonnette*

migraine, a severe headache (not ital.)

mihrab, niche in mosque, *not* me-, mirh-

Mihailovich (**Drazha**), 1893–1946, Yugoslav soldier

MIJ, *use* **MJI**

mijnheer, ordinary Dutch form of **mynheer** as title

mil, one-thousandth of an inch (no point)

mil., military, militia

Milan, Italy; in It. **Milano**

mile/, **-s,** abbr. **m., mls.**; geogr. *or* naut. —, 6,080 feet; **statute** —, 5,280 feet; in Fr. m. *mille*/, **-s**

mileage, number of miles, *not* milage

Milesian, (inhabitant) of anc. Miletus in Asia Minor; (joc.) Irishman

milestone (one word)

milieu/, environment (not ital.), *pl.* **-x.** *See also le juste milieu*

military, abbr. **mil.**

Military Academy, abbr. **MA**

militia, abbr. **M.** *or* **mil.**

milk/ **bar,** — **chocolate** (two words)

milk-float (hyphen)

milkman (one word)

milk/ **pudding,** — **round,** — **run,** — **shake** (two words)

milksop (one word)

Milky Way (astr.) (caps.)

mill, abbr. **m.**

Mill (**James**), 1773–1836, Scottish historian and economist; — (**John Stuart**), 1806–73, his son, English economist and polit. philosopher

Millais (**Sir John Everett**), 1829–96, English painter

Millay (**Edna St. Vincent**), 1892–1950, US poet

mille/ (Fr. m.), a thousand (no *s* in plural), abbr. **M.**; *also* a mile, *pl.* **-s;** — *-feuille* (f.), a confection

millenar/**y,** of a thousand; (celebration of) the thousandth anniversary; adj. **-ian**

millennium/, a thousand years, *pl.* **-s** (two *l*s, two *n*s)

mille passus (Lat.), 1,000 paces of five feet, or the Roman mile; abbr. **MP**

millepede (zool.), *not* -ed, milli-

Millers Dale, Derby. (no apos.)

Milles-Lade, family name of Earl Sondes. *See also* **Mills**

Millet (**Aimé**), 1819–91, French sculptor; — (**Jean-François**), 1814–75, French painter

milli-, prefix meaning one-thousandth, abbr. **m,** e.g. **milliampere, mA**

milliard, a thousand millions

millibar, unit of pressure, one-thousandth of a bar; abbr. **mbar,** *or* **mb** in meteor. usage (no point)

milligram, one-thousandth of a gram, 0.015 grain; abbr. **mg** (no point in scientific and technical work), *not* -gramme

millilitre, one-thousandth of a litre; abbr. **ml** (no point in scientific and technical work)

millimetre, one-thousandth of a metre, 0.039 inch; abbr. **mm.** (*but* no point in scientific and technical work)

millimicron, *use* **nanometre**

million, abbr. **m.** For millions of pounds use, for example, £150m.

millionaire, *not* -onnaire

millipede, *use* **mille-**

Mills, family name of Baron Hillingdon. *See also* **Milles-Lade**

Milngavie, Strathclyde

Milo, It. for **Melos,** Gr. island, esp. in Venus de Milo

M.I.Loco.E., Member of the Institute of Locomotive Engineers

milt, spleen, *not* melt

Milton Keynes, Bucks., 'new town', 1967

Milwaukee, Wis., US

M.I.Mech.E., Member of the Institution of Mechanical Engineers

mimic/, -ked, -king

M.I.Min.E., Member of the Institution of Mining Engineers

MIMM, Member of the Institution of Mining and Metallurgy

min., minim, minimum, mining, minister, ministry, minor; minute(s) (*but* no point in scientific and technical work)

mina/, a starling of SE Asia, *pl.* **-s,** *not* myn-, -ah; anc. Gr. unit of weight and currency, *pl.* **-e**

minac/ious, threatening; **-ity**

minatory, threatening

minauderie, affected, simpering behaviour (not ital.)

mincemeat (one word)

mince pie (two words)

Mindanao, one of the Philippine Islands, *not* -oa

Mindererus/, Latinized name of R. M. Minderer (*c.*1570–1621), German physician; — **spirit,** a diaphoretic

mineralog/y, -ical, abbr. **mineral.,** (of) the study of minerals; *not* minerolog/ical, -y

minever, *use* **mini-**

mingy, mean, *not* -ey

Mini, propr. term for car made by British Leyland (cap.)

mini, in gen. senses, esp. short skirt (not cap.)

mini/bus, -cab, -car (one word)

Minié (Claude Étienne), 1814–79, inventor of rifle, etc.

minikin, a diminutive person or thing. *See also* **manakin, manikin**

minim, a drop, one-sixtieth of fluid drachm; abbr. **min.,** sign ♏

minimize, *not* -ise

minim/um, *pl.* **-a** (not ital.); abbr. **min.**

mining, abbr. **min.**

minion (typ.), name for a former size of type, about 7 pt.

miniscule, *use* **minuscule**

miniskirt (one word)

minister, abbr. **min.**

minium, red oxide of lead

miniver, a fur, *not* -ever

Minn., Minnesota (off. abbr.)

Minneapolis, Minn., US

Minnehaha, wife of Hiawatha

minnesinger, one of a school of medieval German lyric poets (not ital.)

Minnesota, off. abbr. **Minn.** or (postal) **MN**

Minoan, pertaining to the civilization of ancient Crete (from King Minos)

minor, abbr. **min.,** (in Peerage) **M.**

Minorca, Balearic Is.; in Sp. **Menorca**

Minories, a street in London

Minotaur, monstrous offspring of a bull and Pasiphaë, wife of King Minos

Min./ Plen. *or* **Plenip.,** Minister Plenipotentiary; **Res.,** ditto Residentiary

M.Inst.P., Member of the Institute of Physics

Mint (**the**) (cap.)

minthe, *use* **menthe**

minuet, a dance, or its music, *not* -ette

minus/, less, sign − ; a negative quantity, a disadvantage, *pl.* **-es**

minuscule, a small, or lower-case, letter, (adj.) tiny, *not* mini-

minute/, -s, abbr. **m.** *or* **min.** (*but* no point in scientific and technical work), sign ′; — **mark** (′), symbol for feet, minutes, also placed after a syllable on which the stress falls

minuti/ae, small details; *sing.* **-a** (not ital.)

MIOB, Member of the Institute of Building

Miocene (geol.) (cap., not ital.)

miosis, *use* myosis

MIPA, Member of the Institute of Practitioners in Advertising

mirabile/ dictu (Lat.), wonderful to relate; — *visu,* ditto see

Mirat, *use* Meerut

mirepoix (Fr. cook. m.), sautéd chopped vegetables

mirhab, *use* mihrab

mirky, dark, *use* murky

Miró (Joan), b. 1893, Spanish painter

MIRV, multiple independently targeted re-entry vehicle

miry, mirelike, *not* -ey

Mirzapur, Uttar Pradesh, India, *not* -pore

mis/advice, bad counsel; **-advise,** to give it (one word)

misalliance, marriage with a social inferior, not *mésalliance*

misandry, hatred of men

misc., miscellaneous, miscellany

miscegenation, interbreeding between White and non-White

miscellanea, miscellanies, is *pl.*

mischievous, *not* -ious

miscible, capable of being mixed, esp. in scientific and technical contexts. *See also* **mixable**

misdemeanour, *not* -or

*M*ise (Fr.), *marquise* (marchioness), not *Mise*

mise-en-scène (Fr. f.), scenery, stage-effect

misère, declaration of no tricks in cards

Miserere, (musical setting of) the 50th Psalm of the Vulgate

miserere, *properly* **misericord,** bracket on turn-up seat used as support for person standing

misfeasance, a wrongful act; (law) a trespass

mishmash, mixture (one word)

Mishna/h, a collection of Jewish precepts; adj. **-ic**

misle, *use* mizzle

misletoe, *use* mistle-

misogam/y, hatred of marriage; **-ist**

misogyn/y, hatred of women; **-ist**

misprint (typ.), a typographical error

misprision (law), wrong act or omission

Miss (cap. as title, no point)

Miss., Mission, -ary; Mississippi, US (off. abbr.)

missel-thrush, *not* mistle-

misseltoe, *use* mistle-

mis/send, to send incorrectly; **-sent** (one word)

misshape/, -n (one word)

Mission/, -ary, abbr. **Miss.**

missis, slang for **Mrs** (q.v.), *not* -us

Mississippi, river and State, US; off. abbr. **Miss.** or (postal) **MS**

Missolonghi, Greece, *not* Mesolongi

Missouri, river and State, US; off. abbr. **Mo.** or (postal) **MO**

mis/spell, -spend, -spent, -state (one word)

missus, *use* missis

mist (meteor.), abbr. **m.**

mistakable, *not* -eable

Mister, abbr. **Mr**

Mistinguett (pseud. of **Jeanne Marie Bourgeois**), 1875–1956, French entertainer, *not* -e

mistle, *use* mizzle

mistle-thrush, *use* missel-

mistletoe, *not* missel-, misle-

Mistral (Frédéric), 1830–1914, Provençal poet; — **(Gabriela),** 1889–1957, pseud. of **Lucila Godoy de Alcayaga,** Chilean poet

mistral (Fr. m.), cold NW wind in S. France

Mistress, abbr. **Mrs**

M.I.Struct.E., Member of the Institution of Structural Engineers

MIT, Massachusetts Institute of Technology

Mithra/s, Persian sun-god; **-ic, -ism**

Mithridates (VI, 136–63 BC), King of Pontus, *not* Mithra-

Mitilini, *see* **Mytilene**

nitre/-block, -board, -box (hyphens)

nitring, *not* mitreing

nixable, in gen. use, *not* -ible. *See also* **miscible**

nizen/, the aftermost of the fore- -and-aft sails; — **-mast,** the aftermost mast of a three-masted ship; *not* mizz-

Mizen Head, Co. Cork

nizzle, fine rain, *not* misle, mistle

MJI, Member of the Institute of Journalists, *not* MIJ

Mk., identifying a particular design, as Mk. I; mark (Ger. coin), *use* **RM** or **DM**

n.k.s., metre-kilogram-second system

nkt., market

ML, Licentiate in Midwifery, Medieval Latin, Middle Latin, motor launch

Ml., mail

nl., millilitre, -s (no point in scientific and technical work)

MLA, Member of the Legislative Assembly, Modern Languages Association

MLC, Member of the Legislative Council

MLG, Middle Low German

M.Litt., Master of Letters

Mlle, -s, Mademoiselle, Mesde- moiselles (no point)

MLR, minimum lending rate

MLS, Member of the Linnean Society (off. spelling), *not* Linnaean

nls., miles

MM, 2,000, Military Medal, (mus.) Maelzel's metronome, (Their) Majesties, (Fr.) *Mes- sieurs*

mm., millimetre, -s (*but* no point in scientific and technical work)

n.m., *mutatis mutandis* (with the necessary changes)

Mme (Fr.), Madame, *pl.* **Mmes** (no point)

n.m.f., magnetomotive force

M.Mus., Master of Music

MN, Merchant Navy, Minnesota (off. postal abbr.)

Mn, manganese (no point)

Mn., Modern (with names of languages)

Mnemosyne (Gr. myth.), god- dess of memory, mother of Muses

MO, mass observation, Medical Officer, Missouri (off. postal abbr.), money order

Mo, molybdenum (no point)

Mo., Missouri (off. abbr.)

mob/, -bed, -bing

mobilize, *not* -ise

Möbius (August Ferdinand), 1790–1868, German mathema- tician, discoverer of the — **strip** (one-edged)

moccasin, American Indian shoe, N. American venomous snake, *not* the many variants

Mocha, a coffee from Mocha, Arabian port on Red Sea, moss agate, sheepskin

mock turtle soup (three words)

M.o.D., Ministry of Defence, including, from 1964, the Ad- miralty, the War Office, and the Air Ministry

MOD, Ministry of Overseas Development; more usually **ODM**

mod., moderate, modern

mod., moderato (mus.)

mode (Fr. m.), method, (gram.) mood; (f.) fashion

model/, -led, -ler, -ling

moderate, abbr. mod.

moderato (mus.), in moderate time; abbr. *mod.*

Modér/é, fem. -ée, a moderate in French politics

modern, abbr. mod. or **Mn.** (q.v.)

modern face (typ.), a type design with contrasting thick and thin strokes and serifs at right-angles, abbr. **MF**

modern figures (typ.), *see* **nu- merals**

modernize, *not* -ise

modicum/, a small quantity; *pl.*
-s (not ital.)

**modif/y, -iable, -ied, -ier,
-ying**

Modigliani (Amedeo), 1884–
1920, Italian painter and sculptor

modiste, a milliner, dressmaker
(not ital.)

modo praescripto (Lat.), as
directed

Mods., Moderations, the First
Public Examination at Oxford
University

modus/ operandi (Lat.), a plan
of working; — *vivendi,* a way
of living, a temporary compromise

Moët et Chandon, champagne
manufacturers·

mœurs (Fr. f. pl.), manners,
customs

Mogadishu, Somalia

Mogul, a Mongolian, *not*
Mog/hal, -hul, Mughal

MOH, Master of Otter Hounds,
Medical Officer of Health

Mohammed, the Prophet, *use*
Mahomet (trad. Eng. form) or
Muhammad (correct)

Mohammedan, *use* **Muhammadan.** *See also* **Islam, Muslim**

Mohave Desert, *use* **Moj-**

Mohawk, N. Amer. Indian, *not*
Mohock

Mohegan, (member of) eastern
branch of Mahican Indian tribe,
formerly in lower Conn. and
Mass., US

Mohican, (member of) western
branch of Mahican Indian tribe,
formerly on both sides of the
Hudson River, US

Mohocks, London ruffians
(18th c.)

Mohorovičić discontinuity
or **moho** (geol.)

moire (noun), a watered fabric,
usu. silk, orig. mohair (not ital.)

moiré (adj.), (of silk etc.)
watered; (noun) the watered
surface (not ital.)

mois (Fr. m.), month, abbr. *m*
(no point). *See also* **month**

Mojave Desert, Calif., US, *not*
Moh-

mol (chem.), abbr. of **mole,** q.v

molar (chem.), abbr. **M**

molasses, syrup from sugar, *no*
moll-

Mold, Clwyd

Moldavia, a Soviet Socialist Re
public. *See also* **USSR**

mole (chem.), SI unit of amoun
of substance; abbr. **mol**

Molière, pseud. of Jean-Baptiste
Poquelin, 1622–73, French
playwright

moll (Ger. mus.), minor

mollah, *use* **mu-**

mollasses, *use* **mola-**

Molnár (Ferenc), 1878–1952,
Hungarian playwright

Moloch, Canaanite idol to
whom children were sacrificed
(often fig.), *not* -eck

**Moltke (Count Helmuth
Karl Bernard von),** 1800–91
Prussian field marshal; — (**Hel-
muth Johannes Ludwig
von),** 1848–1916, his nephew,
German general in First World
War

molto (mus.), much, very

mol. wt., molecular weight

molybdenum, symbol **Mo**

Mombasa, Kenya, *not* -assa

moment/um, impetus, mass ×
velocity; *pl.* **-a** (not ital.)

Mommsen (Theodor), 1817–
1903, German historian

Mon., Monday

Monaco, principality adjoining
Mediterranean France; adj.
Monégasque

Mona Lisa, portrait by Leonardo da Vinci, also called *La
Gioconda*

mon/ ami (Fr.), *fem.* — *amie,*
my friend

monaural, of reproduction of
sound by one channel. *Also*
monophonic

Mönchen-Gladbach, W. Germany

non cher (Fr. m.), *fem.* **ma chère,** my dear

Monck (Viscount); — **(George, Duke of Albemarle),** 1608–70, English general. *See also* **Monk**

Monckton (Lionel), 1861–1924, English composer; — **(Walter Turner, Viscount),** 1891–1963, British lawyer and politician

Moncreiff (Baron)

Moncreiffe of that Ilk (Sir Rupert Iain Kay, Bt.)

Moncrieff, the more usual spelling

Moncton, NB, Canada

Monday, abbr. **M.** *or* **Mon.**

Mondrian (Pieter Cornelis), 1872–1944, Dutch painter

Monégasque, *see* **Monaco**

Monet (Claude Oscar), 1840–1926, French painter

monetar/y, -ist

money/, *pl.* **-s, -ed,** *not* monied; — **order,** abbr. **MO** (no point). *See also* **numerals**

Monghyr, Bihar, India

Mongol, member of Asian people

mongol/, -ism, (one suffering from) congenital mental deficiency (not cap.)

mongoos/e, an Indian animal, *pl.* **-es,** *not* mun-

Monk Bretton (Baron). *See also* **Monck**

Mono, *see* **Monotype**

mono/, short for monaural; as noun, *pl.* **-s**

monochrom/e, (picture) in one colour (one word); **-atic**

monocle, a single eyeglass

monocoque/, aircraft or vehicle with body of single rigid structure, *pl.* **-s**

monogam/y, marriage with one person; **-ous**

monogyny, marriage with one wife

monophonic, *see* **monaural**

Monophoto (typ.), propr. term

for (computer-driven) filmsetting systems adapted from Monotype principles

monopol/ism, -ist, -istic, -ize, -y

Monotype (typ.), propr. term for a composition system, using separate keyboard and caster, for casting single types; abbr. **Mono**

Monroe Doctrine, that European powers should not interfere in American affairs, from **James Monroe,** 1758–1831, US president

Mons, Belgium

Mons., this abbr. for Monsieur is regarded as incorrect in France

Monseigneur (Fr.), abbr. **Mgr.;** *pl.* **Messeigneurs,** abbr. **Mgrs.** *See also* **Monsignor**

Monserrat, Spain. *See also* **Mont-**

Monsieur/ (Fr.), Mr, Sir, abbr. (to be used in third person only) **M.;** *pl.* **Messieurs,** abbr. **MM.;** — **Chose,** Mr So-and-so. *See also* **Mons.**

Monsignor/, RCC title, abbr. **Mgr.;** *pl.* abbr. **Mgrs.;** It. **-e,** *pl.* **-i**

mons Veneris, rounded mass of fat on woman's abdomen above vulva (one cap.)

Mont., Montana (off. abbr.)

montage, mounting photographs and other objects on a surface to make a design; selecting, cutting, and combining separate 'shots' to make a continuous film

Montagu of Beaulieu (Baron)

Montagu (Lady Mary Wortley), 1689–1762, English writer

Montague, Romeo's family in Shakespeare's *Romeo and Juliet*

Montaigne (Michel Eyquem de), 1533–92, French essayist

Montana, off. abbr. **Mont.** or (postal) **MT**

Mont Blanc (caps.)

mont-de-piété (Fr. m.), Government pawnshop; *pl.* **monts- - - - —** (hyphens)

Monte Cristo, *not* — Christo

Montefiascone, Italian wine, *not* -sco (one word)

Montenegr/o, republic of Yugoslavia; adj. **-in**

Monterey, Calif., US

Monte Rosa, Switzerland (caps., two words)

Monterrey, Mexico

Montesquieu (Charles Louis de Secondat, baron de), 1689–1755, French jurist and philos. writer

Monteverdi (Claudio Giovanni Antonio), 1567–1643, Italian composer

Montevideo, cap. of Uruguay (one word)

Montgomerie, family name of Earl of Eglinton; — **(Alexander),** 1556–1610, Scottish poet

Montgomery, second title of Earl of Pembroke, town and former county in Wales, also towns in Ala. and W.Va., US, and Pakistan

Montgomery of Alamein (Bernard Law, Viscount), 1887–1976, British field marshal in Second World War.

month/, -s, abbr. **m.;** — **(day of the),** to be thus, as 25 Jan., *not* Jan. 25. When necessary, months to be abbreviated thus: Jan., Feb., Mar., Apr., Aug., Sept., Oct., Nov., Dec.; May, June, July to remain in full. In Fr. the names of months do not take caps., as *janvier*

Montpelier, Vermont, US

Montpellier, dép. Hérault, France

Mont-Saint-Michel, dép. Manche, France (caps., hyphens); in Cornwall, **Mount Saint Michael,** *use* **St. Michael's Mount**

Montserrat, Leeward Islands. *See also* **Mons-**

Montyon prizes, of French Academy, *not* Month-

Moodkee, *use* **Mudki**

Moog synthesizer, propr. term for electronic musical instrument (one cap.)

Mooltan, *use* **Mu-moolvi/,** Muslim doctor of the law, *pl.* **-s,** *not* -vee, -vie

moon, cap. only in astronomical contexts and in a list of planets abbr. **m.;** sign for new, ●; first quarter, ☽; full, ○; last quarter ☾

moon/beam, -light, -lit (one word)

moonshee (Ind.), writer or teacher, *not munshi*

moonshine (one word)

Moore (George), 1852–1933, Irish novelist; — **(Sir John),** 1761–1809, British general, killed at Corunna; — **(Thomas),** 1779–1852, 'the bard of Erin'. *See also* **More**

Moosonee (Abp. of), Ontario, Canada

mop/e, -ed, -ing, -ish

Mor., Morocco

moral, (adj.) concerned with principles of right and wrong (e.g. — philosophy), virtuous (opp. to immoral); (noun) a moral lesson or principle, in Fr. f. *morale*

morale, state of mind, in respect of confidence and courage, of groups of combatants or others under stress; in Fr. m. *moral*

moralize, *not* -ise

moratori/um, *pl.* **-ums**

Moray, former Scottish county, *not* Morayshire; — **(Earl of)**

morbidezza (It. art), extreme delicacy

morbilli (med.), measles, *not* -bilia

morbus (cholera) (not ital.)

morceau/ (Fr. m.), a morsel, also a short mus. piece; *pl.* **-x**

mordant, biting; (substance) fixing dye

mordent (mus.), a type of ornament

more (Lat.), in the manner of

More (Hannah), 1745–1833, English religious writer; — (**Sir Thomas**), 1478–1535, English writer, canonized 1935. *See also* **Moore**

moreish, so pleasant one wants more, *not* -rish

morel, an edible fungus, nightshade, *not* -lle

morello/, bitter dark cherry, *pl.* -s

more majorum (Lat.), in the style of one's ancestors

morendo (mus.), dying away

mores (Lat. pl.), social customs and conventions

moresco, an Italian dance (not cap.). *See also* **Morisco**

Moresque, Moorish, *not* Mau-

more suo (Lat.), in his, *or* her, own peculiar way

Moretonhampstead, Devon (one word)

Moreton-in-Marsh, Glos. (hyphens, *not* in-the-Marsh)

morganatic marriage, between royalty and commoner, the children being legitimate but not heirs to the higher rank (not cap.)

morgue, a mortuary

morgue (Fr. f.), haughtiness

Morisco/, Moorish, a Moor, *pl.* -s. *See also* **moresco**

Morison (James), 1816–93, Scottish founder of the Evangelical Union, 1843; — (**James Augustus Cotter**), 1832–88, English hist. writer; — (**Stanley**), 1889–1967, typographer

Morisot (Berthe), 1841–95, French Impressionist painter

Morland (George), 1763–1804, English painter

Mormon, member of Church of Jesus Christ of Latter-day Saints (*not* -an)

morn., morning

Mornay (Duplessis-), 1549–1623, French Huguenot leader

mornay (cook.), sauce flavoured with cheese

Morny (duc de), 1811–65, French statesman

Morocc/o, abbr. **Mor.**; adj. -an; *not* Ma-; indep. 1956

morocco leather (bind.) (not cap.); **french** — —, a low grade with small grain; **levant** — —, high grade with large grain; **persian** — —, the best, usu. finished on the grain side

morph., morphology, the study of forms (bot. or linguistics)

Morpheus, god of sleep, *not* -aeus

morphia, a drug (pop. for **morphine**)

Morrell (Lady Ottoline), 1873–1938, English hostess

Morris (Gouverneur), 1752–1816, US statesman, and his great-grandson, 1876–1953, US writer; — (**William**), 1834–96, English craftsman, poet, and socialist; — (**William Richard, Lord Nuffield**), 1877–1963, English industrialist and philanthropist

morris dance (two words, not cap.)

mortgag/ee, the creditor in a mortgage; -**er**, the debtor; in law, -**or**

mortice, *use* -ise

mortis causa (Lat., Sc. law), in contemplation of death

mortise/, hole for receiving tenon in joint; — **lock**, one recessed in frame etc. (two words), *not* -ice

Morton/ (John), *c.*1420–1500, English statesman, deviser of —**'s Fork**

Morvan (Le), French district

Morven, Grampian

Morvern, Highland

MOS (comp.), metal oxide semiconductor

Mosaic, of Moses

mosaic/, representation using inlaid pieces of glass, stone, etc.;

mosaic (*cont.*)
as verb **-ked, -king; -ist,** maker
of mosaics
Moscow, USSR, in Russ.
Moskva
Moseley, a Birmingham suburb.
See also **Mosl-**
**Moseley (Henry Gwyn Jef-
freys),** 1887–1915, English
physicist
Moselle, river, France–
Germany; in Ger. **Mosel;** a
white wine
Moslem, *use* **Muslim**
**Mosley (Sir Oswald Ernald,
Bt.),** 1896–1980, English politician. *See also* **Mose-**
mosquito/, *pl.* **-es,** *not* mu-
Mossley, Ches., Gr. Manch.,
Staffs.
mosso (mus.), 'moved' (e.g. più
mosso, more moved = quicker)
Most High, as Deity (caps.)
Moszkowski (Moritz), 1854–
1925, Polish composer
MOT, Ministry of Transport,
now Department of Transport,
but still used colloq. in **MOT
test** for motor vehicles
mot (Fr. m.), a word; *mot à
mot,* word for word, abbr. *m.
à m.*
Mother Carey's chicken, the
storm petrel
Mother Hubbard, a character
in a nursery rhyme; a gown
such as she wore
Mother Hubberds Tale, by
Spenser, 1591
Mothering Sunday, fourth in
Lent
mother-in-law, *pl.* **mothers-
-in-law** (hyphens)
motherland (one word)
mother-of-pearl (hyphens)
mother tongue (two words)
motif, dominant idea in artistic
expression; ornament of lace,
etc. (not ital.)
motley, a mixture, *not* -ly
moto/ (It. mus.), motion; —
continuo, constant repetition;
— **contrario,** contrary mo-

tion; — **obbliquo,** oblique motion; — **perpetuo,** a piece of
music speeding without pause
from start to finish; — **prece-
dente,** at the preceding pace;
— **primo,** at the first pace; —
retto, direct or similar motion
motor/ bike, — **boat,** — **car,**
— **coach,** — **cycle** (two words,
hyphens when attrib.)
motorized, *not* -ised
motorway (one word)
mottl/ed, -ing
motto/, *pl.* **-es**
motu proprio (Lat.), of his, *or*
her, own accord
mouezzin, *use* **mue-**
mouflon (Fr. m.), a wild sheep,
not mouff-, muf-
mouillé (Fr.), softened, wet,
(phonet.) palatalized (of consonant)
moujik, Russian peasant, *not*
mujik, muzhik
Moulmein, Burma, *not* Maulmain
moult, a shedding, *not* molt
Mount/, -ain, abbr. **Mt.,** *pl.*
Mts.
Mountain Ash, Mid Glam.
Mount Auburn, Mass., US,
noted cemetery
Mount Edgcumbe (Earl of)
(two words)
Mountevans (Baron) (one
word)
Mountgarret (Viscount) (one
word)
Mount Saint Michael, Cornwall, *use* **St. Michael's Mount,**
in Fr. **Mont-Saint-Michel,**
dép. Manche
Mourne Mountains, Ireland
mousaka, Gr. dish, *not* -ss-
mousetrap (one word)
mousseline/ (Fr. f.) muslin-like
fabric; — **-de-laine,** wool and
cotton muslin; — **-de-soie,**
muslin-like silk
mousseu/x (Fr.), *fem.* **-se,** foam-
ing, as wine
moustache, US mu-
mousy, mouselike, *not* -ey

mouth-organ (hyphen)

mouth/piece, -wash (one word)

mouton (Fr. m.), sheep, mutton

movable, in legal work **moveable**

moyen/ (Fr. m.), medium; — *âge* (*le*), the Middle Ages

moyenne (Fr. f.), average

Mozambi/que, SE Africa, indep. 1975, in Port. **Moç-**; adj. **-can**

Mozart/ (**Wolfgang Amadeus**), 1756–91, Austrian composer; adj. **-ian**

MP, Madhya Pradesh (India), *formerly* Central Provinces, Member of Parliament (*pl.* **MPs**), Metropolitan Police, Military Police, *mille passus* (a thousand paces, the Roman mile)

m.p., melting-point

mp (mus.), mezzo piano (fairly soft)

MPA, Master Printers Association

m.p.g., miles per gallon

m.p.h., miles per hour

M.Phil., Master of Philosophy

MPO, Metropolitan Police Office ('Scotland Yard')

MPS, Member of the Pharmaceutical Society of Great Britain

MR, Master of the Rolls, Municipal Reform

Mr, Mister, *pl.* **Messrs**

MRA, Moral Re-Armament

MRAC, Member of the Royal Agricultural College

M.R.Ae.S., Member of the Royal Aeronautical Society

MRAS, Member of the Royal Asiatic Society

MRBM, medium-range ballistic missile

MRC, Medical Research Council

MRCO, Member of the Royal College of Organists

MRCOG, Member of the Royal College of Obstetricians and Gynaecologists

MRCP, Member of the Royal College of Physicians, London

MRCS, Member of the Royal College of Surgeons, England

MRCVS, Member of the Royal College of Veterinary Surgeons, London

MRE, Microbiological Research Establishment

MRGS, Member of the Royal Geographical Society

MRH, Member of the Royal Household

MRI, Member of the Royal Institution

MRIA, Member of the Royal Irish Academy

MRIC, Member of the Royal Institute of Chemistry, *now* **MRSC**

MRICS, Member of the Royal Institution of Chartered Surveyors

MRINA, Member of the Royal Institution of Naval Architects

Mrs, Missis, Missus, which are corruptions of Mistress

MRSC, Member of the Royal Society of Chemistry

MRSH, Member of the Royal Society of Health

MRSL, Member of the Royal Society of Literature

MRSM, Member of the Royal Society of Medicine, ditto Royal Society of Musicians of Great Britain

MRST, Member of the Royal Society of Teachers

MS, *manuscriptum* (manuscript), *pl.* **MSS** (no point). *See also* **manuscript**

MS, Master of Science, ditto Surgery, multiple sclerosis, (Lat.) *memoriae sacrum* (sacred to the memory of), Mississippi (off. postal abbr.), (It. mus.) *mano sinistra* (the left hand)

Ms, title of woman whether or not married (no point)

m/s, metres per second, SI unit of speed

ms. (Fr.), *manuscrit* (MS)

MSA, Member of the Society of Apothecaries (of London), Mutual Security Agency

M.Sc., Master of Science

MSH, Master of Staghounds

MSI (comp.), medium-scale integration

MSIAD, Member of the Society of Industrial Artists and Designers

m.s.l., mean sea-level

MSM, Meritorious Service Medal

MSS, *manuscripta* (manuscripts). See also **manuscript**

mss. (Fr.), *manuscrits* (MSS)

MT, Masoretic Text (of OT), Mechanical (Motor) Transport, Montana (off. postal abbr.)

Mt., Mount, -ain

MTB, motor torpedo-boat

M.Th., Master of Theology

Mts., Mounts, Mountains

mu, Gr. letter M, μ; **μ** (math.), modulus, (phys.) symbol for magnetic permeability, micron, (as prefix) micro-; **μm,** micrometre; **mμ,** millimicron, *use* **nm**

muc/us, slimy substance; adj. **-ous**

mud/dy (verb), **-died, -dying**

Muddiford, Devon

Mudeford, Dorset

Mudford, Som.

Mudki, E. Punjab, India, *not* Moodkee

muesli, food of cereal and dried fruit, etc. (not ital.)

muezzin, Muslim crier, *not* mou-

muffetee, a wristlet

muflon, *use* **mou-**

mufti (Arab.), a magistrate, not *-tee* (ital.)

mufti (**in**), in civilian dress (not ital.)

Mughal, *use* **Mogul**

mugwump, an Indian chief, one who thinks himself important, (US politics) a politician critical of his party

Muhammad, now correct for **Mohammed** and the traditional English form **Mahomet**

Muhammadan, pertaining to, a follower of, Muhammad; *not* Mahometan, Mohammedan. *See also* **Muslim**

Mühlhausen, Thuringia, E. Germany

mujik, Russian peasant, *use* **moujik**

Mukden, China

mulatto/, offspring of a European and a Black; *pl.* **-s**

mulch, half-rotten vegetable matter, *not* -sh

Mulhouse, Alsace, France; in Ger. **Mülhausen**

mull (bind.), coarse muslin glued to backs of books

mullah (Muslim), a learned man (not cap.), *not* moll-, mool-, -a

mullein, tall yellow-flowered plant, *not* -en

Müller (Friedrich Max-), 1823–1900, English philologist

mulligatawny, a soup, *not* muli-, mulla-

mulsh, *use* **mulch**

Multan, Pakistan, *not* Moo-

Multatuli, pseud. of Edward Dowes Dekker, 1820–87, Dutch writer

multimillionaire, a possessor of several millions (one word)

multinational (one word)

multiple mark (typ.), the sign of multiplication ×

multiplepoinding (Sc. law), a process which safeguards a person from whom the same funds are claimed by more than one creditor (one word)

multiplication point, to be set medially, as m·s, metre-second

multiracial (one word)

multi-storey (hyphen)

multum in parvo (Lat.), much in small compass

mumbo-jumbo/, an object of popular homage, meaningless ritual or language, *pl.* **-s** (hyphen)

Munchausen (Baron), 1720–97, Hanoverian nobleman

whose extravagant adventures are the theme of *Adventures of Baron Munchausen*, collected by Rudolph Eric Raspe; in Ger. **Münchhausen**

München, Ger. for **Munich**

mungoose, use **mon-**

municipalize etc., *not* -ise

Munnepoor, use **Manipur**

munshi (Ind.), use **moonshee**

Munster, Ireland

Münster, W. Germany; Switzerland

muntjak, a S. Asian deer, *not* -jac, -jack

Muntz (metal), alloy of copper and zinc for sheathing ships, etc., *not* Muntz's

Murdoch (Jean Iris), b. 1919, British writer; — (**John**), 1747–1824, friend of Burns

Murdock (William), 1754–1839, inventor of coal-gas lighting, *not* -och

murky, dark, *not* mi-

Murray, the usual spelling, *but see* **Murry**

Mürren, Oberland, Switzerland

murrhine, fluorspar ware; *not* murrine, myrrhine

Murrumbidgee, river in NSW, Australia

Murry (John Middleton), 1889–1957, English hist. writer, *not* -ay; — (**Kathleen**), wife of above, *see* **Mansfield**

Murshidabad, W. Bengal, India

mus., museum, music, -al

musaeo/graphy, **-logy**, use **museo-**

Musalman, use **Mussul-**

Mus.B., *Musicae Baccalaureus* (Bachelor of Music)

musc/a (Lat.), a fly; *pl.* **-ae**

Muscadet, a dry white wine from Brittany. *See also* **muscatel**

Muscat, cap. of Oman, S. Arabia

muscatel, general name for a wine made from the muscat-grape (musk-flavoured)

Muschelkalk (geol.), shell limestone (cap.)

Musc/i, the true mosses; sing. **-us**

muscovado/, unrefined sugar, *pl.* **-s**

Mus.D., *Musicae Doctor* (Doctor of Music)

museo/graphy, museum cataloguing; **-logy**, the science of arranging museums, *not* musae-

Muses (the nine), Calliope, Clio, Erato, Euterpe, Melpomene, Polyhymnia, Terpsichore, Thalia, Urania

museum, abbr. **mus.**

Music (Bachelor of), abbr. **Mus.B., B.Mus.**; — (**Doctor of**), abbr. **Mus.D., D.Mus.**; — (**Master of**), abbr. **M.Mus.** or (Camb.) **Mus.M.**

music/, **-al**, abbr. **mus.**; — **drama** (two words); — **-hall**, **-paper**, **-stool** (hyphens)

Musigny, a red burgundy

musique concrète (Fr. f.), music constructed from recorded sounds

musjid, use **mas-**

Muslim, (a member) of the faith of **Islam**, q.v.; *not* Moslem. Preferred to **Muhammadan** in general usage, as connoting the religion rather than the prophet

muslin-de-lain, use **mousseline-de-laine**

Mus.M., Master of Music (Cambridge)

musquito, use **mos-**

Mussalman, use **Mussul-**

Musselburgh, Lothian, *not* -borough

Mussorgsky (Modest Petrovich), 1839–81, Russian composer

Mussul/man, a Muslim, from Pers. **Musulman**, *not* the many variants; *pl.* **-mans**, *not* -men

mustache, US for **mous-**

Mustafa, Algeria

Mustafabad, Uttar Pradesh, India

mustang (Western US), small, half-wild horse

mustn't, to be printed close up

Musulman, Pers. form of **Mussulman**

mutable, likely to change

mutand/um (Lat.), anything to be altered; *pl.* **-a**

mutatis mutandis (Lat.), with the necessary changes; abbr. *m.m.*

mutato nomine (Lat.), with the name changed

mutton (typ.), colloquial term for an em quad

'mutual' friend, often objected to, though used by Burke, Dickens, Geo. Eliot, and others; the alternative 'common' can be ambiguous

muu-muu (Hawaiian), woman's loose-fitting dress

Muybridge (Eadweard), 1830–1904, English photographer

Muzaffarabad, Kashmir

Muzaffarpur, Bihar, India

Muzak, propr. term for system of piped music (cap.)

muzhik, Russian peasant, *use* **moujik**

MV, motor vessel, (*or* **m.v.**) muzzle velocity

mV (elec.), millivolt (no point)

m.v. (It. mus.), *mezza voce*

MVD, USSR 'Ministry of Internal Affairs', 1953–60. *See also* **KGB, NKVD**

MVO, Member (of fourth or fifth class) of the Royal Victorian Order

M.V.Sc., Master of Veterinary Science

MW, megawatt, Most Worshipful, — Worthy

mW, milliwatt (no point)

Mx, maxwell (no point)

MY, motor yacht

myalism, W. Indian witchcraft; adj. **myal**

myall, wild Australian Aboriginal, Australian acacia

myceli/um (bot.), the thallus of a fungus; *pl.* **-a**

Mycenaean, of Mycenae (anc. Greece) or its civilization

Myddelton Square, Clerkenwell, London. *See also* **Mi-**

Myddleton (Sir Hugh), 1560–1631, London merchant

My dear Sir, (in letters, two caps. only, comma at end)

Myers (Frederic William Henry), 1843–1901, English poet, essayist, spiritualist; — **(Leopold Hamilton),** 1881–1944, English novelist. *See also* **Miers, Myres**

myna (bird), *use* **mina**

mynheer (Du.), Sir, Mr, in Du. *mijnheer*; a Dutchman

myop/ia, shortness of sight; adj. **-ic**

my/osis, contraction of the pupil of the eye, *not* mi-; adj. **-otic**

myosotis, the forget-me-not

Myres (Sir John Linton), 1869–1954, English archaeologist. *See also* **Miers, Myers**

Myriapoda, the centipedes and millepedes, *not* Myrio-

myrobalan, a plum, *not* -bolan

myrrhine, *use* **murrh-**

myrtle, *not* -tel

Mysore, Deccan, India, *not* the more correct Maisur

myst., mysteries

myth., mytholog/y, -ical

mythopoei/a, construction of myths; **-c,** myth-making, *not* -pae-, -pei-

Mytilen/e, Lesbos, Greece; adj. **-aean; Mitylene** in NT; **Mitilini,** the modern city

myxoedema, metabolic disease *not* myxed-

myxomatosis, a contagious and destructive infection of rabbits

N

N, (chess) knight, newton(s), nitrogen, the thirteenth in a series

N., Norse, north, -ern, (Lat.) *nom/en, -ina* (name, -s), *noster* (our)

N.) (naval), navigat/ing, -ion

n, (math.) an indefinite integer, whence (*gen.*) **to the nth (degree)**; (as prefix) nano- (*chem.*), symbol for amount of substance

n., name, nephew, neuter, new, nominative, noon, note, noun

n. (Lat.), *natus* (born), *nocte* (at night)

ñ (Sp.), called '*n* with the tilde', or 'Spanish *n*'; *pron.* as ni in onion; follows *n* without tilde in Spanish alphabetical sequence

NA, National Academ/y, -ician, Nautical Almanac, Naval Auxiliary, North America, -n

Na, *natrium* (sodium) (no point)

n/a (banking), no account; not applicable, not available

NAAFI, Naafi, Navy, Army, and Air Force Institutes

Naas, Co. Kildare; *also* Norway, the home of sloid, q.v.

nabob, Muslim official under Mogul empire, a rich Anglo--Indian (orig. same as **nawab,** but now differentiated)

Nabuchodonosor, *see* Nebuchadnezzar

nach, *use* nautch

nach/ Christi Geburt, — Christo, (Ger.), AD; abbr. *n. Chr.*

Nachdruck (Ger. typ. m.), reprint, pirated edition; **nach- drucken,** to reprint, or pirate

nach und nach (Ger. mus.), little by little

nacre/, mother-of-pearl (not ital.); adj. **-ous,** *not* -rous

nacré (Fr.), like mother-of-pearl (ital.)

---

nadir, the lowest point, opp. to **zenith**

naev/us, a birthmark, *pl.* **-i**

Naga Hills, Assam, India and Burma

Nagaland, Indian state

Nagar, W. Bengal, Mysore, E. Punjab, Kashmir

-nagar (Indian suffix), a town, as Ahmadnagar, *not* -naggore, -nagore, -nugger, -ur

Nagpur, Maharashtra, India, *not* -pore

Nahum, OT, abbr. **Nah.**

naiad/, river-nymph, *not* naid; *pl.* **-s** *or* **-es**

naïf, artless, *use* **naïve;** in Fr. *naï/f, fem.* **-ve**

Naini Tal, Uttar Pradesh, India

nainsook, Indian muslin, *not* -zook

Nairne (Baroness, *née* Caroline Oliphant), 1776–1845, Scottish poet

Nairnshire, former county, Scotland

naïve/, artless, *not* naïv-, naïf, **-ty**

naïveté (Fr. f.), artlessness

NALGO, Nalgo, National and Local Government Officers' Association

Nama (S. Afr.), Hottentot tribesman

namable, *use* **name-**

Namaqualand (off. spelling), S. Africa (one word)

namby-pamby, weakly sentimental (hyphen)

name, abbr. **n.**

nameable, *not* namable

namely, preferred to viz.

N. Amer., North America(n)

namesake (one word)

names of periodicals, *see* **periodicals**

names of persons and places, initial letters to be capitalized, as John Smith, West

---

names of persons and places (*cont.*)

Africa (*see* **capitalization**). They may undergo changes for political or other reasons, as when Saigon becomes Ho Chi Minh City, or Sir Francis Bacon becomes Baron Verulam and then Viscount St. Albans. Names from countries where the Latin alphabet is not used need transliteration, and for most languages there is no universally accepted system; transliterated forms may therefore vary (see, e.g., *Hart's Rules*, p. 130, concerning Russian). The same applies to translated names (Ivan the Terrible or John the Dread). For words derived from proper names, see *Hart's Rules*, pp. 12–13

names of ships (typ.), to be italic

Namibia, *formerly* South West Africa

N. & Q., *Notes and Queries*

nankeen, fabric, *not* -kin

nano-, prefix meaning one--thousand-millionth (10^{-9}), abbr. **n**

nanometre, one-thousandth of a micrometre; abbr. **nm;** use in preference to millimicron ($\text{m}\mu$)

Nansen (Fridtjof), 1861–1930, Norwegian Arctic traveller

Nantasket Beach, Mass., US

Nantucket Island, Mass., US

Nap., Napoleon

nap/, -ped, -ping

nap, a card game, a short sleep, pile in cloth

naphtha, an inflammable oil

naphthal/ene, crystalline substance used in mfr. of dyes, etc.; adj. **-ic**

Napierian logarithms, from **John Napier,** 1550–1617, Scottish mathematician; *not* -perian

Naples, Italy, in It. **Napoli**

napoleon, in Fr. m. **napoléon,** 20-franc coin of Napoleon I

napolitaine (à la) (Fr. cook.), in Neapolitan style (not cap. *n*)

Narbada, Indian river, *not* Ner budda

narciss/us, flower, *pl.* **-i; -ism** admiration of one's own body

narcos/is, stupor induced by narcotics; *pl.* **-es**

narghile, a hookah, *not* the many variants; in Pers. ***nargileh***

Narragansett Bay, RI, US, n -et

narrow measure (typ.), type composed in narrow widths, a in columns

narrow-minded (hyphen)

narwhal, the sea-unicorn, *not* -e, -wal; — **tusk,** *not* horn

NAS, National Association of Schoolmasters, Noise Abatement Society

NASA, National Aeronautics and Space Administration (US

nasalize, *not* -ise

NASD, National Amalgamated Stevedores and Dockers

Nash (Ogden), 1902–71, US poet; — **(Richard,** *or* **'Beau')** 1674–1762, Bath Master of Cer emonies

Nash *or* **Nashe (Thomas),** 1567–1601, English pamphleteer and playwright

Nasirabad, Pakistan

Nasmith (James), 1740–1808 English theologian and antiquary

Nasmyth (Alexander), 1758- 1840, Scottish painter; — **(James),** 1808–90, Scottish en gineer; **Nasmyth hammer,** *not* -th's

nasturtium/ (bot.), *not* -ian, -ion; *pl.* **-s**

Nat., Natal (prov. of S. Africa), Nathanael, -iel, National, -ist

nat., natural, -ist

natch, colloq. for naturally. *See also* **nautch**

nat. hist., natural history

National/, abbr. **Nat.;** — **Aca dem/y, -ician,** abbr. **NA;** — **Assemblies** (*see* **Assemblies**); — **Graphical Association,** abbr. **NGA**

ationalist, lower-case initial, *but* cap. with reference to a particular party or institution, abbr. **Nat.**

ationalize, *not* -ise

National/ Engineering Laboratory, East Kilbride, abbr. **NEL;** — **Physical Laboratory,** Teddington, abbr. **NPL**

National Westminster Bank, Ltd.

ations and places, words derived from, *see* **capitalization** and *Hart's Rules,* pp. 12–13

ation-wide (hyphen)

NATO, Nato, North Atlantic Treaty Organization

at./ ord., natural order; — **phil.,** — philosophy

atrium, sodium, symbol **Na**

NATSOPA, National Society of Operative Printers, Graphical and Media Personnel

attier blue, a soft blue, from **Jean-Marc Nattier,** 1685–1766, French painter

att/y, trim; **-ier, -ily, -iness**

atura (Lat.), nature

atural/, abbr. **nat.; -ism,** moral or relig. system based on nature; **-ist,** one who believes in such; a student of plants or animals. *See also* **naturism**

natural (mus.), sign ♮

aturalize, *not* -ise

atura non facit salt/um, -us (Lat.), Nature makes no leap/, -s

nature, the processes of the material world; cap. only when personified

natur/ism, worship of nature; (advocacy of) (collective) nudism; **-ist.** *See also* **naturalism**

atus (Lat.), born; abbr. *n.*

aught, nothing. *See also* **nought**

Nauheim (Bad), Hesse, W. Germany

Nauru/, W. Pacific, indep. 1968; adj. **-an**

ause/a, sickness; **-ate, -ating, -ous**

Nausicaa, in Homer's *Odyssey* (no accent)

nautch/, Indian dancing entertainment; — **-girl;** *not* nach, natch

nautical/, abbr. **naut.;** — **mile,** (UK) 1,853.18 m., (international) 1,852 m.

Nautical Almanac, abbr. **NA**

nautil/us, a shell; *pl.* **-uses** *or* **-i**

nav., naval, navigation

Navaho/, N. American Indian, *not* Navajo; *pl.* **-s**

navarin (Fr. cook. m.), a stew of mutton or lamb

Navarrese, of Navarre

navigat/ing, -ion, abbr. **(N.)** *or* **nav.**

Navy, Army, and Air Force, in toasts, etc., the Navy precedes, being the senior service

'Navy List' (two words)

nawab, (hist.) Indian nobleman; distinguished Muslim in Pakistan. *See also* **nabob**

Naz/arene, -arite, native of Nazareth, *not* -irite, q.v.

Nazi, member of the German National Socialist Party; **Nazism**

Nazirite, an anc. Hebrew who had taken vows of abstinence, *not* -arite, q.v.

NB, Nebraska (off. postal abbr.), New Brunswick (Canada), North Britain (a name for Scotland, offensive to Scots), *nota bene* (Lat., mark well)

Nb, niobium (no point)

n.b. (cricket), no ball

NBA, net book agreement

NBC, National Book Council, now **NBL;** National Bus Company

n.b.g., no bloody good

NBL, National Book League

NBS, National Broadcasting Service (of New Zealand)

NC, North Carolina (off. abbr.)

NCB, National Coal Board

NCCL, National Council for Civil Liberties

n. Chr. (Ger.), *nach Christo* or *nach Christi Geburt*, AD

NCL, National Carriers, Ltd.

NCO, non-commissioned officer

NCR, no carbon (paper) required

NCU, National Cyclists' Union

NCW, National Council of Women

ND, North Dakota (off. postal abbr.)

N.-D. (Fr.), Notre-Dame

Nd, neodymium (no point)

n.d. (bibliog.), no date given

N. Dak., North Dakota (off. abbr.)

Ndjamena, Chad

NDL, Norddeutscher Lloyd

NE, new edition (also **n/e**), New England, north-east(ern)

Ne, neon (no point)

né (Fr.), *fem.* **née**, born

n/e (banking), no effects. *See also* **NE**

Neal (Daniel), 1678–1743, English Puritan writer; — **(John),** 1793–1876, US writer

Neale (John Mason), 1818–66, English hymnologist

Neanderthal/ man, — skull, names for palaeolithic fossil hominid identified in 1856 in valley of this name near Düsseldorf, W. Germany; **-er,** a Neanderthal man, *not* -tal-

Neapolitan/, (inhabitant) of Naples; — **ice,** ice-cream layered in different colours; — **violet,** kind of sweet-scented viola

neap-tide, tide of smallest range (hyphen)

near, abbr. **nr.; near by** (adv., two words); **nearby** (adj., one word)

Nearctic (zool.), of northern N. America, *not* Neoarctic

neat-house, shed for cattle (hyphen)

neat's-foot oil (one hyphen)

NEB, National Enterprise Board, New English Bible

Nebraska, off. abbr. **Nebr.** or (postal) **NB**

Nebuchad/nezzar, king of Babylon, destroyed Jerusalem in 586 BC (**Nabuchodonosor** in AV Apocr. and in the Vulgate; **-rezzar,** in Jer. 43: 10, etc.)

nebula/, *pl.* **-e**

necess/ary, -ity

Neckar, river, Württemberg, W. Germany

Necker (Jacques), 1732–180 French statesman

nec pluribus impar (Lat.), a match for many (motto of Lou XIV)

NED, *New English Dictionary,* use ***OED*** (*Oxford English Dictionary*

NEDC, National Economic Development Council, colloq. **Neddy**

NEDO, National Economic Development Office

née (Fr. f.), born

needlework (one word)

Neele (Henry), 1798–1828, English poet

neelghau, an antelope, *use* **nilgai**

ne'er/, never; — **-do-well** (hyphens)

ne exeat regno (law), a writ to restrain a person from leaving the kingdom

***nefasti* (*dies*)** (Lat.), blank day

neg., negative, -ly

Negev (the), region of S. Israel *not* -eb

négligé (Fr. m.), women's informal wear, a diaphanous dressing-gown, *not* negligee

negligible, *not* -eable

negotiate, *not* -ciate

Negretti & Zambra, instrument-makers, London

Negrillo/, one of a dwarf Negr people in Cent. or S. Africa; *p* **-s** (cap.)

Negri Sembilan, *see* **Malaya (Federation of)**

Negrito/, one of a dwarf Negroi people in the Malayo-Polynesian region; *pl.* **-s** (cap.)

egritude, quality of being a Negro (cap.)

egr/o, *pl.* **-oes,** *fem.* **-ess** (cap.)

egroid, *not* -rooid

egus, title of Emperor of Abyssinia (Ethiopia)

egus, wine-punch

eh., Nehemiah

ehru (Jawaharlal, 'Pandit'), 1889–1964, Indian nationalist leader, Prime Minister from 1947

.e.i., *non est invent/us, -a, -um* (he, she, *or* it, has not been found)

eice, *use* **niece**

eige (Fr. cook. f.), whisked white of egg

eighbour/, -hood, -ly, *not* -or

eilgherry Hills, S. India, *use* **The Nilgiris**

eill (Alexander Sutherland), 1883–1973, British educationist; — **(Patrick),** d. 1705, first printer in Belfast, of Scottish birth; — **(Patrick),** 1776–1851, Scottish naturalist

either (of two), *is*; neither he nor she *is*; *but* neither he nor they *are*, neither these nor those *are*

ejd ('the plateau'); *see* **Saudi Arabia**

ekton, free-swimming organic life, as opp. to **plankton**

EL, National Engineering Laboratory

ematode (noun *or* adj.), (of) roundworm or threadworm, *not* -oid

em./ con., *nemine/ contradicente;* — *diss.,* — *dissentiente*

emean/ (adj.), of the vale of Nemea in anc. Argolis; — **games,** one of the four principal Panhellenic festivals; — **lion,** slain by Hercules

emes/is (cap. when personified); *pl.* **-es**

emine/ contradicente (Lat.), unanimously, abbr. **nem. con.;** — *dissentiente,* no one dissenting, abbr. **nem. diss.**

emo/ (Lat.), nobody; — *me*

impune lacessit, no one attacks me with impunity (motto of Scotland, and of the Order of the Thistle)

nemophila (bot.), a garden flower, *not* -phyla

N. Eng., New England

ne nimium (Lat.), shun excess

nenuphar, the great white water-lily. *See also* **Nuphar**

neo- is freely added (with hyphen) to the names of philosophies and institutions and to their adjectival forms, to designate their revivals (retain caps. of words which have them): neo-Christianity, neo-classical, neo-colonialism, neo-Freudian, neo-Romantic, etc. An exception is Neoplatonism (q.v.)

Neoarctic, *use* **Nearctic**

neo-colonial/, -ism (hyphen)

Neocomian (geol.) (cap.)

neodymium, symbol **Nd**

neolithic (archaeol.) (not cap.)

neolog/ize, to use new terms, *not* -ise; **-ism,** abbr. **neol.**

neon, symbol **Ne**

Neoplaton/ism, the revived Platonism of the 3rd–6th cc.; adj. **-ic** (one word, one cap.)

Nep., Neptune

Nepal/, kingdom between Tibet and India, *not* -aul; adj. **-ese,** *not* -i

neper, unit for comparing power levels

nephew, abbr. **n.**

ne plus ultra (Lat.), perfection

neptunium, symbol **Np**

ne quid nimis (Lat.), be wisely moderate

Nerbudda, Indian river, *use* **Narbada**

NERC, Natural Environment Research Council

nereid/, a sea-nymph; *pl.* **-s** *or* (with cap.) **-es**

Neri (Saint Philip), 1515–95, founder of the Congregation of the Oratory

nero antico (It.), a black marble

273

Neruda (Jan), 1834–91, Czech writer; — **(Pablo)**, pseud. of **Neftali Ricardo Reyes,** 1904–73, Chilean poet

nerve/-cell, -centre (hyphens)

nerve gas (two words)

nerve-racking, *not* wr-

Nesbit (Evelyn), pseud. of Mrs Edith (Hubert) Bland, 1858–1924, English writer

Nesbitt (Cathleen), b. 1890, English actress

Nessler's reagent (chem.), from **Julius Nessler,** 1827–1905, German chemist

n'est-ce-pas? (Fr.), is it not so?

Nestlé Co., Ltd. (The)

net/, not subject to deduction; — **book,** whose retail price is fixed by publisher; *not* nett

net/, -ted, -ting

net curtain (two words)

Netherlands (The), country, *not* Holland; abbr. **Neth.**

netsuke/, Japanese ornament, *pl.* **-s**

nett, not subject to deduction, *use* **net**

nettle-rash (hyphen)

network (one word)

Neuchâtel, Switzerland, white or red wine from there

Neufchâtel, déps. Aisne and Seine-Maritime, France; a kind of cheese

Neuilly, dép. Seine, France

neume (mus.), (sign indicating) group of notes to be sung with one syllable, *not* neum

neuralgia, pain in a nerve (a symptom, not a disease)

neurasthenia, nervous prostration, chronic fatigue

neurine, a poisonous ptomaine, *not* -in

neuritis, inflammation of nerve--fibres

neurone, nerve-cell and its appendages, *not* -on

Neuropter/a, an order of insects (lacewings, etc.); *sing.* **-on**

neuros/is, a functional derange-

ment through nervous disorde *pl.* **-es**

neurotic/ (adj. *and* noun); **-all■ -ism**

neuter, abbr. **n.** *or* **neut.**

neutralize, *not* -ise

neutrino/ (phys.), an uncharge particle with very small mass; *pl.* **-s**

neutron/ (phys.), an uncharge particle; *pl.* **-s**

Nevada, off. abbr. **Nev.** or (postal) **NV**

névé (Fr. m.), glacier snow (no ital.)

never/-ending, -failing (hyphens)

nevermore (one word)

never-never (colloq.), hire-pur chase (hyphen)

nevertheless (one word)

Nevill, family name of Marque■ of Abergavenny

Neville, family name of Baron Braybrooke

new, abbr. **n.**

Newbery (John), 1713–67, English printer; — **Medal** (US

Newbiggin, Cumbria, Durhan Northumb., N. Yorks.

Newbigging, Strathclyde, Tay side

new/-blown, -born (hyphens)

Newborough (Baron)

New Brunswick, Canada, abbr. **NB**

Newburgh, Fife, Grampian, Lancs.; — **(Countess of)**

New Castle, Ind., Pa., US

Newcastle:, sig. of Bp. of New castle (colon)

Newcastle/ upon Tyne, Tyne & Wear; — **under Lyme,** Staffs. (no hyphens)

newcomer (one word)

Newdigate (Sir Roger), 1719■ 1806, founder of Oxford prize for poem

new edition, abbr. **NE, n/e**

New English Bible (The), firs part (the New Testament) pub■ 1961, complete 1970; abbr. **NEB**

newfangled (one word)
New Forest, Hants (two words)
Newfoundland/ (one word), Canada, abbr. **NF, Nfld.; -er,** an inhabitant
New Guinea, *see* **Papua**
New Hall, Camb. Univ.
Newham, borough of Gr. London (one word)
New/ Hampshire, off. abbr. **NH,** *never* abbr. as — Hants.
New Haven, Conn., US (two words)
Newhaven, Lothian, E. Sussex (one word)
New Hebrides, abbr. **N.Heb.;** *now* **Vanuatu**
New Jersey, US, off. abbr. **NJ**
new-laid egg (one hyphen)
New Mexico, US, off. abbr. **N.Mex.** or (postal) **NM**
New Mills, Ches., Powys
Newmilns, Strathclyde
Newnes (Sir George), 1851–1910, publisher
New Orleans, La., US, abbr. **NO**
new paragraph (typ.), abbr. **n.p.**
New Quay, Dyfed, Essex (two words)
Newquay, Cornwall (one word)
New Red Sandstone (geol.) (caps.)
news agency (two words)
news/agent, -caster, -letter (one word)
New South Wales, Australia (three words, caps., no hyphens); abbr. **NSW**
news/paper, -paperman (one word)
Newspaper Publishers' Association, abbr. **NPA**
newspapers (titles of), *see* **periodicals**
newsprint, paper on which newspapers are printed
news-reader (hyphen)
newsreel (one word)
news/-sheet, -stand (hyphens)
news theatre (two words)
New Style, according to the

Gregorian calendar, adopted in Britain on 14 Sept. 1752 (omitting 3–13 Sept.), and in Russia in 1917; abbr. **NS.** *See also* **Old Style**
new style numerals, *see* **numerals**
newsvendor, *not* -er
newsworthy (one word)
New Testament, abbr. **NT;** for abbr. of books in, *see* each title
newton, SI unit of force; abbr. **N**
Newton Abbot, Devon
Newton-le-Willows, Merseyside, N. Yorks. (hyphens)
Newtonmore, Highland (one word)
Newton Poppleford, Devon
Newton Stewart, Dumfries & Galloway (two words). *See also* **Newtownstewart**
Newtown, Powys, adjacent to site of 'new town', 1967
Newtownabbey, Co. Antrim (one word)
Newtownards, Co. Down (one word)
Newtownbutler, Co. Fermanagh (one word)
Newtowncunningham, Co. Donegal (one word)
Newtownforbes, Co. Longford (one word)
Newtownmountkennedy, Co. Wicklow (one word)
Newtownsands, Co. Limerick (one word)
Newtownstewart, Tyrone (one word). *See also* **Newton Stewart**
New Year's Day (caps.)
New York, the US State, abbr. **NY;** the city, often abbr. **NYC,** but officially **New York, NY**
New Yorker, inhabitant of New York
New Yorker (*The*), magazine (ital.)
nex/us (Lat.), a tie, a linked group; *pl.* **-us** (pedantic) *or* **-uses** (gen.), *not* -i

Ney (Michel), 1769–1815, French marshal

NF, National Front, Newfoundland, New French, Norman-French

Nfld., Newfoundland

NFU, National Farmers' Union

NFWI, National Federation of Women's Institutes

NG, National Giro, National Guard, New Granada

ng, in Welsh a separate letter, following g in alphabetical sequence; not to be divided

n.g., no good

NGA, National Graphical Association

NGO, non-governmental organization

N.Gr., New Greek

NH, New Hampshire (off. abbr.)

N.Heb., New Hebrew, — Hebrides

NHG, New High German

n.h.p., nominal horsepower

NHS, National Health Service

NI, National Insurance, Northern Ireland

Ni, nickel (no point)

Niagara, river separating Ontario, Canada, from New York State, US

Niagara Falls, waterfalls of R. Niagara; town, NY; town, Ontario

Nibelungenlied, German epic, twelfth–thirteenth century, *not Nie-* (one word)

niblick, a golf-club

Nicar., Nicaragua

Nicene Creed, issued in 325 by the Council of Nicaea (in Asia Minor); to be distinguished from the longer **Niceno-Constantinopolitan Creed** of the Thirty-nine Articles

niche, a recess, *not* -ch

Nicholas, Anglicized spelling of the names of five popes, two Russian emperors, and the patron saint of Russia; in Russ. **Nikolai**

Nicholson, the usual spelling, *but see* **Nicolson**

nicht wahr? (Ger.), is it not so?

nick (typ.), a groove cast in the shank of a type

nickel/, symbol **Ni;** — **-plating** (hyphen); — **silver** (two words)

nickel (US), a five-cent coin

nicknack, *use* **knick-knack**

nickname (one word)

Nicobar Islands, Indian Ocean, *not* Nik-

Nicol prism, from **William Nicol,** 1768–1851, Scottish physicist

Nicolson (Hon. Sir Harold George), 1886–1968, British politician and man of letters

Nicomachean Ethics, by Aristotle

Nicosia, Cyprus (Gr. **Levkosia**), Sicily, *not* Nik-

NID, Naval Intelligence Division

nid/us (Lat.), a nest; *pl.* **-i**

Niebelungenlied, use **Nib-**

Niebuhr (Barthold Georg), 1776–1831, German historian and philologist; — **(Karsten)**, 1733–1815, German traveller; — **(Reinhold)**, 1892–1971, US theologian

niece, a relation, *not* nei-

niell/o, Italian metal work; *pl.* **-i** (not ital.)

Niepce (Joseph Nicéphore), 1765–1833, French physicist, originator of photography; — **de Saint-Victor (Claude Marie François)**, 1805–70, his nephew, inventor of heliographic engraving

Niersteiner, a hock (wine)

Nietzsche (Friedrich Wilhelm), 1844–1900, German polit. philosopher

Nièvre, dép. France

Niger (Republic of), W. Cent. Africa, indep. 1960

Nigeria (Republic of), W. coast of Africa, indep. 1960

night/-cap, -clothes, -club, -dress (hyphens)

nightfall (one word)

night/-gown, -life, -light (hyphens)

night-long (adj. *and* adv., hyphen)

nightmarish, *not* -reish

night nurse (two words)

nightshade (bot.) (one word)

night/-shift, -shirt, -spot, -time, -watch, -watchman, -work (hyphens)

nihil/ (Lat.), nothing; — *ad rem,* nothing to the purpose; — *obstat,* no objection is raised (to publication etc.)

Nihon, *use* **Nippon**

Nijinsky (Vaslav), 1890–1950, Russian ballet-dancer; **Nijinska (Bronislava),** 1891–1972, his sister, Russian choreographer

Nijni Novgorod, Russian city (no hyphen), now named **Gorky;** *not* Nizh-

Nikobar Islands, Nikosia, *use* **Nic-**

nil/ (Lat.), nothing; — *admirari,* wondering at nothing (*not* admiring nothing); — *conscire sibi,* to be conscious of no fault; — *desperandum,* despair of nothing, *not* never despair

nilgai, short-horned Indian antelope, *not* neelghau, nylghau

Nilgiris (The), hills, S. India, *not* Neilgherry Hills

ni l'un (*or* *une fem.*) *ni l'autre* (Fr.), neither the one nor the other

nimbo-strat/us (meteor.), low grey layer of cloud, *pl.* -i

nimb/us, a halo, a rain cloud; *pl.* -i *or* -uses; adj. -used

niminy-piminy, affectedly delicate, *not* -i -i -i

n'importe! (Fr.), never mind!

Nin (Anaïs), 1903–77, French-born US writer

nincompoop, a simpleton (one word)

ninepins, game (one word)

ninth, *not* -eth

niobium, symbol **Nb**

nip/, -ped, -per, -ping

Nippon, native name for Japan, *not* Nihon

nirvana, in Buddhism and Hinduism, cessation of sentient existence

nisi/ (Lat.), unless; — *prius,* unless before

Nissen/ (Peter Norman), 1871–1930, British engineer, inventor of — **hut**

nis/us (Lat.), an effort; *pl.* -us

nitrate, a salt of nitric acid

nitrite, a salt of nitrous acid

nitrogen, symbol **N**

nitrogenize, *not* -ise

nitrogenous, *not* -eous

nitro-glycerine, an explosive, *not* -in (hyphen)

nizam, a Turkish soldier, (cap.) title of ruler of Hyderabad; *pl.* same

Nizhni-Novgorod, Russia, *use* **Nijni Novgorod**

NJ, New Jersey (off. abbr.)

NKGB, USSR 'People's Commissariat of State Security', 1943–6. *See also* **NKVD**

Nkrumah (Dr Kwame), 1909–72, Ghanaian politician (President, 1960–6)

NKVD, USSR 'People's Commissariat for Internal Affairs', 1934–43. *See also* **OGPU**

NL, New Latin

n.l. (Lat.), *non licet* (it is not allowed), *non liquet* (it is not clear), (typ.) new line

N. lat., north latitude

NLC, National Liberal Club

NM, New Mexico (off. postal abbr.)

n.M. (Ger.), *nächsten Monats* (next month)

nm, nanometre

N.Mex., New Mexico (off. abbr.)

nn., notes

NNE, north-north-east

NNW, north-north-west

NO, Navigation Officer, New Orleans

NO, n.o. (cricket), not out

No, Jap. drama, *use* **Noh**

No, nobelium (no point)

No., no., from It. *numero* (number); *pl.* **Nos., nos.;** in Fr. **n°,** *pl.* **n^{os}**

no (the negative), *pl.* **noes**

Noachian, pertaining to Noah

Nobel (Alfred Bernard), 1833–96, Swedish inventor of dynamite; — **prizes** (six), awarded annually for Physics, Chemistry, Physiology and Medicine, Literature, Peace, Economics

nobelium, symbol **No**

noblesse/ (Fr. f.) nobility; — *oblige,* — imposes obligations

nobody (one word)

nocte (Lat.), at night; abbr. **n.**

'Noctes Ambrosianae', articles in *Blackwood's Magazine,* 1822–35

nocturn (RCC), part of matins orig. said at night

nocturne, (painting) a night--scene, (mus.) a night-piece (not ital.)

NOD, Naval Ordnance Department

nod/, -ded, -ding

nod/us, difficulty, *pl.* **-i**

Noël, the name

Noël (Fr. m.), Christmas; Eng. **Nowel(l),** obs. except in carols

noetic, of the intellect

Noh, Jap. drama with dance and song, *not* No (cap.)

noisette (Fr. f.), hazel-nut; (cook.) in *pl.* **-s,** small choice pieces of meat

noisette (bot.), a hybrid between China and moss rose

noisome, noxious, ill-smelling (no connection with noise)

noisy, *not* -ey

nolens volens (Lat.), willy--nilly; *pl.* **nolentes volentes**

noli me tangere (Lat.), don't touch me (no hyphens)

nolle prosequi (Lat.), plaintiff's relinquishment of suit; abbr. *nol. pros.*

nolo/ (Lat.), I will not; — *contendere,* I plead guilty; — *episcopari,* I do not wish to be a bishop (formula for avoiding responsible office)

nol. pros., nolle prosequi

nom., nominal, nominative

no man's land, unclaimed territory (apos., three words)

nom de guerre (Fr. m.), assumed name

nom de plume, pseudonym (not in Fr. usage)

nom de théâtre (Fr. m.), stage--name

nom/en (Lat.), a name, *pl.* **-ina,** abbr. **N.;** *nomen/ genericum,* a generic name; — *specificum,* a specific name

nomin., nominative

nominative (gram.), the case of the subject; abbr. **n., nom.,** *or* **nomin.**

non/ (Lat.), not; — *assumpsit,* a denial of any promise

non- is freely added (with hyphen) to adjectives and to nouns of action, condition, or quality to indicate their opposites (retain caps. of words which have them): non-aggression, non-co--operation, non-Euclidean, non--resident, non-toxic, etc. It is also attached to verbs to give sense 'that does not —': non--skid, non-slip, non-stick, etc.

nonce-word, one coined for the occasion (hyphen)

nonchalance, indifference (not ital.)

nonchalant, indifferent

non-commissioned officer/, -s; abbr. **NCO**

non compos mentis (Lat.), of unsound mind

non con., non-content, dissentient

Nonconformist, an English Protestant separated from the Church of England (cap. only in this sense)

non constat (Lat.), it is not clear

non-co-operation (two hyphens)

nones (*pl.*), in Roman calendar the ninth day, counting inclusively, before the ides

non est (Lat.), it is wanting; — — **invent/us, -a, -um,** he, she, *or* it has not been found, abbr. **n.e.i.**

nonesuch (strictly correct), **nonsuch** (usual), person or thing that is unrivalled

Nonesuch Press, founded by Sir Francis Meynell

nonet (mus.), composition for nine performers

none the less (three words)

non-Euclidean (hyphen, one cap.)

non-hero, opposite of a hero (hyphen)

non/ inventus (Lat.), not found; — **libet,** it does not please (me); — **licet,** it is not permitted, abbr. **n.l.;** — **liquet,** it is not clear, abbr. **n.l.;** — **mi ricordo** (It.), I do not remember; — **nobis** (Lat.), not unto us; — **obstante,** notwithstanding, abbr. **non obst.;** — **obstante veredicto,** notwithstanding the verdict

non-lining numerals, *see* numerals

non-net book, one that retailer may sell at less than published price

nonpareil, unequalled; (typ.) name for a former size of type, about 6 pt.

non placet (Lat.), it does not commend itself

nonplus/, to perplex; **-sed, -sing**

non plus ultra (Lat.), perfection

non/ possumus (Lat.), we cannot; — **prosequitur,** he does not prosecute, abbr. **non pros.;** — **sequitur,** it does not follow logically, abbr. **non seq.**

non-resident (hyphen)

non/-skid, -slip, -stick, *see* non-

non-stop (hyphen)

nonsuch, now usual for **nonesuch,** q.v.

nonsuit, stoppage of suit by judge when plaintiff has failed to make a case (one word)

non-toxic (hyphen)

non-U, not characteristic of the upper class (hyphen, no point)

noon, abbr. **n.**

noon/day, -tide (one word)

no one (two words)

n.o.p., not otherwise provided for

no par. (typ.), matter to run on, and have no break

Nor., Norman

Norddeutscher Lloyd (two words); abbr. **NDL**

Nordenfeldt gun

Nordenskjöld (Nils Adolf Erik, Baron), 1832–1901, Swedish Arctic explorer

Norge, Norw. for **Norway**

norm/a (Lat.), a rule or measure; *pl.* **-ae**

normalize, to standardize, *not* -ise

Normanby (Marquess of)

normande (à la) (Fr. cook.), apple-flavoured (not cap.)

Noronha (Fernando de), Brazilian penal colony in S. Atlantic

Norroy and Ulster, the third King-of-Arms

Norse, abbr. **N.**

north/, -ern, abbr. **N.** *See also* **capitalization, compass**

North (Christopher), pseud. of **Prof. John Wilson,** 1785–1854, Scottish poet

Northallerton, N. Yorks. (one word)

North Americ/a, -an, abbr. **NA** *or* **N. Amer.**

Northamptonshire, abbr. **Northants.**

northbound (one word)

North Britain, or Scotland, abbr. **Scot.,** *not* NB

North Carolina, abbr. **NC**

Northd., Northumberland

North Dakota, abbr. **N. Dak.** or (postal) **ND**

North Downs, Kent, etc.

north-east/, -ern, abbr. **NE**

northeaster, wind (one word)

northern, abbr. **N.**

Northern Territory (Australia), abbr. **NT**

Northesk (Earl of)

North-German Gazette (one hyphen)

Northleach, Glos. (one word)

North Pole (caps.)

Northumberland, abbr. **Northumb.** or (postal) **Northd.**

North Wales, abbr. **NW**

north-west/, -ern, abbr. **NW**

Norvic:, sig. of Bp. of Norwich (colon)

Norw., Norway, Norwegian

Norway, in Norw. **Norge**; in Fr. **Norvège.** *See also* **Assemblies**

Norwegian typography, *see* **accents**

Nos., nos., numbers

n^{os} (Fr.), *numéros* (numbers)

nosce teipsum (Lat.), know thyself

nose/bag, -band, -bleed (one word)

nose-cone (hyphen)

nosedive (noun *and* verb, one word)

nose/-piece, -wheel (hyphens)

nosey, *use* **nosy**

noster (Lat.), our, our own; abbr. **N.**

Nostradamus (in Fr., **Michel de Nostredame**), 1503–66, French astrologer

nostrum/, a quack remedy; *pl.* **-s** (not ital.)

nosy, *not* nosey

Nosy Parker, an inquisitive person (two caps.)

nota bene (Lat.), mark well; abbr. **NB**

notabilia (Lat. pl.), notable things

notand/um (Lat.), a thing to be noted; *pl.* **-a**

Notary Public, law officer (caps., two words); *pl.* **-ies** —; abbr. **NP**

notation, *see* **numerals**

note/, -s, abbr. **n.,** *pl.* **nn.** *See als* footnotes, let-in notes, shoulder-notes, side-notes

note/book, -case, -paper, -worthy (one word)

notic/e, -eable, -ing

notice-board (hyphen)

notif/y, -iable, -ied, -ying

notiti/a (Lat.), a list; *pl.* **-ae**

not proven, *see* **proven**

Notre Dame, Ind., US, university town

Notre-Dame (Our Lady), abbr **N.-D.,** French name of many churches, *not* Nô- (hyphen)

Notre-Seigneur (Fr.), Our Lord, abbr. **N.-S.**

Nottinghamshire, abbr. **Notts.**

Nouakchott, Mauritania

n'oubliez pas (Fr.), don't forge

nougat, a confection

nought, the figure zero (o). *See also* **naught**

noumen/on, an object of intellectual intuition, not perceptible by the senses, *opp. to* **phenomenon**; *pl.* **-a**; **-al, -ally**

noun, abbr. **n.**

nouns (collective), if regarded as a whole to be treated as singular, e.g. the army *is*, the committee *meets*. If regarded as a number of units, to be treated as plural, e.g. The French (people) *are* thrifty

nouns (German), all have initial caps. in Ger. usage

nous (Gr.), intellect, shrewdness (not ital.)

nous avons changé tout cela (Fr.), we have changed all that

nous verrons (Fr.), we shall see

nouveau/ riche/ (Fr. m.), parvenu; *pl.* **-x -s**

nouvelles (Fr. f. pl.), news

nouvelle vague (Fr. f.), new wave

Nov., November

nova/ (astr.) new star, *pl.* **-e**

Novalis, pseud. of **Baron
Friedrich von Hardenberg,**
1772–1801, German writer

Nova/ Scotia, Canada, abbr.
NS; — **Zembla,** *use* **Novaya
Zemlya,** Arctic islands (two
words, caps.)

Noveboracensian, of New
York, *not* Nova-

novelette, a small novel

novella, a short novel or narrative

November, abbr. **Nov.;** in Fr.
m. **novembre,** abbr. **nov.** (not
cap.)

noviciate, the state or period of
being a novice, *not* -tiate

novus homo (Lat.), a self-made
man; *pl.* **novi homines**

nowadays (one word)

Nowel (l), *see* **Noël**

nowhere (one word)

noyade (Fr. f.), execution by
drowning

noyau/, a liqueur; also a sweetmeat, *pl.* **-x**

NP, New Providence, Notary
Public

Np, neptunium (no point)

n.p., net personalty, (typ.) new
paragraph, (bibliog.) no place
of publication given

NPA, Newspaper Publishers' Association

NPFA, National Playing Fields
Association

NPL, National Physical Laboratory

NPO, New Philharmonia Orchestra (*earlier and later,* Philharmonia Orchestra)

n.p.p., no passed proof

NR, North Riding

Nr. (Ger.), *Nummer* (number)

nr., near

NRA, National Rifle Association

NRDC, National Research Development Corporation

NS, New Series, New Side, Newspaper Society, New Style (after
1752), Nova Scotia

N.-S. (Fr.), Notre-Seigneur (Our
Lord)

n.s., not specified

n/s (banking), not sufficient

NSB, National Savings Bank

NSPCC, National Society for the
Prevention of Cruelty to Children

NSW, New South Wales

NT, National Theatre, — Trust,
New Testament, (Australia)
Northern Territory, (bridge) no
trumps

NTP, normal temperature and
pressure

n.u., name unknown

nuance, shade of difference (not
ital.)

NUBE, National Union of Bank
Employees

nucle/us, *pl.* **-i**

nuclide (phys.), atom having
specified type of nucleus, *not*
-eide

-nugger, Indian suffix, *use*
-nagar

NUGMW, National Union of
General and Municipal Workers

NUI, National University of Ireland

Nuits-Saint-Georges, a red
burgundy (two hyphens)

NUJ, National Union of Journalists

nulla bona (law), no goods that
can be distrained upon

nullah (Ind.), a dry watercourse
(not ital.)

nulla-nulla, Australian wooden
club, not *-ah -ah*

nullif/y, -ied, -ying

nulli secundus (Lat.), second to
none

Num., Numbers (OT), *not*
Numb.

NUM, National Union of Mine-
workers

num., numeral, -s

number, abbr. **No.** *or* **no.,** *pl.*
Nos. *or* **nos.**

number (**house**), in road or

number (*cont.*)
street, no comma following, as, 6 Fleet Street. *See also* **numerals**
numbskull, *use* **nums-**
numerals (see BS 2961).

I. **Roman**

Use full caps. for Henry VIII, *Gipsy Moth III*, etc., LXX (Septuagint), and in lines of even full caps. give a better appearance to the page, and lower case is used for prelim page numbers, etc. *See also* **authorities**.

II. **Arabic**

Two cuts are used: old style (*or* non-ranging, non-lining) 1234 etc., and new style (*or* modern, ranging, lining) 1234 etc. (used esp. in scientific and technical work). Use numerals, not words, for **ages**, *but* 'he died in his eightieth year'; **bookwork**, rarely, and only those over 100 (*but* in statistical passages numerals are used more freely); **dates**, days and years regularly, months for brevity; **degrees** of heat; **money** (omit the ciphers for cents, pence, etc., when there are none, as e.g. $100, *not* $100.00); **narrow measure** (works of); **numbers** with vulgar or decimal fractions; **races**, distance and time; **scores**, of games and matches; **specific gravity** (relative density); **statistics**; **time of day** when followed by a.m. or p.m.; **votes**; **weights**, when abbreviated units are given.

Numerals are not to be combined with words in one amount (use all numerals, or all words); use commas to separate each group of three consecutive numerals, starting from the right, when there are four or more, except in math. work and pagination; in scientific and foreign-language work thin spaces are used in-

stead of commas (1 234 567 *not* 1,234,567); in pagination, dates, etc., use the least number of numerals possible (42–5, 161–4, 1961–8, 1961–75, 1966/7)—but this does not apply to the numbers 10–19, which represent single words, so 10–11, 16–18, 210–11; dates involving changes of century and all BC dates must be given in full; number of a house in road or street, etc., not to be followed by comma as it does not make the meaning clearer

Use words for: **beginning of sentences**; **degrees** of inclination; **indefinite amounts**, as two or three miles, I have told you a hundred times; **legal work**, *always*; **street names** (numerical), as Fifth Avenue; one-and-twenty, etc. (hyphens).

See also **decimal currency, fractions, lakh**

numér/o (Fr. m.), number, abbr. **n°,** *pl.* **n°ˢ**; *-oter*, to number (e.g. pages); *-oteur,* numbering machine

numis., numismatic, -s, numismatology

Nummer (Ger. f.), number, abbr. **Nr.**

numskull, a dunce, *not* numb-

Nunc Dimittis (Lat.), (musical setting of) the Song of Simeon as a hymn

nuncio/, a papal ambassador; *pl.* **-s**

NUPE, National Union of Public Employees

Nuphar, yellow water-lily genus. *See also* **nenuphar**

NUR, National Union of Railwaymen

Nuremberg, *not* -burg; in Ger. **Nürnberg**

Nureyev (**Rudolf Hametovich**), b. 1939, Russian ballet-dancer

nurl, *use* **knurl**

Nürnberg, Anglicized as **Nuremberg**

nurs/e, -eling, -ing

nursemaid (one word)

nurseryman (one word)

nursery/ rhyme, — school (two words)

NUS, National Union of Seamen; ditto Students

NUT, National Union of Teachers

nut/, -ted, -ting, -ty

nut/crackers; -hatch, a bird; **-shell** (one word)

NUTG, National Union of Townswomen's Guilds

nux vomica, seed of E. Indian tree, source of strychnine (two words, not ital.); abbr. **nux vom.**

NV, Nevada (off. postal abbr.), New Version

NVM, Nativity of the Virgin Mary

NW, north-west(ern), North Wales

NWFP, North-West Frontier Province (Pakistan)

NWT, North-West Territories (Canada)

NY, New York (City or State), US. *See also* **New York**

Nyasaland, Cent. Africa, *not* Nyassa- (one word); *now* **Malawi**

NYC, New York City

nyet (Russ.), no

nylghau, antelope, *use* **nilgai**

nymph/ae (anat.), the *labia minora; sing.* **-a**

Nymphaea (bot.), the white water-lily genus

nymphet, young nymph, sexually attractive girl

nympho/, colloq. for nymphomaniac, *pl.* **-s**

Nynorsk, *see* **Landsmål**

NZ, New Zealand

O

O (exclamation), *see* **Oh**

O, oxygen (no point), (Lat.) *octarius* (a pint), the fourteenth in a series

O., Odd Fellows, Ohio, Order (as DSO), owner, (Fr. m.) *ouest* (west), (Ger. m.) *Osten* (east)

O', Irish name prefix; use apos. as O'Neill, *not* turned comma; **o** or **ó**, or the corresponding caps., followed by thin space, also used, thus: ó Brian, Ó Flannagáin

o., old, (meteor.) overcast

o- (chem.), *ortho-*

o', abbr. for **of**

Ö, ö, in German, Swedish, etc., may *not* be replaced by Oe, oe, (except in some proper names), *O, o*, or Œ, æ

ö (Sw.), ø (Dan.), island

Ø, ø, Danish and Norwegian letter

%, per cent; ‰, per mille

o/a, on account of

oaf/, stupid person, *pl.* -s

Oak-apple Day, 29 May

O. & M., organization and methods

OAP, Old Age Pension(er)

oarweed, seaweed, *not* ore-weed

oas/is, *pl.* -es

OAS, on active service, Organization of American States, Organisation de l'Armée Secrète

oatcake (one word)

Oates (Titus), 1649–1705, false informer of Popish Plot, 1678

oatmeal (one word)

OAU, Organization of African Unity

Oaxaca, Mexico

OB, Old Boy, outside broadcast

ob., oboe

ob., *obiit* (he, she, *or* it died)

Obadiah, abbr. **Obad.**

obbligato (mus.), (required) accompaniment, *not* obl-; abbr. **obb.**

OBE, Officer of the Order of the British Empire

obeah, Negro witchcraft, *not* -ea-eeyah, -i

obeisance, courtesy bow

obelisk, *not* -isc; (typ.) the dagger mark (†); **double obelisk** ‡. *See also* **reference marks**

obel/us, critical mark, *pl.* -i

Oberammergau, Bavaria

obi, witchcraft, *use* obeah

obi, Japanese sash

obiit/ (Lat.), he, she, *or* it died, abbr. **ob.**; — *sine prole*, died without issue, abbr. **ob.s.p.**

obit, date of death, funeral or commemoration service; colloq for obituary (not ital.)

obiter/ dictum (Lat.), a thing said by the way, *pl.* — *dicta*; — *script/um*, ditto written, *pl* — *-a*

object/, -ion, -ionable, -ive, -ively, abbr. **obj.**

object-glass, lens nearest the object

object-lesson (hyphen)

objet d'art (Fr. m.), a work of artistic value; *pl.* **objets** — — (ital.)

objet trouvé (Fr. m.), thing found, *pl.* -s -s

obl., oblique, oblong

obligato, *use* obb-

obligee (law), person to whom another is bound

obliger, one who does a favour

obligor (law), one who binds himself to another

oblique, abbr. **obl.**

oblong, abbr. **obl.**

obo/e (mus.), *not* the many variants; abbr. **ob.**; -ist, *not* -eist

obol, small coin or weight of anc Greece

obol/us, an obol; also used for

284

various small coins in medieval European countries; *pl.* **-i**

O'Brien (Conor Cruise), b. 1917, Irish politician; — **(Edna)**, b. 1932, Irish writer; — **(William Smith)**, 1803–64, and — **(William)**, 1852–1928, Irish patriots

obs., observation, observatory, observed, obsolete

obscurum per obscurius (Lat.), the obscure by means of the more obscure, *not obscurus*

obsequi/es, funeral rites, the *sing.* -y not used; adj. **-al**

obsequious, fawning

observanda (Lat. pl.), things to be observed

observat/ion, -ory, abbr. **obs.**

obsidian (mineral.)

obsolescen/ce, -t, becoming obsolete

obsolete, abbr. **obs.**

ob.s.p., *obiit sine prole* (died without issue)

obstetric/s, midwifery; abbr. **obstet.; -ian**, an expert in it

obstructor, *not* -er

obverse, that side of a coin with the head or main device

OC, Officer Commanding, Officer of the Order of Canada

o.c., *use* **op. cit.**

ocarina, terracotta wind-instrument, *not* och-

Occam/ (William of), 1280–1349, English Franciscan philosopher, deviser of —**'s razor**, principle of fewest possible assumptions; **-ism, -ist**; *not* Ockam

Occident (the), the West, *but* **occidental** (not cap.)

Occleve, *use* **Hoccleve**

occulist, *use* **ocu-**

occur/, -red, -rence, -rent, -ring

Oceania, islands of Pacific Ocean

ocell/us (entom.), simple (as opp. to compound) eye (of insect); *pl.* **-i**

ocharina, *use* **oca-**

ochlocracy, mob-rule

ochone, Anglo-Irish form of **ohone** (q.v.)

ochra, *use* **okra**

ochr/e, yellow pigment, **-eous, -y**, *not* oker

Ockham, Surrey. *See also* **Occam**

o'clock, to be printed close up (not to be abbreviated in print)

O'Connell (Daniel), 1775–1847, Irish statesman

O'Connor (Feargus Edward), 1794–1855, Irish-born Chartist leader; — **(Rt. Hon. Thomas Power)**, 1848–1931, Irish-born journalist and British politician

OCR (comp.), Optical Character Recognition (reading of print etc. by electronic eye for computer input)

ocra, *use* **okra**

Oct., October

octachord, eight-stringed (musical instrument), *not* octo-

octahedr/on, solid figure with eight faces, *pl.* **-ons**; adj. **-al**; *not* octae-, octoe-, ootoh

octaroon, *use* **octo-**

octastyle (archit.), having eight columns, *not* octo-

Octateuch, first eight books of OT, *not* Octo-

octavo/ (typ.), a book based on eight leaves, sixteen pages, to the sheet; *pl.* **-s**, abbr. **8vo** (no point). *See also* **book sizes**

octet (mus.), *not* -ett, -ette

October, abbr. **Oct.;** in Fr. m. *octobre*, abbr. *oct.* (not cap.)

Octobrist (Russian hist.), *not* -erist

octochord, *use* **octa-**

octodecimo, *see* **eighteenmo**

octoedron, -hedron, *use* **octa-**

octopus/, *pl.* **-es**

octoroon, person of one-eighth Negro blood, *not* octa-

octostyle, *use* **octa-**

Octoteuch, use **Octa-**

octroi (Fr. m.), municipal customs duties (not ital.)

OCTU or **Octu,** Officer Cadets Training Unit

oculist, eye-specialist, *not* occ-

Oculi Sunday, third in Lent

ocul/us (Lat.), an eye; *pl.* *-i*

OD, Old Dutch, Ordnance datum

O/D, on demand, overdraft

od, a hypothetical force; (arch.) God, *not* 'od

o.d., outer diameter

odal, use **udal**

odalisque (Turk.), female slave, *not* -isk (not ital.)

O.Dan., Old Danish

odd, used as suffix requires hyphen to avoid ambiguity (e.g. twenty-odd people)

Odd Fellows (Independent Order of) (official, separate words, caps.); abbr. **IOOF**

oddments (typ.), the parts of a book separate from the main text, such as contents, index; a section containing oddments

odd pages (typ.), the right-hand, or recto, pages

Odelsting, lower house of Norwegian Parliament

Odendaalsrus (off. spelling), Orange Free State, *not* -st

Odéon, Paris theatre

Oder–Neisse line, between Poland and E. Germany

odeum/, building for musical performances, *pl.* **-s**

Odeypore, use **Udaipur**

odi profanum vulgus (Lat.), I loathe the common herd

odium, widespread dislike

odium/ aestheticum (Lat.), the bitterness of aesthetic controversy; — *medicum,* ditto medical; — *musicum,* ditto musical; — *theologicum,* ditto theological

ODM, Ministry of Overseas Development

odometer, instrument for measuring distance travelled, *not* ho-

O'Donoghue of the Glens (The) (cap. T)

O'Donovan (The) (cap. T)

odontoglossum/, an orchid; *pl.* **-s;** italics (cap. *O*) as genus

odor/iferous, -ize, -izer, -ous

odour/, -less

Odysseus, Greek hero (Roman Ulysses)

Odyssey, of Homer (ital., not quoted); **Odyssean**

odyssey, a long wandering (no cap., not ital.)

OE, Old English

Oe (phys.), oersted

œ (ligature). *See also* **diphthongs**

OECD, Organization for Economic Co-operation and Development

oecist, the founder of a Greek colony, *not* oik-

oecology, use **ec-**

oecumenic/, -al, *see* **ecumenic/, -al**

Oecumenical Patriarch, Abp. of Constantinople, head of the Orthodox Church

OED, Oxford English Dictionary; NED (New English Dictionary) is now incorrect

oedema/, swelling; **-tous,** *not* ed-

Oedipus/, the Theban hero, *not* Edi-; — **complex** (psych.), a boy's subconscious desire for his mother and hostility to his father (two words); adj. **Oedipal**

OEEC, Organization for European Economic Co-operation, now **OECD,** q.v.

Oehlenschläger (Adam), 1779–1850, Danish writer

œil (Fr. m.), eye, *pl.* **yeux;** **œil-de-bœuf,** a small round window, *pl.* **œils——;** **œil-de-perdrix,** a soft corn, *pl.* **œils——**

œillade (Fr. f.), a glance

oeno/logy, study of wine; **-phile,** lover of wines, *not* en-, oin-

o'er, to be printed close up

oersted (phys.), unit of magnetic field strength, from **Hans Christian Oersted,** 1777–1851, Danish physicist; abbr. **Oe**; in SI units 1000/4π A/m

oesophag/us, the gullet; *pl.* **-i**; **-eal** (not ital.)

Oesterreich (Ger.), Austria, *use* **Ös-**

oestrogen, oestrus (US **e-**)

œuf (Fr. m.), egg; **œufs *à la coque,*** boiled eggs; **— *à la neige,*** whisked eggs; **— *à l'indienne,*** curried eggs; **— *de Pâques,*** Easter eggs; **— *sur le plat,*** fried eggs

œuvre (Fr. f.), work, esp. writer's or artist's work taken as a whole

OF, Odd Fellows, Old French, (typ.) old-face type

O'Faoláin (Seán), b. 1900, Irish writer

off., offic/e, -er, -ial, -inal

Offaly, Ireland, *formerly* King's County

Offa's Dike, between England and Wales (apos.)

off/-beat, -centre (hyphens)

offcut, remnant of paper, wood, etc.; (typ.) a remnant of paper, board, or cloth cut off from a larger piece (one word)

off-day (hyphen)

Offenbach (Jacques), 1819–80, German-born French composer

offer/, -ed, -ing, -tory

offg., officiating

offhand/, -ed, casual, extempore (one word)

offic/e, -er, abbr. **off.**

official, abbr. **off.** *or* **offic.**

officiating, abbr. **offg.**

officina (Lat.), a workshop. *See also* **oficina**

officinal, used in a shop, used in medicine, sold by druggists; abbr. **off.**

off/-licence, -peak (adj.) (hyphens)

offprint (typ.), a separately printed copy, or small edition,

of an article which originally appeared as part of a larger publication

offset (typ.), unwanted transfer of ink from printed sheet to one laid on top of it (now commonly **set-off,** q.v.); a planographic (q.v.) printing process in which ink is transferred on to an intermediate rubber blanket cylinder and then offset on to the paper (*also* **—-litho, —-lithography**)

offshoot (one word)

off shore (adv., two words)

offshore (adj., one word)

offside (sport) (one word)

offspring (one word)

off-stage (adj. *and* adv., hyphen)

offstreet (adj., one word)

oficina (Sp.), a S. American factory. *See also* **off-**

O'Flaherty (Liam), b. 1896, Irish novelist

Oflag, German prison camp for officers

OFM, Order of Friars Minor

OFS, Orange Free State (South Africa)

oft-times (hyphen)

O.Gael., Old Gaelic

ogee (archit.), a moulding; adj. **ogee'd**

og/ham, ancient alphabet, *not* -gam, -um, -hum

ogiv/e, diagonal rib of vault, pointed arch; **-al**

OGPU *or* **Ogpu,** USSR 'United state political administration for struggle against espionage and counter-revolution', 1922–34. *See also* **NKVD**

O'Grady (The) (cap. T)

ogr/e, man-eating giant, *pl.* **-es**; *fem.* **-ess**; adj. **-ish,** *not* -eish

OH, Ohio (off. postal abbr.)

Oh, to be used as an independent exclamation, followed by a comma or exclamation mark. Use **O** to form a vocative, and when it is not separated by punctuation from what follows,

Oh (*cont.*)
as *O mighty Caesar!*, *O for the wings of a dove.* See *Hart's Rules*, p. 30

OHBMS, On Her, *or* His, Britannic Majesty's Service

O. Henry, *see* **Henry**

OHG, Old High German

Ohio, off. postal abbr. **OH**

ohm, SI unit of electrical resistance; symbol Ω; *see also* **mho**; from **Georg Simon Ohm,** 1787–1854, German physicist

OHMS, On Her, *or* His, Majesty's Service

oho! exclamation of surprise, *not* O ho, Oh ho, etc.

ohone, Scottish and Irish cry of lamentation. *See also* **ochone**

oidium, a fungus, *not* oï-

oikist, *use* **oecist**

oil/cake, -can, -cloth (one word)

oil-colour (hyphen)

oilfield (one word)

oil/-fired, -gauge, -painting (hyphens)

oil/skin, -stone (one word)

oil well (two words)

oino/logy, -phile, *use* **oen-**

oi polloi (Gr.), the masses, *use* **hoi** —

O.Ir., Old Irish

Oistrakh (David Fyodorovich), 1908–74, and — **(Igor Davidovich),** b. 1931, his son, Russian violinists

Ojibwa, Wis., US

Ojibway, Amer. Indian tribe

OK, 'all correct'; *not* okay; Oklahoma (off. postal abbr.)

okapi, bright-coloured African ruminant

Okehampton, Devon

O'Kelly (The) (cap. T)

oker, *use* **ochre**

Oklahoma, off. abbr. **Okla.** or (postal) **OK**

okra, pods used for thickening soup; *not* oc(h)-, okro

OL, Old Latin

Ol., Olympiad

Olaf (St.), patron of Norway

old/, abbr. **o.;** **—-clothes-man** (hyphens)

Old Bailey, *see* **Bailey**

Old Believer, *see* **Raskolnik**

Old English (typ.), name for the English style of black letter as 𝔒𝔩𝔡 𝔈𝔫𝔤𝔩𝔦𝔰𝔥

old face (typ.), a type design based on the original roman type of the fifteenth century; abbr. **OF**

old-fashioned (hyphen)

'Old Glory', US stars-and-stripes flag

Old Hickory (US colloq.), Andrew Jackson, President of US 1829–37

'Old Hundredth', hymn-tune, *not* '— Hundred'

old maid, elderly spinster, prissy man, card-game (two words), *but* **old-maidish** (hyphen)

Old Man of the Sea, in *Arab. Nights* (caps.)

Oldmeldrum, Grampian (one word)

Old Pals Act (three words, caps., no apos.)

Old Red Sandstone (geol.) (caps.)

old school tie (three words)

Old Style, according to the Julian calendar, abbr. **OS** (no point). *See also* **New Style**

old style/ (typ.), a type design, being a regularized old face; — — **numerals,** *see* **numerals**; abbr. **OS**

Old Testament, abbr. **OT** *or* **Old Test.**; for abbr. of books in, *see* each title

old-time/ (adj., hyphen); **-r**

old woman, fussy man (two words), *but* **old-womanish** (hyphen)

olefin (chem.), hydrocarbon, *no* -ine

oleiferous, oil-producing, *not* olif-

oleograph, picture printed in oil-colours

O level, examination (no hyphen)

OLG, Old Low German

Olifants River, S. Africa (no apos.)

Oligocene (geol.) (cap., not ital.)

olive/-branch, -oil (hyphens)

Olivetan, an order of monks, *not* -ian

Olivier (Laurence Kerr, Baron), b. 1907, English actor. *See also* **Ollivier**

olla podrida, Spanish national dish, *also* a medley (not ital.)

Ollivier (Olivier Émile), 1825–1913, French statesman and writer. *See also* **Olivier**

olog/y, any science or theory, *pl.* -ies

oloroso/, (drink of) medium- -sweet sherry, *pl.* -s

Olympiad, celebration of Olympic games (anc. *and* mod.), period of four years between celebrations; abbr. **Ol.**

Olympian, of Olympus, the abode of the Greek gods; as noun, dweller in Olympus, superhuman person

Olympic, of Olympia in Greece, *or* of the games held there in antiquity, *or* of the modern games

OM, (Member of the) Order of Merit

o.m., old measurement

omadhaun, Irish term of contempt (*not* the many variants)

Omagh, Co. Tyrone

Omaha, Nebraska, US

Oman/, SE Arabia, adj. **-i**

Omar Khayyám, 1048–1131, Persian poet

ombuds/man, official appointed to investigate complaints against government departments, introduced in Britain in 1965, *pl.* **-men**

Omdurman

omega, the Greek long *o*, last letter of Greek alphabet (ω, Ω); Ω, symbol for ohm

omelette, *not* -et

omicron, the Greek short *o*, (*o*, *O*), *not* omik-

omissions (typ.), *see* **dele, elision, ellipsis**

omit/, -ted, -ting

omnibus/, *pl.* **-es.** See also **bus**

omnium gatherum, a confused medley (not ital.)

ON, Old Norse

ONC, Ordinary National Certificate

oncoming (one word)

OND, Ordinary National Diploma

on dit (Fr. m. sing./pl.), gossip ('people say')

one-and-twenty, etc. (hyphens)

one-eighth, *see* **fractions**

Oneida, socialist community started at Lake Oneida, NY, US, 1847

one-idea'd (hyphen)

O'Neill (Eugene Gladstone), 1888–1953, US playwright; — **(Moira),** pseud. of Agnes Higginson Skrine, *fl.* 1900, Irish poet; — **of the Maine (Baron),** family name of Baron Rathcavan

oneiro/critic, -logy, -mancy, interpreter of, study of, divination by dreams, *not* oniro

oneness (**a,** *not* an) (one word)

one-off, made as one only (hyphen)

oneself is reflexive or intensive; **one's self,** one's personal entity

one-sided (hyphen)

ONF, Old Norman-French

ongoing, continuing (one word)

onirocritic etc., *use* **oneiro-**

onlook/er, -ing (one word)

on ne passe pas (Fr.), no thoroughfare

o.n.o., or near(est) offer

onomastic/, relating to names; **-on,** a vocabulary of proper names

onomatopoe/ia, -ial, -ian, -ic, -ical, -ically, word-formation by imitation of sound; abbr. **onomat.**

onomatopo/ësis, -etic, -etically, *not* -poie-

onrush (one word)

on shore (adv., two words)

onshore (adj., one word)

onside (sport) (one word)

Ontario, Canada; abbr. **Ont.**

on to (two words)

ontolog/y, metaphysical study of the essence of things; **-ize,** *not* -ise

onus/, burden, *pl.* **-es** (not ital.)

onus probandi (Lat.), burden of proof (ital.)

%, per cent

Oodeypore, India, *use* **Udaipur**

oolong, a tea, *not* ou-

oomiak, Eskimo boat (*not* the many variants)

‰, per mille

oopak, a tea, *not* -ack

Oostende, Fl. for **Ostend**

Ootacamund, Madras, India, *not* Utakamand; colloq. abbr. **Ooty**

oozy, muddy, *not* -ey

OP, observation post, *Ordinis Praedicatorum* (of the Order of Preachers, or Dominicans)

op., operation; optime, q.v.

o.p., overproof; (theat.) opposite the prompter's side, or the actor's right; (bibliog.) out of print

op. (Lat.), *opus* (work), *opera* (works)

op art, art in geometrical form giving the illusion of movement (no point)

op. cit., *opere citato* (in the work quoted) (not ital.)

OPEC, Organization of Petroleum Exporting Countries

open-and-shut case (two hyphens)

open-door policy, opportunity for free trade (one hyphen)

open-heart, of surgery (hyphen)

open/-hearted, -mouthed (hyphens)

open-plan (adj., hyphen)

Open Sesame (caps., two words)

open-work/, -ed, -ing (hyphens)

opera, see **opus**

operable, that can be operated (on)

opera buffa (It.), comic opera; in Fr. *opéra bouffe*

Opéra-Comique, Paris theatre (hyphen)

opéra comique (Fr. m.), opera with spoken dialogue

opera/-glass, -house (hyphens)

opera seria (It.), 18th-c. opera on heroic theme

operation, abbr. **op.,** in *pl.* **ops**

opercul/um (biol.), a cover; *pl.* **-a** (not ital.)

opere/ citato (Lat.), in the work quoted, abbr. **op. cit.;** — *in medio,* in the midst of the work

operetta/, light opera, *pl.* **-s**

ophicleide, (mus.) serpent, bass or alto key-bugle, *not* -eid

ophiolatry, worship of serpents

ophiology, study of serpents, *no* ophid-

ophthalmic, of the eye

Opie (Amelia), 1769–1853, English novelist, wife of — (**John**), 1761–1807, English painter

opinion poll (two words)

o.p.n., *ora pro nobis* (pray for us)

opodeldoc, a liniment

opopanax, a perfume

Oporto, Portugal; in Port. **Porto**

opp., opposed, opposite

oppress/or, *not* -er

ops., operations

opt., optative, optical, optician, optics

optical centring (typ.), positioning on page of a title or passage of verse so that it appears to the reader to be centred, although by measurement it is not

optime, at Cambridge, one next in merit to a wrangler (q.v.) (not ital.); abbr. **op.**

optimize, to make the best of, *not* -ise

optim/um, pl. **-a**
opus/ (Lat.), a work, pl. **opera**
(not ital. in mus.), abbr. **op.;**
— **magnum,** a great work, pl.
opera magna. See also **magnum opus**
opuscul/um (Lat.), a small
work, an essay; pl. **-a;** in Eng.
opuscule/, pl. **-s**
opus number (mus.) (two
words)
opus/ operantis (Lat.), the effect of a sacrament resulting
from the spiritual disposition of
the recipient (the Protestant
view); — **operatum,** ditto resulting from the grace flowing
from the sacrament itself (the
RC view)
OR, operational research, Oregon (off. postal abbr.), other
ranks
or, two or more singular subjects
joined by *or* take the verb in the
singular number, e.g. John *or*
William *is* going. Where *or* joins
the last two words of a list, insert
a comma, e.g. black, white, or
green
or (her.), gold
or., oriental
o.r., owner's risk
ora (Lat. pl.), mouths; *see os*
orangeade, *not* -gade
Orange Free State, prov.
of S. Africa (three words), abbr.
OFS
Orange/ism, extreme Irish
Protestantism, *not* -gism; **-man**
(one word, cap.)
Orangeman's Day, 12 July
orange-peel (hyphen)
orang/-utan, *not* ourang-, -out-
ang, -utang (hyphen, not ital.)
**Oranmore and Browne
(Baron)**
ora pro nobis (Lat.), pray for
us; abbr. **o.p.n.**
orat., orator, -ical, -ically
oratio/ obliqua (Lat.), indirect
speech; — **recta,** direct speech.
Manner of reporting speech,
e.g. 'I am glad', he said, 'to see

you again' (*recta*); He said that
he was glad to see me again
(*obliqua*)
oratorio/, pl. **-s;** — **(titles of)**
(typ.), when cited, to be in ital.
orc, a killer whale, *not* ork
Orcadian, of Orkney
orchid, member of a family of
mainly exotic flowering plants;
orchis, (esp. wild) orchid or
one of genus *Orchis*
**Orczy (Baroness Em-
muska),** 1865–1947, Hun-
garian-born English novelist
ord., ordained, order, ordinal,
ordinance, ordinary
Order, abbr. **O.;** when referring
to a society, to be cap., as the
Order of Jesuits
order/, abbr. **ord.** (see also
**botany, zoology); —-book,
-clerk, -form** (hyphens)
ordin/al, -ance (a regulation),
-ary, abbr. **ord.**
ordnance, artillery, mil. stores
**Ordnance/ Survey Depart-
ment** (caps.), abbr. **OSD;** —
datum (one cap.), the standard
sea-level of the Ordnance Survey, abbr. **OD**
ordonnance, the proper disposition of parts of a building or
picture (not ital.)
Ordovician (geol.)
ordre du jour (Fr. m.), agenda
of a meeting
öre, coin of Sweden, Denmark,
and Norway, one-hundredth of
kron/a, -e
Oreg., Oregon (off. abbr.)
oregano, dried leaves of wild
marjoram used for seasoning. See
also **origan**
O'Rell (Max), 1848–1903,
pseud. of Paul Blouet, French
author and journalist
oreo/graphy, -logy, use oro-
Oresteia, trilogy by Aeschylus,
458 BC
ore-weed, use oarweed
orfèvrerie (Fr. f.), goldsmith's
work
orfray, use orphrey

org., organ, -ic, -ism, -ization, -ized

organdie, dress fabric, *not* -di, -dy; in Fr. m. *organdi*

organize, *not* -ise

organ/on, a system of logic, *not* -um

organza, thin transparent dress fabric

org/y, *not* -ie; *pl.* -ies

oriel, small room built out from a wall; its window

Oriel College, Oxford

Orient (the), the East

orient (verb), to site building so as to face east; **orientate** is established usage in fig. senses

oriental/, -ist; abbr. *or. or orient.*

orientalize, *not* -ise (not cap.)

orientate, *see* orient

oriflamme, banner, *not* -flamb

orig., origin, -al, -ally, -ate, -ated

origan, wild marjoram of genus *Origanum.* See also **oregano**

Origen, 185-253, a Father of the Church

original (Sp. typ.), copy

Origin of Species (The), by C. Darwin, 1859

orinasal, of the mouth and nose, *not* oro-

Orinoco, river, S. America. *See also* **oro-**

oriole, a bird

ork, *use* orc

Orkney, region of Scotland

Orkneys, island-group

Orléans (House of, dép. France) (é); **Orlean/ism, -ist** (no accent in Eng.)

Orme's Head (Great and Little), Gwynedd (apos.)

ormolu, a gold-coloured alloy (not ital.)

Ormonde (Marquess of)

Ormuzd, Zoroastrian spirit of good, *not* the many variants

orn/é (Fr.), *fem.* -**ée,** adorned

ornith., ornitholog/y, -ical

Ornithorhynchus, the duck-billed platypus

orogen/esis (geol.), process of formation of mountains; **-etic, -ic**

oro/graphy, -logy, mountain description and science, *not* oreo-

oronasal, *use* ori-

oronoco, a Virginian tobacco, *not* the many variants

Oroonoko, by Aphra Behn, *c.*1678

orotund, magniloquent

Orphe/us (Gr. myth.), a Thracian lyrist; **-an,** like his music; entrancing

Orphic, pertaining to **Orphism,** the mystic cult connected with Orpheus

orphrey, an ornamental border *not* orfray

orpine, purple-flowered plant, *not* -pin

orrery, model of planetary system

orris/, lace or embroidery, iris plant; —**-powder, -root,** used for perfume (hyphens)

Or San Michele, church at Florence, *not* — Saint —

Ortega y Gasset (José), 1883-1955, Spanish philosopher

ortho- (chem. prefix) (ital.)

orthoepy, (correct) pronunciation of words

orthopaedic, concerned with the cure of bone deformities, esp. in the young, *not* -pedic

ortolan, small edible bird (not ital.)

Orvieto, Italian white wine

OS, Old Saxon, — School, — Series, — Side, — Style (before 1752), old style (type), ordinary seaman, Ordnance Survey, (clothing) outsize

Os, osmium (no point)

o.s., only son

o/s, out of stock, outstanding

os (Lat.), a bone; *pl. ossa*

os (Lat.), a mouth; *pl. ora*

OSA, Order of St. Augustine

Osaka, Japan, *not* Oz-

OSB, Order of St. Benedict

Osborn (Sherard), 1822-75,

British rear-admiral and Arctic explorer

Osborne, IW

Osborne, family name of Duke of Leeds; — (**John James**), b. 1929, British playwright

Osbourne (Lloyd), 1868–1947, writer (stepson of R. L. Stevenson)

oscill/ate, to fluctuate; **-ation, -ator, -atory, -ogram, -ograph, -oscope**

oscul/ate, to kiss, adhere closely; **-ant, -ation, -atory**

oscul/um (Lat.), a kiss, *pl.* *-a*; *osculum pacis,* the kiss of peace

OSD, Ordnance Survey Department, Order of St. Dominic

Oset, use **Ossett**

OSF, Order of St. Francis

O.Sl., Old Slavonic

Osler (Sir William), 1849–1919, Canadian physician

Oslo, *formerly* **Christiania,** capital of Norway

Osmanli, of the family of Osman (1259–1326, *not* Othman; founder of the Ottoman empire), *not* -lee, -lie, -ly (not ital.)

osmium, symbol **Os**

o.s.p., *use* **ob.s.p.**

ossa (Lat. pl.), bones; *see* **os**

ossein, bone cartilage, *not* -eine

Ossett, W. Yorks., *not* Oset, Osset

ossia (It. mus.), or

Ossory, Ferns, and Leighlin (Bishop of)

o.s.t. (naut.), ordinary spring tides

Ostend, Belgium; in Fr. **Ostende,** in Fl. **Oostende**

ostensibl/e, outwardly professed; **-y**

ostensive/, indicating by demonstration; **-ly**

osteomyelitis, inflammation of the marrow of the bone (one word)

osteria (It.), an inn

Österreich/ (Ger.), Austria, *not*

Oe-; —**-Ungarn,** Ger. for **Austria-Hungary**

Ostiaks, use **Osty-**

ostinato/ (mus.), repeated melodic figure, *pl.* **-s**

ostrac/on, a potsherd, used in anc. Greece for inscribing and voting, *pl.* **-a** (*not* -k-); **-ize,** to banish dangerously powerful citizen by votes recorded in this way, to exclude from favour, *not* -ise; **-ism**

Ostrovsky (Alexander Nikolaevich), 1823–86, Russian playwright

Ostyaks, of W. Siberia, *not* Osti-

Oswaldtwistle, Lancs., *not* -sle

O.Sw., Old Swedish

Oświęcim, Pol. for **Auschwitz**

OT, occupational therapy, Old Testament

Otaheite, *now* **Tahiti**

OTC, *now* **STC**

O tempora! O mores! (Lat.), what times, what manners!

O.Teut., Old Teutonic

other-world/ly, -liness (hyphen)

Othman, *use* **Osman**

otium/ (Lat.), leisure; — *cum dignitate,* leisure with dignity; — *sine dignitate,* leisure without dignity

otolith, an ear-stone, *not* -lite (not ital.)

otorhinolaryngology, study of ears, nose, and larynx

ottar (of roses), *use* **attar**

ottava rima (It.), stanza of eight lines, as in Byron's *Don Juan*

Ottawa, Canada, *not* Otto-

otto (of roses), *use* **attar**

Ottoman/, a Turk; *pl.* **-s** (cap.)

ottoman/, a sofa, a fabric; *pl.* **-s** (not cap.)

Otway (Thomas), 1652–85, English playwright

OU, Open University, Oxford University

Ouagadougou, Upper Volta

oubliette, a dungeon (not ital.)

Oudenarde, Belgium; battle of —, 1708. In mod. Fr., **Auden-arde,** Fl. **Oudenaarde**

Oudh, India, *not* Oude

OUDS, Oxford University Dramatic Society

Ouida, *see* **Ramée**

ouï-dire (Fr. m. sing./pl.), hearsay

Ouija, propr. term for a board used in spiritualistic seances (cap.)

oukaz, *use* **ukase**

Ouless (Walter William), 1848–1933, British painter

oulong, *use* **oo-**

ounce/, -s, abbr. **oz.,** sign ℥; 437½ grains avoirdupois, 28.35 grams approx. (not now in scientific use)

OUP, Oxford University Press, the whole printing and publishing organization controlled by Oxford University. *See also* **Clarendon Press**

ourang-outang, *see* **orang-utan**

Our/ Father, — Lady, — Lord, — Saviour (caps.)

ours (no apos.)

ousel, *use* **ouz-**

out (typ.), an accidental omission of copy in composition

out-, prefixed to a verb forms a single word unless the verb begins with *t,* as out-talk, out-turn

out-and-out, unreserved(ly) (hyphens)

out/back (noun), **-board, -building** (one word)

outcast, (person) cast out, the gen. word

outcaste (Ind.), (person) with no caste

out/dated, -door (adj.), **-doors** (noun and adv.) (one word)

Outeniqua Mountains, S. Africa

outer (typ.), the side of a sheet containing the signature and first page

outfield (sport) (one word)

outfit/, -ted, -ter, -ting

outgo/, expenditure, *pl.* **-es**

outgrowth (one word)

out-herod (hyphen), should be followed by *Herod*

outhouse (one word)

outl/ie, -ier, -ying

outmanœuvre (one word) (US **outmaneuver**)

outmoded (one word)

out/ of date, — of doors (hyphens when attrib.), **— of sorts,** unwell, (typ.) having any letter in a hand-setting fount all used

out-patient (hyphen)

output/ (verb), **-ting**; past and partic. **output**

outrance (*à*) (Fr.), to the bitter end, not *à l'*-

outr/é (Fr.), *fem.* **-ée,** eccentric

out/rider, -rush, -size (one word)

outspan (S. Afr.), to unyoke; unyoking place

out/standing, -station (one word)

outstrip/, -ped, -ping

out-swinger (sport) (hyphen)

out/-talk, -tray, -turn (hyphens)

outv/ie, -ier, -ying (one word)

outward bound (hyphen when attrib.)

Outward Bound Trust (no hyphen)

outwit/, -ted, -ting

outwork (one word)

ouvert/ (Fr.), *fem.* **-e,** open

ouvri/er (Fr.), a workman; *fem.* **-ère**

ouzel, a bird, *not* **ous-**

ouzo/, Greek drink, *pl.* **-s**

ovenware (one word)

overact (one word)

over-active (hyphen)

overall (noun, adv., and adj., one word)

over-anxious (hyphen)

over/balance, -blown, -board, -book, -burden(some) (one word)

over-careful (hyphen)

overcast (meteor.), abbr. **o.**

over-confiden/ce, -t (hyphen)

overcrowd/, -ed (one word)

Over Darwen, Lancs., *use* **Darwen**

overdevelop/, -ed, -ing (one word)

overemphas/is, -ize, *not* -ise (one word)

Overijssel, province of Netherlands

over/joyed, -land (one word)

over/-large, -many (hyphens)

over/look, -night, -pass (one word)

over/-populate, -produce (hyphens)

over/rate, -reach (one word)

over-react (hyphen)

over/ride, -ripe, -rule (one word)

overrun (typ.), to transfer words from one line to the next (one word)

oversea/, -s (one word)

over/-sensitive, -sexed (hyphens)

oversight (one word)

over/-simplify, -subscribe, -use (hyphens)

over/value, -view, -weight, -work (one word)

Ovid (in Lat., **Publius Ovidius Naso**), 43 BC–AD 17, Roman poet

ovol/o (archit.), a moulding; *pl.* **-i**

ov/um, an egg; *pl.* **-a** (not ital.)

Owens College, Manchester (no apos.)

owner, abbr. **O.**

Ox. *or* **Oxf.,** Oxford

Oxbridge, Oxford and Cambridge Universities regarded collectively

Oxenstjerna (Count Axel), 1583–1654, Swedish statesman

ox-eye/ (hyphen); **— —** **daisy**

Oxfam, Oxford Committee for Famine Relief

Oxford Almanack, not -ac

Oxford English Dictionary (*The*), abbr. *OED,* not *NED*

Oxford hollow (bind.), a flattened paper tube attached to the back of the gatherings and to the spine of the binding; this type of binding

Oxfordshire, abbr. **Oxon.**

Oxford University Press, abbr. **OUP,** q.v.

oxhide (one word)

oxide, *not* -id, -yd, -yde

oxidization, oxidation, *not* oxy-

oxidize etc., *not* -ise, oxy-

Oxon., Oxfordshire, (Lat.) *Oxonia* (Oxford), *Oxoniensis* (of Oxford)

Oxon:, sig. of Bp. of Oxford (colon)

oxtail (one word)

oxy-acetylene (hyphen)

oxychloride, *not* oxi-

oxyd/e, -ation, -ize, *use* oxi-

oxygen/, symbol **O; -ize,** *not* -ise

Oxyrhynchus, Egypt, source of papyri

oxytone (Gr. gram.), (word) with acute accent on last syllable

oyez! hear ye! *not* oyes

oz., ounce, -s (not now in scientific use)

Ozaka, Japan, *use* Os-

P

P, (car) park, (chess) pawn, (as prefix) peta-, phosphorus, (phys.) poise, the fifteenth in a series

P., pastor, post, president, prince, (Fr.) *Père* (Father), (Lat.) *Papa* (Pope), *Pater* (Father), pontifex (a bishop), *populus* (people)

p, penny, pennies, pence (*see* **decimal currency**); (as prefix) pico-

p., page, participle, (meteor.) passing showers, passive, past, per, (Fr.) *passé* (past), *pied* (foot), *pouce* (inch), *pour* (for), (Lat.) *partim* (in part), *per* (through), *pius* (holy), *pondere* (by weight), *post* (after), *primus* (first), *pro* (for)

p- (chem.), *para-*

p, (mech.) pressure; (mus.) piano (softly)

¶ (typ.), the paragraph mark; *see* **reference marks**

PA, Pennsylvania (off. postal abbr.), personal assistant, Press Association, public address, Publishers Association

Pa (phys.), pascal, (chem.) prot-actinium (no point)

Pa., Pennsylvania (off. abbr.)

p.a., per annum (yearly)

Paarl, Cape Province, S. Africa

pabulum, food (not ital.)

PABX, private automatic branch exchange

pace (Lat.), with due respect to (one holding a different view), — ***tua***, by your leave

pacha, *use* **pasha**

pachyderm, a thick-skinned mammal

pachymeter, instrument for measuring small thicknesses, *not* pacho-. *See also* **micrometer**

package/, -s, abbr. **pkg.;** — **holiday,** — **tour** (two words)

pack-drill (hyphen)

packet/, -ed, -ing

pack/-horse, -saddle (hyphens)

packing/-box, -case, -needle, -sheet (hyphens)

packthread (one word)

pad/, -ded, -ding

paddle/-boat, -steamer, -wheel (hyphens)

Paderewski (Ignace Jan), 1860–1941, pianist, first Premier of Polish Rep., 1919

Padishah, a title applied to the Shah of Iran, the Sultan of Turkey, the Great Mogul, and the (British) Emperor of India in Pers. *padshah*

padlock (one word)

padre (colloq.), a chaplain

padre (It., Port., Sp.), father, applied also to a priest

padron/e (It.), a master, employer; *pl.* **-i**

Padua, in It. **Padova**

paduasoy, strong corded silk fabric, *not* the many variants

p. ae., *partes aequales* (equal parts)

paean, a song of triumph. *See also* **paeon, peon**

paedagogy, *use* **ped-**

paederast/, -y, *use* **ped-**

paediatric/ (med.), relating to children (esp. their diseases); **-s** the science; **-ian**

paedo/baptism, -philia, *not* **ped-** (one word)

paella, dish of rice and meat or fish

paeon (Gr. and Lat. prosody), a foot of one long and three short syllables. *See also* **paean, peon**

paeony, *use* **peony**

Paesiello (Giovanni), 1740–1816, Italian composer of opera.

Paganini (Niccolò), 1782–1840, Italian violinist and composer

paganize, *not* -ise

page (typ.), one side of a leaf; type, film, etc., made up for

printing on this; abbr. **p.**, *pl.* **pp.**
See also **pagination**

paginate, to number pages consecutively

pagination (typ.), the numbering of the pages of a book, journal, etc.; may be in headline or at foot of page; generally omitted on opening pages of chapters, main sections, etc. *See also* **preliminary matter**

paging, *not* page-; (comp.), method of presenting text for editing on VDU in static full-screen units. *See also* **scrolling**

Pagliacci (*I*), opera by Leoncavallo, 1892

Pahang, *see* **Malaya (Federation of)**

Pahlanpur, Rajasthan, India, *use* **Pal-**

Pahlavi, language of Persia under Sassanians, *not* Pehlevi

paid, abbr. **pd.**

paillasse (Fr. f.), a straw-mattress, in Eng. **palliasse**

puilles/ (Fr. cook. f.), straws; — *de parmesan,* cheese-straws

Pain (Barry Eric Odell), 1864–1928, English humorous writer. *See also* **Paine, Payn, Payne**

pain (Fr. m.), bread

Paine (Thomas), 1737–1809, English-born political philosopher and American patriot, author of *The Rights of Man. See also* **Pain, Payn, Payne**

painim, a pagan, *use* **pay-**

paint., painting

paintings (titles of), when cited, to be in italic

paintwork (one word)

pair/, -s, abbr. **pr.**

pais/a, coin of India, Pakistan, and Bangladesh, *pl.* **-e**

Paisano/ (Mex.), nickname for a Spaniard; *pl.* **-s**

pajamas, US form of **pyjamas**

Pak., Pakistan, **-i**

pakeha (Maori), a white man (not ital.)

Pakistan/, independent rep. 1956; left Commonwealth 1972; adj. **-i**; abbr. **Pak.** *See also* **Bangladesh**

Pal., Palestine

Palaearctic (zool.), of northern Old World, *not* **Palaeoarctic**

palaeo-, prefix (= ancient), *not* -eo-

palaeography, study of anc. writing; abbr. **palaeog.**

palaeolithic (archaeol.) (not cap.)

palaeology, study of antiquities, *not* paleo-

palaeontology, study of fossils; abbr. **palaeont.**; (typ.) genera, species, and varieties to be italic, other divisions roman. *See also* **botany, zoology**

Palaeozoic (geol.), *not* Paleo- (cap.)

palaestra, a Gr. or Rom. wrestling-school, *not* pale-

palais (Fr. m.), palace

Palanpur, Rajasthan, India, *not* Pahl-

palatable, *not* -eable

palate, roof of the mouth, sense of taste. *See also* **palette, pallet**

palazz/o (It.), palace; *pl.* **-i**

pal/e, -ish

paleo-, prefix, *use* **palaeo-**

Palestin/e, abbr. **Pal.**; adj. **-ian**

palestra, *use* **palaeastra**

paletot), an overcoat (no accent, not ital.)

palette/, artist's thin portable board for colour-mixing; **-knife** (hyphen). *See also* **palate, pallet**

Palgrave (Francis Turner), 1824–97, English anthologist

Pali, language of Buddhist scriptures

palindrom/e, word or phrase reading the same backwards as forwards; adj. **-ic**

Palladian, characterized by wisdom or learning, after **Pallas,** epithet of Gr. goddess Athena; also a Renaissance modification of the classic

Palladian (*cont.*)
Roman style of architecture, from **Andrea Palladio**, 1518–80, Italian architect

palladium, image of Pallas Athena; also metallic element, symbol **Pd**

pallet, mattress, projection on a machine or clock, valve in an organ, platform for carrying loads. *See also* **palate, palette**

palliasse, a straw-mattress; in Fr. f. *paillasse*

palliat/e, to alleviate, minimize; **-ive** (adj. *and* noun)

Pall Mall, London street (two words)

pallor, paleness, *not* -our

Pallottine Fathers, RC society of priests

Palmers Green, London (no apos.)

palmetto/, small palm-tree, *pl.* **-s**

palm/-honey, -oil (hyphens)

Palm Sunday, one before Easter (two words, caps.)

Palomar (**Mount**) (observatory, telescope), Calif., US

palp/us, insect's feeler; *pl.* **-i**

palsgrav/e, count palatinate; *fem.* **-ine**

pam., pamphlet

Pamir, tableland in Cent. Asia, *not* -irs

pampas-grass (hyphen)

Pan., Panama

panacea, a cure-all (not ital.)

Pan/-African, -American (but **Pan American Airways, — — Union**), **-Anglican** (hyphens, caps.)

panakin, *use* **pannikin**

Panama/, Cent. America; abbr. **Pan.**; adj. **-nian**

Pandean pipes, *not* -aean (cap.)

pandect, a treatise covering the whole of a subject; (pl., cap.) the digest of Roman law made under the Emperor Justinian in the sixth century

pandemonium, utter confusion, *not* pandae-

Pandit (**Mrs Vijaya Lakshmi**), b. 1900, Indian diplomat

pandit, *use* **pundit,** *but* **Pandit** (cap.) as Indian title

P. & O., Peninsular and Oriental Steamship Company

p. & p., postage and packing

panegyr/ic, a formal speech or essay of praise; **-ical, -ist, -ize** *not* -ise

panel/, -led, -ling

panem et circenses (Lat.), bread and circus games

paner (Fr. cook.), to dress with egg and breadcrumbs

Panhellen/ic, -ism (one word, cap.)

panic/, -ked, -ky

panikin, *use* **pannikin**

Panislam/, -ic (one word, cap.)

Panizzi (**Sir Antonio**), 1797–1879, Italian-born English librarian

Panjab, *use* **Punjab**

panjandrum, a mock title

pannikin, a little pan, *not* pana-, pani-, -can

pan-pipe, *not* Pan's — (hyphen)

Panslav/ic, -ism, *not* Panscl-

pantagraph, *use* **panto-**

Pantaloon, the lean old man of Italian comedy

panta rhei (Gr.), all things are in a state of flux

pantheon, temple to all gods, (cap.) circular one in Rome

Panthéon, Paris

pantihose, women's tights (one word)

panto/, colloq. for pantomime, *pl.* **-s**

pantograph, instrument for copying to scale, framework for transmitting overhead current to electric vehicle, *not* panta-, penta-

panzer/ (Ger.), armoured; **— division** (two words)

Papa (Lat.), Pope; abbr. **P.**

papal/, -ly

Papal States (the) (hist.) (caps.)

Papandreou (Andreas Georgios), b. 1919, Greek economist and politician; — **(Georgios)**, 1888–1968, his father, Greek statesman

papaw, a small N. American deciduous tree, or its edible fruit; a tropical Amer. evergreen tree, or its edible fruit; *not* pawpaw, papaya

papaya, *use* **papaw**

paperback, a type of book (one word)

paper sizes, *see* **crown, demy, DIN, elephant, foolscap, imperial, medium, post, pott, royal, small royal.** *See also* **book sizes**

papet/ier (Fr.), *fem.* **-ière,** a stationer

Papier (Ger. n.), paper (cap.)

papier (Fr. m.), paper (not cap.)

papier mâché, moulded paper pulp (two words, not ital.)

papill/a, a small protuberance on a living surface, *pl.* **-ae; -ary, -ate, -ose**

papillon (Fr. m.), butterfly

papoose, N. American Indian infant, *not* papp-

Papst (Ger. m.), Pope (cap.)

Papua New Guinea/, indep. 1975, adj. **-n**; abbr. **PNG**

papyr/us, anc. writing-material or MS written on it, *pl.* **-i; -ology**, the study of papyri, **-ologist**

par, equality (of exchange etc.) (no point). *See also* **parr**

par., paragraph, parallel, parish

par (Fr.), by, out of, in, through

Pará, Brazil, *now* **Belém**

para- (chem. prefix) (ital.)

Para., Paraguay

para(s)., paragraph(s)

paradisaical, *not* -iacal

paraffin, *not* -fine

paragon (typ.), name for a former size of type, about 20 pt.

paragraph, a distinct section of type-matter on page; (typ.) in conversation, one for each fresh speaker or interruption. First line usually indented one em. Last line should have more than five letters. Abbr. **par., para.,** *pl.* **pars., paras.,** sign ¶

Paraguay/, S. Amer.; adj. **-an;** abbr. **Para.**

parakeet, *not* paraquet, -oquet, -okeet, parr-

paralipomena, OT books of Chronicles, *not* -leip-

paralips/is, drawing attention to a subject by affecting not to mention it, *pl.* **-es**, *not* -leipsis, -lepsis, -lepsy

parallel/, abbr. **par.; -ed, -ing**

parallelepiped, a solid figure bounded by six parallelograms, *not* -ipiped, -opiped

parallel mark (‖) (typ.), *see* **reference marks**

paralogize, to reason falsely, *not* -ise

paralyse, *not* -ise, -ize, -yze

Paramatta, *use* **Parra-**

Paramecium (zool.), not -aecium, -oecium

parameter (math.), variable factor constant in a particular case; avoid in gen. use

paranoi/a, insanity characterized by delusions, *not* -noea; **-ac, -d**

paraph, the flourish at the end of a signature

paraphernalia (*pl.*), miscellaneous personal belongings

paraquat, a herbicide

paraquet, *use* **parakeet**

parasitize, *not* -ise

paratroops, airborne parachute troops

parbleu! (Fr. colloq.), an exclamation of surprise

parcel/, -led, -ling

parcel post (two words)

parcen/ary (law), joint heirship; **-er**, joint heir

parchment/, inner split sheepskin prepared for writing; — -

parchment (*cont.*)
-paper, imitation parchment (hyphen)
parcimony etc., *use* **parsi-**
parenthes/is, *pl.* **-es,** abbr. **parens.,** the upright curves (). *See also* **brackets, punctuation** X
parenthesize, to insert as a parenthesis, *not* -ise
parerg/on, a subsidiary work; *pl.* **-a** (not ital.)
par/ excellence (Fr.), pre-eminently; — *exemple,* for example, abbr. *p. ex.*
pargana (Ind.), a parish, not *pergunnah*
par hasard (Fr.), by chance, not — *haz-*
parheli/on, a mock sun; *pl.* **-a**
pariah (Ind.), one of low or no caste; a social outcast
pari mutuel (Fr.), a totalizator
pari passu (Lat.), at the same rate
parish/, abbr. **par.;** — **priest,** abbr. **PP**
Parisian, of Paris (cap.); *Parisienne* (Fr. f.), a woman of Paris (cap.)
park, abbr. **P** *or* **pk.**
Park (Mungo), 1771–1806, Scottish traveller in Africa
parka, Eskimo hooded jacket
Parkinson's/ disease, shaking palsy, studied by James Parkinson (1755–1824); — **law,** the law, facetiously expounded by C. N. Parkinson, that work expands to fill the time available for it
Parliament/ (cap.), abbr. **Parl.;** — **House,** Edin., the Scottish Law Courts. *See also* **Assemblies**
Parmesan, a hard cheese made at Parma and elsewhere
Parnass/us (Mount), Greece, sacred to the Muses; adj. **-ian,** poetic
parochialize, *not* -ise
parokeet, *use* **parakeet**
parol (law), oral, not written, *not* -le

parol/e, prisoner's promise not to escape, (mil.) a word of honour; as verb, partic. **-ing;** **-ee,** one paroled
paronomasia, word-play
paroquet, *use* **parakeet**
paroxysm, a fit of pain, passion, laughter
paroxytone (Gr. gram.), (word) with acute accent on last syllable but one
parquet/, wooden flooring; as verb, past and partic. **-ed; -ry;** in Fr. f. *parqueterie*
Parr (Catherine), 1512–48, last wife of Henry VIII; **(Thomas, 'Old'),** ?1483–1635, English centenarian
parr, a young salmon, *not* par
parrakeet, *use* **parakeet**
Parramatta, NSW, *not* -mata, Para-
parramatta, light dress fabric (not cap.)
Parratt (Sir Walter), 1841–1924, English organist
parricid/e, murder(er) of a near relative or of a revered person; adj. **-al.** *See also* **patri-**
Parrish's chemical food
parroquet, *use* **parakeet**
pars., paragraphs
parsec (astr.), a unit of distance, 3.26 light-years, abbr. **pc**
Parsee/, a descendant of the Zoroastrians who fled from Persia to India in the eighth century; *pl.* **-s;** *not* -si
Parsifal, opera by Wagner, 1879
parsimon/y, meanness, adj. **-ious,** *not* parci-
parsnip (hort.), *not* -ep
part, abbr. **pt.**
part., participle
partable (law), *use* **-ible**
parterre, a flower-bed, or garden (not ital.); also area in theatre between orchestra and audience
partes aequales (Lat.), equal parts; abbr. *p. ae.*
part-exchange (noun *and* verb, hyphen)

Parthenon, temple of Athena on Acropolis at Athens

Parthian shot, etc., glance or remark made when turning away (one cap.)

parti (Fr. m.), party (faction), match (marriage), resolution (good or bad)

partible (law), that must be divided, *not* -able

participator, *not* -er

participle, a verbal adjective; abbr. **p.** *or* **part.**

particoloured, variegated, *not* party- (one word)

particularize, *not* -ise

partie/ (Fr. f.), part; **— carrée,** a party of two men and two women

partim (Lat.), in part; abbr. **p.**

parti pris (Fr. m.), foregone conclusion, prejudice

partisan, an adherent of a party (freq. derog.), (mil.) a member of a resistance movement, *not* -zan

partout (Fr.), everywhere

part/-owner, -song (hyphens)

part time (hyphen when attrib.)

party (Conservative, Labour, Liberal, etc.) (not cap. p)

party-coloured, *use* **partic-**

party line (two words)

party-wall (hyphen)

parvenu/, *fem.* **-e,** *pl.* **-s, -es,** an upstart (not ital.)

Pasadena, Calif., US

pascal (phys.), SI unit of pressure, 1 N/m²; abbr. **Pa**

Pas-de-Calais, dép. N. France (hyphens); *Pas de Calais* (Fr. m.), Strait of Dover (three words)

pas de/ chat (Fr. m.), special leap in ballet; **— — deux,** dance for two; **— — quatre,** ditto four; **— — trois,** ditto three

pasha/ (Turk.), a title placed *after* the name, *not* -cha; in Fr. m. *pacha*; **-lic,** pasha's province

Pashto, language of the Pathans, *not* -u, Push-

paso doble (Sp.), ballroom dance in march style

pass., passive

passable, that may be passed. *See also* **passible**

passacaglia (mus.), instrumental piece based on old dance

passbook (one word)

Passchendaele, Belgium, battle (1917)

pass/é (Fr.), *fem.* **-ée,** past, faded; abbr. **p.**

passed master, *use* **past —**

passementerie (Fr. f.), embroidery, *not* passi-

passe-partout, master-key, (picture-frame using) adhesive tape (not ital.)

pas seul (Fr. m.), dance for one person

Passfield, *see* **Webb**

passible, capable of feeling. *See also* **passable**

passim (Lat.), here and there throughout

Passion Week, follows Passion Sunday (fifth in Lent)

passive, abbr. **p.** *or* **pass.**

passkey (one word)

pass-mark (hyphen)

password (one word)

past, abbr. **p.**

pastel/, artist's crayon; **-list.** *See also* **pastille**

Pasternak (Boris Leonidovich), 1890–1960, Russian novelist and poet

pasteurize, to sterilize, *not* -ise

pastiche, medley

pastille, confection, odorizer, *not* -il. *See also* **pastel**

past master, former master in guild, expert, *not* passed —

pastoral/e (mus.), a pastoral composition; *pl.* **-es**

pat., patent, -ed

Pata., Patagonia

patchouli, perfume from Indian plant, *not* -ly

patchwork (one word)

pâté/, paste of meat, fish, etc.

pâté (*cont.*)
(not ital.); — **de foie gras**, a spiced paste of goose-liver

patell/a (anat.), kneecap; *pl.* **-ae**; adj. **-ar**

paten, dish used at Eucharist, circular metal plate, *not* -in, -ine

Patent Office, abbr. **Pat. Off.**

Pater (Lat.), Father; abbr. **P.**

paterfamilias, father of a family

paternoster, the Lord's Prayer (esp. in Lat.), bead in rosary, series (esp. moving compartments)

pater patriae (Lat.), father of his country

Paterson (William), 1658–1719, British founder of the Bank of England

Paterson, NJ, US

path., pathology/y, -ical

Pathétique, name of sonata by Beethoven and symphony by Tchaikovsky

Patiala, E. Punjab, India

patin/, -e, *use* **paten**

patio/, inner court, paved area of garden, *pl.* **-s**

pâtiss/erie (Fr. f.), pastry; *-ier,* *fem.* *-ière,* pastry-cook

Patna, rice (cap.)

patois, a dialect of the common people, jargon, *pl. same* (not ital.)

Patres/ (Lat.), fathers, abbr. **PP;** — *Conscripti,* Conscript Fathers, abbr. **PP C**

patria potestas (Rom. law), father's power over his family

patricide, murder(er) of one's father. *See also* **parricide**

patrol/, -led, -ling

patronize, *not* -ise

Pattenmakers Company, *not* Pattern-

Pattison (Mark), 1813–84, Rector of Lincoln Coll., Oxford

Patton (George Smith), 1885–1945, US general

PAU, Pan American Union

pauca verba (Lat.), few words

pauperize, *not* -ise

Pausanias (*fl. c.* AD 150), Greek traveller, author of 'Description of Greece', abbr. **Paus.**

pavan, stately dance, *not* -ane; in Fr. f. *pavane*

pavé (Fr. m.), pavement, jewellery setting with stones close together

pavilion, *not* pavill-

pavior, one who lays pavements, *not* -ver, -vier, -viour; **Paviors Company**

Pavlova (Anna), 1885–1931, Russian ballet-dancer

pawaw, *use* **powwow**

pawn (chess), abbr. **P**

pawpaw, *use* **papaw**

PAX, private automatic exchange

pax vobiscum! (Lat.), peace be with you!

paxwax, the neck cartilages (*not* the many variants)

pay/-bed, -claim, -day (hyphens)

PAYE, pay-as-you-earn, method of tax-collection

payload (one word)

paymaster, abbr. **paymr.** *or* **PM** (one word)

Paymaster-General (caps., hyphen); abbr. **PMG**

payment, abbr. **pt.**

Payn (James), 1830–98, English novelist and editor. *See also* **Pain, Paine, Payne**

Payne (Edward John), 1844–1904, English historian; — **(John Howard),** 1791–1852, US playwright, wrote 'Home, Sweet Home'. *See also* **Pain, Paine, Payn**

paynim, a pagan, *not* pai-

pay-off (noun, hyphen)

pay-packet (hyphen)

pay phone (two words)

payroll (one word)

paysage (Fr. m.), landscape (painting) (ital.); **paysagist,** landscape-painter (not ital.)

Pays-Bas (Fr. m. pl.), the Netherlands (caps., hyphen)

ᵖB, *Pharmacopoeia Britannica*, Plymouth Brethren, Prayer Book

ᵖb, *plumbum* (lead) (no point)

ᵖBI (mil. slang), (poor bloody) infantry(man)

ᵖBX, private branch exchange

ᵖC, Panama Canal, Parish Council, -lor, Police Constable (q.v.), Privy Council, Counsellor

ᵖc (astron.), parsec

ᵖ/c, petty cash, prices current

ᵖ.c., per cent, police constable, postcard

ᵖCC, Parochial Church Council

ᵖCS (Sc.), Principal Clerk of Session

ᵖD, *Pharmacopoeia Dublinensis*, Postal District (London), *privat-docent*

ᵖd, palladium (no point)

ᵖd., paid

ᵖ.d., per diem, (elec.) potential difference

ᵖ.d.q., pretty damn quick

ᵖDSA, People's Dispensary for Sick Animals

ᵖE, *Pharmacopoeia Edinburgensis*, physical education, Port Elizabeth (Cape Province), potential energy, Protestant Episcopal

ᵖ.e., personal estate

ᵖ/e, price/earnings ratio

ᵖEA, Physical Education Association of Great Britain and Northern Ireland

peaceable, *not* -cable

peace/maker, -time (one word)

peace/-offering, -pipe (hyphens)

pea-chick, young peafowl (hyphen)

peach Melba, ice-cream and peaches with liqueur, *not* pêche —

pea/cock, -fowl, -hen (one word)

pea-green (hyphen)

pea-jacket, sailor's overcoat (hyphen)

pearl (in knitting), *use* **purl**

pearl (typ.), name for a former size of type, about 5 pt.

Pearl Harbor, US naval base, Hawaii, *not* — Harbour

Pears (Sir Peter), b. 1910, English tenor

Peary (Robert Edwin), 1856–1920, US Arctic explorer, first at N. Pole, 1909

pease-pudding (hyphen)

peat/bog, -moss, -reek (one word)

peau-de-soie, smooth silky fabric (not ital.)

pebbl/e, -y

p.e.c., photoelectric cell

peccadillo/, a trifling offence; *pl.* **-es**

peccary, S. American mammal, *not* -i

peccavi, confession of guilt (not ital.)

pêche (Fr. f.), fish/ery, -ing; peach; **pêche Melba,** *use* **peach Melba**

peckoe, peco, *use* **pekoe**

pecorino/, It. cheese made from ewes' milk, *pl.* **-s**

peculat/e, -or, embezzl/e, -er

Ped. (mus.), pedal

pedagog/ue, schoolmaster (arch. and derog.); **-y, -ics,** science of teaching, *not* pae-

pedal/, -led, -ling

pederast/, -y, *not* paed-

pedlar, travelling vendor of small wares, *not* -er

pedo/baptism, -philia, *use* **paedo-**

Peeblesshire, former Sc. county, abbr. **Peebles.**

peel (hist.), small square tower, *not* pele

Peele (George), ?1558–?97, English playwright

peep/-hole, -show (hyphens)

peeping Tom, voyeur (one cap.)

peer group (two words)

peewit, the lapwing, *not* pewit

Peggotty, family in Dickens's *David Copperfield*

Pehlevi, *use* **Pahlavi**

PEI, Prince Edward Island, Canada

peignoir (Fr. m.), a woman's loose dressing-gown

peine forte et dure (Fr. f.), severe punishment, a medieval judicial torture

Peirce (**Charles Sanders**), 1839–1914, US mathematician, founder of philosophical pragmatism

Peking/, China, *not* Peip-, -kin; in Pinyin **Beijing; -ese,** (inhabitant) of Peking; **Pekinese,** a small dog, colloq. **peke**

pekoe, a black tea, *not* peckoe, peco (not cap.)

pele, *use* **peel**

pell-mell, confusedly

Peloponnese, the modern Morea, S. Greece

pemmican, dried meat for travellers, *not* pemi-

Pen., Peninsula

penalize, *not* -ise

pen-and-ink (adj., hyphens)

Penang, *see* **Malaya** (**Federation of**)

penates (*pl.*), Roman household gods. *See also* **lar**

pence, *see* **penny**

penchant, bias (not ital.)

pencil/, -led, -ling

PEN Club, an international association of writers (= *P*oets, *P*laywrights, *E*ditors, *E*ssayists, *N*ovelists)

pendant, anything hanging

pendent (adj.), suspended

pendente lite (Lat.), during the trial

pendulum/, *pl.* **-s**

penetrable, that may be penetrated

penetralia (*pl.*), innermost recesses (not ital.)

pen-feather, quill-feather. *See also* **pin-feather**

pen-friend (hyphen)

penguin, a bird. *See also* **pinguin**

penholder (one word)

penicillin (med.)

Penicillium (bot.), a genus of fungi, mould (ital.)

Penicuik, Lothian

peninsula/ (noun), *pl.* **-s; adj. -r**

Peninsular/ Campaign, SE Virginia, 1862, in American Civil War; — **War,** Spain and Portugal, 1808–14, in Napoleonic Wars

penis/, the male organ, *pl.* **-es**

penknife (one word)

Penmaenmawr, Gwynedd

penman/, -ship (one word)

pen-name (hyphen)

pennant (naut.), a piece of rigging, a flag

pennon (mil.), a long narrow flag

penn'orth (colloq.), a pennyworth

Pennsylvania, off. abbr. **Pa.** or (postal) **PA,** *not* Penn., Penna.

penny, *pl.* **pennies** (number of coins), **pence** (sum of money) abbr. *s.* and *pl.* **d.**; **new penny** *pl.* **new pence** (*not* pennies), abbr. **p** (no point). *See also* **decimal currency**

penniless, *not* penny-

penny-a-liner (hyphens)

pennyroyal (bot.), a kind of mint (one word)

pennyweight, 24 grains (approx. 1.55 g.); abbr. **dwt.**

pennyworth (one word)

penology, science of punishment, *not* poe-

pen-pal (hyphen)

Penrhyn, Gwynedd, also **Baron —**

Penryn, Cornwall

pensée (Fr. f.), thought, maxim; *also* pansy

pension/ (Fr. f.), a boarding--house, -school; *en —,* on boarding terms; *-nat* (m.), a boarding-school

penstemon, *use* **pentstemon**

Pent., Pentecost

pentagon/, figure or building with five sides; (cap.) headquarters of US defence forces; adj. **-al**

pentagram, five-pointed star

pentagraph, *use* **panto-**

entameter (prosody), verse of five units

entateuch/, first five books of the OT; -al (cap.)

entecost, Whit Sunday; abbr. Pent.

entecostal (not cap.)

enthouse (one word)

entstemon (bot.), *not* pens-

enumbr/a (astr.), lighter shadow round dark shadow of an eclipse; *pl.* -ae

eon, a servant. *See also* paean, paeon

eony, a flower, *not* pae-

EP, Political and Economic Planning

epo/, fleshy fruit of melon, etc., *pl.* -s

epsin, an enzyme in gastric juice, *not* -ine

epys (Samuel), 1633–1703, diarist and civil servant

er., period

er (Lat.), by, for; abbr. p.

*ERA, Production Engineering Research Association of Great Britain

erai, *use* piranha

erak, *see* Malaya (Federation of)

er annum, yearly, abbr. p.a. *or* per ann.

er caput, for each person, preferable to per capita

erceiv/e, -able, -er

er cent, for each hundred (two words, no point), symbol %

ercentage (one word)

erceptible, *not* -able

erceval, one of King Arthur's knights; — (Spencer), 1762–1812, English statesman

erch, (rod or pole) 5.029 m., (area) 25.29 sq. m.; do not abbreviate

erchance, perhaps

ercolat/e, -ing, -or

er/ contra (Lat.), on the other hand; — *curiam,* by the court

erdendosi (mus.), dying away

er diem, daily; as noun, daily allowance

perdu/ (Fr.), *fem.* -e, concealed, lost

Père (Fr. m.), RCC father; abbr. P.

père (Fr. m.), father, as Dumas père (not ital.)

Père Lachaise, Paris cemetery (two words, caps.)

perf., perfect, (stamps) perforated

perfect, abbr. perf.

perfect binding, unsewn binding, q.v.

perfecter, one who perfects, *not* -or; (typ.), a printing press which prints both sides of the paper at one pass

perfectib/le, -ility, *not* -ab-

perfecting (typ.), printing the second side of a sheet

perforated, abbr. perf.

perforce, of necessity (one word)

Pergamon Press, Ltd., publishers

Pergamum, city and kingdom in anc. Asia Minor

Pergolesi (Giovanni Batista), 1710–36, Italian composer

pergunnah (Ind.), a parish, *use* pargana

periagua, *use* piragua

peridot, gem, *not* -te

perig/ee (astr.), abbr. perig.; adj. -ean

Périgord (Fr. cook.), cooking based on truffles

perimeter, circumference

per incuriam (Lat.), by oversight

perine/um (anat.), *pl.* -ums; adj. -al

per interim (Lat.), in the mean time

period, abbr. per.; in typ. called the full point or point; *see* punctuation V

periodicals (titles of), when cited, to be italic; as a rule, the definite article should be in roman lower case, except in *The Economist* and *The Times*

peripatetic, walking about; (cap.) Aristotelian (school of philosophy)

peripeteia, sudden reversal of fortune, *not* -tia (not ital.)

periphras/is, circumlocution, *pl.* **-es;** adj. **-tic**

perispomen/on (Gr. gram.), (word) with circumflex on last syllable, *pl.* **-a**

peristyle (archit.), row of columns round temple

periton/eum (anat.), *pl.* **-eums;** **-eal,** **-itis**

perityphlitis (path.)

periwig, **-ged,** *not* perri-

periwinkle, plant, and mollusc

Perlis, *see* **Malaya (Federation of)**

perm, colloq. for permanent wave, permutation, and corresponding verbs (no point)

permeable, that may be permeated

per/ mensem, for each month; — **mille,** for each thousand, symbol ‰

permis de séjour (Fr. m.), residence permit

permissible, *not* -able

permit/, **-ted,** **-ting**

Perón (Juan Domingo), 1895–1974, pres. of Argentina 1946–55, 1973–4

per pais (Norman Fr.), by jury (= by the county)

perpetuum mobile (Lat.), something never at rest

per/ procurationem (Lat.), by agency of, *now often* on behalf of, *abbr.* **per pro.** *or* **p.p.**

perriwig, *use* peri-

Pers., Persia, -n

pers., person, -al, -ally

per/ saltum (Lat.), at a leap; — *se,* by himself, herself, itself, *or* themselves

Perse, *use* **Leger**

Persia/, **-n,** abbr. **Pers.;** *but use* **Iran/,** **-ian,** of the modern state

Persian, usually printed in Arabic alphabet with slight modifications

Persian/ carpet, — **cat,** — **ru** (caps.), *but* **persian morocc** (bind.) (no caps.)

persiflage, banter (not ital.)

persil (Fr. m.), parsley

persimmon (bot.), the date-plum, *not* -simon

persist/ence (in Fr. f. *persist ance*), **-ency,** **-ent**

persona/, perceived characteristics of personality, *pl.* **-e** (not ital.)

persona/ grata (Lat.), an acceptable person; — *gratissima,* a most acceptable person — *ingrata,* — *non grata,* unacceptable person; *pls.* **-ae** **-ae**

personalize, *not* -ise

personalty (law), personal estate, *not* -ality

personnel, staff of persons employed in any service, member of armed forces; — **carrier,** — **manager** (two words)

persp., perspective

Perspex, propr. term for a transparent plastic (cap.)

perspicac/ity, clearness of understanding; adj. **-ious**

perspicu/ity, clearness of statement; adj. **-ous**

per stirpes (Lat.), by the number of families

persuadable, *use* with ref. to a particular instance, in gen. *use* next

persuas/ible, open to persuasion; **-ive,** able to persuade

PERT, programme evaluation and review technique

per thousand, symbol ‰

pertinac/ity, persistence; adj. **-ious**

peruke, a wig, *not* -que

Peruv., Peruvian

per viam (Lat.), by way of

pes (Lat.), a foot; *pl.* **pedes**

peseta, Spanish unit of currency abbr. **pta.**

Peshawar, W. Pakistan, *not* -u

peso/, unit of currency in some Latin-American countries, *pl.* -

Pestalozzi (**Johann Heinrich**), 1746–1827, Swiss educationist

Pesth, *see* **Budapest**

Pet. (**1, 2**), First, Second Epistle of Peter (NT)

peta-, prefix meaning 10^{15}, abbr. **P**

Pétain (**Henri Philippe**), 1856–1951, French Marshal

petal/, -led

peterel, *use* **petrel**

Peterlee, Co. Durham, 'new town', 1948

Peter Schlemihl, a well-meaning unlucky fellow (title of a novel by Chamisso, 1814)

Peter's pence (hist.), an annual tax of one penny per household paid to the papal see, discontinued in England under Henry VIII; from 1860 a world-wide voluntary contribution to the papal treasury on St. Peter's Day

petit/ (Fr.), *fem.* **-e,** small; — **bourgeois,** member of lower middle class; — **déjeuner,** breakfast; — **four,** a small fancy cake (*pl.* **-s fours**)

petitio principii (Lat.), begging the question

petit/ mal (Fr. m.), mild form of epilepsy; — **point,** embroidery in small stitches

petits/ pois (Fr. m. pl.), green peas; — **soins,** little attentions

petit verre (Fr. m.), a glass of liqueur

Petrarc/h (**Francesco**), 1304–74, Italian scholar and poet, in It. **-a**; adj. **-an**

Petre (**Baron**)

petrel, a bird, *not* -erel

Petriburg:, sig. of Bp. of Peterborough (colon)

Petrograd, name given to St. Petersburg during the First World War; *now* **Leningrad**

petrol, refined petroleum used as fuel

petroleum, unrefined oil

petrology, abbr. **petrol.**

Pettie (**John**), 1839–93, Scottish painter

pettifog/, to cavil in legal matters; **-ger, -gery, -ging**

petty/ bourgeois; — **cash,** abbr. **p/c;** — **officer,** abbr. **PO** (two words)

peu à peu (Fr.), little by little

Peugeot, Fr. make of car

peut-être (Fr.), perhaps

pewit, *use* **peewit**

p. ex. (Fr.), *par exemple* (for example)

PF, Patriotic Front, Procurator-fiscal

pF, picofarad (no point)

Pf. (Ger.), *Pfennig*

pf (mus.), più forte (a little louder) *or* piano forte (soft, then loud)

Pfd. (Ger.), *Pfund* (pound)

pfennig, a small German coin, $\frac{1}{100}$ of a mark; Ger. m., cap., *Pfennig*

Pfitzner (**Hans**), 1869–1949, German composer

Pfleiderer (**Edmund**), 1842–1902, and his brother — (**Otto**), 1839–1908, German philosophers

Pfund (Ger. n.), pound (cap.); abbr. **Pfd**

PG, paying guest

pg, picogram (no point)

PGA, Professional Golfers' Association

PGM (Freemasonry, Odd Fellows), Past Grand Master

pH, measure of hydrogen ion concentration

ph, phot

Phaedon, b. *c.*417 BC, pupil of Socrates

phaenomenon etc., *use* **phenomenon**

Phaethon (Gr. myth.), son of Helios, *not* Phaë-, -ton

phaeton, a carriage

phalan/x, line of battle, a compact body of men; *pl.* **-xes;** (biol.) a bone or stamen-bundle, *pl.* **-ges**

phall/us, *pl.* **-uses,** (bot., med., relig.) **-i**

phanariot, one of Greek official class in Constantinople under Ottoman rule

phantasize, use **fan-**

phantasmagor/ia, a shifting scene of real or imagined figures, *pl.* **-ias**; adj. **-ic,** *not* fa-

phantasy, use **fan-**

phantom, *not* f-

Phar., Pharmacopoeia

Pharao/h, *not* -oah; adj. **-nic**

Pharis/ee, one of an anc. Jewish sect, strict observers of religious forms, hence a self-righteous person, *pl.* **-ees**; **-aism,** adjs. **-aic, -aical**

pharm., pharmaceutical, pharmacy

pharmacol., pharmacology

Pharmacopoeia, a book describing drugs, abbr. **P.** (*see* BP) *or* **Phar.**; *Pharmacopoeia/ Dublinensis* (of Dublin), abbr. **PD**; — *Edinburgensis* (of Edinburgh), abbr. **PE**; — *Londiniensis* (of London), abbr. **PL**

pharos, lighthouse, (cap.) the one at Alexandria or the island on which it stood

pharyn/x, the cavity behind the larynx; *pl.* **-ges**; **-gal, -geal, -gitis**

phas/e, -ic, -ing

Ph.B., *Philosophiae Baccalaureus* (Bachelor of Philosophy)

Ph.D., *Philosophiae Doctor* (Doctor of Philosophy)

Phebe, *see* **Phoebe**

Phèdre, by Racine, 1677

Pheidias, use **Phidias**

Phenician, use **Phoe-**

phenix, use **phoe-**

phenomen/on, an appearance; *pl.* **-a**; adj. **-al,** *not* phae-

Phidias, 5th c. BC, Athenian sculptor, *not* Pheid-

Phil., Philadelphia, Philharmonic, Philippine, Epistle to Philippians (NT)

philabeg, use **filibeg**

Philadelphia, abbr. **Phil.** *or* **Phila.**

philatel/y, stamp-collecting; **-ic, -ically, -ist**

Philem., Philemon (NT)

philemot, use **fi-**

philharmonic, fond of music (cap. as part of name of orchestra or society)

philhellen/e, (one) friendly to the Greeks; **-ic, -ism, -ist**

philibeg, use **filibeg**

Philip (George) & Son, map publishers

Philippe, Kings of France

Philippians, Epistle to, abbr. **Phil.**

Philippic, (*pl.*) speeches of Demosthenes against Philip of Macedon; also those of Cicero against Antony; (not cap.) a bitter invective

philippina, a game of forfeits, *not* the many variants

Philippine Islands, abbr. **PI**

Philippines (Republic of the), indep. 1946, inhabited by **Filipinos**

Philipps, family name of Viscount St. Davids and Baron Milford. See also **Phill-**

Philips (Ambrose), 1675–1749, English writer of pastoral and nursery poems, hence 'namby-pamby'; — **(John),** 1676–1709, parodist of Milton; — **(Katherine),** 1631–64, English poet. See also **Phill-**

Philister (Ger.), a townsman, a non-student, *pl. same* (cap.)

Philistin/e, an inhabitant of anc. Palestine; (not cap.) a person indifferent to culture; **philistinism**

phillipina, use **philipp-**

Phillipps (James Orchard), see **Halliwell-Phillipps**; — **(Sir Thomas),** 1792–1872, English book-collector. See also **Philipps**

Phillips (Sir Claud), 1846–1924, English art critic; — **(Edward,** 1630–94, lexicographer, and **John,** 1631–1706), Milton's nephews; —

(**John Bertram**), b. 1906, English Bible translator; — (**Stephen**), 1864–1915, English poet; — (**Wendell**), 1811–84, US abolitionist. *See also* **Philips**

Phillpotts (**Eden**), 1862–1960, English novelist and dramatist; — (**Henry**), 1778–1869, Bp. of Exeter. *See also* **Philpott**

philopoena, *use* **philippina**

philosophers' stone, substance thought to change base metals to gold, *not* 's

philosophize, *not* -ise

philosophy, abbr. **philos.**

Philpott (**Henry**), 1807–92, Bp. of Worcester. *See also* **Phillpotts**

Phil. Soc., Philological Society of London, Philosophical Society of America

Phil. Trans., the *Philosophical Transactions of the Royal Society of London*

philtre, aphrodisiac drink. *See also* **filter**

Phiz, illustrator of Dickens, see **Browne** (**Hablot Knight**)

phiz (arch. *and* colloq.), the face, *not* phizz

phlebit/is, inflammation of the veins; adj. **-ic**

Phnom Penh, Cambodia

phobia/, a morbid fear; *pl.* **-s**

Phoebe (*but* **Phebe** in *As You Like It*)

Phoebus, 'bright', epithet of Apollo

Phoenician, *not* Phe-

phoenix, myth. bird that rose rejuvenated from its own ashes, *not* phen-

phon, unit of loudness

phon., phonetics

phone, short for telephone, *not* 'phone

phone-in, broadcast incorporating listeners' telephoned views (hyphen)

phon/ey, false, *not* phony; **-ier**, **-ily**, **-iness**

phonol., phonology

phosphor/us, symbol **P**; adj. **-ous**

phosphuretted, *not* -eted, -oretted

phot, unit of illuminance, in SI units 10 klx; abbr. **ph**

photo/, photograph (no point); *pl.* **-s**

photo/composition, -setting (one word) (typ.), setting copy on to film etc. instead of in metal type

photocop/y, -ier (one word)

photoelectric (one word)

photo finish (two words)

photofit, Identikit (q.v.) picture using photographs of facial features (one word)

photog., photograph/y, -ic

Photograph/ (Ger. m.), photographer; **-ie** (f.), photograph, -y

photograph/e (Fr. m. and f.), photographer; **-ie** (f.), photograph, -y

photogravure (typ.), intaglio printing process; shortened as **gravure**

photolithography (typ.), the photographic processes for making a plate for printing by offset lithography; colloq. **photolitho**

photom., photometr/y, -ical

photomicrograph, a photograph of a minute object taken with a microscope. *See also* **microphotograph**

photomontage, a montage of photographs

photosetting, see **photocomposition**

phr., phrase

phren., phrenology

phrenetic, *use* **fren-**

phrenitic, affected with **phrenitis**, inflammation of the brain

phthalic (chem.), derived from naphthalene

phthisis, tuberculosis

Phyfe (**Duncan**), 1786–1854, Scottish-born US furniture--maker

phyl/um, a main division of the animal or vegetable kingdom; *pl.* **-a**

phys., physical, physician, physics

physic/, -ked, -king, -ky

physico-chemical (hyphen)

physiol., physiolog/y, -ical, -ist

physique, constitution (not ital.)

physique (Fr. f.), physics (natural philosophy)

PI, Philippine Islands

pi (typ.), *use* **pie**

pi, Gr. Π, π; $\prod$ (math.) product; π, ratio of circumference to diameter of circle, or $3.14159265 \ldots$

pianissimo (mus.), very soft; abbr. *pp*

pianississimo (mus.), as softly as possible; abbr. *ppp*

piano/ (mus.), softly, abbr. *p*; instrument formally called **pianoforte,** *pl.* **-s**; — **-player, -stool** (hyphens)

Pianola, propr. term for automatic piano-player (cap.)

piano nobile (archit.), the main floor of a building (ital.)

piassava, fibre from palm-trees, *not* -aba

piastre, small coin of Middle Eastern countries, *not* -er

piazz/a, an open square, *pl.* **-as**

pibroch, an air on the bagpipe, *not* the bagpipe itself

pica (typ.), size of letters on some typewriters; the standard for typographic measurement, equals 12 pt. or about one-sixth of an inch; also name for a former size of type, about 12 pt.

picaninny, *use* **picca-**

picaresque, of a style of fiction describing the life of an (amiable) rogue (not ital.)

Picasso (Pablo), 1881–1973, Spanish painter

picayune, a small US coin pre- -1857, (colloq.) any trifling coin, person, or thing; (adj.) trifling, mean

piccalilli, a pickle

piccaninny, Negro or Austral. Aboriginal infant, *not* pica-, picka-

piccolo/, the smallest flute; *pl.* **-s**

pi character (comp.), special sort: *see* **pi plaque**

pick (typ.), blob of ink or dirt stuck to sort; proneness of paper surface fibres or coating to become detached on to the offset blanket

pick-a-back, *use* **piggy-back**

pickaxe (one word)

pickelhaube (Ger. hist. f.), infantry helmet

picket/, pointed stake, person stationed at site of strike, (mil.) body of troops watching for enemy, *not* picq-, piq-; as verb, **-ed, -ing**; — **line** (of industrial pickets) (two words)

pick-me-up/, a tonic (hyphens) *pl.* **-s**

pickpocket (one word)

picnic/, -ked, -ker, -king

pico-, prefix meaning 10^{-12}, abbr. **p**

picquet, *use* **picket**

pictures (titles of), when cited, to be in italic

pidgin English, Chinese jargon from *pidgin* (corruption of *business*), (two words, one cap.). *See also* **pigeon**

pi-dog, *use* **pye-dog**

pie (typ.), composed type which has been jumbled, *not* pi, pye

piebald, of two colours in irregular patches, usu. of horses, usu. black and white; *not* pye-. *See also* **skewbald**

pièce de résistance (Fr. f.), the principal dish at a meal, the most remarkable item, *pl.* *pièces* —

piecemeal, one portion at a time (one word)

piece/-rate, -work (hyphens)

pied/ (Fr. m.), a foot, abbr. **p.**; — **-à-terre,** an occasional

residence, *pl.* **pieds-** — - —
(hyphens)

pie-dog, *use* **pye-dog**

Pierce (Franklin), 1804–69,
President of US 1853–7

Pierce the Ploughman's
Crede, anon. about 1394. *See*
also **Piers**

Pierian Spring, the fountain of
the Muses in Thessaly

Pierides, the nine Muses

pierr/ot, seaside entertainer derived from French pantomime;
fem. **-ette**

Piers Plowman (The Vision of
William concerning), by W.
Langland, first ed. 1362. *See also*
Pierce

pietà, representation of dead
Christ in his Mother's arms

pietas (Lat.), respect due to an
ancestor, etc.

Pietermaritzburg, S. Africa

pietr/a dur/a (It.), a stone
mosaic; *pl.* **-e -e**

piezo-electric (hyphen)

pigeon (**not my**), not my affair.
See also **pidgin**

pigeon-hol/e, -ed (hyphen)

piggy-back, *not* pick-a-back

pigheaded (one word)

pig-iron (hyphen)

pigm/y, -aean, *use* **pygm-**

pig/skin, -sty, *pl.* **-sties, -swill,**
-weed (one word)

Pike's Peak, Rocky Mountains,
US (apos.)

pikestaff (one word)

Piketberg (off. spelling), Cape
Province, *not* Piquet-

pilaff, dish of rice with meat or
fish, spices, etc., *not* the many
variants

pillule, *use* **pilule**

pilot/, -ed, -ing

pilot/-fish, -house, -light (hyphens)

Piloty (Karl von), 1826–86,
German painter

Pilsen (Czechoslovakia); in
Czech **Plzeň**

Pilsener, a light beer (cap.)

Piłsudski (Józef), 1867–1935,

Polish general, first President of
Poland, 1918

pilule, a small pill, *not* pill-

pimento/, allspice, sweet pepper, *pl.* **-s,** *not* pimi-

pinakothek (Ger. from Gr.),
picture-gallery

pin-ball (hyphen)

pince-nez, spring eyeglasses, *pl.*
same (not ital., hyphen)

pincushion (one word)

Pindar, 518–438 BC, Greek poet;
abbr. **Pind.**

Pindar (Peter), 1738–1819,
pseud. of Dr J. Wolcot

pineapple (one word)

pine-cone (hyphen)

Pinero (Sir Arthur Wing),
1855–1934, English playwright

pin-feather, small feather. *See*
also **pen-feather**

ping-pong (hyphen)

pinguin, W. Indian plant or
fruit. *See also* **penguin**

pin/-head, -hole, -money,
-point (hyphens)

pinprick (one word)

pinscher, a breed of dog, *not*
pinch-. *See also* **Dobermann**
pinscher

pin-stripe (hyphen)

pint/, -s, abbr. **pt.**, sign ℺

pin-table (hyphen)

pintado/, petrel, mackerel-like
fish, *pl.* **-s**

Pinturicchio, nickname of **Ber-**
nardino di Betto, 1454–1513,
Italian painter

pinxit (Lat.), painted this; abbr.
pnxt. or *pinx.*

Pinyin, *see* **Chinese**

pion (phys.), subatomic particle

Pippa Passes, by R. Browning,
1841

pi plaque (comp.), set of special
sorts called into use when required

pipy, like, or having, pipes, *not*
-ey

piquan/t, sharp, **-cy** (not ital.)

piqu/e, resentment; a score in
piquet; as verb, to irritate, **-ed,**
-ing

piqué, a thick cotton fabric (not ital.)

piqué (Fr.), (of wine) slightly sour; (mus.) short, detached; (cook.) larded

piquet, a card-game, *not* picq-. See also **picket**

Piquetberg, *use* **Piket-**

PIRA, Pira, Research Association for the Paper and Board, Printing and Packaging Industries

Piraeus, port of Athens

piragua, S. American dug-out canoe, *not* the many variants

Pirandello (Luigi), 1867–1936, Italian playwright

Piranesi (Giambattista), 1720–78, Italian architect and etcher

piranha, ferocious S. American fish, *not* perai, piraya

pis aller (Fr. m.), a second best

piscin/a, a fish-pond, (eccl.) stone basin for disposing of water used in washing the chalice, etc.; *pl.* **-ae**

pisé (Fr. m.), rammed earth (ital.)

Pissarro (Camille), 1831–1903, French painter

pistachio/, a nut; *pl.* **-s**; *not* -acho

piste (Fr. f.), ski-track

pitch-and-toss (hyphens)

pitch/blende, -fork, -stone (one word)

Pitman (Sir Isaac), 1813–97, English inventor of shorthand

Pitti (Palazzo), art gallery, Florence

Pittsburgh, Pa., US, *not* -burg

più (It. mus.), more

pius (Lat.), holy; abbr. **p.**

pivot/, -ed, -ing

pix, *use* **pyx**

pix/ie, a small fairy, *not* -y; **-ilated,** bewildered, *not* pixy--led

pixie/ hat, — hood (two words)

Pizarro (Francisco), 1475–1541, Spanish *conquistador*

pizz/a (It. cook.), layer of dough

covered with cheese, ham, olives, etc. and baked, *pl.* **-as; -eria,** place where pizzas are made

pizzicato (mus.), pinched, plucked; abbr. **pizz.**

PJ, presiding judge, Probate Judge

PK, psychokinesis

pk., park, peak

pkg., package, -s

PL, Paymaster Lieutenant, *Pharmacopoeia Londiniensis*, Poet Laureate, Primrose League

P/L, Profit and Loss

Pl., Plate, -s

pl., place, plate, platoon, plural

PLA, Port of London Authority

place (Fr. f.), square in a town

place aux dames! (Fr.), ladies first!

placebo/, opening antiphon of vespers for the dead; (med.) a medicine given to humour the patient; *pl.* **-s**

place-name (hyphen)

placent/a, organ nourishing foetus in womb, *pl.* **-ae**

placet (Lat.), it pleases, permission granted

plafond (Fr. m.), a ceiling, esp. decorated (ital.)

plagiarize, use another's writings as one's own, *not* -ise

plagu/e, -esome, -ily, -y

Plaid Cymru, Welsh nationalist party

plainchant (one word)

plain sailing, (fig.) easy work (two words). See also **plane sailing** (naut.)

plain/sman, -song (one word)

plaintiff, abbr. **plf.**

Plaisterers (Worshipful Company of)

planchet, a coin-blank

planchette, small board on castors and a pencil-point, used to trace letters spontaneously at spiritualistic seances

Planck's/ constant, h (6.6262×10^{-34} J s); **— Law of Radiation,** in the quantum theory, from **Max Planck**, 1858–1947, German physicist

plane, short for aeroplane, *not* 'plane

plane sailing (naut.), calculation of a ship's position on the assumption that it is moving on a plane surface (two words). *See also* **plain sailing**

planetari/um, *pl.* **-ums**

planetary signs, *see* **astronomy**

plankton, microscopic drifting organic life found in water. *See* **benthal** *and* **nekton**

planographic (typ.), of a printing process based on a flat surface, on which the image areas are made greasy so as to accept ink and the rest wet to reject it

Plantagenet, family name of the English sovereigns from 1154 to 1399

Plantin (**Christophe**), *c.*1520–89, French printer

Plasticine, propr. term for modelling substance (cap.)

plasticy, plastic-like, *not* -cky

plastron (Fr. dress. m.), a bodice front

plat (Fr. cook. m.), a dish; — *du jour,* special dish of the day

Plate/, -s, abbr. **Pl.**

plate, (typ.) planographic (q.v.) or typographic (q.v.) printing surface of complete page or sheet cast or etched in metal or polymerized resin; halftone etc. illustration; (photog.) **whole** —, 8½ × 6½ in., **half-**—, 6½ × 4¼ in., **quarter-**—, 4¼ × 3¼ in.

plateau/, an elevated plain; *pl.* **-x** (not ital.)

plate glass (two words)

plate-rack (hyphen)

platinize, to coat with platinum, *not* -ise

platinum, symbol **Pt**

Platon/ic (cap. when in hist. or philos. contexts), *but* **platonic love** (not cap.); **-ism, -ist, -ize,** *not* -ise

Platt-Deutsch, Low German

platypus/, the Australian duck-bill; *pl.* **-es**

platyrrhine, broad-nosed (of monkeys), *not* -yrrhine

plausible, *not* -able

Plautus (Titus Maccius), *c.*254–*c.*184 BC, Roman playwright; abbr. **Plaut.**

play-act/or, -ing (hyphens)

play/bill, -boy, -goer, -ground, -mate (one word)

play/-off (noun), **-pen** (hyphens)

plays (titles of), when cited to be in italic

play/time, -wright (one word)

plaza/ (Sp.), a public square; *pl.* **-s** (not ital.)

PLC, Public Limited Company

pleasur/e, -able

plebeian, commoner in anc. Rome, vulgar, common, *not* -bian

plebiscit/e, a vote of the people (no accent, not ital.); adj. **-ary**

plebiscit/um (Lat.), a law passed by the *plebs*; *pl.* **-a**

plebs (Lat.), the populace

plectr/um, *pl.* **-a**

Pleiad (Gr. myth.), one of the **Pleiades,** the seven daughters of Atlas; any brilliant group of seven, such as the French poets of the late sixteenth century (in Fr. f. *pléiade*); **Pleiades,** a star-group

plein air (Fr. m.), the open air

pleinairist, painter reproducing effects of atmosphere and light (not ital.)

Pleistocene (geol.) (cap., not ital.)

Plen., plenipotentiary

pleno jure (Lat.), with full authority

plethora, (med.) excess of red corpuscles in the blood; (gen.) unhealthy repletion

pleur/a, a membrane lining the thorax or enveloping the lungs; *pl.* **-ae; -isy,** inflammation of this; adj. **-itic**

pleuro-pneumonia (hyphen)

Plexiglas, propr. term for a transparent plastic (cap.)

plexus/, a network (of nerves, etc.); *pl.* **-es**

plf., plaintiff

Plimsoll line (naut.)

plimsolls, rubber-soled canvas shoes

Plinlimmon, *use* **Plynlimon**

Pliny/ the Elder (in Lat., **Gaius Plinius Secundus**), 23–79, and his nephew — **the Younger** (**Gaius Plinius Caecilius Secundus**), *c.*61–*c.*112, Roman writers

plissé (Fr. m.), gathering, kilting, or pleating

PLO, Palestine Liberation Organization

plod/, -der, -ding

PL/1, a computer programming language; caps. or even s. caps.

PLP, Parliamentary Labour Party

PLR, Public Lending Right

plumb, vertical

plumbum, lead, symbol **Pb**

plummy, abounding in plums, plumlike

plum pudding (hyphen when attrib.)

plumy, plumelike, *not* -ey

plung/e, -ing

Plunket, family name of Baron Plunket

Plunkett, family name of Earl of Fingall, and of Barons Dunsany and Louth

pluperfect, abbr. **plup.**

plural/, abbr. **pl.**; **-ize,** *not* -ise

plurals of abbreviations, as MPs, B.Litt.s (no apos.)

plurals of letters, as *A*s, *a*s, (italic letter, roman s, no apos.); *but note* **p's and q's**

plus/, with the addition of, sign +; an additional amount, an advantage, *pl.* **-es**

Plutarch, *c.*50–*c.*120, Greek philosopher and biographer, abbr. **Plut.**

Pluto (Rom. myth.), the god of

the underworld; (astr.), planet, symbol **P**

plutonium, symbol **Pu**

Plutus (Gr. myth.), personification of riches

pluviometer, rain-gauge, *not* pluvia-

Plymouth Brethren, religious sect; abbr. **PB**

Plynlimon, Welsh mountain, *not* Plin-, -limmon

Plzeň, Czech for **Pilsen**

PM, paymaster, postmaster, post-mortem, Prime Minister, Provost-Marshal

Pm, promethium (no point)

pm, picometre (no point)

pm., premium, premolar

p.m. (Lat.), *post meridiem* (after noon) (lower case, points)

PMG, Paymaster-General, (*formerly*) Postmaster-General

p.m.h., production per man-hour

PMRAFNS, Princess Mary's Royal Air Force Nursing Service

p.n., promissory note

PNdB, perceived noise decibels

PNEU, Parents' National Educational Union

pneumatic, relating to air or gases, abbr. **pneum.**

pneumon/ic, pertaining to the lungs; **-ia**

PNG, Papua New Guinea

pnxt., pinxit (painted this)

PO, petty officer (RN), pilot officer (RAF), postal order, post office

Po, polonium (no point)

po/, colloq. for chamber-pot, *pl.* **-s**

pocket-book (hyphen)

pocket handkerchief (two words)

pocket/-knife, -money, -piece (hyphens)

poco/ (It.), a little; — *a poco,* little by little

pococurant/e, apathetic, careless (one word, not ital.); **-ism,** *not* -eism

POD, pay on delivery, Post Office Department

POD, *Pocket Oxford Dictionary*

podestà (It. m.), municipal magistrate

podi/um, base or pedestal, *pl.* **-a**

Podsnappery, British philistinism, from Mr Podsnap in Dickens's *Our Mutual Friend*

podzol/, acid sandy soil deficient in humus, *not* -sol; **-ization, -ize,** *not* -isation, -ise

Poe (Edgar Allan), 1809–49, US writer

poems (titles of), when cited, to be roman quoted; but titles of poems long enough to form a separate publication should be in italic

poenology, *use* **pen-**

poet., poetic, -al, poetry

poeticize, *not* -ise

Poet Laureate (caps., two words); abbr. **PL,** *see* **Austin, Betjeman, Bridges, Cibber, Lewis, Masefield, Southey, Tennyson**

poetry (typ.), to be centred on longest line, unless such line is disproportionately long, in which case use optical centring (q.v.); turnover lines to be one em more than greatest indentation of poem; a grave-accented *è* may be used to show that an otherwise mute syllable is to be separately pronounced, as *raisèd*; when lines are numbered, no point after the figure

Poets' Corner, Westminster Abbey (caps.)

po-faced, solemn (hyphen)

pogrom, an organized massacre, esp. of Jews in Russia

poignard, *use* **poniard**

poikilothermic, cold-blooded

poilu (Fr. m.), private soldier

Poincaré (Raymond), 1860–1934, French statesman, president 1913–20

poiniard, *use* **poniard**

poinsettia (bot.), *not* point-

point (typ.), all marks of punctuation, especially the full stop. *See also* **compass, point system, punctuation** V

point-blank (hyphen)

point/ d'appui (Fr. m.), a base of operations; — *d'attaque,* base of offensive operations

Point de Galle, *see* **Galle**

point-device, extremely precise (hyphen)

point-duty (hyphen)

point et virgule (Fr.), the semicolon

pointsettia, *use* **poinsettia**

point system (typ.), the Anglo--American standard by which the bodies of all types shall be multiples, or divisions, of the twelfth of a pica, which is theoretically the sixth of an inch (72 metric points = 1 inch); abbr. **pt.**

pois (Fr. m. sing. and pl.), pea

poise (phys.), unit of dynamic viscosity, in SI units 10^{-1} Pa s; abbr. **P**

poisson (Fr. m.), fish; — *d'avril,* April fool

poivre (Fr. m.), pepper

poky, small and cramped, *not* -ey

Pol., Poland, Polish

Poland, abbr. **Pol.**

polarize, to restrict vibrations of light or other electromagnetic waves to one plane; (elec., mag., chem.), to separate positive and negative charges; (fig.) to give special meaning or unity of direction to; *not* -ise

pole, *see* **perch**

pole-axe (noun *and* verb, hyphen)

polecat (one word)

pol. econ., political economy

pole/-jump, -vault (hyphens)

Police (Ger.), *police* (Fr.), policy of insurance. *See also* **Polizei**

police/ constable, — **sergeant** (two words), abbr. **p.c., p.s. (PC, PS,** when used before a name)

policeman (one word)

police-officer (hyphen)

police state, a totalitarian one controlled by political police (two words)

police station (two words)

policewoman (one word)

polichinelle (Fr. m.), puppet, buffoon

poliomyelitis, infantile paralysis; colloq. **polio/,** *pl.* **-s**

Polish, abbr. **Pol.;** (typ.) has 24 letters as in Eng. without *q* and *v*; *see* **accents.** In dividing words the letters *ch, cz, dz (dż), rz,* and *sz* should not be separated

polit., political, politics

Politburo, executive committee of Communist Party (USSR and elsewhere)

politesse (Fr. f.), politeness

Politian (in It., **Angelo Ambrogini Poliziano**), 1454–94, Italian humanist

political economy, abbr. **pol. econ.**

politicize, *not* -ise

politico/, one devoted to politics, *pl.* **-s**

polity, organized society, form of civil government

Polizei (Ger. f.), police (cap.). *See also* **Police**

polka dot (hyphen when attrib.)

pollack, sea-fish, *not* -ock

pollen/ analysis, — **count** (two words)

poll/ex, thumb, *pl.* **-ices**

Pollok (Robert), 1798–1827, Scottish poet

Pollok/shaws, -shields, Glasgow

pollster, sampler of public opinion

poll-tax (hyphen)

Pollyanna, cheerful optimist

polonium, symbol **Po**

poly/, colloq. for polytechnic, *pl.* **-s**

polyandr/y, marriage to more than one husband; adj. **-ous**

polyanthus/ (bot.), *not* -os; *pl.* **-es**

Polybius, *c.*200–*c.*118 BC, Greek historian; abbr. **Polyb.**

polyethylene, polymerized ethylene

polygam/y, marriage to more than one wife or husband; adj. **-ous**

polygen/y, derivation of man from independent pairs of ancestors; **-ism, -ist**

polyglot/, -tal, -tic

polygyn/y, marriage to more than one wife; adj. **-ous**

polyhedr/on, *not* polye-; *pl.* **-a**

Polyhymnia (Gr.), muse of rhetoric

Polynesia, islands in central Pacific (a bishopric, not a political unit)

Polyolbion, by Drayton, 1613–22

polyp/ (zool.), *pl.* **-s**

polyp/us (path.), *pl.* **-i**

polytechnic/, (institution, college) concerned with many technical subjects; (cap. when part of name); colloq. **poly**

polythene, commercial name for a polyethylene

poly/urethane, -vinyl chloride, synthetic resins or plastics

polyzoan, aquatic animal forming colonies

pomade, preparation for the hair, *not* pomm-

pomelo/, a fruit, the shaddock, *pl.* **-s,** *not* pomm-, pu-

Pommard, a burgundy, *not* Pomard

pommel, knob, saddle-bow. *See also* **pummel**

pommelo, *use* **pomelo**

pommes/ (Fr. f. pl.), apples; — **(de terre),** potatoes

pompano/, W. Indian and N. Amer. food-fish, *pl.* **-s**

Pompeian, of Pompeii

pom-pom, automatic gun. *See also next*

pompon, ornamental tuft, *not* pom-pom (q.v.)

poncho/, S. Amer. cloak or one of similar design, *pl.* **-s**

pondere (Lat.), by weight; abbr. **p.**

Pondicherry, India; in Fr. **Pondichéry**

pongo/, ape, *pl.* **-s**

poniard, a dagger, *not* poign-, poin-

Poniatowski (Joseph Antony), 1762–1813, Polish soldier; **(Stanisław August),** 1732–98, last king of Poland

pons/ (Lat.), a bridge, *pl.* *pontes*; — *asinorum,* bridge of asses, Euclid, i. 5

Pontacq, a white wine

pontif/ex, (eccl.) a bishop, (Rom. hist.) member of priestly college, *pl.* **-ices; Pontifex Maximus,** head of college

pont/iff, a bishop, esp. the Pope; **-ifical**

Pont-l'Évêque, dép. Calvados, France; a cheese

Pontypridd, Mid Glam.

Pooh-Bah, holder of many offices at once

pooh-pooh, to scorn, ridicule, *not* poo-poo

Pool, Cornwall, W. Yorks.

Poole, Dorset

Poole (William Frederick), 1821–94, US librarian, compiler of *Poole's Index*

Poona, India, *not* -ah

poor-box (hyphen)

poorhouse (one word)

Poor Law (caps. when hist.)

poor-rate (hist.), assessment for relief of poor (hyphen, no caps.)

pop, colloq. for popular (music, etc.); — **art,** that based on popular culture (two words)

pop., popular, population

Pope (the) (cap.)

Pope Joan, a card-game (two words)

pop group (two words)

Popocatepetl, volcano, Mexico

popularize, *not* -ise

population, abbr. **pop.**

populus (Lat.), people; abbr. **P.**

porc (Fr. m.), pork, pig

Porchester Terrace, London. *See also* **Port-**

Porsche, Ger. make of car

Porson/, a Greek typeface, supposedly derived from the Greek hand of — **(Richard),** 1759–1808, English Greek scholar

Port., Portug/al, -uese

portament/o (mus.), a continuous glide (not break) from one note to another of different pitch, *pl.* **-i**

Portarlington (Earl of)

Port au Prince, Haiti

Port aux Basques, Newfoundland

Portchester, Hants. *See also* **Porch-**

Porte (hist.), the Turkish Court and Government; more fully **the Sublime —**

porte-cochère (Fr. f.), a carriage entrance

Port Elizabeth, Cape Province; abbr. **PE**

portent/, -ous, *not* -ious

Porter (William Sydney), *see* **Henry (O.)**

portfolio/, *pl.* **-s**

Port Glasgow, Strathclyde

Porthmadog, Gwynedd

portico/, *pl.* **-es, -od**

portière (Fr. f.), door-curtain, a portress, carriage door, *or* window

portmanteau/, *pl.* **-s** (not ital.)

Porto, Port. for **Oporto**

Porto Bello, Panama, *use* **Puerto Bello**

Portobello/, Lothian; — **Road,** London (one word)

Port of Spain, Trinidad

Porto Rico, *use* **Puerto Rico**

Portpatrick, Dumfries & Galloway (one word)

portrait (typ.), a book, page, or illustration of which the depth is greater than the width. *See also* **landscape**

Port-Royal, monastery, Versailles and Paris (hyphen)

Port Royal, Jamaica; *also* SC, US (no hyphen)

Port Salut, a cheese (no hyphen)

Portsmouth:, sig. of Bp. of Portsmouth (colon)

Portugal, abbr. **Port.** *See also* **Assemblies**

Portuguese, abbr. **Port.;** (typ.) alphabet has 25 letters as in Eng., without *w*. *See also* **accents**

pos., positive

pos/e, -ed, -ing

Posen, Poland, use **Poznań**

pos/er, a problem; **-eur,** *fem.* **-euse,** one who behaves affectedly, *pl.* **-eurs, -euses**

posey, *use* **posy**

posit/, -ed, -ing

posology (med.), study of dosages

poss., possess/ion, -ive

posse/, body of men, *pl.* **-s;** — **comitatus** (hist.), county force of men over 15

possessive case (see *Hart's Rules,* pp. 31–2):

  1. The apostrophe for this must be used only for proper and common nouns; *not* for the pronouns hers, its (it's = it is), ours, theirs, yours.

  2. In nouns (singular or plural) that end in a letter other than *s,* the apostrophe must *precede* the added *s,* as in President's house, men's hats, fox's earth.

  3. In nouns in the singular number that end in *s,* the possessive is usually formed by adding the *'s,* as in octopus's tentacles.

  4. In nouns in the plural number that end in *s,* the apostrophe must follow the *s,* as in boys' clothing, octopuses' tentacles.

  5. When the added *s* would be silent *in speech,* it is generally omitted, as in for goodness' sake.

  6. In English names and surnames add *'s,* as in Burns's poems, St. James's Street; but euphony often demands omission of a further *s,* e.g. Bridges'

poems (be guided by ease of pronunciation).

  7. Ancient names ending in *s* usually omit a further *s* in the possessive, as Venus' rites, Moses' law, Jesus' love.

  8. French names ending in silent *-s* or *-x* add an *'s,* e.g. Crémieux's.

  9. For place-names involving possessives, e.g. Earls Court, King's Cross, see *Hart's Rules,* p. 32, and individual entries.

  10. Abbreviations add *'s* in the singular (MP's), and *s'* in the plural (MPs')

post, former size of paper with several variations: —, $19 \times 15\frac{1}{4}$ in.; **pinched** —, $18\frac{1}{2} \times 14\frac{1}{2}$ in.; **large** —, $16\frac{1}{2} \times 21$ in.

post (Lat.), after; abbr. **p.**

post, abbr. **P.**

postage stamp (two words)

postal code, *use* **post-code** (q.v.)

postal order, abbr. **PO**

post-box (hyphen)

postcard (one word); abbr. **p.c.**

post-classical (hyphen)

post-code (hyphen). *See also* **capitalization**

post-date (verb *and* noun, hyphen)

poste restante, Post Office dept. where letters remain till called for

post-exilic, subsequent to exile of Jews in Babylon (hyphen)

postgraduate (one word)

post-haste (hyphen)

post hoc (Lat.), after this. *See also propter hoc*

posthorn (mus.), one word

posthumous, occurring after death, *not* postu-

Posthumus, in Shakespeare's *Cymbeline*

postilion, one who guides post- or carriage-horses, riding the near one, *not* -llion

Post-Impressionism (hyphen, caps.)

post litem motam (Lat.), after litigation began

post/man, -mark (one word)

postmaster, abbr. **PM**; *also* a scholar at Merton Coll., Oxford (one word)

Postmaster-General, former ministerial post (hyphen, caps.); abbr. **PMG**

post meridiem (Lat.), after noon; abbr. **p.m.**

postmistress (one word)

post mortem (Lat.), after death

post-mortem (adj. *and* noun), abbr. **PM**

postnatal (one word)

post-obit, taking effect, a bond payable, after death (hyphen, not ital.)

Post Office (The) (caps., no hyphen), public corporation, replaced the govt. dept. in 1969

post office (not caps., no hyphen), a local office of the above, abbr. **PO**

post-paid, abbr. **p.p.**

postposit/on, word or particle placed after another; **-ve,** designating this (one word)

postprandial, after dinner (one word)

postscript (one word); abbr. **PS**, *pl.* **PSS**

post terminum (Lat.), after the conclusion

postumous, *use* **posth-**

post-war (hyphen)

posy, a nosegay, *not* -ey

pot., potential

potage (Fr. m.), soup

potassium, symbol **K**

potato/, *pl.* **-es**

pot-au-feu (Fr. cook. m.), a meat broth

poteen, illicit whisky, *not* pott-, poth-

potential, abbr. **pot.**

pot-hole/, -r (hyphens)

pot-pourri, a medley

potsherd, a piece of broken earthenware, *not* -ard

pott, former size of paper, $12\frac{1}{2} \times$

$15\frac{1}{2}$ in.; — **4to,** $7\frac{3}{4} \times 6\frac{1}{4}$ in; — **8vo,** $6\frac{1}{4} \times 3\frac{7}{8}$ in. (untrimmed)

potteen, *use* **poteen**

potting-shed (hyphen)

potto/, W. Afr. lemur, *pl.* **-s**

pouce (Fr. m.), an inch; a thumb; abbr. **p.**

pouding (Fr. cook. m.), pudding

poudr/é (Fr.), *fem.* **-ée,** powdered

Poughkeepsie, New York State, US

Pouilly/, dép. Saônc-ct-Loire, France, source of — **-Fuissé** white wine; — **-sur-Loire,** dép. Nièvre, France, source of — **-Fumé** white wine

poularde (Fr. f.), fat pullet

poule (Fr. f.), a hen

poulet (Fr. m.), a young chicken

poulette (Fr. f.), young hen

POUNC, Post Office Users' National Council

pound, avoirdupois, approx. 453 g.; abbr. **lb.,** *pl. same* (not now in scientific use)

pound mark (money), **£** (*sing.* and *pl.*), q.v. at **l**

pour/ (Fr.), for, abbr. **p.**; — *ainsi dire,* so to speak

pourboire (Fr. m.), gratuity, tip

pourparler (Fr. m.), preliminary discussion (one word, ital.)

pousse-café (Fr. m.), a liqueur (after coffee)

poussette, a dance with hands joined, to do this (not ital.)

Poussin (Nicolas), 1594–1665, French painter

poussin, a very young chicken (not ital.)

pou sto (Gr.), standing-place

POW, Prince of Wales, prisoner of war

powder keg (two words)

powder/-magazine, -puff, -room (hyphens)

Powell (Anthony Dymoke), b. 1905, British author

powerhouse (one word)

power-station (hyphen)

Powis (Earl of), family name Herbert

powwow, N. American Indian conference, meeting compared to this, *not* pawaw

Powys, family name of Baron Lilford; Welsh county

Poznań, Poland

PP, parish priest, Past President, (Lat.) *Patres* (Fathers)

pp., pages

p.p., past participle, post paid, (Lat.) *per procurationem* (q.v.)

pp (mus.), pianissimo (very soft) *or* più piano (softer)

p.p.b., parts per billion (i.e. 10^9)

PP C (Lat.), *Patres Conscripti* (Conscript Fathers)

PPE, Philosophy, Politics, and Economics (Oxford degree subject)

p.p.m., parts per million

PPP, Psychology, Philosophy, and Physiology (Oxford degree subject)

ppp (mus.), pianississimo (as softly as possible)

PPS, Parliamentary *or* principal private secretary; *post-postscriptum* (further postscript)

PQ, Parliamentary Question, previous (*or* preceding) question; Province of Quebec, Canada; (Fr.) *Parti Québécois* (Quebec Party)

PR, prize ring, Proportional Representation, Public Relations, Puerto Rico, (Lat.) *Populus Romanus* (the Roman people)

Pr., priest

Pr, praseodymium (no point)

pr., pair, -s, price

PRA, President of the Royal Academy

praam, a boat, *use* **pram**

Prachtausgabe (Ger. typ. f.), de luxe edition

practice (noun; US also verb)

practise (verb)

praemunire, a writ (not ital.)

praenomen, Rom. first name, *not* pren-

praepostor, a school prefect, *not* prep-, -itor

praetor/, a Roman magistrate, adj. **-ian,** *not* pre-

Praga, suburb of Warsaw

Prague, Czechoslovakia; Eng. and Fr. for Czech **Praha; Ger Prag**

prahu, *use* **proa**

praseodymium, symbol **Pr**

Prayer Book (caps., no hyphen); abbr. **PB**

Prayer of Manasses (Apocr.) abbr. **Pr. of Man.**

PRB, Pre-Raphaelite Brotherhood (group of artists), 1848

pre-adam/ic, -ite (hyphens, nc cap.)

Préault (Antoine Auguste), 1809–79, French sculptor

preb., prebend, -ary

prec., preceding, precentor

Pre-Cambrian (geol.) (caps., hyphen)

precede, *not* -eed

precent/or, controller of cathedral music, *fem.* **-rix;** abbr. **prec.**

precept/or, teacher, *fem.* **-ress**

preces (Lat.), prayers

precession (astr.), earlier occurrence (of the equinoxes); (dyn.), the rotation of the axis of a spinning body about another body

pre-Christian (hyphen, cap. C)

Précieuses ridicules (*Les*), play by Molière, 1659

précieu/x (Fr.), an affected man; *fem.* **-se**

précis/, (to write) a summary, *pl. same* (not ital.); as verb, **-ed, -ing**

precisian, one who is rigidly precise, esp. (hist., in hostile sense) a Puritan

precisionist, one who makes a practice of precision

pre-Columbian (hyphen, cap. C)

preconce/ive, -ption (one word)

precursor, forerunner, *not* -er

pred., predicative, -ly

predecease (noun *and* verb, one word)

predicant, a preacher, esp. Dominican

predictor, *not* -er

predikant (Afrik.), a preacher of the Dutch Protestant Church, esp. in S. Africa

predilection, partiality

predispos/e, -ition (one word)

pre-elect/, -ion (hyphens)

pre-eminen/t, -ce (hyphens)

pre-empt/, -ion, -ive (hyphens)

pre-exilic, prior to exile of Jews in Babylon (hyphen)

pre-exist/, -ence (hyphens)

pref., preface, preference, preferred, prefix, -ed

prefab, colloq. for prefabricated building (no point)

prefabricate (one word)

preface, the introductory address of the author to the reader, in which he explains the purpose and scope of the book. Abbr. **pref.** *See also* **preliminary matter**

préfecture/ (Fr. f.), county hall in a French town; — *de police,* office of commissioner of police

prefer/, -able, -ably, -ence, -red, -ring, abbr. **pref.**

préfet (Fr. m.), prefect

prefix/, -ed, abbr. **pref.**

preform (one word)

pre-heat (hyphen)

prehistoric (one word)

Preignac, a white wine

prejudge/, -ment (one word)

prelim., preliminary

preliminary matter (typ.), material giving *identification* and any *explanations* desirable for bibliographical and trade purposes, or to 'prepare' the reader of a book; the order should be: series title; publisher's announcements, e.g. list of other titles in the same series; book half-title; frontispiece; title-page, naming work, author, publisher, place and date of

publication; title verso, giving copyright notice, publisher's agencies, ISBN, impression lines enumerating (at least) current impression/edition, and stating any geographical limitation on sales, country of origin, printer's imprint (this is sometimes printed elsewhere, e.g. on the last page of the book); dedication; acknowledgements; foreword (introducing the author, and usually written by someone else); preface (written by the author, introducing the book and stating its purpose); contents, list of illustrations/maps/tables/graphs/etc.; introduction. Colloq. **prelims**

pre-makeready (typ.), careful preparation of the forme before it goes to the machine

pre/-marital, -menstrual (hyphens)

premi/er (Fr.), first, *fem.* -ère, abbr. 1er

première, (to give) first performance of a film or play; as verb, **-d** (not ital.)

première danseuse (Fr. f.), principal fem. dancer in a ballet

premise (verb), to say or write by way of introduction

premises (noun, pl.), foregoing matters, (law) aforesaid houses or land, (gen.) building with grounds etc. *See also* next

premiss/ (logic), a proposition; *pl.* -es

premium/, abbr. **pm.**; *pl.* -s

premolar, a tooth; abbr. **pm.**

prenomen, *use* **praenomen**

preoccup/y, -ation (one word)

pre-ordain (hyphen)

prep., preparat/ion, -ory, preposition

pre/paid, -pay (one word)

prepositor, *use* **praepostor**

pre/possess, -prandial (one word)

Pre-Raphaelite/ (hyphen, caps.); — **Brotherhood,** abbr. **PRB** (q.v.)

prerequisite (one word)

Pres., president

présalé (Fr. m.), salt-marsh sheep or mutton (one word); pl. *prés-salés* (hyphen)

Presb., Presbyterian

presbyop/ia, a failing of near sight in the elderly; adj. **-ic**

Presbyterian (cap.); abbr. **Presb.**

prescribable, *not* -eable

pre-select (hyphen)

preselect/ive (of motor-car gears), **-or**, *not* -er (one word)

Preseli, district of Dyfed

preses (Sc.), president, *or* chairman, *pl. same*

pre-shrunk (hyphen)

president, abbr. **P.** *or* **Pres.** *See also* **capitalization**

pre-Socratic (philos.), prior to Socrates (hyphen, one cap.)

Press (**the**), newspapers etc. (cap.)

press agent (two words)

Press Association, abbr. **PA**

press-box, shelter for reporters at outdoor functions (hyphen)

press-button (hyphen)

press conference (two words)

Pressensé (**Edmond Dehaut de**), 1824–91, French theologian and statesman

press-gallery, esp. in House of Commons (hyphen)

press-gang (hyphen)

pressman (one word)

press/-mark, that which shows the place of a book in a library (hyphen); *now usu.* **shelf-mark**; **— -proof**, the last one examined before going to press or plate-making (**machine revise**, after **makeready**, may follow)

press-stud (hyphen)

presswork (typ.), the preparation for and control of the printing-off of composed material; the work thus produced

pressure (mech.), symbol *p*

Presteigne, Powys

Prester John, mythical medi-

eval priest-king of Cent. Asia or Abyssinia; *Prester John*, novel by John Buchan, 1910

prestissimo (mus.), very quickly

presto (mus.), quickly

Prestonpans, Lothian (one word)

pre-stressed (hyphen)

Prestwich, Greater Manchester

Prestwick, Northumberland, Strathclyde

presum/e, -able, -ably, -ing

presuppos/e, -ition (one word)

pret., preterite

pre-tax (adj., hyphen)

preten/sion, -tious

preterite, past tense, *not* -it; abbr. **pret.**

preternatural etc. (one word)

pretor, *use* **prae-**

pretzel (Ger.), a salted biscuit (not ital.)

Preussen, Ger. for **Prussia**

preux chevalier (Fr. m.), a brave knight

prevail/, -ed, -ing

preventive, *not* -tative, -titive

previous, abbr. **prev.**

Prévost (**Marcel**), 1862–1941, French novelist

Prévost d'Exiles (**Antoine François**), 1679–1765, French novelist, known as **Abbé Prévost**

prévôt (Fr. m.), provost

pre-war (hyphen)

Prez (**Josquin des**), *c.*1440–1521, French composer

PRI, President of the Royal Institute of Painters in Water--Colours

PRIBA, President of the Royal Institute of British Architects

Pribilof Is., Bering Sea

price, abbr. **pr.**

price/-list, -ring, -tag (hyphens)

prices current, abbr. **p/c**

pricey, *not* -cy

prie-dieu, kneeling stool (hyphen, not ital.)

priest, abbr. **Pr.**

prig/, -gery, -gish, -gism

prim., primary, primate, primitive

prima (typ.), pron. *i* as in *bite,* the page of copy on which a new take (q.v.) of proof begins; also mark on copy where this begins or where reading is to be resumed after interruption

prima/ ballerina, *pl.* — **ballerinas;** — **donna,** *pl.* — **donnas** (not ital.)

primaeval, *use* **primeval**

prima facie (adv., hyphen as adj.)

Prime Minister (two words, caps.); abbr. **PM**

primer (typ.), *see* **great —, long** — (sizes of type); *pron.* primmer

primeval, of the first age of the world; *not* -aeval

primigravid/a, woman in first pregnancy, *pl.* **-ae**

primipar/a, woman bearing first child, *pl.* **-ae**

primitive, abbr. **prim.**

primo/ (mus.), upper part in a duet, *pl.* **-s**

primo (Lat.), in the first place; abbr. **1°**

Primus, propr. term for kind of portable stove (cap.)

primus/ (Lat.), first, abbr. **p.;** — **inter pares,** first among equals

prin., principal

prince, abbr. **P.; Prince/ of Glory, Life,** *or* **Peace,** as Deity (caps.); — **of Wales/,** abbr. **POW;** — — — **Island,** off. Penang

princeps (Lat.), the first, *pl.* **principes.** *See also* **editio princeps**

Princes Town, Trinidad

Princeton University, NJ, US

Princetown, Devon, Mid Glam.

principal, (adj.) chief; (noun) the chief person (cap. when the title of an office); abbr. **prin.**

principle, a fundamental truth, moral basis

print., printing

print (in), still on sale; — **(out of),** new copies no longer obtainable, abbr. **o.p.**

printani/er (Fr. cook.), *fem.* **-ère,** with early spring vegetables

printer, abbr. **pr.; King's** *or* **Queen's Printer/ of Bibles and Prayer Books,** may print Bibles (AV) and Prayer Books (1662) to the exclusion of all other English presses, except those of the Universities of Oxford and Cambridge; — — **of the Acts of Parliament,** has the duty of providing an authoritative printing of all Acts of Parliament etc., and controls their copyright

printer's error, *pl.* **print/er's,** *or* **-ers', errors** (use hyphen to avoid ambiguity: bad printers'- -errors)

printing, abbr. **ptg.** *or* **print.**

Prinz/ (Ger. m.), prince (usu. of the blood royal), *pl.* **-en;** *fem.* **-essin,** *pl.* **-essinnen** (cap.)

prise, to force open, *not* -ize

priv., privative

privat docent, German university teacher paid only by students' fees (hyphen, not cap.); abbr. **PD**

Private (mil.), cap. as title, abbr. **Pte.**

privative, denoting the loss or absence of something; esp. in gram.; abbr. **priv.**

priv/y, secret, **-ily**

Privy Coun/cil, -sellor (two words, caps.), abbr. **PC; Privy Seal, PS**

prix/ (Fr. m.), prize, price; — **fixe,** fixed price

prize, to force open, *use* **-ise**

prize/-fight, -fighter, -money, -ring, -winner (hyphens)

p.r.n., *pro re nata* (as occasion may require)

PRO, Public Record Office, — Relations Officer

pro/, colloq. for professional, *pl.* **-s** (no point)

pro (Lat.), for; abbr. **p.**

proa, Malay vessel, *not* the many variants

pro-am, involving professionals and amateurs (no points)

pro and con, for and against; *pl.* **pros and cons** (not ital., no points)

prob., probab/le, -ly, problem

pro bono publico (Lat.), for the public good

proboscis/, long flexible trunk or snout; *pl.* **-es**

Proc., proceedings, proctor

proceed, *not* -ede

procès (Fr. m.), lawsuit

process blocks (typ.), those made by photographic and etching processes, for printing illustrations by letterpress

procès-verb/al (Fr.), official report, minutes; *pl.* **-aux** (not ital.)

proconsul (one word)

procrustean, using force to impose conformity, like Procrustes (Gr. myth.) (not cap.)

Procter (Adelaide Anne), 1825–64, English poet; — **(Bryan Waller,** *not* Walter), 1787–1874, pseud. 'Barry Cornwall', English poet

proctor, a university official, an attorney in spiritual courts; abbr. **Proc.; King's** *or* **Queen's Proctor,** an official who can intervene in divorce cases

Proctor (Richard Anthony), 1837–88, British astronomer

Procurator-fiscal, Scottish law officer (cap. P, hyphen), abbr. **PF**

prodrome, a preliminary treatise or symptom, *not* -dromus

producible, *not* -able

product (math.), symbol ∏

proem/, preamble, adj. **-ial**

pro et contra (Lat.), for and against. *See also* **pro and con**

Prof(s)., professor(s)

professoriate, *not* -orate

proffer/, **-ed, -ing**

profit/, **-ed, -ing**

profiterole, confection of pastry and sweet filling (not ital.)

Pr. of Man., Prayer of Manasses

pro forma (adv. *and* adj.), (done) as a matter of form; (noun) an account showing the market prices of specified goods; *pl.* **-s**

progr/amme, but **-am** (noun *and* verb) in computing

pro hac vice (Lat.), on this occasion, *not* — hâc —

projector, *not* -er

Prokofiev (Sergei Sergeyevich), 1891–1953, Russian composer

prolegomen/a, preliminary remarks; *sing.* (rare) **-on**

proleps/is, anticipation, esp. of adjectives in gram., *pl.* **-es**

proletari/at, the poorest class in a community, *not* -te; adj. **-an,** verb **-anize,** *not* -ise

prologize etc., to deliver a prologue, *not* -ise, -uize

PROM (comp.), programmable read-only memory

prom, a promenade (at seaside), a promenade concert (no point); **The Proms,** the Henry Wood Promenade Concerts

prom., promontory

Promethean, of or like Prometheus (Gr. myth.)

promethium, symbol **Pm**

promissory note, abbr. **p.n.**

promoter, one who or that which promotes, *not* -or

pron., pronominal, pronoun, pronounced, pronunciation

proneur, flatterer, in Fr. *prôneur*

pronoun, abbr. **pron.;** when relating to Deity, l.c. unless caps. specified

pronounc/e, -eable, -ed, -ement, -ing; abbr. **pron.**

pronunciamento/, a manifesto; *pl.* **-s** (not ital.) ; *pronunciamiento* (Sp.) does not have this meaning

pronunciation, *not* -nounc-
proof (typ.), a trial printing of
composed matter, taken for correction; **author's —**, a clean
proof as corrected by the compositor, supplied to the author,
and later returned by him with
his corrections, abbr. **a.p.;**
clean —, one having very few
printer's errors; **first —**, the
first taken after composition,
usu. for in-house correction
only; **foundry —**, the final one
taken from type prepared for
plating; **galley** *or* **slip —**, a
proof taken before the matter is
made up into pages: usu. about
18 in. long; **page-on-galley** *or*
slip-page —, one having type
made up into pages but not yet
imposed; **page —**, *or* **— in
sheets,** one made up into pages;
slip — (*see* **galley,** above); **—
marks** (*see* **proof-correction
marks**); **— paper,** that used
for taking proofs; **plate —**, one
taken from a plate; **press —**,
the final one passed by author,
editor, or publisher, for the
press; **— -reader,** one who
reads and corrects printers'
proofs; **rough —**, one taken
without special care
proof-correction marks.
Two systems are in current use:
for a comprehensive treatment
see BS 1219 and 5261 part 2.
One or other must be used
consistently throughout a given
book. The second system is recommended when dealing with
a foreign author or printer; a
list of its marks is given in *Hart's
Rules,* pp. 34–5. The first is still
in wide use in UK, and the
following are its commonest
marks:
bold, print in bold face the sorts
with wavy underline in proof-
-text.
cap, change to capital letters
those trebly underlined (or
ringed where space is tight).

eq #, justify word-spacing of
line.
ital, change to italic letters those
singly underlined.
lc, change to lower-case letters
(small, not caps. or s. caps.)
those ringed.
lig, use ligatured sort for those
ringed.
np, begin a new paragraph
with the word after the mark
[*or* ⌋ .
press, print off.
Qy, *or* **?,** added by reader to
mark something about which
he is uncertain.
revise, submit another proof.
rom, change to roman letters
those ringed.
run on, and a line drawn from
the last word of the first paragraph to the first word of the
second, no new paragraph.
sc, change to small capital
letters those doubly underlined.
stet, let the cancelled word
which is dotted underneath
remain.
trs, transpose letters, words,
etc., marked in thus proof-
-text.
wf, wrong fount, area or letter
ringed is set in wrong size or
cut of type.
⌀ , delete what is crossed
through.
⋏ , insert preceding words etc.
at point marked thus in proof-
-text.
□ indent matter 1 em.
□ indent matter 1 en.
replace sort by, or insert, space
at point in proof-text marked
by / or ⋏.
⟳ turn ringed sort other way
up.
⌒ remove space, close up (same
mark in text and margin).
⋁ or ⋏ to be put under all
superior sorts required in corrections located by / or ⋏ in
proof-text.

proof-correction marks (*cont.*)

⊥ press down sort ringed in proof-text.

⌡ align (same mark in text and margin).

= straighten words having a horizontal line above and below them in proof-text.

× sorts ringed in proof-text are battered or blotched and must be cleaned or replaced.

⊙ full point needed where indicated, ⊙ colon ditto.

/ a stroke as this is to be put after each correction (except additional matter which ends with ∧) in the margin to show that it is concluded, to separate it from others, and to call attention to it.

Write corrections clearly in the margin by the line to which they refer, in the same left-to--right sequence as they occur there. Use red for printer's errors, black or blue for author's changes. Never correct in pencil or crayon. Mark the precise point in the proof-text where correction is required. Ring round all words not to be printed (e.g. stet, lig, etc.)

Prop., Propertius

prop, colloq. for (stage) property, propeller (no point)

prop., proposition

propaganda, an activity for the spread of a doctrine or practice, is *singular*

proparoxytone (Gr. gram.), (word) with acute accent on last syllable but two

pro patria (Lat.), for one's country

propel/, -led, -ler, -ling

propell/ant (noun) ; **-ent** (adj.)

properispomen/on (Gr. gram.), (word) with circumflex on last syllable but one, *pl.* **-a**

Propertius (**Sextus**), *c.*50–*c.*14 BC, Roman poet; abbr. **Prop.**

prophe/cy (noun) ; **-sy** (verb)

proportional (math.), sign ∝

proposition, abbr. **prop.**

Proprietary (Austral., NZ, S. Afr.), after name of company, abbr. **Pty.**

proprietary terms, *see* p. xiv *and* **capitalization**

proprio motu (Lat.), of his, *or* her, own accord

propter hoc (Lat.), because of this. *See also* **post hoc**

propylae/um (Lat. from Gr.), entrance to temple, *pl.* **-a** ; **Propylaea,** entrance to Acropolis at Athens (pl., cap.)

propylon/ (Gr.), entrance to temple, *pl.* **-s**

pro rata, in proportion (not ital.) ; *pro re nata,* as occasion may require (ital.), abbr. **p.r.n.**

pros., prosody

pros and cons, *see* **pro and con**

prosceni/um, the front part of the stage ; *pl.* **-ums**

proselyt/e, a religious convert ; **-ize,** to convert, *not* -ise

prosit! (Lat.), your good health! (used by German students and others)

prosod/y, the laws of poetic metre ; abbr. **pros.** ; **-ic, -ist**

prosopograph/y, hist. enquiry into social and family connections ; **-er, -ical**

prosopopoeia (rhet.), introduction of pretended speaker

prospector, *not* -er

prospectus/, *pl.* **-es**

pros/y, commonplace, *not* -ey ; **-ily**

Prot., Protectorate, Protestant

protactinium, symbol **Pa**

protagonist, leading character in a play, novel, or cause (*not* the opp. of antagonist)

pro tanto (Lat.), to that extent

protean, assuming different shapes, like Proteus (Gr. myth.), *not* -ian

protector, *not* -er

Protectorate, abbr. **Prot.**

protég/é, one under the protec-

tion of a patron, *fem.* **-ée,** *pl.* **-és,
-ées** (not ital.)
pro tempore (Lat.), for the time
being, colloq. **pro tem** (no
point, not ital.); abbr. ***p.t.***
Protestant/, abbr. **Prot.; -ism**
(cap.)
protester, *not* -or
'Prothalamion', by Spenser,
1596
protocol/, first draft of a diplomatic document, diplomatic
etiquette; as verb, to record in
protocol, **-led, -ling**
protonotary, a chief clerk,
esp. to some lawcourts, *not*
protho-
prototype, an original model
protozo/on, single-cell form of
life, *pl.* **-a** (not ital.)
protractor, drawing instrument, *not* -er
Proudhon (Pierre Joseph),
1809–65, French socialist
Prov., Proven/ce, -çal, Proverbs,
Province, Provost
prov., proverbially, provincial,
provisional
prov/e, -able, -ed (lit., Sc., and
US **-en:** *see* next), **-ing**
proven (*not*) (Sc. law), a verdict
pronouncing evidence insuffi-
cient to determine guilt or innocence
provenance, (place of) origin,
not -venience (not ital.)
provençale (à la) (Fr. cook.),
with garlic or onions (not cap.)
Proven/ce, S. France (no cedilla), **-çal;** abbr. **Prov.**
provenience, *use* provenance
Proverbs, OT, abbr. **Prov.**
provinc/e, abbr. **Prov.; -ial,**
abbr. **prov.**
Province of Quebec, Canada,
abbr. **PQ**
Provincetown, Mass., US
proviso/, stipulation; *pl.* **-s** (not
ital.)
Provo/, member of IRA, *pl.* **-s**
Provost/, abbr. **Prov.; — -
-Marshal** (caps., hyphen),
abbr. **PM**

proxime/ accessit (Lat.), he, *or*
she, came nearest (to winning a
prize, etc.); abbr. ***prox. acc.;*** *pl.*
— accesserunt
proximo, in, *or* of, the next
month; better not abbreviated
PRS, President of the Royal
Society (London)
PRSA, President of the Royal
Scottish Academy
PRSE, President of the Royal
Society of Edinburgh
prud'homme (Fr. m.), (*formerly*) good and true man; (*now*)
expert, umpire
Prudhomme, *see* **Sully Prudhomme**
Prufrock (J. Alfred), in T. S.
Eliot's *Love Song,* 1917
prunella, a throat affliction; a
strong silk or worsted material;
plant of self-heal genus (as
genus-name, cap. and ital.)
Pruss., Prussia, -n
Przemyśl, Poland
PS, permanent secretary, Police
Sergeant (*see also* **police constable**), *postscriptum* (postscript)
(*pl.* **PSS**), private secretary,
Privy Seal, (theat.) prompt
side
Ps., *see* **Psalm**
p.s., police sergeant
Psalm, abbr. **Ps.; Psalms** (*pl.*),
abbr. **Pss.; Psalms (Book of,**
OT), abbr. **Ps.**
Psalmist (the) (cap.)
Psalter (the), the Book of
Psalms, separately printed
(cap.)
psaltery, an anc. and medieval
plucked instrument
p's and q's (apos., no points)
psepholog/y, study of patterns
in voting; **-ical, -ist**
pseud (colloq.), false (person)
pseudepigraph/a (*pl.*), spurious writings; **-ic, -ical**
pseudo/, false (person), *pl.* **-s**
pseudonym, an assumed name;
abbr. **pseud.**
pshaw! an exclamation, *not*
psha, -h

psi, Greek letter ψ, used to denote parapsychological factors

p.s.i./, pounds per square inch; — **g.,** ditto gauge

psittacosis, a contagious disease

PSS, *postscripta* (postscripts) (no point)

Pss., *see* **Psalm**

PSV, public service vehicle

psych., psychic, -al

psychedelic, (of drugs) inducing fantastic mental pictures, (gen.) having comparable effect (esp. of colour), *not* psycho-

psycho/ (colloq.), psychotic (person), *pl.* **-s**

psychoanaly/sis, -se, -st, -tic, -tical (one word)

psychokinesis, movement by psychic agency; abbr. **PK**

psychol., psycholog/y, -ical

psycho/sis, *pl.* **-ses,** adj. **-tic**

psychotherapy (one word)

PT, physical training, post town, pupil teacher

Pt, platinum (no point)

Pt. (geog.), Point, Port

pt., part, payment, pint, -s, (math. & typ.) point

p.t., *pro tempore* (for the time being)

PTA, Parent-Teacher Association

pta., peseta

Ptah (Egypt. religion), creator of the world

Pte. (mil.), Private

pterodactyl, *not* -le

ptg., printing

ptisan, medicated drink, *not* tisane

PTO, please turn over, Public Trustee Office

ptomaine poisoning, former name for 'food poisoning'

Pty. (Austral., NZ, S. Afr.), Proprietary

Pu, plutonium (no point)

pub/ (colloq.), public house; — **-crawl** (hyphen)

pub./, public, -an, publish, -ed, -er, -ing; — **doc.,** public document

public address (hyphen when attrib.), abbr. **PA**

public house (two words)

publicity agent (two words)

publicize, *not* -ise

public school (two words)

publish/, -ed, -er, -ing, abbr. **pub.**

Publishers Association (The) (no apos.); abbr. **PA**

publisher's binding (bibliog.) standard binding in which edition is supplied to bookseller

pucka, *use* **pukka**

pudend/um, usu. *pl.* **-a,** genital (usu. female)

Puebla, Mexico

Pueblo, Colorado, US

pueblo/, town or village in Latin America, *pl.* **-s**

Puerto/ Bello, Panama; — **Rico,** W. Ind. island, adj. — **Rican;** *not* Porto —

puff/-adder, -ball (hyphens)

puff pastry (two words)

pug/-dog, -nose(d) (hyphens); adjs. **pugg/ish, -y**

puggaree, a hat-scarf, *not* the many variants

Pugwash/, NS, Canada, site of first — **Conference**

puîn/é (Fr.), *fem.* **-ée,** younger; opposed to **aîné,** senior

puisne (law), subsequent (to)

pukka, *not* pucka, pukkah. *See also* **sahib**

pul/e, to whine; **-ing**

Pulitzer (Joseph), 1847–1911, US journalist, endowed — **prizes** for various genres of writing

pull (typ.), a proof

pull a proof (typ.), print a few copies of galley etc. for correction purposes

Pullman, railway carriage or motor coach (cap.)

pull-out, (something) that pulls out (hyphen)

pullover, knitted garment (one word)

pulque, Mexican drink

pulsar (astr.), source of pulsating radio signals

pulse-rate (hyphen)

pulsimeter, a pulse measurer

Pulsometer, propr. term for a pumping engine

pulverize, *not* -ise

pumelo, *use* **pomelo**

pumice/, -stone (hyphen); adj. **-ous**

pummel/, -led, -ling, to pound with the fists. *See also* **pommel**

pummelo, *use* **pomelo**

pumpernickel, German rye bread

Punchinello (cap.), principal character in Italian puppet-show; hence **Punch** (no point)

punctatim (Lat.), point for point

punctilio/, (scrupulous observance of) a point of behaviour; *pl.* **-s**

punctuation

I. **General purpose**

Punctuation in the written word corresponds to pauses, inflexions, and emphases in the spoken word, the aim being to make the sense clear. Within the framework of a few basic rules, an author's choice of punctuation is as personal as his choice of words, and editors, unless asked to impose their own style upon the copy, must use great caution in emendation. The punctuation in the copy should be followed by the printer when so ordered, and always when setting legal texts, or extracts or quotations from any source.

II. **comma** (,)

This is the least emphatic of the separating marks. It may be used:

1. To separate main clauses when the second is not closely identified with the first, e.g. 'Cars will turn here, and coaches will go straight on.'

But cf. 'He turned and ran.'

2. When, without the comma, the eye or tongue would run on and momentarily mistake the sense, e.g. 'In the valley below, the villages looked very small.'

3. When the sentence would mean something different without the comma, e.g. 'He did not go to church, because he was playing golf.'

4. Between adjectives which each qualify a noun in the same way, e.g. 'a cautious, eloquent man' (*but* 'a distinguished foreign author').

5. To separate items (including the last) in a list of more than two items, e.g. 'potatoes, peas, and carrots', 'potatoes, peas, or carrots', 'potatoes, peas, etc.'.

6. After salutations and vocatives if they do not end a sentence, e.g. My Lord, Dear Sir, O God, and also before them if they do not begin the sentence.

7. To mark the beginning and end of a parenthetical word or phrase, e.g. 'I am sure, however, that it will not happen', 'Infidels, i.e. Christians, may not enter', 'Fred, who is bald, complained of the cold' (*but* 'Men who are bald wear hats').

8. Before a quotation. Note the increasing weight of the break before the quotation in the following: 'You say "It cannot be done." I boldly cried out, "Woe to this city!" Then he wrote these words: "I have named none to their disadvantage."'

9. In numbers of four or more figures, to separate each three consecutive figures, starting from the right, e.g. 10,135,793. For exceptions *see* **numerals.**

III. **semicolon** (;)

This separates those parts of

punctuation (*cont.*)

a sentence between which there is a more distinct break than would call for a comma, but which are too closely connected to be made into separate sentences. Ideally, it should separate clauses or phrases that are similar in importance and in grammatical construction, e.g. 'To err is human; to forgive, divine.'

See also **Greek.**

IV. **colon** (:)

This is used:

1. When the preceding part of the sentence is complete in sense and construction, and the following part naturally arises from it in sense though not in construction, e.g. 'The universe would turn to a mighty stranger: I should not seem part of it.'

2. To lead from introduction to main theme, from cause to effect, or from premiss to conclusion, e.g. 'Country life is the natural life: it is there that you will find real friendship.'

3. To introduce a list of items (a dash should not be added), and after such expressions as 'namely', 'for example', 'to resume', 'to sum up', 'the following'.

V. **period, full point, full stop** (.)

This is used:

1. At the end of all sentences which are not questions or exclamations. The next word should generally begin with a capital letter.

2. After many abbreviations (q.v.) and initials. If such a point closes the sentence, it also serves as the sentence's full point, e.g. '. . . cats etc.' *but* '(. . . cats etc.).'

3. Do not use a further full point at the end of a sentence that concludes with a quotation

itself ending with a full point, question mark, or exclamation mark, e.g. 'He cried "Be off!"'' But where the quoted sentence is a short statement, and the introducing sentence has much more weight, the point is put outside the quotation marks, e.g. 'He took many years considering the implications of the simple phrase "Thou art the man".'

See also **decimal currency, ellipsis, time of day.**

VI. **question mark, note of interrogation** (?)

1. This should follow every question which expects a separate answer. It is not used after indirect questions, e.g. 'He asked why I was there.' The next word should generally begin with a capital letter.

2. It may be placed before a word etc. whose accuracy is doubted, e.g. 'T. Tallis ?1505–85'. *See also* XI. 1.

3. ? is used in chess reports to denote a bad move.

See also **Greek, quotation marks, Spanish.**

VII. **exclamation mark, note of exclamation** (!)

This is used:

1. After an exclamatory word, phrase, or sentence expressing absurdity, command, contempt, disgust, emotion, enthusiasm, pain, sorrow, a wish, wonder.

2. In mathematics, as the factorial sign; in chess reports, to denote a good move.

3. In square brackets after a quotation, to express the editor's amusement, dissent, or surprise.

See also **quotation marks, Spanish.**

VIII. **apostrophe** (')

This is used:

1. To show the possessive case, q.v.

2. To show an omission, e.g. e'er, tho', we'll, he's (he is, he has), it's (it is), '69 (1969, or whichever century the context demands).

3. In Irish names, such as O'Connor; *but see* **Mac.**

4. As the 'closing quote'; *see* **quotation marks.**

IX. **turned comma** (')
This is used:

1. As the 'opening quote'; *see* **quotation marks.**

2. In Scottish names, such as **M'Gregor**; *see* **Mac.**

X. **parentheses** ( )
These enclose:

1. Interpolations and remarks made by the writer of the text himself, e.g. 'He is (as he always was) a rebel.'

2. An authority, definition, explanation, reference, or translation.

3. In the report of a speech, interruptions by the audience.

4. Reference letters or figures, e.g. (1), (*a*). These do not need a point as well as the parentheses. *See also* XV below.

XI. **brackets**

1. *square* [ ]. These enclose comments, corrections, explanations, interpolations, notes, question marks, or translations, which were not in the original text, but have been added by subsequent authors, editors, or others, e.g. 'He [Bloggs] fell down.'

2. *angle* ⟨ ⟩ (*narrow*) < > (*wide*). (*a*) In text-critical editions, pairs of these (preferably narrow) enclose words conjecturally supplied at a defective or illegible point in the source. (*b*) In scientific, mathematical, philological, etc. work, a single bracket (preferably wide) indicates relative size of entities, direction or derivation of argument, etc.

3. *Curly* (or *hooked*) brackets

{ }, *double square* brackets ⟦ ⟧, together with square and angle brackets, have a special significance in epigraphical work: *see* A. G. Woodhead, *The Study of Greek Inscriptions.*

XII. **dashes**

(*a*) The **en rule** (–) is used:

1. To join pairs wherever movement or tension, rather than co-operation or unity (for which use hyphen) is felt, e.g. '1914–18 war' (*but* 'from 1914 to 1918'), 'current–voltage characteristic', 'the Fischer–Spassky match', 'the London–Horsham–Brighton route', 'the Marxist–Trotskyite split' (*but* 'the Marxist-Leninist position' (hyphen)). Note also 'Franco--Prussian War' (hyphen, because 'Franco-' is a prefix which cannot stand alone).

2. For joint authors (hyphen would lead to confusion with a single double-barrelled name), e.g. 'the Lloyd–Jones hypothesis' (two men), 'the Lloyd--Jones hypothesis' (one man), 'the Lloyd-Jones–Scargill talks' (two men).

(*b*) The **em rule** (—) is used:

1. Instead of the parentheses in X. 1 above.

2. In informal contexts, to replace the colon in IV. 1, 2 above.

3. To indicate pauses in hesitant speech, or the ending and resumption of a statement interrupted by an interlocutor. If the sentence is not interrupted but abandoned, use the ellipsis (q.v.).

4. In dictionaries, indexes, etc., to represent the headword, and so save space.

5. To replace an omitted word.

(*c*) The **two-em rule** (——) is used in bibliography entries cued in by author's name, to avoid the repetition of the same

punctuation (*cont.*)

name(s) in successive entries. One 2-em rule serves for the complete name (in whatever form) of each person.

XIII. **hyphen** (-)

This is used:

1. To join two or more words so as to form a single expression, e.g. ear-ring, get-at-able, and words having a syntactical relationship which form a compound, as weight-carrying (objective), punch-drunk (instrumental); and in a compound used attrib., to clarify the unification of the sense, e.g. a blood-red hand, the well--known man, *but* prettily furnished rooms, the man is well known (predic.).

2. To join a prefix to a proper name, e.g. anti-Darwinian.

3. To prevent misconceptions by linking words, e.g. a poor--rate collection, a poor rate--collection.

4. To prevent misconceptions by separating a prefix from the main word, e.g. recover, re--cover (an umbrella); (a footballer) resigns, re-signs.

5. To separate two similar consonant or vowel sounds in a word, as a help to understanding and pronunciation, e.g. sword-dance, Ross-shire, co--operate (the hyphen here is preferable to a **diaeresis,** q.v.).

6. To represent a common second element in all but the last word of a list, e.g. two-, three-, or fourfold.

7. At the end of a line of print, to indicate that the last word has been divided; *see* **division of words.**

See also XII (*a*) above, **fractions,** and, in general, *Hart's Rules*, pp. 76–81.

XIV. **brace** (})

This is used (usu. vertically)

to connect words, lines, staves of music, etc., and in mathematical and tabular work. When used to show that one thing comprises several others the brace should point toward the single item, which should be centred on it, e.g.

$$\text{Biology} \left\{ \begin{array}{l} \text{Botany} \\ \text{Zoology.} \end{array} \right.$$

XV. **duplication of points**

This should be avoided where possible. A comma should not precede or follow a dash, nor a full point an exclamation or question mark. If a sentence which is a statement ends with quoted matter which is itself a statement, the full point before the final 'quote' will be sufficient, unless this is itself closed by a parenthesis. *See also* V. 2, 3 above. There are many complications (see *Hart's Rules,* pp. 44–8), but, as in all punctuation, clarity is the object, and logic and common sense the touchstones.

See also **quotation marks;** and for a detailed treatment of the whole topic of this article, *Hart's Rules*, pp. 38–49

punctus (Lat.), a point; *pl. same*

pundit, Hindu sage, (gen.) expert. *See also* **pandit**

Punica fides (Lat.), Punic faith, treachery

Punjab, India and Pakistan, *not* -aub, Panjab

Punjabi, (inhabitant or dialect) of Punjab, *not* -bee

punkah (Anglo-Ind.), a large fan, *not* -a

Punkt (Ger. typ. m.), point, dot, a full stop (cap.); *punktieren,* to point, dot, or punctuate; *Punktierung* (f.), punctuation

PUO (med.), pyrexia (fever) of unknown origin

pup/a (entom.), a chrysalis; *pl.* **-ae** (not ital.)

pupillage (two *l*s)

pur, *use* **purr**

pur (Ind.), a city, as Nagpur, Kanpur

Purana, Sanskrit sacred poem

Purchas (Samuel), 1577–1626, English compiler of voyages

purchas/able, *not* -eable; **-er,** *not* -or

purdah (Ind.), (curtain for) seclusion of women of rank, *not* -a

purée, pulped and sieved vegetables etc.

purgatory, place of temporary suffering and purification of souls of the dead (not cap.)

Purim, Jewish festival

puritan/, one strict in religion or morals (cap. in hist. use); **-ical, -ism, -ize,** *not* -ise

purl, in knitting, *not* pearl

Purleigh, Essex

Purley, Berks., London

purlicus, surroundings

purpose/, -ful, -less, -ly

purr, of or as a cat, *not* pur

pur sang (Fr. m.), pure blood; adj., thoroughbred, total

purslane (bot.), a salad herb, *not* -lain

purveyor, one whose business is to supply meat or meals, *not* -er

push/-bike, -button, -chair (hyphens)

push-over (colloq.), a difficulty easily overcome (hyphen)

Pushtu, *use* **Pashto**

put, in golf, *use* **putt**

putre/fy, to go rotten, *not* -ify; **-scent,** in process of rotting; **-scible,** liable to rot

putsch (Ger. m.), a revolutionary attempt (ital.)

putt, in golf, *not* put

puttees, strips of cloth worn round the lower leg for protection, *not* -ies

putt/o (It.), figure of child in Renaissance art, *pl.* **-i**

Puy-de-Dôme, dép. France (hyphens, two caps.). In phrases, *puy* (Fr. m.), a small volcanic cone, to have lower-case *p* and no hyphens, as *le puy de Dôme*

PVC, polyvinyl chloride

Pwllheli, Gwynedd

PWR, pressurized water reactor

pwt., pennyweight, *use* **dwt.**

pyaemia (med.), a type of blood-poisoning, *not* pyem-

pye (typ.), *use* **pie**

pyebald, *use* **pie-**

pye-dog, an oriental ownerless dog, *not* pi-, pie-

pygm/y, *not* pigmy; adj. **-aean,** *not* -mean

pyjamas, *not* the many variants; US **pa-**

pyknic, (person) of short, squat stature

pyorrhoea, discharge of pus

Pyrenees Mts., France and Spain

Pyrénées/-Atlantiques; Hautes--; --Orientales, déps. France

pyro/lysis, decomposition from heat, *pl.* **-lyses; -lyse,** *not* -lyze

pyrotechnics, the art of fireworks; abbr. **pyrotech.**

pyrrhic/ dance, anc. Gr. war-dance; **— foot** (prosody), foot of two short syllables; **— victory,** one won at great cost, like that of Pyrrhus, King of Epirus, over Romans in 279 BC

Pyrrhon/, *c*.300 BC, Greek sceptic philosopher; **-ism**

Pytchley Hunt (the)

Pythian, of Delphi, anc. Greece

pyx (trial of the), at the Royal Mint, *not* pix (not cap.)

Q

Q, quartermaster, Quarto (Shakespeare), (chess) queen (no point), the sixteenth in a series

Q., pseud. of Sir Arthur Thomas **Quiller-Couch,** q.v.

Q., Queen('s), question

q., quaere, query, quintal, quire, -s, (meteor.) squalls

QAB, Queen Anne's Bounty

Qaisar-i-Hind, *use* **K-**

Qantas, Queensland and Northern Territory Aerial Service (the Australian Commonwealth airline)

QARANC, Queen Alexandra's Royal Army Nursing Corps

QARNNS, Queen Alexandra's Royal Naval Nursing Service

Qatar/, Persian Gulf; adj. **-i**

QB, Queen's Bench; (chess) queen's bishop

QBD, Queen's Bench Division

Q-boat (First World War), a merchant-vessel with concealed guns

QBP (chess), queen's bishop's pawn

QC, Queen's, *or* Queens', College, Queen's Counsel

q.e., quod est (which is)

QED, *quod erat demonstrandum* (which was to be demonstrated)

QEF, *quod erat faciendum* (which was to be done)

QEI, *quod erat inveniendum* (which was to be found out)

QF, quick-firing

qibla, use **kiblah**

QKt (chess), queen's knight; **QKtP,** queen's knight's pawn

q.l., quantum libet (as much as you please)

Qld, Queensland (Australia) (no point)

QM, Quartermaster

q.m., *quomodo* (by what means)

QMC, Queen Mary College, London

Q.Mess., Queen's Messenger

QMG, Quartermaster-General

QMS, Quartermaster-Sergeant

QN (chess), queen's knight; **QNP,** queen's knight's pawn

Qom, *use* **Qum**

QP (chess), queen's pawn

q.pl., quantum placet (as much as seems good)

QPM, Queen's Police Medal

QPR, Queen's Park Rangers

qq.v., quae vide (which see; refers to plural)

QR (chess), queen's rook

qr., quarter (28 lb.), quire

QRP (chess), queen's rook's pawn

qrs., quarters

QS, quarter sessions, Queen's Scholar

q.s., quantum sufficit (as much as suffices)

QSO, quasi-stellar object, quasar

qt., quantity, quart, -s

q.t., (colloq.), (on the) quiet

qu., question

qua (Lat.), in the character of, *not* -à, -â

quad (colloq.), quadrangle; quadraphony; a quadruplet; (typ.) in metal-setting, a piece of spacing material, usually of size en, em, 2-em, 3-em, or 4-em; in photosetting, the action of spacing out a line; (paper) a size of printing paper four times (quadruple) the basic size; (no point in any sense)

quad., quadrant

Quadragesima, first Sunday in Lent

quadraphon/y, sound reproduction using four transmission channels; **-ic, -ically,** *not* quadri-, quadro-; (colloq.) **quad**

quadrenni/um, a period of four years; *pl.* **-ums; -al;** *not* quadrie-

334

quadroon, offspring of White and mulatto, with quarter--Negro blood

quaere (Lat. imperative), inquire; (as noun) a question; abbr. **q.**

quaestor, anc. Rom. magistrate. *See also* que-

uae vide (Lat.), which see; refers to plural; abbr. **qq.v.**

uag/, -gy

quagga (S. Afr.), extinct zebra--like animal

quai (Fr. m.), quay, railway platform

quaich, Scots drinking vessel, *not* -gh

Quai d'Orsay, Paris, the French Foreign Office

Quaker, a member of the Society of Friends

uand même (Fr.), notwithstanding, all the same

quango/, '*quasi* non-governmental *organization*', *pl.* **-s**

quantity, abbr. **qt.**

quantize, *not* -ise

quant/um (not ital.), a natural minimum quantity of an entity; *pl.* **-a;** *hence* **-um theory,** theory of the emission and absorption of energy in finite amounts. *See also* **Planck**

quant/um (Lat.), a concrete quantity, *pl.* **-a;** *quantum/ libet,* as much as you please, abbr. *q.l.;* — *meruit,* as much as he, or she, deserved; — *placet,* as much as seems good, abbr. *q.pl.;* — *sufficit,* as much as suffices, abbr. *q.s.* or *quant. suff.;* — *vis,* as much as you wish, abbr. *q.v.*

Qu'Appelle/, river; — (**Bp. of**), Rupert's Land, Canada

quark (phys.), one of orig. three hypothetical components of elementary particles, from 'three quarks for Muster Mark' in James Joyce's *Finnegans Wake*

Quarles (Francis), 1592–1644, English poet

quarrel/, -led, -ler, (*not* -lor), **-ling, -some**

quart, two pints, abbr. **qt.;** position in fencing, *not* carte; sequence in piquet

quart., quarterly

Quart (Ger. n.), quarto (cap.)

quarter/, -s, abbr. **qr., qrs.**

quarter-binding, the spine in a different material from the rest of the case

quarter-day (hyphcn)

quarterdeck (onc word)

quarter/-final, -light (hyphens)

quartermaster (one word), abbr. **QM** *or* **Q; Quartermaster/-General** (hyphen, caps.), abbr. **QMG;** —**-Sergeant** (hyphen, caps.), abbr. **QMS**

quarter sessions (two words, not caps.), abbr. **QS**

quartet, *not* -ette

quartier/ (Fr. cook. m.), quarter; — *d'agneau,* — of lamb; *quartier-général,* mil. headquarters; *Quartier Latin,* the Latin Quarter of Paris

quarto/ (typ.), a book based on four leaves, eight pages, to the sheet; *pl.* **-s;** abbr. **4to** (no point). *See also* **book sizes**

quartz, form of silica

quasar, a quasi-stellar object; abbr. **QSO**

quasi-, prefix with sense 'seeming(ly)', 'almost'

Quasimodo/, first Sunday after Easter (one word, cap.); character in Hugo's *Notre-Dame de Paris;* — (**Salvatore**), b. 1901, Italian poet

quass (Russ.), rye-beer; *use* **kvass**

quater/centenary, (celebration of) four-hundredth anniversary; **-nary,** (set) of four parts; *not* quart-

quater-cousin, *use* **cater-**

quatrain, stanza of four lines usu. with alternate rhymes

Quatre-Bras (battle of), 1815

quattrocento, 1400–99, art and literature of the early Renaissance period

quay, a wharf, *not* key

QUB, Queen's University, Belfast

Quebec/, Canada, abbr. **PQ** *or* **Que.**; adj. and n. **-ois,** *pl. same*

Quechua, Peruvian Indian, *not* Qui-

Queen, abbr. **Q.**; (chess, no point)

Queen Anne's Bounty, formerly for augmenting C. of E. livings (apos.); abbr. **QAB**

Queenborough, Sheppey, Kent. *See also* **Queens-**

Queen Mary College, London, *not* Mary's, abbr. **QMC**

Queen's, abbr. **Q.**

Queens, borough of New York City (no apos.)

Queensberry/ (Marquess of); — **Rules,** in boxing

Queensborough, Co. Louth. *See also* **Queenb-**

Queensbury, London, W. Yorks.

Queen's College (The), Oxford (named after one queen)

Queens' College, Camb. (named after two queens)

Queen's Counsel, abbr. **QC**

Queen's County, Ireland, now **Laois**

Queensland (Australia), abbr. **Qld**

Queen's Printer, *see* printer

Quelimane, Mozambique

quelque chose (Fr. m.), something, a trifle

quenelle (Fr. cook. f.), a forcemeat ball (ital.)

Quentin Durward, by Sir W. Scott, 1823

query, abbr. **q., qy,** *sign* **?**

Quesnay (François), 1694–1774, French economist

question, abbr. **Q.** *or* **qu.**; — **mark (?),** *see* **capitalization, punctuation** VI

questionnaire, a formulated series of questions (not ital.)

questor, official of RCC *or* French Assembly. *See also* **quæ**

Quételet (Lambert Adolphe Jacques), 1796–1874, Belgian mathematician and astronomer

Quetta, Pakistan

quetzal, beautiful bird of Central America, currency unit of Guatemala

Quetzalcoatl, trad. culture-hero of the Aztecs

queu/e, persons in line, to form this, **-ed, -ing,** *not* cue

queue/ (Fr. f.), tail; — *de bœuf,* oxtail

quiche, open flan (not ital.)

Quichua, *use* **Quechua**

quick/lime, -sand, -set, -silver, -step (one word)

quicumque vult (Lat.), whosoever will (first words of the Athanasian Creed)

quid (Lat.), that which a thing is

quidam (Lat.), an unknown person, *pl. same*

quidnunc, a gossip (not ital.)

quid pro quo, something in return, an equivalent (not ital.)

¿Quién sabe? (Sp.), Who knows? (turned question mark before, unturned after)

quietus, a settlement, final riddance (not ital.)

Quiller-Couch (Sir Arthur Thomas), 1863–1944, English novelist and essayist (hyphen), pseud. **Q.**

quin, colloq. for quintuplet (no point)

quincentenary, (celebration of) five-hundredth anniversary (one word)

quincun/x, five arranged as on dice; *pl.* **-xes**; adj. **-cial**

quinine, medicine in treating malaria and fevers, *not* -in

Quinquagesima, the Sunday before Lent

quinquenni/um, a period of five years; *pl.* **-ums**

quinquereme, anc. galley with five banks of oars

uins/y, tonsillitis, *not* -cy, -sey, -zy; **-ied**

uint, sequence of five cards in piquet. *See also* **quinte**

uint/al, 100 kg., 220½ lb., 1.968 cwt., *not* kentle, kintle; *pl.* **-als** (in French **-aux**); abbr. **q.**

uinte, position in fencing. *See also* **quint**

uintet (mus.), *not* -ette

uintillion, cardinal number, 1 with 30 noughts; (US) 1 with 18 noughts

uiproquo (Fr. m.), mistake

uipu, the anc. Peruvian language of knotted cords, *not* -po, -ppo, -ppu

uire, 24 or 25 sheets of paper; abbr. **q.** or **qr.,** *pl.* same. *See also* **choir**

Qui s'excuse s'accuse (Fr.), to excuse oneself is to accuse oneself

uisling, traitor, from **Vidkun Quisling,** 1887–1945, pro--Nazi leader in Norway

uit/, -ted, -ter (*not* -tor), **-ting**

Quito, cap. of Ecuador

Qui va là? (Fr.), Who goes there?

Qui vive? (Fr.), Who goes there?

ui vive (on the), on the alert (*not* ital.)

uixotic (not cap.), extravagantly romantic and visionary, like Don Quixote, hero of the romance (1605) by Cervantes

uiz/, to interrogate, a competition by interrogation, **-zed, -zer, -zes, -zing**

Qum, river, city, and former province, central Iran, *not* Qom

Qumran, Jordan, site associated with Dead Sea Scrolls

quoad/ (Lat.), as far as; **— hoc,** to this extent

quod erat demonstrandum etc., *see* **QED** etc.

quod est (Lat.), which is; abbr. **q.e.**

quodlibet, (mus.) a medley; (hist.) (exercise on) subject of philos. or theol. disputation

quod vide (Lat.), which see; refers to sing.; abbr. **q.v.**

quoins (typ.), wedges or expanding devices which secure type, blocks, and furniture in the chase; *pron.* coins

quoits, a game, *not* coits

quo jure? (Lat.), by what right?

quomodo (Lat.), by what means; abbr. **q.m.**

quondam (adj.), former, from Lat. *quondam* (adv.), formerly

quor/um, the number of members whose presence is needed to make proceedings valid; *pl.* **-ums** (not ital.); adj. **-ate**

quot., quot/ation, *-ed*

quota/, a share; *pl.* **-s**

quotation marks (typ.), in English, one turned comma at the beginning and one apostrophe at the end; colloq. **quotes** (no point). The apostrophe at the end of the quotation should come before all punctuation marks unless these form part of the quotation itself; *see also* **punctuation** V. 3. Quotes and roman type are to be used when citing titles of articles in magazines, series titles, chapters of books, essays, poems (*but see* **poems, titles of**), and songs. They are *not* to be used for the titles of the books of the Bible; where the substance only of an extract is given; or where the tense or person has been altered. For further guidance, see *Hart's Rules*, pp. 44–9, and (for foreign languages) 100–1, 109, 129, 133

quotations, all extracts in the exact words of the original, if set run on in the text matter, to have 'quotes' at the commencement, and at the beginning of each paragraph (*not* each line, except in special cases), and at the end of the quotation. If the extract is set broken off from the main text, quotation marks are not required, except in conversational matter. Punctuation of the extract to be *exactly* as in the

quotations (*cont.*)
original. The concluding point
to be outside the last quotation
mark if not in the original. *See
also* **ellipsis**

**quotations within quo-
tations** to have double
quotation marks within the
single. Quotations within the
double quotation to be single-
-quoted

quote (typ.), to enclose within
quotation marks

quotes (typ.), quotation marks,
q.v.

quot homines, tot sententiae
(Lat.), there are as many
opinions as there are persons

quo vadis? (Lat.), where are yo
going?

Qur'ān, *see* **Koran**

q.v., *quantum vis* (as much as you
wish), *quod vide* (which see; refe
to sing.)

qwerty keyboard, of typewriter or computer input (no
caps.; acronym from the first
letters on top row of this keyboard)

qy, query (no point)

R

R, radius, rand, *retarder* (on timepiece regulator), roentgen, (chess) rook, the seventeenth in a series; rupee, *use* **Re**

R (elec.), symbol for resistance

R., rabbi, Radical, railway, rector, regiment, registered, reply, (US) Republican, River, Royal, (Fr.) *Rue* (street), (Ger.) *Recht* (law), (Lat.) *regina* (queen), *respublica* (commonwealth), *rex* (king), (naut.) run (deserted), (theat.) right (from actor's point of view), (thermom.) Réaumur

Ⓡ, registered trade mark

℞, *recipe* (take)

℟, response (to a versicle)

r., rare, recto, residence, resides, right, rises, rouble, -s, (meteor.) rain

r months, Sept. to Apr.

RA, Rear-Admiral, Royal/ Academy, — Academician, — Artillery, (astr.) right ascension

Ra, radium (no point); Egyptian god of the sun

RAA, Royal Academy of Arts

RAAF, Royal Auxiliary Air Force

Rabat, cap. of Morocco

Rabb., Rabbinic

rabbet/, -ed, -ing, groove in woodwork; *also* **rebate;** *not* rabbit

rabbi/, Jewish expounder of the law; *pl.* **-s;** abbr. **R.** (cap.); **Chief Rabbi** (caps.)

rabbin/, *usually pl.,* **-s,** Jewish authorities on the law, mainly of 2nd–13th cc.; **-ical, -ism, -ist**

Rabbinic (cap.) (noun), late Hebrew; abbr. **Rabb.**

rabbit/, a rodent; to hunt rabbits, **-ed, -ing;** — **-warren** (hyphen). *See also* **rabbet, rebate**

rabdomancy, *use* **rhab-**

Rabelais/ (François), 1483– 1553, French writer; adj. **-ian** (cap.)

rabscallion, *use* **raps-**

RAC, Royal Agricultural College, — Armoured Corps, — Automobile Club

raccoon, *use* **racoon**

racecourse (one word)

race/goer, -horse (one word)

race relations (two words)

race-riot (hyphen)

racey, *use* **racy**

RACS, Royal Arsenal Co-operative Society

rach/is (bot., zool.), *more usual than* rhachis; *pl.* (incorrectly formed) **-ides**

rachiti/s, rickets; **-c**

Rachmaninov (Sergei Vasilyevich), 1873–1943, Russian composer

Rachmanism, slum landlordism (from 1963)

rack (and ruin). But use wrack for destruction, seaweed

racket/, *not* racquet; **-eer, -s** (game), **-y**

rackett (mus.), Renaissance woodwind instrument, *not* racket, rank-

racont/eur, *fem.* **-euse** (not ital.)

racoon, bushy-tailed animal, *not* racc-

racquet, *use* **racket**

rac/y, *not* -ey; **-ily**

RAD, Royal Academy of Dancing

rad, former unit of absorbed radiation dose, = 0.01 Gy (*now use* **gray**); radian

rad., radix (root)

Rad., Radical

RADA, Royal Academy of Dramatic Art

radar, *ra*dio *d*etection *a*nd *r*anging (not cap.)

339

RADC, Royal Army Dental Corps

Radcliffe (Ann), 1764–1823, English writer; — **(John),** 1650–1714, English physician; — **College,** Mass., US; — **Camera, Infirmary, Science Library,** and **Observatory,** Oxford. *See also* **Ratradian,** SI unit of plane angle (approx. 57.296°), abbr. **rad**

radiator, *not* -er

Radical (polit.), abbr. **R.** *or* **Rad.**

radical (chem.), *not* -cle; — **sign** (math.), √

radicle (bot.)

radio/, *pl.* **-s**

radioactiv/e (one word); **-ity**

radio/-carbon, -cobalt, etc. (hyphens)

radiograph, X-ray photograph

radio-isotope (hyphen)

radium, symbol **Ra**

rad/ius, *pl.* **-ii,** abbr. **R**

rad/ius vect/or, *pl.* **-ii -ores**

rad/ix, a root; *pl.* **-ices;** abbr. **rad.**

radon, symbol **Rn**

RAE, Royal Aircraft Establishment(s)

Rae (John), 1813–93, Scottish Arctic traveller. *See also* **Ray, Reay**

Rae Bareli, Uttar Pradesh, India, *not* Ray Bareilly

Raeburn (Sir Henry), 1756–1823, Scottish painter

RAEC, Royal Army Education Corps

R.Aero.C., Royal Aero Club of the United Kingdom

R.Ae.S., Royal Aeronautical Society

Raetia, *use* Rh-

RAF, Royal Air Force

RAFA, Royal Air Forces Association

raffia, a palm-fibre, *not* rafia, raphia

RAFRO, Royal Air Force Reserve of Officers

RAFVR, Royal Air Force Volunteer Reserve

rag/, -ged, -ging

rag-and-bone man (two hyphens)

rag-bag (hyphen)

rag doll (two words)

ragee, a coarse Indian grain, *not* ragg-, -i

ragout, a rich meat stew; in Fr. m. *ragoût*

ragtag (one word) **and bobtail** (one word)

ragtime (mus.) (one word)

Ragusa, It. for **Dubrovnik** (Yugoslavia); *also* a town in Sicily

RAI, Royal Anthropological Institute

raie (Fr. f.), skate (fish)

Raikes (Robert), 1735–1811, English originator of Sunday Schools

rail/car, -head, -road, -way (abbr. **R.** *or* **rly.**), **-wayman** (one word)

rain/, abbr. **r.; -bow, -coat, -drop, -fall** (one word)

rain forest (two words)

rainhat (one word)

rain-water (hyphen)

raison/ de plus (Fr.), all the more reason; — *d'état,* a reason of State; — *d'être,* purpose of existence

raisonn/é (Fr.), *fem.* *-ée,* reasoned out

raj (Hindi), sovereignty

raja/, Indian title, Malay or Javanese chief, *not* -h (not cap.). *See also* **ranee**

Rajagopalachari (Chakravarti), 1878–1972, last Gov.--Gen. of India, 1948–50

Rajasthan, Indian state

Rajput/, member of Hindu soldier caste, *not* -poot; **-ana,** Indian region

Rajshahi, Bangladesh, *not* Rajeshaye

râle/, noise made in difficult breathing; *pl.* **-s**

Ralegh (Sir Walter), 1552–1618, English courtier, colonizer, soldier, and writer; -eigh usual but erroneous

Raleigh (Prof. Sir Walter Alexander), 1861–1922, English writer. *See also* **Rayleigh**

Ralfs (John), 1807–90, English botanist

rallentando/ (mus.), (passage performed) with decreasing pace, *pl.* -**s**; abbr. **rall.**

RAM, (comp.) random-access memory; Royal Academy of Music (London)

ram/, -med, -ming

Ramadan, Muslim fast, also ninth Muslim month; Pers. and Turk. **Ramazan**

Raman (Sir Chandrasekhara Venkata), 1888–1970, Indian physicist; — **effect**, change of frequency in scattering of radiation

Ramayana, Hindu epic

Rambouillet, dép. Seine-et-Oise, France

RAMC, Royal Army Medical Corps

ramchuddar, Indian shawl

Ramée (Marie Louise de la), 1839–1908, French writer of English novels; pseud. **Ouida**

ramekin, (a cheesecake served in) a small mould, *not* -quin

Rameses, name of kings of ancient Egypt, *not* Ramses

Ramillies (battle of), 1706

Rampur, Uttar Pradesh, India

Ramsay (Allan), 1686–1758, Scottish poet; — **(Allan)**, 1713–84, Scottish painter; — **(Sir William)**, 1852–1916, Scottish chemist

Ramses, *use* **Rameses**

Ramsey, Cambs., Essex, I.o.M.

Ramsey (Sir Alfred), b. 1920, English footballer and manager; — **(Arthur Michael)**, b. 1904, Bp. of Durham 1952–6, Abp. of York 1956–61, Abp. of Canterbury 1961–74; — **(Ian**

Thomas), 1915–72, Bp. of Durham 1966–72

ram/us (Lat.), a branch; *pl.* -**i**

rancher/o (Sp.), *fem.* -**a**, a small farmer

ranc/our, spite, *but* -**orous**

Rand (colloq.), the Witwatersrand, S. Africa

rand, S. Afr. unit of currency, *pl.* same; abbr. **R**

R. & A., Royal and Ancient (Golf Club)

R. & B., rhythm and blues

R. & D., research and development

Randolph-Macon College, Ashland and Lynchburg, Va.

ranee, Hindu queen, wife or widow of raja (q.v.), *not* rann-, rani

Ranelagh Gardens, Chelsea (1742–1804)

rang/e (typ.), to align sorts vertically and/or horizontally; -**ing numerals**, lining numerals, q.v.

Rangoon, Burma, *not* -un

rani, *use* **ranee**

Ranjit Singh, 1780–1839, founder of Sikh kingdom

Ranjitsinhji (Kumar Shri), 1872–1933, Indian cricketer; Maharaja Jam Sahib of Nawanagar 1906–33

Ranke (Leopold von), 1795–1886, German historian

ranket(t), *use* **rackett**

rannee, *use* **ranee**

Ransom (John Crowe), 1888–1974, US poet

Ransome (Arthur), 1884–1967, English writer

ranuncul/us, a buttercup, *pl.* -**uses**; (bot.) the genus including buttercups (cap. and ital.)

ranz-des-vaches, Swiss alpenhorn melody

RAOB, Royal Antediluvian Order of Buffaloes

RAOC, Royal Army Ordnance Corps

rap/, -ped, -ping

RAPC, Royal Army Pay Corps

Raphael/ (in It. **Raffaello Sanzio**), 1483–1520, Italian
painter; adj. **-esque**
raphia, use **raffia**
raphi/s (bot., zool.), a needle-
-like crystal; pl. **-des;** not
rha-
Rappahannock, river,
Virginia
rapparee, a seventeenth-
-century Irish freebooter
rappee, a coarse snuff
rappel/, (make) descent of rock-
-face using doubled rope, **-led,**
-ling (not ital.)
rapport, fruitful communication or relationship (not ital.)
rapporteur, one who prepares
account of proceedings (not
ital.)
rapprochement (Fr. m.), establishment or renewal of friendly
relations
rapscallion, rascal, not rabs-
rar/a av/is (Lat.), a prodigy,
literally a rare bird; pl. **-ae -es**
rare, abbr. **r.**
rarebit, see **Welsh rabbit**
rarefaction, not -efication
rarefy, not rari-
rarity, not -ety
RARO, Regular Army Reserve
of Officers
Rarotonga, Cook Islands, New
Zealand
RAS, Royal Agricultural,
Asiatic, or Astronomical Society
RASC, Royal Army Service
Corps, now **RCT**
rase, to destroy, use **raze**
Raskolnik/ (Russ.), a dissenter
from the Orthodox Church, pl.
-i; now commonly called **Old
Believer**
Rasoumoffsky, see **Rasumovsky**
Rasselas, by S. Johnson, 1759
Rastafarian, member of Jamaican sect
Rasumovsky Quartets (the),
by Beethoven
rat/, **-ted, -ting**
ratable, use **rateable**

ratafia, liqueur, biscuit, not -afie,
-ifia
ratan, use **rattan**
ratany (bot.), use **rh-**
ratatouille (Fr. f.), vegetable
stew
ratchet/, -ed
Ratcliff Highway, not -e, now
The Highway. See also **Rad-**
Ratcliffe College, Leicester
rateable, not ratable
ratepayer (one word)
Rathaus (Ger. n.), town hall
ratifia, use **ratafia**
ratio/, arith. relation; pl. **-s**
rationale, logical cause, reasoned explanation
rationalize, not -ise
Ratisbon, Bavaria, use **Regensburg**
ratline (naut.), the ladder-rope
on the shrouds, not -in, -ing
rat race (two words)
rattan, a cane, not ratan
rattlesnake (one word)
**Raumer (Friedrich Ludwig
Georg von),** 1781–1873, German historian; — **(Rudolf
von),** 1815–76, German philologist
RAVC, Royal Army Veterinary
Corps
ravel/, -led, -ling
ravioli (It. cook.), small squares
of pasta filled with meat (not
ital.)
Rawalpindi, Pakistan, not
Rawul-
raw sienna, brownish-yellow
pigment, not — siena
ray (mus.), not re
Ray (John), 1627–1705, English naturalist, spelt Wray till
1670. See also **Rae, Reay**
Ray Bareilly, use **Rae Bareli**
Rayleigh, Essex
**Rayleigh (John William
Strutt, Baron),** 1842–1919,
English physicist. See also **Raleigh**
raze, to destroy, to erase, not rase
Razumovsky, see **Rasumovsky**

razzmatazz, humbug, *not* the many variants

Rb, rubidium (no point)

RBA, Royal Society of British Artists

RBS, Royal Society of British Sculptors

RC, Red Cross, reinforced concrete, (theat.) right centre, Roman Catholic; — **paper** (typ.), resin-coated paper (for photosetting)

RCA, Royal College of Art

RCC, Roman Catholic Church

RCDS, Royal College of Defence Studies

RCM, Royal College of Music (London)

RCN, Royal College of Nursing

RCO, Royal College of Organists

RCOG, Royal College of Obstetricians and Gynaecologists

RCP, Royal College of Physicians

RCS, Royal College of Science, ditto Surgeons, Royal Commonwealth Society, Royal Corps of Signals

RCT, Royal Corps of Transport

RCVS, Royal College of Veterinary Surgeons

RD, Royal Dragoons, Royal Naval Reserve Decoration, Rural Dean

R/D, refer to drawer (of a cheque)

Rd., road

RDC, Royal Defence Corps, Rural District Council

RDF, Radio Direction-Finding

RDI, Royal Designer for Industry (Royal Society of Arts)

RDS, Royal Drawing Society

RE, Reformed Episcopal, Right Excellent, Royal Engineers, — Exchange, — Society of Painter-Etchers and Engravers

Re, Reynolds number, rhenium, rupee (no point)

re (mus.), *use* ray

Re (*il*) (It.), the King (no accent)

re (Lat.), with regard to

re- (prefix), when followed by *e* and separately sounded, to have hyphen, as re-echo, re-entry. It also has a hyphen when forming a compound to be distinguished from a more familiar one-word form, as re-form (to form again), re-sign (to sign again)

react, produce response (one word)

re-act, to act again (hyphen)

Read (**Sir Herbert Edward**), 1893–1968, English critic

readdress (one word)

Reade (**Charles**), 1814–84, English novelist. *See also* **Read, Rede, Reed, Reid**

reader (typ.), a corrector of the press; one who reports on MSS to a publisher (not cap.)

Reader, a university teacher, in some universities intermediate between Lecturer and Professor (cap. as title)

Reader's Digest Association Ltd., publishers

readers' marks, *see* proof--correction marks

reading/-desk, -lamp, -light, -room (hyphens)

readjourn, readjust, re-admission, readmit, -ted, -ting (one word)

read-out (noun, in computing, hyphen)

ready/-made, -to-wear (adjs., hyphens)

reafforest/, *not* reforest (one word); **-ation**

reagent (one word)

real/, former Portuguese and Brazilian coin, *pl. reis*; —, Spanish and Mexican coin, current in US (a 'bit', value 12½ cents), *pl. -es*

realiz/e, -able, *not* -is-

realpolitik (Ger. f.), practical politics (ital., not cap.)

real tennis, *see* **tennis**

realty (law), landed property

ream of paper, 500 (formerly also 480, 504, or 516) sheets; abbr. **rm.**

reanimate, reappear, re-appoint, reapprais/e, -al (one word)

Rear-Admiral, abbr. **RA** or **Rear-Adm.** (hyphen, caps.)

rearguard (one word)

rearm/, -ament (one word)

rearrange, reascend, reassemble, reassert/, -ed, -ing, reassur/e, -ance (one word)

Réaumur (René Antoine Ferchault de), 1683–1757, inventor of thermom. scale; the scale itself, abbr. **R.** or **Réaum.** See also **Raumer**

reawake etc. (one word)

Reay (Donald James Mackay, Baron), 1839–1921, Scottish (Dutch-born) Governor of Bombay and first president of the Brit. Academy. See also **Rae, Ray**

rebaptize, not -ise (one word)

rebate, to reduce, a reduction. See also **rabbet**

rebec (mus.), medieval stringed instrument, not -eck

rebel/, -led, -ling

rebind, to bind (a book) again, but **re-bound** (hyphen)

rebound, to bound, act of bounding, back (one word)

rebus/, a puzzle; pl. **-es**

rebut/, -tal, -ted, -ting

rec., receipt, recipe, record, -ed, -er

Récamier (Jeanne Françoise Julie Adélaïde, Madame), 1777–1849, a leader of French Society

recap/, colloq. for recapitulate, **-ped, -ping**

recast (one word)

recce/, slang for reconnaissance, to reconnoitre, pl. **-s**

recd., received

recede, to withdraw (one word)

re-cede, to cede back (hyphen)

receipt, abbr. **rec.**

receivable, not -eable

Rechabite, a total abstainer (cap.)

réchauffé (Fr. m.), a warmed-up dish, (fig.) a rehash

recherché, chosen with care, far-fetched (not ital.)

recidiv/ist, one who relapses into crime; **-ism**

recipe/, pl. **-s,** abbr. **rec.**

réclame (Fr. f.), notoriety by advertisement; (journ.) editorial announcement

recogniz/ance, a bond given to a court, not -sance; **-ant,** not -sant

recognize, not -ise

recollect, to remember (one word)

re-collect, to collect again (hyphen)

recommit/, -ted, -ting

recompense (noun and verb)

reconcilable, not -eable

reconciler, not -or

reconnaissance, preliminary survey, not reconnoi- (not ital.)

reconnoitre, to make a preliminary survey, not -er

record/, -ed, -er, abbr. **rec.**

record-player (hyphen)

recount, to narrate (one word)

re-count, to count again, a further count (hyphen)

recoup, to recompense, recover, make up for

recover, to regain possession of, to revive (one word)

re-cover, to cover again (hyphen)

recreat/e, to refresh; **-ion,** amusement, pastime (one word)

re-creat/e, to create again; **-ion, -or,** not -er (hyphen)

rect., rectified

rectif/y, -iable, -ied, -ier, -ying

recto/ (typ.), the right-hand page, usu. having an odd page number, 1, 3, 5, etc.; abbr. **ʳ, r.** (not ital.); pl. **-s**

rector/, in Ch. of England, incumbent of a parish, originally one whose tithes were held by the parson; in RCC head priest of church etc.; head of academic or relig. institution; (cap.)

elected representative of students on governing body of Scottish university; abbr. **R.**; adj. **-ial**

rectri/x, strong tail-feather, *pl.* **-ces**

rect/um (anat.), straight part of intestine, *pl.* **-ums**

rect/us (anat.), straight muscle, *pl.* **-i**

reçu (Fr. m.), a receipt

recueil (Fr. m.), a literary compilation

reculer pour mieux sauter (Fr.), to withdraw to await a better opportunity

recur/, -red, -ring

rédact/eur (Fr.), editor; *fem.* **-rice**

rédaction (Fr. f.), editing, editorial department

Redakteur (Ger. m.), editor (cap.)

Redbourn, Herts.

Redbourne, Humberside

redbreast, a robin (one word)

redbrick, new (university) (one word)

Redbridge, Gr. London

redd/en, -ish, -y

Reddish, Gr. Manchester

Redditch, Hereford & Worc.; 'new town', 1964

Redeless (Ethelred the), King of England 978–1016

Rede Lecture, Camb. Univ. *See also* **Read, Reade, Reed, Reid**

Redemptionists, an order of Trinitarian friars devoted to the redemption of Christian captives from slavery

Redemptorists, an order of missionaries founded in 1732 by Alfonso Liguori

Redgauntlet, by Sir W. Scott, 1824 (one word)

red-handed (hyphen)

redhead (one word)

red-hot (hyphen)

redivivus, restored to life

red lead (two words)

red-letter day (one hyphen)

redoubt, mil. outwork, *not* -out

redress, to remedy (one word)

re-dress, to dress again (hyphen)

redskin (one word)

red tape, excessive (use of) formalities (two words)

reducible, *not* -eable

reductio ad/ absurdum (Lat.), disproof by reaching an obviously absurd conclusion; — — *impossibile,* ditto an impossible conclusion

red-water, cattle disease (hyphen)

reebok (Afrik.), an antelope, *not* rhe-

re-echo (hyphen)

Reed (Alfred German), 1847–95, English actor; — **(Edward Tennyson),** 1860–1933, English caricaturist; — **(Talbot Baines),** 1852–93, English writer of boys' books; — **(Walter),** 1851–1902, US army surgeon (mosquitoes and yellow fever). *See also* **Read, Reade, Rede, Reid**

re/-edit, -educate (hyphens)

reef-knot (hyphen)

Reekie (Auld), Old Smoky, that is, Edinburgh

re/-elect, -election, -eligible, -embark (*not* reim-), **-emerge, -emergence, -enact, -enactment, -enforce** (to enforce again), **-enslave, -enter, -enthrone, -entrant, -entry, -establish** (hyphens)

reeve (naut.), to thread or fasten rope, past **rove.** *See also* **ruff**

re/-examine, -exchange, -exhibit, -export (hyphens)

Ref., the Reformation

ref., referee, reference, referred, reform/ed, -er

Ref. Ch., Reformed Church

refer/, -able, -ence, -red, -rer, -ring

refer/ee, -eed, abbr. **ref.**

reference marks (typ.), may be used (esp. in math. setting) as an alternative to superior figures for footnote references in the

reference marks (*cont.*)
order * (not used in math.
works) † ‡ § ¶ ‖, repeated in
duplicate as ** etc. if necessary.
See also **footnotes**

references, *see* **authorities**

referend/um, referring to the
electorate on a particular issue;
pl. **-ums**

referrible, *use* **-erable**

refit/, **-ted, -ting**

refl., reflect/ion, -ive, reflexive

reflectible, *use* **reflexible**

reflection, *not* -exion; abbr.
refl.; in Fr. f. *réflexion*

reflective, (of surfaces) giving
back a reflection, abbr. **refl.**;
(of people) meditative

reflector, *not* -er

reflet (Fr. m.), lustre on
pottery

reflexible, able to be reflected,
not reflect-

reflexion, *use* **reflection**

reflexive (gram.), implying that
the action is reflected upon the
doer, abbr. **refl.**

reforest/, -ation, *use* **reaffor-
est/, -ation**

reform/, to improve, correct;
-ation (one word)

re-form/, to form again; **-ation**
(hyphen)

Reformation (**the**) (cap.);
abbr. **Ref.**

Reform Bills, 1832, 1867,
1884-5

reform/ed, -er, abbr. **ref.**

refractor, a type of lens or
telescope

refrangible, that can be re-
fracted

refuse, to say no (one word)

re-fuse, to fuse again (hyphen)

Reg., Regent, *regina* (queen)

reg., regis/ter, -trar, -try, regular,
-ly

regalia, insignia, is *plural*

regd., registered

regenerator, *not* -er

Regensburg, Bavaria, *not* Ratisbon, -berg

Regent, abbr. **Reg.**

Regent's Park/, London
(apos.); —— **College,** Oxford

reggae, W. Indian style of music
with strong subsidiary beat

Reg.-Gen., Registrar-General

Régie, governmental control of
articles paying duty in Austria,
France, Italy, Spain, and Turkey

regime (not ital., no accent). *See
also ancien régime*

regimen/, *pl.* **-s** (not ital.)

regiment, abbr. **regt.** *or* **R.**

regina (Lat.), queen; abbr. **R.** *or*
Reg.

régisseur, ballet or theatre director (not ital.)

register, abbr. **reg.**

register (binding), a bookmar-
ker; (typ.) when pages back one
another exactly, or when the
separate colour printings of an
illustration superimpose exactly, they are said to be 'in
register'

registered, abbr. **regd.**

register marks (typ.), crosses
used to help achieve good
register

Register Office, off. name, *not*
Registry ——

registrable, *not* -erable

registr/ar, -y, abbr. **reg.**

Registrar-General (hyphen,
caps.) abbr. **Reg.-Gen.**

Registrary, official of Cambridge University (*not* Regis-
trar)

Registry Office, *see* **Register
Office**

regium donum (Lat.), a royal
grant

Regius Professor, abbr. **Reg.
Prof.**

règle (Fr. f.), a rule

Reg. Prof., Regius Professor

**regret/, -ful, -fully, -table,
-tably, -ted, -ting**

regt., regiment

regul/a (Lat.), a book of rules;
pl. **-ae**

regular/, -ly, abbr. **reg.**

regulator, *not* -er

rehash (noun *and* verb, one word)

re-heat (noun *and* verb, hyphen)

Reich, the German State and Commonwealth (cap.); **Reichs/anstalt,** German off. institution (1871–1945); *Reichsanzeiger,* German Gazette; **-kanzler,** German Chancellor; **-mark,** abbr. **RM; -tag,** German legislative body (1871–1933)

Reid (Sir George), 1841–1913, Scottish painter; — **(Capt. Thomas Mayne),** 1818–83, British novelist; — **(Thomas),** 1710–96, Scottish metaphysician; — **(Whitelaw),** 1837–1912, US journalist and diplomat. *See also* **Read, Reade, Rede, Reed**

Reikiavik, *use* **Reykj-**

Reilly (the life of), *use* **Riley**

reimbark etc., *use* **re-embark** etc.

reimburse, to pay for loss or expense (one word)

réimpression (Fr. f.), a reprint

Reims, dép. Marne, France, *not* Rh-

Reine Claude (Fr. f.), greengage; *pl.* **Reines —**

re infecta (Lat.), with the object not attained

reinforce/, to strengthen; **-ment.** *See also* **re-enforce**

Reinhardt (Max), 1873–1943, German theatrical producer

reinstate (one word)

reis, pl. of *real,* a former Portuguese and Brazilian coin

Reis Effendi (Turk.), title of former Secretary of State for Foreign Affairs

rejectamenta (mod. Lat. pl.), wasted matter (not ital.)

rel., relative, -ly, religion, religious, (Fr.) *relié* (bound), (Lat.) *reliquiae* (relics)

relat/er, one who relates; **-or** (law), one who lays information before the Attorney-General

relation, a relative; a narrative; the way in which one person or thing relates to another

relationship, the state or fact of being related; kinship

relative/, -ly, abbr. **rel.**

relativity, the state of being relative, Einstein's theory of space-time

relativize, *not* -ise

relator, *see* **relater**

relay, to arrange in relay(s), to pass on (one word)

re-lay, to lay again (hyphen)

releas/er, one who releases; **-or** (law), one who grants a release

re-let/, to let again; **-ting** (hyphen)

relev/é (Fr.), *fem.* **-ée,** exalted, noble; (cook.) highly seasoned, not *ré-*

reliable, *not* -y-

relié (Fr. bibliog.), bound

relief-printing (typ.), printing from surfaces raised to contact ink and paper

relievo, use **rilievo**

religieuse/ (Fr. f.), a nun; *pl.* **-s**

religieux (Fr. m.), a monk; *pl.* same

relig/ion, -ious, abbr. **rel.**

religionize, *not* -ise

religious/ denominations, as Baptist, Protestant, to have caps.; — **marks,** *see* **ecclesiastical signs**

reliquiae (Lat. pl.), relics; abbr. **rel.**

rel/y, -ied, -ying

REM, rapid eye-movement

rem, roentgen equivalent man

rem., remarks

remainder (typ.), that part of an edition which is unsaleable at its original price

Remarque (Erich Maria), real name **Krämer,** 1898–1970, German-born US novelist

remblai/, earth used to form rampart, *pl.* **-s**

Rembrandt/ (in full, — **Harmenszoon van Rijn**), 1606–69, Dutch painter; adj. **-esque**

REME, Royal Electrical and Mechanical Engineers

Remembrance Sunday, nearest to 11 Nov.

remerciment (Fr. m.), thanks, not *-iement*

rem/ex, wing quill-feather; *pl.* **-iges**

Reminiscere Sunday, the second in Lent

remise (law), to make over, *not* *-ize*

remissible, capable of being forgiven

remit/, -tal, -tance, -tee, -tent, -ter

remonstrator, *not* -er

rémoulade or rémolade (Fr. f.), salad-dressing, kind of sauce

removable, *not* -eable

Rémusat (Charles François Marie, comte de), 1797–1875, French politician and writer; — **(Jean Pierre),** 1788–1832, French Chinese scholar

Renaissance (the), *not* -ascence (cap.)

Renard, the fox, *use* **Rey-**

Renault, French make of car

rendezvous/, *sing.* and *pl.* (one word, not ital.); as verb **-es, -ed, -ing;** Fr. *rendez-vous* (hyphen)

renege, renounce, abandon, *not* -gue

Renoir (Auguste), 1841–1919, French painter; — **(Jean),** 1894–1979, his son, French film-director

renounceable, *not* -cable

renouncement, *use* **renunciation**

renovator, *not* -er

rent-a/-car, -crowd, etc. (hyphens)

rentes/ (Fr. f.), independent income, *also* government stocks; — *sur l'État,* interest on government loans

rent/ier (Fr.), *fem.* *-ière,* one whose income is derived from investments

renunciation, preferable to renouncement

reoccup/y, -ation (one word)

reometer etc., *use* **rhe-**

re/open, -order, -organize, *no* -ise (one word)

reorient/, -ate (one word); *see* **orient**

Rep., Representative

rep, a fabric, *not* repp; colloq. fo representative, (theat.) repertory (no point)

rep., report, -er, republic, -an

rep/air, -airable (of material things), **-aration** (amends)

reparable (of loss, etc.), that ca be made good

repartee, (the making of) witty retorts (not ital.); in Fr. f. *repartie*

repêchage (Fr. m.), contest between runners-up

repel/, -led, -lent, -ling, -ler

repertoire (not ital.)

repertorium (Lat.), a catalogue

repetatur (Lat.), let it be repeated; abbr. *repet.*

répétiteur (Fr. m.), one who rehearses opera-singers (ital.)

repetitorium (Lat.), a summary

replaceable, *not* -cable

replica/, a copy (esp. one made by the artist); *pl.* **-s** (not ital.)

réplique (Fr. f.), a reply

repl/y, -ies, -ier, -ying

report/, -er, abbr. **rep., rept.**

repoussé, (ornamental metal work) hammered from the reverse side (not ital.)

repp, a fabric, *use* **rep**

repr., representing, reprint

reprehensible, open to rebuke, *not* -able

representable, *not* -ible

Representative, abbr. **Rep.**

Representatives (House of), lower division of US Congress (caps.)

repress/, -ible, -or, *not* -er

reprint (typ.), a second or new impression of any printed work (only minor corrections); a reimpression (no corrections at all); printed matter taken from some other publication for reproduction; also printed 'copy' (one word); abbr. **RP** *or* **repr.** *See also* **edition**

reprisal, act of retaliation, *not* -izal

reprise, (law) a yearly charge or deduction, (mus.) repeated passage, *not* -izc

reprize, to prize anew

repro., reproduction; — **pull** (typ., no point), a good proof made for camera use

reproducible, *not* -eable

reproof (noun), a rebuke

reproof (verb), to make waterproof again

reprove, to rebuke

rept., report

republic/, -an, abbr. **rep.**

Republican, abbr. (US) **R.**

République française (l.c. *f*), French Republic; abbr. **RF**

repudiator, *not* -er

reputable, respectable

requiem/, the Mass for the dead, its mus. setting; *pl.* -**s**

requies/cat in pace (Lat.), may he, *or* she, rest in peace, abbr. **RIP**, *pl.* -*cant* — —; -*cit* — —, he, *or* she, rests in peace

reredos/, ornamental screen or panelling behind altar, *not* the many variants; *pl.* -**es**

re-route/ (hyphen); -**ing**

rerun (noun and verb, one word)

res., research, reserve, resid/ence, -es, resigned

res/ (Lat.), a thing or things; — *adjudicata,* a matter already decided; — *angusta domi,* scanty means at home

re/sale, -sell (one word)

rescuable, *not* -ueable

research, abbr. **res.**

reserve, that which is kept back, restraint, to keep back (one word); abbr. **res.**

re-serve, to serve again (hyphen)

Reserve (Army) (caps.)

reservist, *not* -eist

res gestae (Lat. pl.), things done, matters of fact

resid/ence, -es, abbr. **r.** *or* **res.**

residence permit (two words)

residu/um, *pl.* -**a**

resigned, abbr. **res.**

resin, gum from trees or synthetically produced. *See also* **rosin**

resist/ance, -ant, *not* -ence, -ent

resist/er, person; -**or,** thing

res/ judicata (Lat.), a thing already decided; — *nihili,* a nonentity

resoluble, that can be resolved (one word)

re-soluble, that can be dissolved again (hyphen)

resolv/able, -er

resonator, an instrument responding to a certain frequency of vibrations

resort, have recourse to, expedient, holiday-spot (one word)

re-sort, to sort again (hyphen)

resource, in Fr. f. *ressource*

resp., respondent

respecter, *not* -or

Respighi (Ottorino), 1879–1936, Italian composer

respirator, *not* -er

respondent, abbr. **resp.**

response mark (typ.), ℞

responsible, in Fr. *responsable*

restaurateur, restaurant-keeper, *not* -rant-

rest/-cure, -day, -home (hyphens)

restor/able, -er

Restoration (the) (cap.)

restrain, to check (one word)

re-strain, to strain again (hyphen)

rest-room (hyphen)

resum/e, -able

résumé, a summary (not ital.)

resurgam (Lat.), I shall rise again

resuscitat/e, -or

Reszke (Édouard de), 1856–1917, Polish bass; — **(Jean de)**, 1850–1925, his brother, Polish tenor

ret., retired, returned

retable, shelf or framed panels above back of altar

retd., retired, returned

R. et I. (Lat.), *Rex et Imperator*, king and emperor, *Regina et Imperatrix*, queen and empress

retin/a, inner membrane of the eyeball; *pl.* (anat.) **-ae**, (gen.) **-as**

retired, abbr. **ret.** *or* **retd.**

retraceable, *not* -cable

retractable, *not* -ible

retractor, *not* -er

retree, slightly defective paper

retrial (one word)

retriever, a dog

retrochoir (archit.) (one word)

retroflex (adj.), turned backwards (one word)

retro-rocket (hyphen)

retrouss/é (Fr.), *fem.* **-ée,** turned up

retry (one word)

returned, abbr. **ret.** *or* **retd.**

Reuben (Bib.), son of Jacob

Reubens (Peter Paul), *use* **Rubens**

reunion, a social gathering; in Fr. f. *réunion*

Réunion (Île de), Indian Ocean

re/unite, -urge, -usable, -use (one word)

Reuters Ltd., reporting agency

Rev., Book of Revelation (NT), Review. *See also* **Reverend**

rev., revenue, reverse, revis/e, -ed, -ion, revolution, -s

Reval (Estonian SSR), *use* **Tallinn**

revalorize, to establish a new value, *not* -ise

Revd, *see* **Reverend**

réveil (Fr. m.), an awaking, a morning call

reveille, morning call to troops (no accents, not ital.)

revel/, -led, -ler, -ling

Revelation (Book of) (NT), *not* -ions; abbr. **Rev.**

revenons à nos moutons (Fr.), let us return to our subject

revenue/, abbr. **rev.**; **— office — tax** (two words)

reverberator, *not* -er

Reverend, abbr. **(The) Revd;** *pl.* **Revds; Very Revd** (dean, provost, or former moderator), **Right Revd** *or* **Rt. Revd** (bishop or moderator), **Most Revd** (archbishop or Irish RC bishop); *not* Rev.

rever/end, deserving reverence; **-ent**, showing it

reverie, day-dream, *not* -y

revers, the front of a garment turned back showing the inner surface, *pl. same*

reverse, side of a coin other than obverse (q.v.), abbr. **rev.**

reversed block (typ.), a design in which the illustration or wording appears in white against a background; block whose contents are transposed left-to-right for offset printing

reversible, *not* -able

rêveu/r (Fr.), *fem.* **-se,** (day-)dreamer

Review, abbr. **Rev.**

Reviews (titles of) (typ.), when cited, to be italic

revise (typ.), second or subsequent proof; abbr. **rev.**

revis/e, *not* -ize; **-able, -ing**

revis/ed, -ion, abbr. **rev.**

Revised Standard Version (of Bible) (caps.); abbr. **RSV**

Revised Statutes, abbr. **RS** *or* **Rev. Stat.**

Revised Version (of Bible) (caps.); abbr. **RV**

reviv/er, one who revives; (law) **-or,** a proceeding to revive a suit

revoir (*à*) (Fr.), to be revised. *See also* **au revoir**

revo/ke, -cable, -cation

Revolution (the), Amer. 1775–83, Chin. 1911–12, Eng. 1688–9, Fr. 1789–95, 1830, 1848, 1870, Russ. 1905, 1917 (cap.)

revolution/, -s, abbr. **rev.,** but cf. **r.p.m., r.p.s.**

revolutionize, *not* -ise

Rev. Stat., Revised Statutes

revue, a theatrical entertainment (not ital.)

rex (Lat.), king; abbr. **R.**

Rexine, propr. term for a kind of imitation leather (cap.)

Reykjavik, Iceland, *not* Reiki-

Reynard, the fox, *not* Ren- (cap.)

Reynolds News (no apos.)

Reynolds number (two words, no apos.), abbr. **Re**

rez-de-chaussée (Fr. m.), ground floor (hyphens)

Rezniček (Baron Emil Nikolaus von), 1860–1945, Austrian composer

RF, *République française* (French Republic)

Rf, rutherfordium (no point)

r.f., radio frequency

rf., see **rinforzando**

RFA, Royal Field Artillery, Royal Fleet Auxiliary

RFC, Rugby Football Club

RGS, Royal Geographical Society

RH, Royal Highness

Rh, rhesus, rhodium (no point)

r.h., right hand

RHA, Regional Health Authority, Road Haulage Association Ltd., Royal Hibernian Academy, — Horse Artillery

rhabdomancy, divination by rod, *not* ra-

Rhadamanth/ine, stern, like **-us** (Gr. myth.), judge of the dead

Rhaet/ia, -ian, -ic, (of) Austrian Tyrol, *not* Rae-, Rhe-

rhaphis, *use* **ra-**

rhapsodize, to be enthusiastic, *not* -ise

rhatany (bot.), S. American shrub with astringent root, *not* rat-

rheebok, *use* **ree-**

Rheims, *use* **Reims**

Rhein (Ger. m.), the Rhine

Rheingold (Das), first opera in Wagner's *Ring des Nibelungen,* 1869

Rhemish, of Rheims

Rhenish, (wine) of the Rhine

rhenium, symbol **Re**

rheology (phys.), science of flow

rheo/meter, -stat, -trope, instruments for (respectively) measuring, regulating, reversing electric current, *not* reo-

rhesus, Indian monkey; **rhesus factor** (in the blood), abbr. **Rh-factor** (hyphen, no point); **Rh-positive,** reacting to blood tests like rhesus monkeys; **Rh-negative,** not ditto

rhet., rhetoric

Rhetia, *use* **Rhae-**

RHF, Royal Highland Fusiliers

RHG, Royal Horse Guards (the Blues)

Rhine/, European river, in Fr. **Rhin,** in Ger. **Rhein; -land** (one word)

rhinestone, imitation diamond

rhinoceros/, *pl.* **-es;** colloq. **rhino/,** *pl.* **-s** (no point)

R.Hist.S., Royal Historical Society

Rhode Island, off. abbr. **RI**

Rhod/es, island in Aegean Sea, and its capital; adj. **-ian**

Rhodes Scholar (at Oxford)

Rhodesia, Africa, UDI Nov. 1965, republic by unilateral declaration, Mar. 1970, renamed Zimbabwe-Rhodesia 1979, Zimbabwe 1980

rhodium, symbol **Rh**

rhododendron/, *pl.* **-s**

rhodomontade, *use* **rodo-**

rhomb/us (geom.), *pl.* **-uses**

Rhondda, river and town, Mid Glam.

Rhône, Fr. dép. and river, *not* Rhone

RHS, Royal Historical, Horticultural, *or* Humane Society

rhumba, *use* **rumba**

rhumb-line (naut.), line cutting all meridians at same angle (hyphen)

rhym/e (noun and verb), *not*
rime; **-er** (**-ester** is usu. derog.)

rhythm/, -ic

RI, Rhode Island (off. abbr.),
Royal Institute of Painters in
Water Colours, Royal Institu-
tion

RI, = *R. et. I.* (q.v.)

RIA, Royal Irish Academy

Rialto, Venice (cap.)

riant, laughing, cheerful, pleasant (not ital.)

rib/, -bed, -bing

RIBA, Royal Institute of British
Architects

riband, *see* ribbon

ribband, a light spar used in
shipbuilding

ribbon, *not* riband, except in
sport and heraldry

RIC, Royal Institute of Chemistry, *now* part of Royal Society of
Chemistry

Ricard/o (**David**), 1772–1823,
English economist; adj. **-ian**

Riccio, *use* Rizzio

ricercar (mus.), type of instrumental composition

Richelieu (**Armand Jean du
Plessis, duc de, Cardinal**),
1585–1642, French statesman

Richepin (**Jean**), 1849–1926,
French writer

Richter (**Hans**), 1843–1916,
German conductor of Wagnerian opera and the Hallé
Orchestra; — (**Johann Paul
Friedrich**), 1763–1825, German writer, pseud. Jean Paul;
— (**Karl**), 1926–81, German
conductor; — (**Sviatoslav**),
b. 1915, Russian pianist

Richthofen (**Baron Manfred
von**), 1892–1918, German
flying ace

rick, sprain or strain, *not* wrick

rickets, a bone disease

rickettsi/a, a micro-organism
such as typhus, *pl.* **-ae**; from
Howard Taylor Ricketts,
1871–1910, US pathologist

rickety, shaky, *not* -tty

rickrack, *use* ricrac

rickshaw, light two-wheeled
vehicle, orig. a shortening of the
now disused word **jinricksha**,
not -sha

ricochet/, to skip or rebound
(of projectile), **-ed, -ing** (not
ital.)

ricrac, zigzag braid trimming,
not rickrack

RICS, Royal Institution of Chartered Surveyors

rid/, -dance, -ded, -ding

rid/e, -able, -den, -ing

ridge/-piece, -pole, -tile, -tree
(hyphens)

ridgeway (one word)

ridgy, *not* -ey

Riemann/ (**Georg Friedrich
Bernhard**), 1826–66, German
mathematician; adj. **-ian**

Riesling, a white wine

Rievaulx Abbey, N. Yorks., *no.*
Riv-

rifaciment/o (It.), a remaking;
pl. **-i**

Riff (**The**), mountainous region
in Morocco, *not* Rif

riff-raff (hyphen)

rifl/e, -ing

rigadoon, lively dance, *not* ri-
gaudon

right/ (theat., from actor's point
of view), abbr. **R.**; — **centre,**
abbr. **RC**

right angle, symbol ∟

right ascension (astr.), abbr.
RA

righteous

right/-hand, -handed, adjs.
(hyphens)

right-hand pages (typ.), the
recto pages, usu. with odd page
numbers

Right Reverend (for bishops
and moderators in Presbyterian
churches and Church of Scotland), abbr. (**The**) **Right
Revd, Rt. Revd**

right wing (two words), *but*
right-winger (hyphen)

Rigi, Switzerland, *not* -hi

rigor (med.), a shivering-fit (*not*
ital.)

rigor mortis (Lat.), stiffening after death (not ital.)

rigour, severity, *but* **rigorous**

Rigsdag, the former Danish Parliament

Rig-Veda, Sanskrit religious book (caps., hyphen)

RIIA, Royal Institute of International Affairs

Riksdag, the Swedish Parliament

Riksmål, the literary language of Norway, now called **Bokmål**

Riley (**the life of**), slang for carefree existence, *not* Reilly;
— (**Bridget Louise**), b. 1931, British painter; — (**James Whitcomb**), 1849–1916, US poet

riliev/o (It.), raised or embossed work, not *re-*; *pl.* -*i*

Rilke (**Rainer Maria**), 1875–1926, German poet

rill, small stream

rille, narrow valley on moon

rim/, -med, -ming

rim/a (It.), verse, *pl.* -*e*

rim/e, hoar-frost; -**y.** *See also* **rhyme**

Rimsky-Korsakov (**Nicholas Andreievich**), 1844–1908, Russian composer

RINA, Royal Institution of Naval Architects

rinderpest, pleuro-pneumonia in cattle (one word)

rinforz/ando, -ato (It. mus.), with more emphasis; abbr. *rf.* or *rinf.*

Ring des Nibelungen (*Der*), cycle of operas by Wagner, 1869–76

ring/leader, -master (one word)

Rio de Janeiro, Brazil, *not* Rio Janeiro

Rio Grande, river, Brazil and Mexico

RIP, *requiescat* (or *-ant*) *in pace* (may he, she (*or* they) rest in peace)

RIPA, Royal Institute of Public Administration

R.I.P.H. & H., Royal Institute of Public Health and Hygiene

ripien/o (mus.), player or instrument additional to leader or soloist, *pl.* -*i*

Ripman (**Prof. Walter**), 1869–1947, English educationist

rippl/e, -y

Rip Van Winkle, by Washington Irving, 1820 (three caps.)

ris de veau (Fr. cook. m.), sweetbread

rises, abbr. **r.**

risible, provoking laughter; *not* -able

Risorgimento, Italian unification movement in 19th c.

risott/o, rice with meat and onions, etc., *pl.* -**os**

risqu/é (Fr.), *fem.* -*ée*, risky, indelicate

rissole, fried cake of minced meat (not ital.)

rissolé (Fr. cook.), well-browned

ritardand/o (mus.), holding back, *pl.* -**os**; abbr. **rit.** *or* **ritard.**

Ritchie-Calder (**Baron**) (hyphen)

ritenuto (mus.), held back; abbr. **riten.**

ritornell/o (mus.), a short instrumental passage in a vocal work; return of full orchestra after a solo passage in a concerto; *pl.* -**os**

ritualist/, one devoted to ritual (adj. -**ic**); (cap. with hist. reference)

ritualize, *not* -ise

Riv., River

rival/, -led, -ling

Rivaulx, *use* **Rie-**

River, when with name to have cap., as Yellow River, River Dart; abbr. **R.** *or* **Riv.**

river (typ.), apparent tracks of white twisting down printed page, caused by bad word-spacing

Rivera y Orbaneja (Miguel Primo de), 1870–1930, Spanish general, dictator 1923–30

river/-god, -mouth (hyphens)

riverside (one word)

Rivesaltes, a French wine

rivet/, -ed, -er, -ing

Riviera, the coast of France and Italy from Nice to La Spezia; **riviera,** similar region elsewhere

Rivière (Briton), 1840–1920, English painter

rivière (Fr. f.), river, (of diamonds) collar

Rivingtons, Ltd., publishers

Riyadh, cap. of Nejd, and one of the federal capitals of Saudi Arabia. *See also* **Mecca**

riz (Fr. m.), rice

Rizzio (David), 1540–66, Italian musician, favourite of Mary Q. of Scots, assassinated, *not* Ricc-

RL, Rugby League

RLO, Returned Letter Office (formerly Dead — —)

R.L.S., Robert Louis Stevenson

RLSS, Royal Life Saving Society

rly., railway

RM, Reichsmark, Resident Magistrate, Royal Mail, — Marines

rm., ream, room

RMA, Royal Military Academy (Sandhurst, *formerly* Woolwich)

RMCS, Royal Military College of Science

R.Met.S., Royal Meteorological Society

RMP, Royal Military Police

RMS, Royal/ Mail Service, — Mail Steamer, — Microscopical Society, — Society of Miniature Painters

r.m.s., root-mean-square

RMSM, Royal Military School of Music

RN, Royal Navy

Rn, radon (no point)

RNA, ribonucleic acid

RNC, Royal Naval College

RNCM, Royal Northern College of Music

RNIB, Royal National Institute for the Blind; **RNID,** ditto Deaf

RNLI, Royal National Life-boat Institution

RNR, Royal Naval Reserve

RO, Receiving Office, -r, Relieving Officer, Returning —, Royal Observatory

ro., rood

Road, *after* name to be cap., as Fulham Road; abbr. **Rd.**

road-block (hyphen)

road fund licence (three words)

road/-hog, -map (hyphens)

road/side, -stead, -way (one word)

road-works (hyphen)

roadworthy (one word)

roan (bind.), a soft and flexible sheepskin, often imitating morocco

Roanoke, Virginia, US

roast (to rule the), *use* **roost**

Robarts (Mr), in Trollope's *Framley Parsonage*

Robbe-Grillet (Alain), b. 1922, French writer

Robben Island, S. Africa

Robbia (Luca della), 1399–1482, Italian sculptor. *See also* **Della-Robbia**

Robbins Report, 1963, on Higher Education (no apos.)

Robens (Alfred, Lord), b. 1910, Chairman of the National Coal Board 1961–71

Robespierre (Maximilien François Marie Isidore de), 1758–94, French revolutionary

Robin Goodfellow, a sprite (caps.)

Robin Hood/, hero of medieval legend; — —'s **Bay,** a town, N. Yorks. (apos., three words)

robin redbreast (two words)

Robinson Crusoe, by Defoe, 1719

Rob Roy, Robert ('the Red') Macgregor, 1671–1734, Scottish outlaw (caps., two words)

ROC, Royal Observer Corps

roc, a fabulous bird, *not* the many variants

roccoco, *use* **rococo**

Roch (St.)

Rochefoucauld, *see* **La —**

roches moutonnées (geol.), *pl.*, a glaciated type of rock-surface (not ital.)

rochet, a surplice-like linen garment, *not* rotch-, -ette

Rocinante, Don Quixote's steed, usu. Anglicized as **Rosinante**

Rock (the), Gibraltar

Rockefeller/ (John Davison), 1840–1937, head of family of US capitalists; — **(Nelson Aldrich),** 1908–79, US statesman; — **Center,** nexus of office skyscrapers, NY; — **Foundation,** 1913, philanthropic; — **Institute,** 1901, for medical research, a university since 1965

rock-garden (hyphen)

Rockies (the) (N. Amer.), the Rocky Mountains

'Rock of Ages' (caps.)

rock-plant (hyphen)

rock salmon (two words)

rococo/, ornamental style of decoration in 18th c. Europe, also in mus.; *pl.* **-s,** *not* rocc-

rod, *see* **perch**

Rod (Édouard), 1857–1910, French writer

Rodd (James Rennell, first Baron Rennell), 1858–1941, English diplomat and scholar

rodeo/, *pl.* **-s**

Roderic, d. 711, last king of the Visigoths, *not* -ick

Rodin (Auguste), 1840–1917, French sculptor

rodomontade, bragging talk, *not* rh-

roebuck (one word)

roe-deer (hyphen)

roentgen/, former unit of ionizing radiation dose, *now* expressed in coulomb per kilogram; abbr. **R;** **-ography,** photography with X-rays;

-ology, study of X-rays. *See also* **Röntgen**

ROF, Royal Ordnance Factory

Roffen:, sig. of Bp. of Rochester (colon)

Rogation Sunday, that before Ascension Day

Rogers (Bruce), 1870–1951, US typographer, designer of Centaur type

Roget (Peter Mark), 1779–1869, compiled *Thesaurus*

rognons (Fr. m.), kidneys

rogues' gallery, *not* rogue's

Rohilkhand, Uttar Pradesh, India, *not* Rohilc-, -und

Rohmer (Eric), b. 1920, French film-director; — **(Sax),** pseud. of Arthur Sarsfield Ward, 1883–1959, British mystery-writer

ROI, Royal Institute of Oil Painters

roi fainéant (Fr.), do-nothing king

roinek, *use* **rooinek**

roisterer, noisy reveller, *not* roy-

rok, a fabulous bird, *use* **roc**

Rokitansky (Karl, Baron von), 1804–78, Czech-born Austrian anatomist

role, actor's part, in Fr. m. *rôle*

roll-call (hyphen)

roller-coaster, switchback (hyphen)

roller-skate (noun and verb, hyphen)

rolling-pin (hyphen)

rollmop (one word)

rollock, *use* **rowlock**

roll-on (adj., hyphen)

Rolls-Royce (hyphen)

Rölvaag (Ole Edvart), 1876–1931, Norw.-US novelist

roly-poly, a pudding, *not* the many variants

ROM (comp.), read-only memory

Rom/, a male gypsy, *pl.* **-a** (cap.)

Rom., Roman, Romance, Romans (Epistle to the)

rom., roman type

Romaic, modern Greek

romaika, a national dance of mod. Greece

Romain (**Jules**), *c.* 1499–1546, Italian architect, *but use* **Giulio Romano**

Romains (**Jules**), 1885–1972, French writer

romaji, Roman alphabet used to transliterate Japanese

roman (typ.), ordinary upright letters as distinct from bold or italic; abbr. **rom.**

Roman Catholic/ (caps.), abbr. **RC**; —— **Church** (caps.), abbr. **RCC**

roman-à-clef (Fr. m.), novel about real people under disguised identities, *pl.* *romans-à--clef*

Roman de la Rose, thirteenth--cent. French allegorical verse romance, source of *Romaunt of the Rose,* attributed to Chaucer

Romanée-Conti, a red burgundy wine

Romanes, gypsy language

Romanes lectures, at Oxford Univ.

Romanesque, style of architecture between classical and Gothic

roman-fleuve (Fr. m.), long novel about same characters over a period, *pl.* *romans--fleuves*

Romania/, **-n,** the official Romanian spelling, now preferred to the traditional English Roum- or Rum-

Romanize, *not* -ise (*not* cap. with reference to type)

roman numerals, *see* **numerals**

Romanov, Russian dynasty, 1613–1917, *not* -of, -off

Romans (**the Epistle to the**); abbr. **Rom.** (NT)

Romansh, Rhaeto-Romanic, dial. of E. Switzerland, *not* Rou-, Ru-, -ansch, -onsch

Romany, a gypsy, *not* -ncy, -mmany (cap.)

Romaunt of the Rose, see Roman de la Rose

Romeo/, a young lover, *pl.* **-s** (cap.)

Romney (**George**), 1734–1802, English painter

Romney Marsh, Kent

romneya, poppy-like shrub

Romsey, Hants

Ronaldsay, Orkney; **Ronaldshay** (**Earl of**), Marquess of Zetland's heir

Ronaldsway Airport, I.o.M.

Ronda, Spain

rondeau/, a form of poem; *pl.* **-x**

rondel, a special form of the rondeau

rondo/ (mus.), a movement with recurring main theme; *pl.* **-s**

rone, a water-pipe

Roneo/, propr. term for machine used to duplicate documents; as noun *pl.* **-s**; as verb **roneo**/, **-es, -ed, -ing**

Röntgen (**Julius**), 1855–1934, Dutch composer; — (**Wilhelm Konrad von**), 1845–1923, German physicist, discoverer of Röntgen (or X-) rays. *See also* **roentgen**

rood, abbr. **ro.**

roof/**-garden, -rack, -top** (hyphens)

rooinek (Afrik.), an Englishman, *not* roi-

rook (chess), abbr. **R,** *also* a bird

Rooke (**Sir George**), 1650–1709, English admiral

room, abbr. **rm.**

rooming-house (hyphen)

room-mate (hyphen)

room service (two words)

Roosevelt (**Franklin Delano**), 1882–1945, US President 1933–45; — (**Theodore**), 1858–1919, US President 1901–9

roost (**to rule the**), to dominate, *not now* roast

root, to rummage, to give support, to search out. *See also* **rout**

root (math.), sign √

rop/e, -ing, -y

rope/-dancer, -ladder, -walk (hyphens)

Roquefort, a Fr. cheese

rorqual, a whale

Rorschach test (psych.) (with ink-blots)

rosaceous, of the rose family

Rosalind, in Shakespeare's *As You Like It,* and Spenser's *Shepheardes Calendar*

Rosaline, in Shakespeare's *Love's Labour's Lost, Romeo and Juliet*

rosary, (RCC) (beads for) set of devotions. *See also* **rosery**

Roscommon, Ireland

rose (Fr. m.), pink colour; (f.) a rose; *couleur de* —, roseate, attractive

rosé (Fr. m.), pink wine

Rosebery (Earl of), *not* -berry, -bury

rose/-bowl, -bud, -bush (hyphens)

rosella, Austral. parakeet

Rosencrantz and Guildenstern, in Shakespeare's *Hamlet*

Rosenkranz (Johann Karl Friedrich), 1805–79, German metaphysical philosopher

Rosenkreuz, *see* **Rosicrucian**

roseola (med.), rosy rash, German measles

rose-red (hyphen)

rosery, a rose-garden. *See also* **rosary**

rose-water (hyphen)

rosewood (one word)

rosey, *use* **rosy**

Rosh Hashanah, the Jewish New Year's Day

Rosicrucian, (a member) of an order devoted to occult lore, founded by Christian Rosenkreuz, 1484

rosin, a solid residue from turpentine distillation, used esp. on strings of musical instruments. *See also* **resin**

Rosinante, *see* **Rocinante**

Roskilde/, Denmark; —

(Treaty of), 1658, between Denmark and Sweden

ROSLA, raising of school-leaving age

Roslin, Lothian

Rosny (Joseph Henry), joint pseud. of French novelist brothers Joseph Henri Boëx, 1856–1940, and Séraphin Justin François Boëx, 1859–1948

rosoli/o, a sweet cordial of S. Europe, *pl.* **-os,** *not* -oglio, -oli (not cap.)

RoSPA, Royal Society for the Prevention of Accidents

Ross (Sir James Clark), 1800–62, English Arctic explorer; — **(Sir John),** 1777–1856, his uncle, Scottish explorer; — **(Sir Ronald),** 1857–1932, English physician (malaria-mosquito)

Rosse (Earl of)

Rossetti (Christina Georgina), 1830–94, English poet; — **(Dante Gabriel),** 1828–82, her brother, English painter and poet; — **(Gabriele),** 1783–1854, Italian poet and liberal, father of the other three; — **(William Michael),** 1829–1919, his son, English author and critic

Rosslyn (Earl of). *See also* **Roslin**

Ross-shire (hyphen); *now* the district of **Ross and Cromarty**

Rostand (Edmond), 1868–1918, French playwright

roster, a list of persons, showing rotation of duties

Rostropovich (Mstislav), b. 1927, Russian cellist and conductor

rostr/um, speaker's platform, *pl.* **-a** (not ital.)

ros/y, *not* -ey; **-ily**

rot/, -ted, -ting

rota/, a roster; *pl.* **-s**

Rotarian, a member of a Rotary Club (cap.)

rotary, revolving, *not* rotatory;

rotary (*cont.*)
(typ.) printing machine in which the plate(s) are mounted on a cylinder

Rotary Club, a branch of the world-wide Rotary Movement, aiming at service to humanity (caps.)

rotator, a revolving part, (anat.) a muscle that rotates a limb

rotatory, *use* **rotary**

Rotavator, propr. term for a rotary cultivator (cap.)

rotchet, garment, *use* **rochet**

rote, mechanical memory or performance

Rothamsted, Herts., agric. station for soil research

Rothe (Richard), 1799–1867, German theologian

Rothes, Grampian; — **(Earl of)**

Rothschild, European family of bankers

rôti (Fr. m.), roast meat

rotifer/, minute aquatic animal, *pl.* **-s; Rotifera,** phylum containing them

rotogravure (typ.), printing from photogravure cylinders on a web-fed rotary press

rotor (mech.), a revolving part

rottenstone, decomposed siliceous limestone (one word)

Rottingdean, E. Sussex (one word)

Rottweiler, breed of tall black dog

rotund/a, a domed circular building or hall, *pl.* **-as**

roturi/er (Fr.), *fem.* **-ère,** of low birth

Rouault (Georges), 1871–1958, French painter

Roubiliac (Louis François), 1695–1762, French sculptor, *not* -lliac

rouble, Russian coin and monetary unit, *not* ru-; abbr. **r.**

rouche, *use* **ruche**

roué, a debauchee (not ital.)

Rouge/Croix, — Dragon, pursuivants

rouge-et-noir, a game of chance (hyphens, not ital.)

Rouget de l'Isle (Claude Joseph), 1760–1836, composer of *La Marseillaise*

rough, trump, *use* **ruff**

rough-and-/-ready, -tumble (hyphens)

roughcast (one word)

rough-/dry, -hew (hyphens)

rough house (hyphen as verb)

roughneck (one word)

rough pull (typ.), proof pulled by hand on inferior paper for correction purposes

rough-rider (hyphen)

roughshod (one word)

roulade (mus.), a florid passage (not ital.)

rouleau/, a roll of money; *pl.* **-x** (not ital.)

Roumania/, -n, *use* **Romania/, -n,** q.v.

Roumansh, *use* **Rom-**

Roumelia/, -n, *use* **Rum-**

round about (adv. and prep., two words), *but* **roundabout** (noun and adj., one word)

round-house (hyphen)

rounding (bind.), shaping back of book into a convex curve

round robin, a petition (two words)

roundworm (one word)

Rousseau (Henri), 1844–1910, French painter, called *Le Douanier*; — **(Jean Baptiste),** 1670–1741, French poet; — **(Jean Jacques),** 1712–72, Geneva-born French philosopher; — **(Pierre Étienne Théodore),** 1812–67, French painter

Roussillon, Fr. province, a red wine

rout (verb), to put to flight; (noun) a rabble, a disorderly retreat. *See also* **root**

route/, to send by a particular route, **-ing**

Routledge & Kegan Paul, Ltd., publishers

roux (cook.), mixture of fat and flour

rowlock (naut.), *not* roll-, rull-
Roxburgh, Borders
Roxburgh Club, exclusive
club for bibliophiles
Roxburghe (Duke of)
Royal/ (cap.), abbr. **R.; —**
Academ/y, — -ician, — Ar-
tillery, abbr. **RA; — High-**
ness, RH
royal, former size of paper, 20 ×
25 in.; — **4to,** $12\frac{1}{2}$ × 10 in., —
8vo, 10 × $6\frac{1}{4}$ in. (untrimmed).
See also **book sizes**
Royal Society, abbr. **RS**
Royal/ Welch Fusiliers, —
Welch Regiment (*but* **Welsh**
Guards)
roysterer, *use* **roi-**
RP, read for press, received pro-
nunciation, Reformed Presby-
terian, reply paid, reprint, (Fr.)
Révérend Père (Reverend
Father), Royal Society of Por-
trait Painters
RPC, Royal Pioneer Corps
RPE, Reformed Protestant Epis-
copal
RPI, retail price index
RPM, resale price maintenance
r.p.m., revolutions per minute.
In scientific and technical work
rev/min (no points) is pre-
ferred
RPO, Royal Philharmonic
Orchestra
r.p.s., revolutions per second. *See*
also **r.p.m.**
RPS, Royal Philharmonic Soci-
ety, Royal Photographic Society
RQMS, Regimental Quarter-
master-Sergeant
RRC, (Lady of the) Royal Red
Cross
RRE, Royal Radar Establish-
ment, *now* **RSRE**
RS, Revised Statutes, Royal
Scots, Royal Society
Rs, rupees
r.s., right side
RSA, Royal Scottish Academ/y,
-ician, Royal Society of Arts
RSAA, Royal Society for Asian
Affairs

RSC, Royal Shakespeare Com-
pany, Royal Society of Chemis-
try
RSD, Royal Society of Dublin
RSE, Royal Society of Edin-
burgh
RSFSR, Russian Soviet Federa-
tive Socialist Republic, one (and
by far the largest) of the repub-
lics forming the Union of Soviet
Socialist Republics (**USSR,**
q.v.)
RSH, Royal Society of Health
R. Signals, Royal Corps of
Signals
RSL, Royal Society of Literature
RSM, Regimental Sergeant-
-Major, Royal School of Mines,
Royal Society of Medicine,
Royal Society of Musicians of
Great Britain
RSMA, Royal Society of Marine
Artists
RSO, railway sub-office, — sort-
ing office
RSPB, Royal Society for the
Protection of Birds
RSPCA, Royal Society for the
Prevention of Cruelty to Ani-
mals
RSRE, Royal Signals and Radar
Establishment
RSV, Revised Standard Version
(of Bible)
RSVP, *répondez, s'il vous plaît*
(please reply) (not to be used
in writings in the third
person)
RT, radio-telegraphy, radio-
-telephony, received text
rt., right
Rt. Hon., Right Honourable
RTO, Railway Transport Offi-
cer
RTR, Royal Tank Regiment
Rt. Revd, Right Reverend (of a
bishop or moderator)
RTS, Royal Toxophilite Society
RTYC, Royal Thames Yacht
Club
RU, Rugby Union
Ru, ruthenium (no point)
Ruanda, *use* **Rwanda**

rub-a-dub/, make sound of drum, **-bed, -bing** (hyphens)

Rubáiyát (*The*), by Omar Khayyám; in Persian poetry *Rubá'iyát, pl.*, are four-lined stanzas

rubat/o (mus.), (performed at) tempo varied for expressive effect, *pl.* **-os**

rubber stamp (hyphen as verb)

rubef/y, to make red, *not* -ify; **-acient, -action**

rubella (med.), German measles

Rubens (**Peter Paul**), 1577–1640, Flemish painter, *not* Reu-

rubeola (med.), measles

Rubicon (**to cross the**), to take an irretraceable step (cap.); **rubicon** (noun and verb, in piquet, not cap.)

rubidium, symbol **Rb**

Rubinstein (**Anton Gregor**), 1829–94, Russian composer and pianist; — (**Artur**), b. 1886, Polish pianist

ruble, *use* **rouble**

rubric, instruction in liturgical book, orig. printed in red; special passage or heading

ruby (typ.), name for a former size of type, about 5½ pt.

ruc, -ck, -kh, a fabulous bird, *use* **roc**

RUC, Royal Ulster Constabulary

ruch/e, a quilling or frilling, **-ing**, *not* rou-

rudd, a fish, *not* rud

Rüdesheimer, a Rhine wine

rue, to regret; **rueful, ruing**

RUE (theat.), right upper entrance

ruff, a bird, *fem.* **reeve**; a fish (*not* ruffe); a frill worn round the neck; (cards) to trump, a trumping

rug/a (Lat.), a wrinkle; *pl.* **-ae**

Rugbeian, member of Rugby School

Rugby football (one cap.), colloq. **rugger** (not cap.)

Ruhmkorff/ (**Heinrich**

Daniel), 1803–77, German electrician; — **coil**

RUKBA, Royal United Kingdom Beneficent Association

rule (typ.):
  dotted, ...;
  double, ⹀;
  em, —;
  en, –;
  French, ⟶ ;
  milled, ----;
  parallel, ⹀;
  single, ——— ;
  spread *or*
  swelled, ⬥——— ;
  total, ⬥——— ;
  wavy, 〰〰〰

'**Rule, Britannia!**' (comma)

rullock, *use* **rowlock**

Rumania/, -n, *use* **Romania/, -n,** q.v.

Rumansh, *use* **Rom-**

rumba, *not* rhumba

rumb-line, *use* **rh-**

Rumelia/, an area of the Balkans, mainly in Bulgaria; **-n;** *not* Rou-

rumen/, ruminant's first stomach, *pl.* **-s**

rumin/ant, an animal that chews the cud; **-ator,** one who ponders, *not* -er

Rumpelstiltskin, a dwarf in German folklore

run (naut.), deserted; abbr. **R.**

runabout, small light vehicle (one word)

runaway (noun and adj., one word)

Runeberg (**John Ludwig**), 1804–77, Finnish poet

running/ headline *or* — **title** (typ.), *see* **headlines**

Runnymede, meadow on R. Thames where King John signed Magna Charta, 1215

runoff (noun, one word)

run/ on (typ.), continued operation of press subsequent to first (or stated amount of) copy; (matter) to be set without break or paragraph; — **round,** set text etc. round three or more sides of block etc.

runway (one word)

rupee, abbr. **Re,** *pl.* **Rs**

Rupert's Land (Bp. of), Canada

RUR, Rossum's Universal Robots (title of play by K. Čapek, 1920)

ruralize, *not* -ise

RUSI, Royal United Services Institute for Defence Studies

rus in urbe (Lat.), the country within a town

Russ., Russia, -n

Russel (Alexander), 1814–76, editor of the *Scotsman*

Russell cord, a ribbed fabric (two words)

Russell (Bertrand Arthur William, Earl), 1872–1970, English philosopher and mathematician; — **(Baron, of Liverpool);** *also* family name of Duke of Bedford and Barons Ampthill and De Clifford; — **(George),** *see* **A.E.;** — **(Jack),** breed of dog

Russia/, -n, abbr. **Russ.;** (typ.) 36 letters, of which 4 were abolished in the New Orthography of 1918; for details see *Hart's Rules,* p. 119

rut/, -ted, -ting, -ty

Ruth, not to be abbr.

ruthenium, symbol **Ru**

Rutherford (Ernest, Baron), 1871–1937, New Zealand scientist; — **(Mark),** 1831–1913, pseud. of **William Hale White,** English writer

rutherfordium, symbol **Rf**

Ruthven (Baron)

Ruwenzori, mountain, Uganda and Zaïre

Ruy Lopez, chess opening

Ruysbroeck (Jan van), 1293–1387, Flemish mystic

Ruysdael (Jakob van), 1628–82, and — **(Salomon van),** 1600–70, Dutch painters

RV, Revised Version (of Bible)

RW, Right Worshipful, — Worthy

Rwanda/, Cent. Africa, indep. 1962, *not* Ruanda; adj. **-n**

RWF, Royal Welch Fusiliers

RWS, Royal Society of Painters in Water Colours

Rx., tens of rupees

Ry., railway

RYA, Royal Yachting Association

Rye House Plot, 1682–3 (no hyphens, three caps.)

RYS, Royal Yacht Squadron (a club)

Ryukyu Islands, Japan, *not* the many variants

Ryun (Jim), b. 1948, US runner

S

S, siemens, sulphur, (on timepiece regulator) slow, the eighteenth in a series

S., Sabbath, Saint, school, series, *Signor*, Socialist, Society, soprano, south, -ern, sun, Sunday, surplus, (Fr.) *saint* (saint), (Ger.) *Seite* (page), (Lat.) *sepultus* (buried), *socius* or *sodalis* (Fellow)

s., second, -s (of time or angle) (no point in scientific and technical work), section, see, sets, shilling, -s, sign, -ed, singular, (meteor.) snow, solo, son, spherical, stem, (meteor.) stratus cloud, substantive, succeeded, (Fr.) *siècle* (century), *sud* (south), (Ger.) *siehe* (see), (Lat.) *semi* (half), (It.) *sinistra* (left)

s/ (Fr.), *sur* (on), e.g. Boulogne s/M = sur-mer

's, abbr. for Du. *des* (of the), as 's--Gravenhage (The Hague)

'S., *see* **dal segno**

$ or $, the dollar mark; to be *before*, and close up to, the numerals

ʃ, italic form of long s (q.v.), (math.) sign of integration

SA, the Salvation Army, sex appeal, South Africa, South America, South Australia, (Fr.) *Société Anonyme* (Co.), (Ger.) *Sturm-Abteilung* (Nazi storm--troops)

sa., sable

s.a., *sine anno* (without date)

Saar/, river and region, **-land,** province, W. Germany (one word)

Sabaean, of ancient Yemen

Sabah, part of Malaysia

Sabaism, star-worship in anc. Arabia and Mesopotamia (cap.)

Sabaoth (Lord (God) of) (Scrip.), Lord of Hosts (cap.). *See also* **sabbath**

Sabatini (Rafael), 1875–1950, Anglo-Italian author

Sabbatarian/, -ism (caps.)

sabbath/ (day) (not cap.); — **day's journey** (Heb.), about ⅔ mile. *See also* **Sabaoth**

Sabian, member of a sect mentioned in the Koran

Sabine, of an anc. Italian tribe

sable (her.), black

sabra, native-born Israeli (not cap.)

sabretache, bag for cavalry, *not* -tash, -tasche

sabre-toothed/ lion, — - — tiger (one hyphen)

SAC, Senior Aircraftman

sac (med. and biol.), a baglike cavity

saccharimeter, instrument for measuring sugar content by means of polarized light, *not* -ometer

saccharin, sugar-substitute

saccharine (adj.), sugary

saccharometer, instrument for measuring sugar content by hydrometry, *not* -imeter

Sacheverell (Henry), ?1674–1724, English ecclesiastic and politician

Sachs (Hans), 1494–1576, German poet; — **(Julius),** 1832–97, German botanist. *See also* **Sax, Saxe**

Sachsen, Ger. for **Saxony**

sackbut, medieval trombone

sackcloth (one word)

sack-race (hyphen)

Sackville-West (Victoria Mary), 1892–1962, English author

sacr/é (Fr.), *fem.* **-ée,** sacred

sacrileg/e, -ious, *not* sacre-, -ligious

SACW, Senior Aircraftwoman

saddle/-bag, -cloth, -horse (hyphens)

saddle-stitch (bind.), stitching through the back of all pages

362

of a pamphlet etc. placed open on a saddle-shaped support

sadhu, Indian holy man

Saʿdī, ?1184–1291, Persian poet, real name Mushh ud-Din

sadism, a sexual perversion marked by love of cruelty, from **Donatien Alphonse François, 'marquis' de Sade,** 1740–1814, French soldier and writer

Sadleir (Michael), 1888–1957, English author and publisher, son of following

Sadler (Sir Michael Ernest), 1861–1943, English educationist

sado-masochism (hyphen)

s.a.e., stamped addressed envelope

safari/, hunting expedition, *pl.* **-s**

safe/ conduct, — **deposit** (hyphen when attrib.)

safety/-belt, -catch (hyphens)

safety/ curtain, — **film,** — **lamp,** — **match,** — **net,** — **razor** (two words)

safety/-pin, -valve (hyphens)

Saffron Walden, Essex (no hyphen)

S. Afr., South Africa, -n

saga/, Norse (spec. Icelandic) prose narrative; *hence* any long or heroic story; *pl.* **-s**

sagesse (Fr. f.), wisdom

Saghalien Island, *use* **Sakhalin** —

sago/, palm or starch produced from it, *pl.* **-s**

sahib (Ind.), European master, gentleman; (cap.) an honorific affix, as Smith Sahib; fem. **memsahib; pukka sahib,** perfect gentleman

Saidpur, India, *not* Sayyid-

saignant/, fem. **-e** (Fr. cook.), underdone

Saigon, Vietnam, *now* **Ho Chi Minh City**

sailcloth (one word)

sailed, abbr. **sld.**

sail/er, ship of specified power; **-or,** seaman

sailing/-boat, -ship (hyphens)

Sailors' Home, *not* -'s

sailplane (one word)

sainfoin (bot.), a leguminous fodder-plant, *not* saint-

Saint, abbr. **S.** *or* **St.,** *pl.* **SS** *or* **Sts.;** in alphabetic arrangement always place under Saint, *not* under St-. In Fr. small *s* and space after if relating to the person of a saint, as saint Jean, but cap. *S* and hyphen if relating to the name of a place or person or saint's day: as Saint-Étienne, Sainte-Beuve, la Saint-Barthélemy. In Fr. abbr. **S.,** *fem.* **Ste,** for the persons of saints; **St-,** *fem.* **Ste-,** in names of places, of persons other than saints, or of saints' days. In Ger. **Sankt,** abbr. **St.** In It. & Sp. **San, Santo, Santa,** according to gender and form of name; similarly in Port. **São, Santa**

St. Abbs, but **St. Abb's Head,** Borders; **St. Albans,** Herts. (no apos.); **St. Albans (Bp.** *and* **Duke of);** **St. Albans:,** sig. of the Bp. (colon); **St. Alban's Head,** Dorset (apos.); **St. Aldwyn (Earl); St. Andrew's Cross,** ×; **St. Andrew's Day,** 30 Nov.; **St. Andrews, Dunkeld, and Dunblane (Bp. of) St. Andrews (University),** Fife (no apos.); **St. Anne's College,** Oxford; **St. Anne's Day,** 26 July; **St. Anne's on Sea,** Lancs. (no hyphens, apos.); **St. Anthony's cross,** T; **St. Anthony's fire,** erysipelas; **St. Antony's College,** Oxford; **St. Arvans,** Gwent (no apos.); **St. Aubin,** Jersey, *not* n's; **St. Barnabas's Day,** 11 June; **St. Bartholomew's Day,** *see* **Bartholomew Day,** 24 Aug.; **St. Bees (Head, School)** (St. Begha's), Cumbria (no apos.); **St. Benet's Hall,** Oxford; **St. Bernard** (dog or Pass); **St.**

Saint *(cont.)*

Boswells, Borders (no apos.);
St. Catharine's College,
Cambridge; **St. Catherine's
College,** *formerly* **Society,** Oxford; **St. Clears,** Dyfed (no
apos.); **St. Clement's Day,** 23
Nov.; **St. Crispin's Day,** 25
Oct.; **St. Cross Church** *and*
College, Oxford; **St. Davids
(Head),** Dyfed, *and* **Viscount
St. Davids** (no apos.); **St.
David's Day,** 1 Mar.; **St.
Denis's Day,** 9 Oct.; **St. Dunstan's Day,** 19 May; **St. Edm.
and Ipswich,** sig. of present
Bp. of St. Edmundsbury and
Ipswich; **St. Edmund Hall,**
Oxford; **St. Edmund's
House,** Cambridge; **St.
Elmo's fire,** an electric discharge; **Saint Émilion,** a
claret; **Saint Estèphe,** a claret;
Saint Étienne, dép. Loire,
France; **Saint-Exupéry (Antoine Jean Baptiste Marie
Roger de),** 1900–44, French
aviator and writer; **St. Fillans,**
Tayside (no apos.); **Saint-Gall**
or **St. Gallen,** Switz.; **St.
George's Channel** (apos.); **St.
George's Cross,** red + on
white; **St. George's Day,** 23
Apr.; **St. Germans (Bishop**
and **Earl of); St. Giles',** Oxford street; **St. Gotthard,**
Switz., *not* Goth-; **St. Gregory's Day,** 12 Mar.; **St. Helens,** Cumbria, IW, Merseyside, *and* **Baron St. Helens** (no
apos.); **St. Ive,** Corn.; **St. Ives,**
Cambs., Corn., Dorset (no
apos.); **St. James's,** the British
court; **St. James's Day,** 25
July; **St. James's Palace,
Park, Square, Street,** London; **St. John,** as proper name
pron. sinjun (with stress on first
syllable); **Saint John,** New
Brunswick; **St. John Ambulance Association, Brigade,**
not John's; **St. John's,** Antigua,
Newfoundland, Quebec; **St.**

John's College, Oxford, Cambridge, Durham; **St. John the
Baptist's Day,** 24 June; **St.
John the Evangelist's Day,**
27 Dec.; **St. John's Wood,**
London; **St. John's wort,** hypericum; **Saint-Julien,** a
claret; **St. Just,** Cornwall, *and*
**Baron St. Just; St.-Just-in-
-Roseland,** Cornwall (hyphens); **Saint-Just (Louis Antoine Léon de),** 1767–94,
French revolutionary; **St.
Kitts** (no apos.), St. Christopher Island, WI; **St. Lambert's Day,** 17 Sept.; **St.
Lawrence River,** Canada; **St.
Lawrence's Day,** 10 Aug.; **St.
Leger,** a race; as surname *pron.*
sillinger *or* sentlejer (with stress
on first syllable); **St. Leonards,**
Bucks., Dorset, Lothian, Strathclyde, *and* **Baron St. Leonards**
(no apos.); **St. Leonards-on-
-Sea,** E. Sussex (hyphens, no
apos.); **St. Lucia,** WI, indep.
1979; **St. Luke's Day,** 18 Oct.;
St. Luke's summer, mid--Oct.; **St. Margaret's at
Cliffe,** Kent; **St. Margaret's
Day,** 20 July; **St. Mark's Day,**
25 Apr.; **St. Martin-in-the-
-Fields,** London church, *not*
Martin's (hyphens); **St. Martin's Day,** 11 Nov.; **St. Martin's summer,** mid-Nov.; **St.
Mary Abbots,** Kensington (no
apos.); **St. Mary Church,** S.
Glam. (three words); **St. Marychurch,** Devon (two words);
St. Mary Cray, London (no
's); **St. Matthew's Day,** 21
Sept.; **St. Matthias's Day,** 24
Feb. (s's); **St. Mawes,** Corn.;
**St. Michael and All Angels'
Day,** 29 Sept.; **St. Michael's,**
Azores, *use* **São Miguel; St.
Michael's Mount,** Corn.;
Mont-Saint-Michel, Fr. (hyphens); **St. Neot,** Corn.; **St.
Neots,** Cambs. (no apos.); **St.
Nicholas,** patron of Russia,
town in Belgium; **St. Nicho-**

las's clerks, thieves; **St. Olaf,** patron of Norway; **St. Patrick's Day,** 17 Mar.; **St. Paul,** Minn.; **St. Paul de Loanda,** W. Africa, *use* **Luanda;** **St. Paul's,** London; **St. Paul's Cray,** London; **St. Paul's Day,** 25 Jan.; **St. Peter Port,** Guernsey, *not* Peter's; **St. Peter's,** Rome; **St. Peter's Day,** 29 June; **St. Petersburg** (Petrograd), *now* **Leningrad;** **St. Philip and St. James's Day,** 1 May; **Saint-Pierre,** a claret; **Saint-Pierre (Jacques Henri Bernardin de),** 1737–1814, French author; **Saint-Saëns (Charles Camille),** 1835–1921, French composer; **St. Sepulchre (Church of); Saint-Simon (Claude Henri de Rouvroy, comte de),** 1760–1825, founder of French socialism, whence **Saint-Simonism,** a form of socialism; **Saint--Simon (Louis de Rouvroy, duc de),** 1675–1755, French diplomat and writer; **St. Simon and St. Jude's Day,** 28 Oct.; **St. Stephen,** Corn.; **St. Stephen's,** the Houses of Parliament; **St. Stephen's Day,** 26 Dec.; **St. Swithun,** *see* **Swithun; St. Thomas's Day,** 21 Dec.; **St. Valentine's Day,** 14 Feb.; **St. Vitus's dance,** chorea (s's)
Sainte-Beuve (Charles Augustin), 1804–69, French literary critic
Sainte-Claire Deville (Henri Étienne), 1818–81, French chemist (one hyphen)
saintfoin, *use* **sainfoin**
saintpaulia, African violet
Saintsbury (George Edward Bateman), 1845–1933, English literary critic
Sakandarabad, Hyderabad, India, *use* **Secunder-;** Uttar Pradesh, India, *use* **Sikandarabad**

sake, a Japanese fermented liquor, *not* -ké, -ki
Sakhalin, island, E. Asia, *not now* Saghalien, Karafuto
Sakharov (Andrei Dimitrievich), b. 1921, Russian physicist and dissident
saki/, a S. American monkey, *pl.* -s
Saki, pseud. of Hector Hugh Munro, 1870–1916, English writer
Śākyamuni, title of Buddha
salaam, oriental salutation, *not* -lam (not ital.)
salable, *use* **sale-**
salad days (two words)
salad/-dressing, -oil (hyphens)
salade (Fr. f.), salad
salami/, *pl.* -s, Anglicized form of It. *salam/e, pl.* -i, a highly seasoned sausage
Salammbô, by Flaubert, 1862
salariat, the salaried class, *not* -ate
saleable, *not* salable
Salem, Madras; Mass. (US); a Nonconformist chapel
Salempur, Uttar Pradesh, India
Salesian, (member) of an order founded by Don Bosco in honour of St. Francis de Sales
Salic law, allegedly limiting succession to certain lands among the Salian Franks to males, *not* -ique
salicylic acid
Salinger (Jerome David), b. 1919, US novelist
salle/ (Fr. f.), hall; — *à manger,* dining-room; — *d'attente,* waiting-room
sallenders, dry eruption on horse's hind leg, *not* sellanders
Sallust (in Lat. **Gaius Sallustius Crispus**), 86–?34 BC, Roman historian; abbr. **Sall.**
Sally Lunn, a teacake (caps., two words)
salmagundi, a medley; also a seasoned dish

salmi, a ragout, esp. of game; in Fr. m. *salmis*

salmon trout (two words)

salon (Fr.), reception-room, room where hairdresser etc. receives clients, (cap.) annual exhibition of living artists' pictures (not ital.)

Salonika, Greece

Salop, *now* Shropshire

Salpêtrière (La), hospital for the aged or insane, Paris

salpicon (Sp. m.), cold minced meat, not *-çon*

salsify (bot.), purple goat's--beard, *not* -afy

SALT, Strategic Arms Limitation Talks

saltarello/, It. and Sp. dance, *pl.* -s

salt-cellar (hyphen)

salt lake (two words)

Salt Lake City, Utah, US (caps., three words)

salt-mine (hyphen)

Saltoun (Baron)

saltpetre, *not* -peter (one word)

saltus (Lat.), a jump; *pl. same*

salt water (hyphen when attrib.)

saluki, Arabian gazelle-hound

salutary, beneficial, *not* -ory

salutatory, welcoming, *not* -ary

Salvador/, *properly* **El Salvador,** rep. of Cent. America; adj. **-ean**

salvage, rescue of, to rescue, ship or contents from shipwreck, goods from fire, etc.; the property salvaged or the payment made

salver, a tray. *See also* **salvor**

salvo/, simultaneous discharge of guns, bombs, cheers, *pl.* **-es**; reservation, excuse, *pl.* **-s**

salvo jure (Lat.), reserving the right

sal volatile, ammonium carbonate, smelling-salts

salvor, one who salvages. *See also* **salver**

Salzkammergut, Austria

SAM, surface-to-air missile

Sam., Samaritan

Sam. (1, 2), Samuel, First, Second Book of (OT)

samarium, symbol **Sm**

Samarkand, Uzbekistan, USSR, *not* -cand, -quand

sambo/, half-breed esp. of Negro and Indian or European blood, (cap., derog.) a Negro, *pl.* **-s**

sambok (Afrik.), *use* **sjambok**

Sam Browne (belt), officer's, with shoulder strap

S. Amer., South America, -n

samizdat, secret publication of banned matter

Samoyed, a Mongolian of NW Siberia, a breed of dog, *not* -oied, -oide, -oyede

sampan, Chinese boat, *not* san-

Sampson (Dominie), in Scott's *Guy Mannering*

Samuel, First, Second Book of (OT), abbr. **1 Sam., 2 Sam.**

samurai, Japanese mil. class (*sing.* and *pl.*)

Sana'a, N. Yemen

sanatorium/, *not* sanitarium; *pl.* -s

sanatory, healing. *See also* **sanitary**

sanbenito/, penitential garment, *pl.* -s

Sancho Panza, Don Quixote's squire

Sancho-Pedro, card-game (caps., hyphen)

sanctum/, a retreat, *pl.* **-s**; — **sanctorum** (Lat.), holy of holies in Jewish temple, a special retreat, *pl.* **sancta sanctorum** (not ital.)

Sand (George, *not* Georges), pseud. of Madame Amandine Aurore Lucile Dupin, baronne Dudevant, 1804–76, French novelist

sandal/, -led

sandarac, realgar, resin, *not* -ach

Sandars Reader, Camb.

sand/bag, -bank (one word)

sand-blast (noun and verb, hyphen)

Sandburg (Carl), 1878–1967, US poet

sand-castle (hyphen)

sandhi (gram.), change in form of word due to its position

sand/paper, -piper (bird) (one word)

sand-pit (hyphen)

sand/stone, -storm (one word)

sandwich-board (hyphen)

sandwich course (two words)

sang-de-bœuf (Fr. m.), a deep-red colour

sang-froid, composure (hyphen, not ital.)

sangría, Spanish drink (ital.)

Sanhedrin, supreme council in anc. Jerusalem, *not* -im

sanitarium, *use* **sanatorium**

sanitary, healthy, *not* -ory. *See also* **sanatory**

sanitize, *not* -ise

Sankt (Ger.), saint; abbr. **St.**

sannup (Amer.-Ind.), husband of a squaw, *not* -op

sannyasi, Hindu religious mendicant

sanpan, *use* **sam-**

sans (Fr.), without

San Salvador, cap. of El Salvador

sans/ appel (Fr.), without appeal; — *cérémonie,* informally; — *changer,* without changing

Sanscrit, *use* **Sansk-**

sansculott/e, in Fr. Rev., a man of the lower classes; a strong republican or revolutionary, *pl.* **-es; -erie, -ism, -ist** (one word); in Fr. m. *sans-culott/e* (hyphen), *-isme*

sans doute (Fr.), no doubt

sanserif (typ.), any typeface without serifs, *not* sans serif

Sansevieria (bot.), a genus of lily

sans/ façon (Fr.), informally; — *faute,* without fail; — *gêne,* free and easy (hyphen if used as a noun)

Sanskrit, *not* -crit, abbr. **Skt.;** (typ.) printed in Devanagari

alphabet with some 50 letters and various added vowel-marks; many ligatures exist

sans/ pareil (Fr.), unequalled; — *peine,* without difficulty; — *peur et sans reproche,* fearless and blameless; — *phrase,* without circumlocution

sans serif (typ.), *use* **sanserif**

Sanssouci, palace of Frederick II at Potsdam (one word)

sans/ souci (Fr.), without cares; — *tache,* stainless

Santa (It., Sp., Port.), female saint; abbr. **Sta**

Santa Claus, *not* — Kl-

Santa Fe, NM, US, and Argentina (no accent)

Santander, N. Spain (one word)

Santayana (George), 1863–1952, Spanish-born US philosopher

Santenot, a burgundy wine

Santo Domingo, former name of the Dominican Republic, WI; name of its capital, *formerly* Ciudad Trujillo

Santos-Dumont (Alberto), 1873–1932, Brazilian aeronaut

São (Port.), saint, e.g. in place-names, as **São Miguel,** Azores, **São Paulo,** Brazil

Saône/, French river; **Haute--—,** dép. (hyphen); — **-et--Loire,** dép., abbr. **S.-et-L.** (hyphens)

São Tomé and Príncipe, W. Africa, indep. 1975

Sapho, novel by Daudet, opera by Gounod, 1851

Sapper, abbr. **Spr.**

Sapph/o, *c.*600 BC, Greek poetess of Lesbos; adj. **-ic,** but **sapphic/ metre,** — **verse** (not cap.); **sapphics,** verse in sapphic stanzas; **sapphism,** homosexual relations between women

Sar., Sardinia, -n

saraband, stately Sp. dance, music for this, *not* -bande

Saragossa, Spain; in Sp. **Zara-goza**

sarai, *use* **serai**

Sarajevo, Yugoslavia, *not* Se-

sarape, *use* **serape**

Sarawak, part of Malaysia

sarcenet, *use* **sars-**

sarcom/a (path.), a tumour; *pl.* **-ata**

sarcophag/us, stone coffin; *pl.* **-i**

Sardinia, in It. **Sardegna;** abbr. **Sar.**

Sardou (Victorien), 1831–1908, French playwright

saree, *use* **sari**

sargasso/, seaweed, *pl.* **-s**

Sargent (John Singer), 1856–1925, US painter; — **(Sir Harold Malcolm Watts),** 1895–1967, English conductor

sari, Indian female garment, *not* -ee

SARL (Fr.), *société à responsabilité limitée* (Co. Ltd.)

sarong, a Malay or Javanese long garment for man or woman

sarsaparilla, (med.) dried root of a tropical American smilax

sarsenet, a fabric, *not* sarc-

Sartor Resartus, 'the tailor re-tailored', by Carlyle, 1833–4

Sartre (Jean-Paul), 1905–80, French philosopher

Sarum:, sig. of Bp. of Salisbury (colon)

SAS, Special Air Service

SASC, Small Arms School Corps

sash/-cord, -window (hyphens)

Saskatchewan, Canada; abbr. **Sask.**

sassafras (bot. and med.), N. American tree; an infusion from its bark, *not* sasse-

Sassan/ian, -id, (member) of Persian dynasty ruling AD 211–651, *not* Sasa-

Sat., Saturday

Satan (cap.)

satanic, devilish (not cap. unless directly referring to Satan)

SATB (mus.), soprano, alto, tenor, bass

sateen, a shiny fabric, *not* satt-

sati, *use* **suttee**

satinet, a thin satin, *not* -ette

satire, literary work holding up folly or vice to ridicule. *See also* **satyr**

satirize, *not* -ise

satrap/, anc. Persian viceroy; **-y,** his province or office

sat sapienti (Lat.), sufficient for a wise man

satsuma, kind of orange, (cap.) Japanese pottery

satten, *use* **sateen**

Saturday, abbr. **Sat.**

Saturnalia (Lat. pl.), anc. Rom orgiastic festival of Saturn; (gen., not cap.) time or occasion of wild revelry (in this sense commonly sing.); adj. **saturnalian** (not cap.)

Saturnian, of the god or planet Saturn

satyr/, (Gr. and Rom. myth.) a woodland deity with human appearance but horse's ears and tail (goat's in Rom. myth.); a lascivious man; adj. **-ic,** type of Greek drama with chorus of satyrs. *See also* **satire**

Sauchiehall Street, Glasgow (two words)

saucisse (Fr. f.), fresh pork sausage

saucisson (Fr. m.), large highly seasoned sausage

Saudi Arabia, *formerly* Hejaz, Nejd, and Asir

sauerkraut (Ger.), chopped pickled cabbage

Saumur, a champagne

sauna bath (two words)

sausage/-dog (colloq.), **-meat, -roll** (hyphens)

saut/é, (food) lightly fried; to cook in this way; **-éd,** *not* -éed

Sauterne, a white Bordeaux wine from **Sauternes,** dép. Gironde, France

sauve qui peut (Fr.), let him save himself who can

avannah, a treeless plain of subtropical regions

avannah, river and town, Ga., US; town, Tenn., US

avant/, man of learning, *pl.* **-s;** *fem.* **-e,** *pl.* **-es** (not ital.)

avigny, a red burgundy

avile/, family name of Earl of Mexborough; — (**Baron**); — **Club,** — **Row,** London; (**Sir Henry**), 1549–1622, founder of Savilian chairs at Oxford, *not* -ille

avin, kind of juniper; its dried tips used as a drug, *not* -ine

avings/ bank, — **certificate** (two words)

avoir/-faire (Fr.), skill, tact; — **-vivre,** good breeding (hyphens)

avonarola (**Girolamo**), 1452–98, Italian religious reformer

avory, the herb

avoury, appetiz/er, -ing

avoyard, of Savoy (region or theatre)

aw/fish, -mill, -tooth (one word)

ax, colloq. for saxophone; slater's chopper, *not* zax

ax., Saxon, Saxony

ax (**Adolphe**), 1814–94, Belgian inventor of saxhorn and saxophone. *See also* **Sachs, Saxe**

axe (Saxony), in Ger. **Sachsen**

axe (**Hermann Maurice, comte de**), 1696–1750, French marshal; — (**John Godfrey**), 1816–87, US poet and humorous writer. *See also* **Sachs, Sax**

axe/-Altenburg, — **-Coburg- -Gotha,** — **-Meiningen,** — **-Weimar** (hyphens), former duchies in E. Germany, incorporated in Thuringia (Coburg in Bavaria), 1919; in Ger. **Sachsen-**

axhorn, brass wind instrument with long winding tube and bell opening

Saxon/, -y, abbr. **Sax.**

saxophone, brass wind instrument with keys and reed like that of clarinet, colloq. **sax**

SAYE, save as you earn

Saye and Sele (**Baron**)

Sayyidpur, India, *use* **Saidpur**

Sb, *stibium* (antimony) (no point)

sb, stilb

sb., substantive

s.b., single-breasted

SBN, Standard Book Number, now **ISBN,** International ditto

SC, South Carolina, Special Constable, Staff College, — Corps, Supreme Court, (Lat.) senatus consultum (a decree of the Senate), (law) same case, (paper) super-calendered

Sc, scandium (no point)

Sc., Scotch, Scots, Scottish

sc., scene, scruple, (Lat.) scilicet (namely)

s.c. (typ.), small capitals (without points as a proof-correction mark)

sc., sculpsit (carved or engraved this)

Sca Fell, English mountain, Cumbria, *not* Scaw — (two words)

Scafell Pike, highest English mountain (two words)

scagliola, imitation marble, *not* scal-

scal/a (anat.), a canal in the cochlea; *pl.* **-ae**

Scala (**La**), theatre, Milan

scalable, *not* -eable

scalar (math.), a non-vector number

scald, anc. Scand. composer and reciter, *use* **skald**

scaler, one who, *or* that which, scales

Scaliger (**Joseph Justus**), 1540–1609, French philologist; — (**Julius Caesar**), 1484– 1558, his father, Italian-born scholar

scaliola, *use* **scagl-**

scallop/, a shellfish, also used in cook. and dressmaking; **-ed, -ing,** *not* sco-, escallop

scallywag, ill-fed animal, scamp, *not* scalla-

scampi, (*pl.*) large prawns, (*sing.*) a dish of these

scan/, to analyse metre (of verse), to be metrically sound, to survey intently or quickly, **-ned, -ning, -sion**

Scand., Scandinavia, -n

scandalize, *not* -ise

scandal/um magnatum (Lat.), defamation of high personages, *pl. -a magnatum*; abbr. *scan. mag.*

Scandinavia/, -n, abbr. **Scand.**

scandium, symbol **Sc**

SCAPA, Society for Checking the Abuses of Public Advertising

s. caps. (typ.), small capitals; in proof-reading *use* **sc**

scar, craggy part of cliff etc., *not* scaur

Scarborough, N. Yorks.

Scarbrough (Earl of)

scar/f, *pl.* **-ves**

scarlatina, scarlet fever, *not* scarlet-

Scarlatti (Alessandro), 1660–1725, and — **(Domenico),** 1685–1757, his son, Italian composers

Scarlett, family name of Baron Abinger

scaur, *use* **scar**

Scaw Fell, *use* **Sca Fell**

Sc.B., *Scientiae Baccalaureus* (Bachelor of Science)

SCC, Sea Cadet Corps

Sc.D., *Scientiae Doctor* (Doctor of Science)

SCE, Scottish Certificate of Education

scen/a (It., Lat.), scene in a play or opera; It. *pl.* **-e,** Lat. *pl.* **-ae**

scenario/, outline of a ballet, play, or film; *pl.* **-s**

scène (Fr. f.), scene, stage; *en* —, on the stage

Scenes of Clerical Life, by George Eliot, 1858, *not* from

scent-bottle (hyphen)

scep/sis, philosophic doubt;

-tic, one inclined to disbelieve adj. **-tical,** *not* sk-

SCGB, Ski Club of Great Britai

sch., scholar, school, schooner

schadenfreude (Ger. f.), malicious delight in others' misfo tune

Schadow (Friedrich), 1789– 1862, German painter; — **(Jo hann Gottfried),** 1764–1850 German sculptor, father of Ru dolph and Friedrich; — **(Rudolph),** 1786–1822, German sculptor

Schäfer (E. A.), *see* **Sharpey- -Schafer**

schako, *use* **shako**

schallot, schalom, *use* **sh-**

Schaumburg-Lippe, W. Germany (hyphen)

Scheele/ (Karl Wilhelm), 1742–86, German chemist; — **'s green** (apos.)

Scheherazade, the relater in *The Arabian Nights*

Schelde, river, Belgium-Holland; in Fr. **Escaut;** (*not* -dt)

Schelling (Friedrich Wilhelm Joseph von), 1775– 1854, German philosopher

schem/a (Gr.), an outline; *pl.* **-ata**

schematize, *not* -ise

schemozzle, *use* **sh-**

Schenectady, NY, US

scherzando/ (mus.), in a playfu or lively manner; as noun, *pl.* **-**

scherzo/ (mus.), a playful or vigorous piece or movement; *p* **-s**

Schiedam (cap.), Hollands gin schnapps

Schiehallion, Mt., Tayside

Schiller (Johann Christoph Friedrich von), 1759–1805, German poet

schilling, Austrian coin

Schimmelpenninck (Mary Anne, Mrs), 1778–1856, Eng lish writer

schipperke, a breed of dog

schizanthus (bot.), the butter- fly-flower

schizo   Schwyz

schizo/, colloq. for schizo-
phrenic, *pl.* **-s**
Schleswig-Holstein, W. Germany (hyphen)
Schliemann (Heinrich),
1822–90, German archaeologist
schlieren, forming patterns of
light to show variation of density
etc.
schmaltz, sentimentalism, *not*
sh-, -alz (not ital.)
schnapps (Ger.), strong spirit,
not -aps
schnauzer, German breed of
dog
schnitzel (Ger.), veal cutlet;
Wiener —, this fried in breadcrumbs (one cap.)
Schnitzler (Arthur), 1862–
1931, Austrian playwright
schnorkel, *use* **snorkel**
Schobert (Johann), *c.*1720–67,
German composer
Schoenberg (Arnold), 1874–
1951, Austrian-born composer
scholi/um (Lat.), marginal
note in a manuscript; *pl.* **-a;**
-ast
**Schomberg (Friedrich Her-
mann, Duke of)**, 1615–90,
German born French marshal
and English mercenary officer
**Schomburgk (Sir Robert
Hermann)**, 1804–65, German
traveller
Schönbrunn, Vienna
school, abbr. **S.**
school/-bag, -board, -book
(hyphens)
school/boy, -child, -girl (one
word)
school bus (two words)
school-days (hyphen)
schoolhouse, building for
school (one word); **school
house,** headmaster's or house-
master's house at boarding-
-school (two words)
school-leaver (hyphen)
**school/master, -mistress,
-room, -teacher** (one word)
schooner, abbr. **sch.**
Schopenhauer (Arthur),

1788–1860, German philo-
sopher
**Schreiner (Olive Emilie
Albertina)**, 1855–1920, S. Af-
rican writer
Schrödinger (Erwin), 1887–
1961, Austrian physicist
**Schubart (Christian Fried-
rich Daniel)**, 1739–91, Ger-
man poet
Schubert (Franz Peter),
1797–1828, Austrian composer
Schuman (Robert), 1886–
1963, French statesman; —
(William Howard), b. 1910,
US composer
**Schumann (Robert Alex-
ander)**, 1810–56, German com-
poser
Schütz (Heinrich), 1585–1672,
German composer
Schutzstaffel (Ger. f.), Nazi élite
corps, abbr. **SS**
Schuyler (Eugene), 1840–90,
US writer; — **(Philip John)**,
1733–1804, US statesman and
soldier
Schuylkill, town and river, Pa.,
US
schwa (phonetics), indistinct
vowel sound, *not* sheva, shwa
schwärmerei (Ger. f.), a senti-
mental enthusiasm
Schwarzerd, *see* **Melanch-
thon**
Schwarzkopf (Elisabeth), b.
1915, Polish-born soprano
Schwarzwald (Ger.), the Black
Forest
Schweinfurt/, Bavaria; —
blue, — green, etc.
**Schweinfurth (Georg
August)**, 1836–1925, German
traveller in Africa
Schweitzer (Albert), 1875–
1965, German philosopher,
theologian, organist, med.
missionary at Lambaréné, Fr.
Equatorial Africa (now Gabon)
Schweiz (die), Ger. for
Switzerland
Schwyz, canton and its capital
in Switz.

sci., scien/ce, -tific

scia/graphy, art of the perspective of shadows, X-ray radiography; **-gram, -graph, -graphic; -machy,** fighting with shadows; *not* scio-, skia-

science, abbr. **sci.**

science fiction (two words); colloq. **sci-fi,** abbr. **SF**

scienter (Lat.), knowingly

scientific, abbr. **sci.**

scientology, religious system based on study of knowledge (not cap.)

scilicet (Lat.), namely; abbr. **sc.**

Scill/y (Isles of), off Cornwall; adj. **-onian**

scimitar, oriental sword, *not* the many variants

Scind, Pakistan, *use* **Sind**

scintilla/, a spark, trace; *pl.* **-s**

sciography etc., *use* **scia-**

scion, *not* cion

scirocco, *use* **si-**

Sclavic etc., *use* **Sl-**

sclerom/a, *pl.* **-ata, scleros/is,** *pl.* **-es** (med.), hardening

SCM, State Certified Midwife, Student Christian Movement

Sc.M., *Scientiae Magister* (Master of Science)

Scofield (Paul), b. 1922, British actor

scollop, *use* **sca-**

Scone, Tayside

SCONUL, Standing Conference of National and University Libraries

score (bind.), to break the surface of board to help folding

score, three-, four-, five-, etc. (one word)

score/-board, -card, -sheet (hyphens)

scori/a, slag; *pl.* **-ae**

Scot, native of Scotland

Scot., Scotch, Scotland, Scottish

Scotch, in Scotland preferred only for phrases — (whisky), — broth, — mist, etc.; otherwise, **Scots** *or* **Scottish** is preferred, hence also often outside Scotland; abbr. **Sc.** *or* **Scot.**

Scotchman, *see* **Scotsman**

scot-free (hyphen)

scotice, *use* **scottice**

Scotic/ism, -ize, *use* **Scott-**

Scotism, the doctrine of Duns Scotus

Scots/, *see* **Scotch;** abbr. **Sc.;** — **Guards** (no apos.)

Scotsman, in Scotland and increasingly elsewhere preferred to Scotchman

scottice (Lat.), in Scots dialect (no accent), *not* *scotice*

Scotticism, a Scottish expression, *not* Scoti-

Scotticize, make or become like the Scots, *not* -ise, Scoti-

Scottie (colloq.), Scottish person or terrier

Scottish, in Scotland and increasingly elsewhere preferred to **Scotch,** q.v.; *not* Scotish; abbr. **Sc.** *or* **Scot.**

scouse (colloq.), native or dialect of Liverpool (not cap.)

Scout, *not now* Boy Scout (cap.)

scow, a flat-bottomed boat, *not* skow

SCR, Senior Common Room; (Camb.) — Combination Room

scr., scruple, -s

screen (typ.), a fine grating on film or glass which breaks an illustration with various tones into dots of the appropriate size

screenplay, script of a film (one word)

screen printing, *see* **silk screen**

Scriabin(e), *use* **Skriabin**

scribes and Pharisees (cap. P only)

scrips/it (Lat.), wrote this; *pl.* **-erunt** *or* **-ere**

script (typ.), type resembling handwriting

script., Scriptur/e, -al

scriptori/um, a writing-room; *pl.* **-a**

Scriptures (the) (cap.)

scrolling (comp.), VDU presentation in which text appears to move vertically up-screen until arrested by operator for editing. *See also* **paging**

Scrooge, a miser, from character in Dickens's *Christmas Carol* (cap.)

scrot/um (anat.), *pl.* **-a**

scruple, 20 grains; abbr. **sc.,** sign ℈

scrutator, a scrutineer

scrutinize, *not* -ise

scuba/, acronym for *self-*contained *u*nderwater *b*reathing *a*pparatus, *pl.* **-s**

sculduggery, *use* **skul-**

sculp., sculpt/or, -ural, -ure

sculps/it (Lat.), engraved, or carved, this; *pl.* **-erunt** *or* **-ere**; abbr. **sc.** or ***sculps.***

sculptures (**titles of**) (typ.), when cited, to be in italic

scurry (noun and verb), *not* sk-

Scutari, Albania, *not* Sk-

scutcheon, *use* **escut-**

Scylla and Charybdis (class. myth.), personified rock and whirlpool in Str. of Messina

scymitar, *use* **scim-**

SD, South Dakota (off. postal abbr.)

sd. (books), sewed

s.d., shillings & pence, (Ger.) *siehe dies* (= q.v.), (Lat.) *sine die* (indefinitely)

S. Dak., South Dakota (off. abbr.)

SDLP, Social Democratic and Labour Party

SDR, Special Drawing Right

SE, south-east(ern), (Fr.) *Son Excellence* (His *or* Her Excellency)

S/E, Stock Exchange

Se, selenium (no point)

Sea, when with name, to be cap., as North Sea, Sea of Marmara

sea-bed (hyphen)

seaboard (one word)

sea/-borne, -breeze (hyphens)

SEAC, South-East Asia Command

sea change (two words)

sea/farer, -faring, -food (one word)

Seaford, E. Sussex

Seaforde, Co. Down

sea/going, -gull (one word)

sea-horse (hyphen)

Seal, Kent

Seale, Surrey

sea/-legs, -level (hyphens)

sealing-wax (hyphen)

sealskin (one word)

seamstress, sewing-woman, *not* semp-

Seanad Éireann, Upper House of Irish Parliament

seance, a sitting (not ital., no accent); in Fr. f. ***séance***

sea/plane, -port, -scape (one word)

sear, to scorch, wither(ed). *See also* **sere**

searchlight (one word)

search/-party, -warrant (hyphens)

sea/ serpent, — shell (two words)

sea-shore (law), the land between high and low water (hyphen)

sea/sick, -sickness, -side (one word)

season-ticket (hyphen)

SEATO, South-East Asia Treaty Organization

sea-urchin (hyphen)

sea-way (hyphen)

sea/weed, -worthy (one word)

sebaceous, fatty, *not* -ious

Sebastopol, the spelling generally used in contemporary accounts of the Crimean War. *But see* **Sevastopol**

sec, (math.) secant (no point); colloq. for second

sec (Fr.), *fem.* ***sèche,*** dry (ital.)

sec., second, -s (of time or angle, no point in scientific and technical work), secondary, secretary

*sec., see **secundum***

secateurs (pl.), pair of pruning clippers (no accent, not ital.)

secco/, painting on dry plaster, *pl.* **-s**

Sec.-Gen., Secretary-General

Sechuana, language of the Bechuanas, *use* **Tswana**

Secker & Warburg, Ltd., publishers

second (adj.), abbr. **2nd**

second/, -s, abbr. **s.** *or* **sec.** (*but* no point in scientific and technical work); sign *"*; — **mark** (*"*), symbol for inches or seconds. *See also* **secund**

seconde, a fencing parry

second-hand (adj., hyphen)

secondo (It. mus.), lower part in a duet

Second World War, 1939–45 (caps.); *also* **World War II**

secrecy, *not* -sy

sec. reg., secundum regulam (according to rule)

secretaire, a writing-table (not ital.)

secretariat, a secretary's office, members of an administration, *not* -ate

Secretary, head of State department (cap.)

secretary, person who deals with correspondence, typing, etc.; a literary assistant; abbr. **sec.**

secret/e, to hide, to form and separate (blood, sap, etc.); **-ion, -or, -ory**

section (typ.), a chapter subdivision, abbr. **s., sect.,** symbol §; — **mark,** §, fourth ref. mark for footnotes; *pl.* §§

secularize, *not* -ise

secund (biol.), on one side only. *See also* **second**

Secunderabad, Hyderabad, India, *not* Sak-, Sek-; *but* **Sikandarabad,** Uttar Pradesh, India

secundo (Lat.), in the second place; abbr. **2°**

secundum/ (Lat.), according to, abbr. **sec.;** — *artem,* ditto art, abbr. **sec. art.;** — *naturam,* naturally, abbr. **sec. nat.;** — *quid,* in some respects only; —

regulam, according to rule, abbr. **sec. reg.**

SED, Scottish Education Department

Sedbergh, Cumbria

se defendendo (Lat.), in defending himself, *or* herself

Seder, Jewish Passover ritual

sederunt (Lat.), a meeting, *or* sitting (not ital.)

sede vacante (Lat.), when the see is vacant

Sedgemoor, Somerset; — (**battle of**), 1685 (one word)

sedil/e, stone seat for priest in chancel of church, usu. one of three, *pl.* **-ia** (not ital.)

seduc/er, *not* -or; **-ible,** *not* -eable

séduisant/ (Fr.), *fem.* **-e,** bewitching

See (the Holy), the Papacy

see (verb), often italicized in indexes and reference books to distinguish from words being treated, abbr. **s.**

Seefried (Irmgard), b. 1919, German soprano

Seeley (Sir John Robert), 1834–95, English hist. writer

Seely, family name of Baron Mottistone

see-saw (hyphen)

seethe, to boil, *not* -th

sego/, N. American lily, *pl.* **-s**

segue (It.), follows

seguidilla/, Sp. dance, music for this, *pl.* **-s**

seicento, 1600–99, and art-style of that century

seiche, a fluctuation in the level of a lake

seigneur/, a feudal lord, *not* -ior; adj. **-ial** (not ital.)

seigniorage, superior's prerogative, right of Crown to revenue from bullion, *not* seignor-

seigniory, lordship

Seine/-et-Marne, abbr. **S.-et-M., — -Maritime, — - -St.-Denis,** French déps. (hyphens)

374

seise (law), to put in possession of. *See also* **seize**

seisin (law), (taking) possession by freehold, *not* -zin

seismo/graph, instrument to record force of earthquakes; **-logy,** study of earthquakes

Seit/e (Ger. f.), a page, abbr. **S.,** *sing.* and *pl.*

seize, to grasp. *See also* **seise**

seizin, *use* **seisin**

Sejm, Polish Parliament

séjour (Fr. m.), sojourn

Sekunderabad, *see* **Sec-**

sel., selected

Selangor, *see* **Malaysia (Federation of)**

Selassie (Haile), *see* **Haile Selassie**

Selborne (Earl of)

Selborne (Natural History of), by Gilbert White, pub. 1789

selector, one who, or that which, selects, *not* -er

selenium, symbol **Se**

self- is freely added (with hyphen) as a reflexive prefix to nouns, adjectives, and participles, as self-abuse, self-evident, self-made, etc. *But note* selfsame

self-conscious (hyphen), *but* **unselfconscious** (one word)

Seljuk, (member) of Turkish dynasties ruling from 11th to 13th cc., *not* -ouk

sellanders, *use* **sallenders**

selle de mouton (Fr. f.), saddle of mutton

seller's option, abbr. **s.o.**

Sellindge, Hythe, Kent

Selling, Sittingbourne, Kent

Sellotape, propr. term for transparent adhesive tape (cap.)

Selsey, Glos., E. Sussex, *not* -sea

seltzer, a German mineral water

selvage, an edging of cloth, *not* -edge

Sem., Semitic

semant/ic (adj.), concerning the meaning of words; **-ics,** the

branch of philology concerned with this

semeio-, *use* **semio-**

semester, a college or university course of half a year (esp. in Germany and US)

semi/, colloq. for semi-detached house, *pl.* **-s**

semi (Lat.), half; abbr. *s.*

semi-barbar/ian, -ic, -ism, -ous (hyphens)

semi-bold (typ.), a weight between light roman and bold

semi/circle, -circular (one word)

semicolon (one word); *see* **punctuation** III

semi/conductor, -final (one word)

semi-monthly, twice a month (hyphen)

Seminole, *sing.* and *pl.,* Amer. Indian

semi-official (hyphen)

semi/ology, -otics, branch of linguistics concerned with signs and symbols; (path.) science of symptoms, *not* semeio-

semiprecious (one word)

Semite, *not* Sh-

Semitic, abbr. **Sem.**

Semitize, *not* -ise (cap.)

semi/transparent, -tropical (one word)

semi-weekly, twice a week (hyphen)

semp. (It. mus.), sempre

semper/ eadem (Lat. f. sing., and n. pl.), always the same; **— fidelis,** always faithful, *pl.* **— fideles; — idem** (m. and n. sing.), always the same

semplice (mus.), in simple style

sempre (mus.), always; abbr. *semp.*

sempstress, *use* **seams-**

SEN, State Enrolled Nurse

Sen., Senat/e, -or, Seneca, senior

sen., senior

Senat/e, -or, abbr. **Sen.**

Senatus (Lat.), the Senate

Senatus Academicus, the governing body in Scottish universities (not ital.)

senatus consultum, a decree of the Senate; abbr. **SC**

Seneca (Lucius Annaeus), *c.*3 BC–AD 65, Roman Stoic philosopher and statesman; abbr. **Sen.**

Senegal/, W. Africa, rep. 1959, indep. 1960; adj. **-ese**; in Fr. **Sénégal**

Senhor/ (Port.), Mr; *-a,* Mrs; *-ita,* Miss

senior, abbr. **sen.** or **Senr.**

seniores priores (Lat.), elders first

Sennacherib, d. 681 BC, King of Assyria

se non è vero, è ben trovato (It.), if it is not true, it is well invented

Señor/ (Sp.), Mr, abbr. **Sr.**; *-es,* Messrs; *-a,* Mrs, abbr. **Sra.**; *-ita,* Miss, abbr. **Srta.**

Senr., Senior, use **sen.**

sensitize, make sensitive, *not* -ise

sensori/um (Lat.), grey matter of brain and spinal cord; *pl.* **-a**

sensory, pertaining to the senses or sensation

sensual, concerned with gratification of the senses, carnal

sensualize, *not* -ise

sensu/ lato (Lat.), in the wide sense; — *stricto,* in the narrow sense

sens/um (philos.), sense-datum, *pl.* **-a**

sensuous, pertaining to or affected by the senses in aesthetic terms

sentimentalize, *not* -ise

Senussi, member of fanatic Muslim sect; *pl.* **same**

senza (It. mus.), without

Seoul, cap. of S. Korea, *not* Seul, Soul

separate, abbr. **sep.**; (typ.), a reprint of one of a series of items

separator, *not* -er

Sephardi/, Jew of Spanish or Portuguese descent, *pl.* **-m** (cap.)

sepoy (hist.), Indian soldier in European service (not cap.)

seppuku (Jap.), hara-kiri

Sept., September, Septuagint, *not* Sep.

September, abbr. **Sept.**; in Fr. m. *septembre,* abbr. *sept.* (not cap.)

septemvir/ (Lat.), one of a committee of seven; *pl.* **-i**

septet (mus.), *not* -ett, -ette

septfoil, seven-lobed ornamental figure

septicaemia, blood-poisoning, *not* -cemia

septime, a fencing parry

septuagenarian, person in seventies

Septuagesima Sunday, the third before Lent

Septuagint, Greek version of OT, *c.*270 BC, abbr. **Sept.** or **LXX**

sept/um (biol.), a partition; *pl.* **-a**

Sepulchre (Church of St.), *not* -'s

sepultus (Lat.), buried; abbr. **S.**

seq. (sing.), *sequens* (the following), *sequente* (and in what follows), *sequitur* (it follows)

seqq. (pl.), *sequentes, sequentia* (the following), *sequentibus* (in the following places)

sequel/a (path.), a symptom following a disease; *pl.* **-ae**

sequen/s, -te, see **seq.**

sequent/es (m. f.), *-ia* (n.), *-ibus,* see **seqq.**

sequitur (Lat.), it follows; abbr. **seq.**

ser., series

serac/, a castellated mass of ice in a glacier; *pl.* **-s** (not ital.)

seraglio/, a harem, a Turkish palace, *pl.* **-s**

serai, a caravanserai, *not* sar-, -ay

Serajevo, use **Sa-**

serape, Sp.-Amer. shawl, *not* sar-, zar-

seraph/, a celestial being; *pl.* **-s**; Heb. *pl.* **-im,** *not* -ims

Serb., Serbian

sere, catch of gun-lock; (ecol.) sequence of animal or plant communities. *See also* **sear**

sere/cloths, -ments, *use* **cere-**

serein (Fr. m.), rain from cloudless sky

serge, large candle, *use* **cierge**

sergeant (mil.; but **-j-** in some official contexts), abbr. **Sgt.** (cap. as title). *See also* **serjeant**

seriatim, serially, point by point

series, *sing.* and *pl.*; abbr. **S.** *or* **ser.**

serifs (typ.), short lines across the ends of arms and stems of letters, *not* ceriphs

Serinagar, *use* **Sr-**

Seringapatam, Mysore, India

serio-comic (hyphen)

serjeant (law). *See also* **sergeant**

serra (Port.), sierra, mountain range

ser/um, the fluid that separates from clotted blood, antitoxin, *pl.* **-a**

serviceable, *not* -cable

serviceman (one word)

serviette, table-napkin (not ital.)

servitor, an attendant, *not* -er

servo/, mechanism for powered automatic control of larger system, *pl.* **-s**; as comb. form (with hyphen) denoting machine with this function, e.g. servo-motor

Sesotho, Bantu language, *not* Sesuto

sesquicentenary, (celebration of) one-hundred-and-fiftieth anniversary

sesquipedalian, (of word) one and a half (metrical) feet long; many-syllabled

sess., session

Session (Court of), supreme Scottish Court, *not* — — Sessions; — (**Parliamentary**) (caps.)

sesterce, Roman coin, in Lat. **sesterti/us,** *pl.* **-i,** symbol **HS** (no point)

sesterti/um, 1,000 sesterces, *pl.* **-a**

sestet, last six lines of a sonnet. *See also* **sextet**

Sesuto, *use* **Sesotho**

set, badger's burrow, shoot for planting, alignment in weaving, paving-block, (typ.) width of sort, *not* sett

set-back (noun, hyphen)

S.-et-L., dép. France, Saône-et-Loire; **S.-et-M.,** dép. France, Seine-et-Marne

set-off (typ.), unwanted transfer of ink from one sheet to another

set/ piece, — point (two words)

sets, abbr. **s.**

sett, *use* **set**

Settlement (Stock Ex.) (cap.)

settler, one who settles; **settlor** (law), one who makes a settlement

set-up (adj. and noun, hyphen)

Seul, Korea, *use* **Seoul**

Sevastopol, Crimea, USSR, *but see* **Sebastopol**

Sevenoaks, Kent (one word)

Seventh-day Adventist (two caps., one hyphen)

Seven Years War, 1756–63 (caps., no apos.)

Sévigné (Marie de Rabutin-Chantal, Madame de), 1626–96, French writer

Sèvres porcelain

sewage/, the refuse that passes through sewers; — **farm,** — **works** (two words)

sewerage, a system of sewers

sewin, salmon trout, *not* -en

sewing (bind.), the individual sewing of each book-section to its neighbours. *See also* **stitching**

sexagenarian, person in sixties

Sexagesima Sunday, the second before Lent

sex/ appeal, — change, — maniac (two words)

sextet (mus.), (a work for) a group of six performers, *not* -ett, -ette. *See also* **sestet**

sexto (typ.), a book based on 6 leaves, 12 pages, to the sheet; abbr. **6to**; **sextodecimo** *or* **sixteenmo**, a book based on 16 leaves, 32 pages, to the sheet; abbr. **16mo**

sexualize, to attribute sex to, *not* -ise

Seychelles (**Republic of**), Indian Ocean, indep. 1976

Seymour (**Jane**), ?1509–37, 3rd wife of Henry VIII of England

SF, San Francisco, science fiction

s.f., sub finem (towards the end)

SFA, Scottish Football Association

Sforza/, Milanese ducal family, notably — (**Ludovico**), 1451–1508; — (**Carlo, Count**), 1873–1952, leader of anti-Fascist opposition in Italy

sforz/ando, -ato (mus.), with sudden emphasis on a chord or note; abbr. *sfz*

sfumato (It. paint.), with indistinct outlines

S.-G., Solicitor-General

SGA, Society of Graphic Artists

Sganarelle, name of characters in Molière's comedies

sgd., signed

s.g.d.g. (Fr.), *sans garantie du gouvernement* (without government guarantee)

sgraffit/o, decorative work in which different colours are got by removing outer layers; *pl.* -**i.** *See also* **graffito**

's-Gravenhage, formal Dutch for **The Hague**

Sgt., Sergeant

sh., shilling, -s

shadoof, Egyptian water-raising apparatus; in Egypt. Arabic *šādūf*

shagreen, untanned leather from skin of fish, sharks, etc.

Shahabad, Bengal, Hyderabad, Punjab, Uttar Pradesh (India)

shaikh, *use* **sheikh**

Shakespear/e (**William**), 1564–1616, English playwright, abbr. **Shak.**; **-ian, -iana,** *not* the many variants

Shakespeare Society (**the**), *but* the **New Shakespere Society**

shako/, mil. head-dress, *not* sch-; *pl.* **-s**

shaky, unsteady, *not* -key

shallop, a small boat or dinghy, *not* shalop, -oop

shallot, a kind of onion, *not* esch-, sch-, shallott

shalom, Jewish salutation, *not* sch-

Shalott (*The Lady of*), by Tennyson, 1833

shamefaced (one word)

shammy-leather, *use* **chamois- —**

shandrydan, a rickety vehicle, *not* -dery-

Shanghai, China; **shanghai/** (not cap.), to make drunk and ship as a sailor; **-ed**

Shangri-La, earthly paradise (hyphen, two caps.)

shan't, to be printed close up, one apos. only

shanty, hut, rough dwelling, sailor's song, *not* ch-; — **town** (two words)

shapable, *not* shape-

SHAPE, Supreme Headquarters, Allied Powers, Europe

shareholder (one word)

share-list (hyphen)

share-out (noun, hyphen)

sharif, *use* **sherif**

sharp (mus.), sign ♯; **double —,** sign ×

Sharp (**Becky**), in Thackeray's *Vanity Fair*; — (**Cecil James**), 1859–1924, English collector of folk-songs; — (**Granville**), 1735–1813, English abolitionist; — (**James**), 1613–79, Scottish prelate; — (**William**), 1855–1905, Scottish poet and novelist, pseud. **Fiona Macleod**

Sharpe (Charles Kirk-patrick), 1781–1851, Scottish antiquary and artist; — **(Samuel)**, 1799–1881, English Egyptologist and biblical scholar

Sharpeville, S. Africa

Sharpey-Schafer (Sir Edward Albert), né **Schäfer**, 1850–1935, English physiologist

sharpshoot/er, -ing (one word)

shaving/-brush, -cream, -soap, -stick (hyphens)

Shaw (George Bernard), 1856–1950, Irish-born English playwright; adj. **Shavian**; — **(Aircraftman)**, see **Lawrence (T. E.)**

shaykh, use **sheikh**

shchi, Russian cabbage soup, not the many variants

sheaf, pl. **sheaves**

shealing, Highland hut, use **shiel-**

shear/, to cut, past **-ed**, partic. **shorn**, but **-ed** with reference to mechanical shears (see also **sheer**); — **-hulk**, — **-legs**, use **sheer-**

shearwater, a bird (one word)

sheath (noun)

sheathe (verb)

shebeen (Ir.), unlicensed house selling spirits

Shechinah, use **Shekinah**

sheep/-dip, -dog, -fold, -run (hyphens)

sheepshank, a type of knot (one word)

sheepskin, a rug, a bookbinding leather, a parchment (one word)

sheer, (adj.) mere, vertical, (adv.) quite, vertically, (verb) to deviate, swerve. See also **shear**

sheer/-hulk, a dismasted ship; **-legs**, hoisting apparatus for masts (hyphens)

Sheer Thursday (Maundy Thursday), not Shere —

sheet (typ.), a piece of paper of a definite size; a signature of a book

sheet/ metal, — music (two words)

sheets (in) (typ.), not folded, or, if folded, not bound

sheet work (typ.), printing the two sides of a sheet from two formes

Sheffield:, sig. of Bp. of Sheffield (colon)

sheikh (Arab.), a chief, not sha-, -ik, -yk

sheiling, use **shieling**

Shekinah, visible glory of God, not Shech-

sheldrake, a water-bird, not shell-; fem. and pl. **shelduck**, not sheld duck

shelf-ful/, the quantity that fills a shelf, pl. **-s** (hyphen)

shelf-mark (hyphen)

shellac/, a gum, not -ack, shelac, shelack; as verb, **-ked, -king**

Shelley (Percy Bysshe), 1792–1822, English poet

shellfish (one word)

shell-shock (hyphen)

Shelta, a cryptic jargon used by tinkers

sheltie, a Shetland pony, not -y

shelv/e, -ing, -y

Shemite, use **Semite**

shemozzle, rumpus, not sch-

Shepheardes Calendar (The), by Spenser, 1579

Shepheard's Hotel, Cairo

Shepherd Market, London

Shepherd's Bush, London

shepherd's/ needle, — pie, — purse (apos.)

Sheppard (Jack), 1702–24, highwayman

Sheppey (Isle of), Kent

Shepton Mallet, Somerset

Sheraton (Thomas), 1751–1806, English furniture designer

Sherborne, Dorset, Glos.; — **(Baron)**

Sherbourne, Warwick

Sherburn, Dur., N. Yorks.

Sherburn in Elmet, N. Yorks.

Shere, Surrey

Shere Thursday, use **Sheer —**

sherif, Muslim leader, *not* -eef, sharif

sheriff, county officer

Sheringham, Norfolk

Sherpa, one of a Himalayan people from borders of Nepal and Tibet, *not* -ah

Sherpur, Bangladesh

Sherrington (Sir Charles Scott), 1857–1952, English physiologist

's Hertogenbosch, Netherlands

Shetland/, region of Scotland; — **Islands,** or -s, the main group of islands in it

sheva, *use* **schwa**

shew, *use* **show** except in Sc. law, and biblical and Prayer Book citations

Shewbread (Scrip.), loaves displayed in Jewish temple, *not* Show-

sheyk, *use* **sheikh**

Shiah, sect of the Shiite Muslims

shibboleth, test word, *not* shiboleth

shieling, Highland hut or sheep-shelter, *not* sheal-, sheil-

Shifnal, Shropshire

shih-tzu, breed of dog

Shiite, (member) of Muslim sect

shikaree (Ind.), a hunter, *not* the many variants

Shikarpur, Pakistan

shillelagh, Irish cudgel, *not* the many variants

shilling/, -s, abbr. **s.** *or* **sh.**; symbol /-; — **mark** (typ.), /

shilly-shally/, -ing

Shinto/, -ism, Japanese religion

shiny, *not* -ey

'ship (typ.), *see* **companion-ship**

ship/builder, -building, -mate, -owner, -shape (one word)

shipping/-agent, -office (hyphens)

ship's names to be italic

Shipston on Stour, War. (three words)

Shipton, Glos., N. Yorks., Shropshire

Shiptonthorpe, Humberside

Shipton under Wychwood, Oxon. (three words)

ship/wreck, -wright, -yard (one word)

Shire/brook, Derby; **-hampton,** Avon; **-newton,** Gwent (one word)

shirt-sleeve (hyphen)

shirtwaister, dress with bodice like shirt (one word)

shish kebab (two words)

Shiva, *use* **Siva**

shmaltz, *use* **schmaltz**

shock/therapy, — treatment (two words)

shock troops (two words)

shoe/, -ing

shoe/black, -horn, -maker (one word)

shoe/-string, -tree (hyphens)

Sholokhov (Mikhail Aleksandrovich), b. 1905, Russian novelist

Shooters Hill, London

shop assistant (two words)

shop/-boy, -front, -girl (hyphens)

shopkeeper (one word)

shop/-lifter, -walker (hyphens)

shop steward (two words)

short-change (verb, hyphen)

short circuit (noun); **short-circuit** (verb)

short/mark, that placed over short vowel, breve; — **ton,** 2,000 lb.; abbr. **s.t.**; — **vowel,** ă, ĕ, etc.

shorty (colloq.), short person or garment, *not* -ie

Shostakovich (Dmitri Dmitrievich), 1906–75, Russian composer

shot-gun (hyphen)

shoulder (typ.), platform on the shank of a sort from which the face rises

shoulder-notes (typ.), marginal notes at the top outer corner of the page

shouldst, to be printed close up, no apos.

shovel/, -led, -ler, -ling, -ful,
pl. -fuls
shovelboard (one word)
show, *see* **shew**
Showbread, *use* **Shew-**
show business (two words),
colloq. **showbiz** (one word)
show/girl, -man (one word)
show house (two words)
show-piece (hyphen)
showroom (one word)
show-through (typ.), the degree to which printing is visible
on the other side of the paper
s.h.p., shaft horsepower
shrievalty, the office of sheriff
shrill/y, -ness
shrivel/, -led, -ling
Shropshire, *formerly* **Salop**
Shrovetide, from the Saturday
evening before to Ash Wednesday morning
Shrove Tuesday, day before
Ash Wednesday
shumac, *use* **sumac**
'shun!, colloq. for attention!
Shute (Nevil), pseud. of N. S.
Norway, 1899–1960, English
novelist
s.h.v., *sub hac voce* or *hoc verbo*
(under this word)
shwa, *use* **schwa**
shy/er, -est, -ly, -ness, *not* shi-
SI, Système International
d'Unités, a coherent system of
scientific units based on the
metre, kilogram, second, ampere, kelvin, mole, and candela;
adopted by the General Conference of Weights and Measures
(CGPM) and endorsed by the
International Organization for
Standardization (ISO); Sandwich Islands, Star of India,
Staten Island (NY)
Si, silicon (no point)
si (mus.), *use* **te**
SIAD, Society of Industrial Artists and Designers
sialagogue, an agent inducing
a flow of saliva, *not* sialo-
Siam, *use* **Thailand**
Siamese/, *use* **Thai,** except in

set phrases such as — **cat,** —
twins
Sib., Siberia, -n
sibilant, (adj.) hissing; (noun)
letter(s) sounded with a hiss (e.g.
s, sh)
sibyl/, a prophetess; adj. -**line;
the Sibylline books** (one
cap.), anc. Rom. collection of
oracles. *See also* **Sybil**
Sic., Sicil/y, -ian
sic (Lat.), thus, so (print usu. in
brackets, as parenthetical comment on quoted words)
sice, the six on dice. *See also*
syce
Sicilian Vespers, massacre of
the French in Sicily, 1282; subject of opera by Verdi, *Les Vêpres
siciliennes,* 1855
sick/-bed, -leave, -pay, -room
(hyphens)
sicut ante (Lat.), as before
sic vos non vobis (Lat.), thus
you labour, but not for your-
selves
side/-arms, -bet (hyphens)
side/-board, -burns (one word)
**side/-car, -chapel, -dish,
-door, -drum, -effect** (hyphens)
side/-light, -line, -long (one
word)
side-notes (typ.), those in margin, generally outer (hyphen)
**side/-road, -saddle, -show,
-slip, -step, -street, -table,
-track** (hyphens)
side valve, of engine (hyphen
when attrib.)
sideways (one word)
Sidgwick & Jackson, Ltd.,
publishers
Sidney (Sir Philip), 1554–86,
English soldier, courtier, writer.
See also **Syd-**
siècle (Fr. m.), century; abbr. **s.**
Siegfried/, a *Nibelungenlied* hero,
eponym of third opera in Wag-
ner's *Ring,* 1876, *not* Sig-; —
Line, German fortified line on
Franco-German border before
1939

siehe (Ger.), see; abbr. **s.**

siehe/ dies (Ger.), see this (=q.v.), abbr. **s.d.**; —
oben, = see above, abbr. **s.o.**;
— **unten,** = see below, abbr. **s.u.**

Siemens (Werner von), 1816–92, German founder of the electrical firm; — (**Sir William, Karl Wilhelm**), 1823–83, his brother, German-born British engineer

siemens (elec.), unit of conductance, abbr. **S**

Sienkiewicz (Henryk), 1846–1916, Polish novelist

Sienna, Italy; in It. **Siena**

sierra/ (Sp.), a mountain-chain, *pl.* **-s**

Sierra/ Leone, W. Africa, indep. 1961, adj. — **Leonean;** — **Madre,** Mexican mountain-chains; — **Nevada,** mountain-chain in E. California, US, and in Spain

siesta/ (Sp.), afternoon rest; *pl.* **-s**

Sieyès (Emmanuel-Joseph), 1748–1836, French statesman

Sig., *Signor, -i*

sig., signature

Sigfried, *use* **Sieg-**

sight-read/, -er, -ing, of music (hyphens)

sightsee/r, -ing (one word)

sigill/um (Lat.), a seal; *pl.* **-a**

sigl/um, sign denoting source of text, *pl.* **-a**

sigma, Gr. letter (Σ, σ, final ς); $\sum$ (math.) sum

sign, abbr. **s.**

signal/, -ize, *not* **-ise, -led, -ler, -ling, -ly**

signatory, one who has signed, *not* -ary

signature (mus.), the key or time sign at beginning of the stave; (typ.) complete part of a book, usu. 4, 8, 16, 32, or 64 pp.; these are given letters of the alphabet at foot of first page; text usu. begins with B, omits J, V, W; when the alphabet is

exhausted, the letters are duplicated as AA, AAA, etc.; a, b, c, etc. are used for prelims.; abbr. **sig.**

signed, abbr. **s.** *or* **sgd.**

Signor/ (It.), Mr, *pl.* **-i,** abbr. **Sig.;** *-a,* Mrs, *pl.* **-e, -ina,** Miss, *pl.* **-ine**

signs, *see* **astronomy, ecclesiastical, logic symbols, proof-correction marks**

Sikandarabad, Uttar Pradesh, India; *but* **Secunderabad,** Hyderabad, India

Sikes (Bill), in Dickens's *Oliver Twist, not* Sy-

Sikkim, E. Himalayas

Sikorski (Gen. Władysław), 1881–1943, Polish Prime Minister

silhouette, shadow-outline

siliceous, of silica, *not* -ious

silicon, chem. element, symbol **Si**

silicone, organic polymer containing silicon

siliqu/a, long narrow seed-vessel; *pl.* **-ae**

silk screen (typ.), process based on a stencil supported on a fine mesh

silkworm (one word)

sill (of door, window), *not* cill

sillabub, milk or cream curdled with wine, *not* sy-

Sillanpää (Frans Eemil), 1888–1964, Finnish novelist

Sillery, a champagne

silo/, airtight chamber for storing grain; *pl.* **-s**

silvan, of woods, *not* sy-

silver, symbol **Ag** (*argentum*) (no point)

silvicultur/e, -ist, *not* sylvi-

s'il vous plaît (Fr.), if you please; abbr. **s.v.p.**

simile/, a (literary) likening of one thing to another; *pl.* **-s,** *not* -ies; *adv.* (mus.), in the same manner

similia similibus curantur (Lat.), like cures like

similiter (Lat.), in like manner

simitar, *use* **scimitar**

Simla, cap. of Himachal Pradesh, India

Simonstown, Cape Province, S. Africa (one word)

simoom, a hot desert wind, *not* -oon

simpatico (It.), congenial

simpliciter (Lat.), absolutely, without qualification

simulacr/um, an image, a deceptive substitute; *pl.* **-a**

simultaneous, *not* -ious

simurg, monstrous bird of Persian myth, *not* -org, -urgh

sin (math.), sine (no point)

Sinaitic, of Mount Sinai or the Sinai peninsula (cap.)

Sind, Pakistan, *not* -e, -h, Scinde

Sindbad, the sailor, *not* Sinb-

Sindi/, Sind native; *pl.* **-s**

sine (math.), abbr. **sin**

sine/ (Lat.), without; — *anno,* without the date, abbr. *s.a.;* — *cura,* without office; — *die,* without a day (being named), abbr. *s.d.;* — *dubio,* — doubt; — *invidia,* — envy; — *legitima prole,* — lawful issue, abbr. *s.l.p.;* — *loco, anno, vel nomine,* — place, year, or name, abbr. *s.l.a.n.;* — *loco et anno,* — place and date (of books without imprints, abbr. *s.l.e.a.*); — *mascula prole,* — male issue, abbr. *s.m.p.;* — *mora,* — delay; — *nomine,* — (printer's) name, abbr. *s.n.;* — *odio,* — hatred; — *prole (superstite),* — (surviving) issue, abbr. *s.p.(s.);* — *qua non,* an indispensable condition, *not* — quâ —

sinfon/ia, (It. mus.), an overture or symphony, *pl.* **-s; -ietta,** a small-scaled symphony or orchestra, *pl.* **-iettas**

sing., singular

Singapore/, island off south end of Malay Peninsula, indep. 1965; adj. **-an**

sing/e, to scorch, **-ed, -eing**

Singh, Indian title, as Ranjit Singh, *not* -ng

singsong (adj. and noun, one word)

singular, abbr. **s.** *or* **sing.**

sinh, hyperbolic sine (no point)

Sinha (**Baron**)

Sinhalese, (member of major population group or language) of Sri Lanka, *not* Sing-, Cing-

sinister (her.), the shield-bearer's left, the observer's right, opp. **dexter**

sinistra (It.), left-hand side, abbr. **s.;** — *mano* (mus.), the left hand, abbr. **SM**

Sinn Fein/, the movement for Irish independence; **-er, -ism**

sinus/ (anat.), cavity in bone or tissue; *pl.* **-es**

Sion, *use* **Zion**

Siou/x, N. American Indian; *pl. same,* adj. **-an**

siphon, *not* sy-

si quis (Lat.), if anyone, first words of notice of ordination

Sirach, *use* **Ecclesiasticus**

sircar (Ind.), head, the Government, *not* -kar

sirdar (Ind. etc.), leader, one in command

siren, a sea-nymph, a temptress; (device which emits) warning sound, *not* sy-

Sirenia, order of aquatic mammals

siringe, *use* **sy-**

sirocco/, wind from Sahara reaching Italy and S. Europe; *pl.* **-s;** *not* sci-

sirup, *use* **sy-**

sirvente (Fr. m.), medieval Provençal narrative poem

SIS, Secret Intelligence Service (US)

sister/-german, having same parents; — **-in-law,** *pl.* **sisters-in-law;** — **-uterine,** having same mother only (hyphens)

Sistine Chapel, in the Vatican, *not* Six-

sisyphean, condemned to eternal punishment, like Sisyphus (Gr. myth.) (not cap.)

Sitapur, Uttar Pradesh, India

sitting-room (hyphen)

situs (Lat. *sing.* and *pl.*), a site

Sitwell (Dame Edith), 1887–1964, English poet; — (**Sir Osbert**), 1892–1969, English poet, novelist, essayist; — (**Sir Sacheverell**), b. 1897, English poet and art critic

sitz-bath, a hip-bath (hyphen)

Siva, Hindu god, *not* Shiva

Siwalik Hills, India, *not* Siv-

Six Day War, 1967

Six Mile Bottom, Cambs. (three words)

Sixmilebridge, Co. Clare, Co. Limerick (one word)

Sixmilecross, Tyrone (one word)

Six Road Ends, Co. Down (three words)

sixte, a fencing parry

sixteenmo (typ.), *see* **sexto-decimo**

Sixtine, *use* **Sistine**

sixty-fourmo (typ.), a book based on 64 leaves, 128 pages, to the sheet (hyphen); abbr. **64mo**

sizable, *use* **sizea-**

sizar, an assisted student

sizeable, *not* -zable

sizes of type, for the names of former sizes *see* **bourgeois, brevier, brilliant, canon, diamond, double pica, emerald, English, gem, great primer, long primer, minion, nonpareil, paragon, pearl, pica, ruby, small pica**; the **point system** (q.v.) is now used

SJ, Society of Jesus (Jesuits)

SJAA, St. John Ambulance Association

SJAB, St. John Ambulance Brigade

sjambok (Afrik.), a hide whip, *not* sambok

SJC (US), Supreme Judicial Court

Skagerrak, arm of the North Sea between Denmark and Norway (one word), *not* Skager Rack

skald, anc. Scand. composer and reciter, *not* scald

Skara Brae, Orkney

skateboard (one word)

skean-dhu (Sc.), dirk worn in the stocking, *not* skene-

skein, of silk etc., *not* -ain

Skelmersdale, Lancs., 'new town', 1962; — (**Baron**)

skep, basket, *not* skip

skepsis etc., *use* **sc-**

skewbald, of two colours in irregular patches, usu. of horses, usu. white and a colour other than black. *See also* **piebald**

ski/, *pl.* **-s;** as verb, **-'d, -ing** (no hyphen)

skiagraphy etc., *use* **scia-**

skier, person using skis. *See also* **skyer**

skiey, *use* **skyey**

skiing, *see* **ski**

ski-joring, sport in which skier is towed (hyphen)

skilful, *not* skill-

skilless, without skill (two *l*s)

skinflint, a miser (one word)

skinhead, youth with close--cropped hair (one word)

Skinner (Burrhus Frederic), b. 1904, US psychologist

skip, basket, *use* **skep**

skiver (bind.), binding-leather split from the grain side of sheepskin

Skopje, Yugoslavia

skow, *use* **scow**

Skriabin (Aleksandr Nikolayevich), 1872–1915, Russian composer, *not* Scr-, Skry-

Skt., Sanskrit

skulduggery, trickery, *not* scul-, skull-

skull-cap (hyphen)

skull-less (hyphen, three *l*s)

skurry, *use* **scurry**

Skutari, *use* **Sc-**

skyer, high ball at cricket. *See also* **skier**

Skye terrier (cap. S)

skyey, *not* skiey

sky/jack, -lark, -light, -line, -scraper (one word)

SL, serjeant-at-law, solicitor-at--law

SLADE, Society of Lithographic Artists, Designers, Engravers, and Process Workers

slaked lime, *not* slack- - -

slalom, ski-race down zigzag course, *not* slallom

Slamannan, Central

s.l.a.n., sine loco, anno, vel nomine (without place, year, or name)

slapdash (one word)

slap-happy (hyphen)

slapstick (one word)

S. lat., south latitude

slaty, like, or made of, slate, *not* -ey

Slav/ic, -onian, -onic, of the Slavs, *not* Sc-; abbr. **Slav.**

SLBM, submarine-launched ballistic missile

sld., sailed

s.l.e.a., sine loco et anno (without place or date)

sledge-hammer (hyphen)

sleeping/-bag, -pill, -suit (hyphens)

sleight, of hand, *not* sli-

slenderize, *not* -ise

Slesvig, W. Germany, *use* **Schleswig**

sleuth-hound (hyphen)

S level, examination (no hyphen)

slew, to turn, *not* slue

slily, *use* **slyly**

slip proofs, *see* proof (**galley** *or* **slip**)

slipshod (one word)

slivovitz, Balkan plum brandy

Sloane (Sir Hans), 1660–1753, English naturalist

sloe/, (fruit of) blackthorn; **-gin,** liqueur of sloes (hyphen); **-worm,** *use* **slow-**

sloid, system of manual training, esp. by wood-carving, *not* -jd, -yd

sloot, *use* **sluit**

sloping fractions (typ.), those with an oblique stroke as $^1/_2$. *See also* **fractions**

sloth, laziness; *also* (zool.) an animal

Slough, Berks.

slough, to shed

slowcoach (one word)

slow-worm, *not* sloe- (hyphen)

sloyd, *use* **sloid**

s.l.p., sine legitima prole (without lawful issue)

slue, *use* **slew**

slug (typ.), a line of type cast as a single piece

slug-abed, lazy person (hyphen)

sluit (Afrik.), a narrow channel, *not* sloot

slyly, *not* sli-

slype, a passage between walls, esp. between south transept of cathedral and chapter house

SM, Sergeant- or Staff-Major, (mus.) short metre, (Fr.) *Sa Majesté,* (It.) *Sua Maestà,* (Sp.) *Su Magestad* (His, *or* Her, Majesty), (Ger.) *Seine Majestät* (His Majesty, It. mus.) *sinistra mano* (the left hand)

Sm, samarium (no point)

sm., small

small capitals, *see* **capitals** (**small**) *and* **proof correction marks**; abbr. **sc** *or* **s. caps.**

small pica (typ.), name for a former size of type, about 11 pt.

smallpox (one word)

small royal, former size of paper, 25×19 in.; — **4to,** $12\frac{1}{2} \times 9\frac{1}{2}$ in.; — **8vo,** $9\frac{1}{2} \times 6\frac{1}{4}$ in. (untrimmed); basis for size of **metric royal,** 1272×960 mm. *See also* **book sizes**

Smalls, Oxford 'Responsions' Examination

smart alec, *not* -eck, -ick (two words)

SMD (mus.), short metre double

SME, *Sancta Mater Ecclesia* (Holy Mother Church)

Smelfungus, Sterne's name for Smollett (one *l*)

smelling/-bottle, -salts (hyphens)
smell-less (hyphen, three *l*s)
smelt, past of smell, *not* smelled
smelt, a small sea-fish. *See also* **smolt**
SMI (Fr.), *Sa Majesté Impériale* (His, *or* Her, Imperial Majesty)
Smith (Sydney), 1771–1845, English clergyman and wit; — **(Sir William Sidney),** 1764–1840, English admiral, defender of St. Jean d'Acre
Smithsonian Institution, Washington, DC, US; abbr. **Smith. Inst.;** founded 1846 from funds left by **James Smithson,** 1765–1829, English chemist
SMM, *Sancta Mater Maria* (Holy Mother Mary)
smok/e, -able, -y
smolder, *use* **smoulder**
Smollett (Tobias George), 1721–71, Scottish physician and novelist
smolt, a young salmon. *See also* **smelt**
smooth/ (verb), **-s,** *not* -e, -es
smorgasbord, Scandinavian hors-d'œuvres (not ital.), in Sw. *smörgåsbord*
smorzando (It. mus.), gradually dying away
smoulder, *not* smol-
s.m.p., *sine mascula prole* (without male issue)
Smyrna, in Turk. **İzmir**
Smyrniot, (native) of Smyrna
Smyth (Dame Ethel), 1858–1944, English composer; — **(John),** 1586–1612, founder of the English Baptists
Smythe (Francis Sydney), 1900–49, English mountaineer
Sn, *stannum* (tin) (no point)
s.n., *sub nomine* (under a specified name)
s.n., *sine nomine* (without name)
snapdragon, a Christmas game; (bot.) antirrhinum (one word)

snipe/, a gamebird, *pl.* **-s** and (collect.) *same*
snivel/, -led, -ler, -ling
snoek (Afrik.), a sea-fish
snorkel/, device for underwater swimming etc., **-ling,** *not* schn-
Snorri Sturluson, 1178–1241, Icelandic historian and poet. *See also* **Heimskringla**
snow (meteor.), abbr. **s.**
snowball (one word)
snow-drift (hyphen)
snow/drop, -fall, -flake, -man (one word)
snow-shoe (hyphen)
snowstorm (one word)
SNP, Scottish National Party
Snr., *use* **sen.**
SO, Staff Officer, Stationery Office, sub-office
so (mus.), *use* **soh**
s.o., seller's option, substance of *s.o.* (Ger.), *siehe oben* (= see above)
so-and-so/, particular person not specified, *pl.* **-'s; Mr So--and-so** (one cap.)
Soane's (Sir John) Museum, London
SOAS, School of Oriental and African Studies
sobriquet, nickname, *not* sou- (not ital.)
Soc., Socialist, Society, Socrates
socage, feudal tenure of land, *not* socc-
so-called (adj., hyphen)
Socialist (cap. when name of polit. party); abbr. **S.** *or* **Soc.**
socialize, *not* -ise
société/ (Fr. f.), society; — **anonyme,** public company; **Société des Bibliophiles françois,** founded 1820, *not* français (l.c. *f*); abbr. **S^{té}**
Society, abbr. **S.** *or* **Soc.**
socio/-cultural, -economic, -linguistic (hyphens)
sociol., sociology
socius/ (Lat.), Fellow, Associate, abbr. **S.;** — **criminis,** associate in crime
socle (archit.), base or plinth supporting column, statue, etc.

386

Socotra, Indian Ocean, *not* -ora, Sok-

Socrat/es, 469–399 BC, Greek philosopher, *not* Sok-; adj. **-ic**; abbr. **Soc.**

sodium, symbol **Na** (*natrium*)

sodomize, *not* -ise

Sodor and Man (**Bp. of**); **Sodor & Man:,** his sig. (one colon)

SOE, Special Operations Executive

SOED, Shorter Oxford English Dictionary

sœur/ (Fr. f.), sister, nun; — *de charité,* a Sister of Mercy

Sofar, *s*ound *f*iring *a*nd *r*anging (under water). *See also* **Sonar**

soffit (archit.), under-surface of arch

Sofi, *use* **Sufi**

Sofia, Bulgaria

S. of S., Song of Solomon

softa, Muslim student of sacred law (not ital.)

soft copy (comp.), transient reproduction of keyboard input on VDU (two words). *See also* **hard copy**

S. of III Ch., Song of the Three Children (Apocr.)

softie, silly weak person, *not* -y

SOGAT *or* **Sogat,** Society of Graphical and Allied Trades

sogenannt (Ger.), so-called, abbr. **sog.**

soh (mus.), *not* so

soi-disant (Fr.), self-styled (hyphen)

soigné/ (Fr.), *fem.* -*e,* well-groomed

soirée/, an evening party; — **dansante,** ditto with dancing; — **musicale,** ditto with music (not ital.)

Sokotra, Sokrates, *use* **Soc-**

Sol (Lat.), the sun (cap.)

Sol., Solomon

sol., solicitor, solution

sola, *see* **solus**

solan goose, the gannet. *See also* **solen**

solarize, *not* -ise

solati/um (Lat.), compensation; *pl.* **-a**

sola topi (Ind.), a sun-helmet, *not* solar —

solecis/m, a blunder in speaking, writing, *or* behaviour; adj. **-tic**

solecize, *not* -ise

solemnize, *not* -ise

solen, a mollusc. *See also* **solan**

solenoid, cylindrical wire coil as magnet

sol-fa (mus.) (hyphen); **sol-faed**

solfegg/io (mus.), a sol-fa exercise for the voice; *pl.* **-i**

solicitor, abbr. **sol.** *or* **solr.**

Solicitor-General (caps., hyphen); abbr. **S.-G.** *or* **Sol.-Gen.**

solicitude, anxiety, concern; in Fr. f. *sollicitude*

solid (typ.), text matter set without extra spacing between lines or letters

solid/us (Lat.), shilling; *pl.* **-i,** abbr. **s.;** (typ.) the shilling stroke, /, used to denote alternatives (and/or) and ratios (miles/day). *See also* **fractions**

soliloqu/y, speech made regardless of hearers, *pl.* **-ies; -ize,** *not* -ise

solmization (mus.), sol-fa system, *not* -sation

sol/o, abbr. **s.;** *pl.* **-os,** It. mus. *pl.* **-i**

Solon, ?638–?558 BC, Athenian lawgiver

solr., solicitor

soluble, that can be dissolved *or* solved

sol/us (theat.), *fem.* **-a,** alone

solution, abbr. **sol.**

Solutrean, a palaeolithic culture, *not* -ian

solvable, that can be solved, *not* -eable, -ible

solvent, *not* -ant

solvitur ambulando (Lat.), the question settles itself naturally

Solzhenitsyn (**Alexander Isayevich**), b. 1918, Russian writer

Som., Somerset
Somalia, NE Africa, indep. 1960; adj. **Somali**
Somaliland (one word)
sombrero/ (Sp.), broad--brimmed hat; *pl.***-s**
somebody, some person
some body, unspecified group of persons
some/how, -one (one word)
some one (person, thing), when each word to retain its meaning (see *Hart's Rules*, p. 77)
Somerby, Leics., Lincs.
somersault, *not* -set, summer-
Somersby, Lincs., birthplace of Tennyson
Somerset, English county, abbr. **Som.**
Somers Town, London (two words)
Somerstown, W. Sussex (one word)
some/thing, -time, -what, -when, -where (one word)
Son (the), as Deity (cap.)
son, abbr. **s.**
Sonar, *s*ound *na*vigation and ranging. *See also* **Asdic, Sofar**
sonata/ (It. mus.), pl. **-s**
Sondes (Earl, *not* of)
son et lumière (Fr.), night entertainment with sound and light effects
song/bird, -book (one word)
song cycle (two words)
Song/ of Solomon, abbr. **S. of S.**; — **of Songs,** same as preceding; — **of the Three Children** (Apocr.), abbr. **S. of III Ch.**
songs (titles of) (typ.), to be roman quoted when cited
son-in-law (hyphens); *pl.* **sons--in-law**
Son of/ God, —— Man (caps.)
soochong, *use* sou-
Soofee, *use* **Sufi**
Sop., soprano, -s
Sophocles, 495–406 BC, Greek playwright; abbr. **Soph.**
sophomore (US), second-year student, *not* sophi-

Sophy, a Persian ruler in the 16th and 17th cc. *See also* **Sufi**
sopra (It.), above
sopranino/ (It. mus.), instrument higher than a soprano, *pl.* **-s**
sopran/o (It. mus.), *pl.* **-os**; It. *pl.* **-i**; abbr. **S.** *or* **Sop.**
Sorbonne, medieval theological college of Paris, now comprising three of the Paris universities with faculties of science and arts also
sordin/o (It. mus.), a mute, *pl.* **-i**
sorghum, tropical cereal grass, *not* sorgum
sortes/ (Lat.), (divination by) lots; — *Biblicae or Sacrae, Homericae, Vergilianae,* (divination from) random passages of Scripture, Homer, Virgil
sorts (typ.), pieces of type; **special** —, *see* **accents.** *See also* **height to paper**
SOS, signal for help, *pl.* **SOSs**
so-so, passable (hyphen)
sostenuto/ (It. mus.), (passage played) in sustained manner, *pl.* **-s**
Sotheby, Wilkinson, & Hodge (Messrs.), auctioneers, *now* **Sotheby Parke Bernet & Co.**
sotto/ (It.), under; — *voce*, in an undertone
sou, former French coin of small value
Soubirous (St. Bernadette), 1844–79, of Lourdes, French nun
soubrette, maid or other pert fem. part in (musical) comedy
soubriquet, *use* sob-
souchong, a black tea, *not* soo-
Soudan/, -ese, *use* Sud-
soufflé (Fr.), light spongy dish using egg-whites
souffleur (Fr. m.), theat. prompter
souk, a Muslim market-place, *not* suk, sukh, suq
Soul, Korea, *use* **Seoul**

Soult (Nicolas Jean de Dieu), 1769–1851, French marshal, Duke of Dalmatia

sound barrier (two words)

sound/-track, -wave (hyphens)

soupçon, a very small quantity (not ital.)

souper (Fr. m.), supper, not *-pé*

souteneur (Fr. m.), man living off prostitute

south, abbr. **S.** *See also* **capitalization, compass**

South/ Africa, -n, abbr. **S. Afr.** *or* **SA; — America/, -n,** abbr. **S. Amer.** *or* **SA: South Australia/, -n,** abbr. **SA; — Carolina,** US, abbr. **SC; — Dakota,** US, off. abbr. **S. Dak.** or (postal) **SD**

southbound (one word)

Southdown, sheep and mutton (one word)

South Downs, Hants etc.

south-east(ern), abbr. **SE**

southeaster, wind from south-east (one word)

southern, abbr. **S.**

Southern Africa, geog. term, incl. Zimbabwe and other countries as well as Republic of South Africa

southernwood, species of wormwood

Southern Yemen, *see* **Yemen (South)**

Southesk (Earl of)

Southey (Robert), 1774–1843, Poet Laureate 1813–43

southpaw, left-handed (person) (one word)

South Pole (caps.)

South Wales, abbr. **SW**

Southwark, *or* 'The Borough', formerly in Surrey; **Southwark:,** sig. of Bp. of Southwark (colon)

Southwell:, sig. of Bp. of Southwell (colon)

south-west(ern), abbr. **SW**

South West Africa (three words)

southwester, wind from south-west (one word)

sou'wester, sailor's hat

sovereign/, -s, abbr. **sov., sovs.**

soviet (Russ.), an elected council, the basis of Russian governmental machinery since 1917; adj. (cap.), as in Soviet Russia. *See also* **USSR**

Soweto, Black township near Johannesburg

Soxhlet (chem.), an extraction apparatus

Sp., Spain, Spanish

sp., species (*sing.*), specimen, spelling, spirit

s.p., self-propelled, starting price

s.p., sine prole (without issue)

S.p.A. (It.), *Società per Azioni* (joint-stock company)

space/craft (*sing.* and *pl.*), **-man, -ship, -suit** (one word)

spaces (typ.), blanks between words or letters. *See also* **proof-correction marks, letter-spacing**

space/ station, — vehicle, — walk (two words)

space-time (hyphen)

spacial, *use* **spatial**

spaghetti, strings of pasta (not ital.)

spahi, Turkish horse soldier, Algerian French cavalryman; in Turk. **sipahi**

Spain, abbr. **Sp.**

Spalato (Yugoslavia); *now called* **Split**

spandrel (archit.), space between curve of an arch and enclosing mouldings, *not* -il

Spanish, abbr. **Sp.;** (typ.) alphabet consists of 27 letters including *ch, ll,* and *ñ,* but does not include *k* and *w*; *see* **accents**; *ch, ll,* and *rr* must not be divided. The portion carried over to begin with a consonant. Question and exclamation marks are inverted before and upright after their phrases. For further information see *Hart's Rules*, pp. 130–3

Spanish n (ñ), 'curly n', or 'n with the tilde'; *pron.* as ni in onion

Spanish Town, Jamaica

spatial, *not* spacial

SPCK, Society for Promoting Christian Knowledge

SPE, Society for Pure English

spearhead (one word)

spec, colloq. for specification, speculation (no point)

spec., special, -ly, specific, -ally, -ation, specimen

spécialité (Fr. f.), a speciality (ital.)

speciality, in gen. senses, *not* specialty (q.v.)

specialize, *not* -ise

special sorts, *see* **accents**

specialty (law), a contract under seal

specie, coin, as distinct from paper money

species, *sing.* and *pl.*; abbr. **sp.,** *pl.* **spp.** *See also* **botany, zoology**

specific gravity, abbr. **sp. gr.**

specimen, abbr. **sp.** *or* **spec.**

Spectaclemakers Society (no hyphen, no apos.)

specs, colloq. for **spectacles**

spectr/um, *pl.* **-a**

speculat/e, -or

specul/um, a mirror, (surg.) instrument for viewing body cavities, (ornith.) bright patch on a bird's wing; *pl.* **-a**

speech-day (hyphen)

speedboat (one word)

spele/an, cave-dwelling (adj.); **-ologist, -ology,** *not* -ae-

spel(l)ican, *use* **spillikin**

spell, past and partic. **spelled** *or* **spelt**

spellbound (one word)

spencer, short double-breasted overcoat, woollen jacket (not cap.)

Spencer (Earl, *not* of)

Spencer (Herbert), 1820–1903, English philosopher; — **(Sir Stanley),** 1891–1959, English painter

Spengler (Oswald), 1880–1936, German philosopher of history

Spen/ser (Edmund), 1552–99, English poet; **-serian,** of the poet, his style, *or* his stanza

Spetsai, Greece, *not* Spezzia

spew, to vomit, *not* spue

Speyer, W. Germany, *not* Spires

Spezzia, *use* **Spetsai** (Greece) *or* **La Spezia** (Italy)

sp. gr., specific gravity

spherical, abbr. **s.**

sphinx/, *not* sphy-; *pl.* **-es**

spianato (It. mus.), made smooth

spick and span (three words)

spicy, *not* -ey

spiegando (It. mus.), 'unfolding', becoming louder

spifflicate (joc.), to trounce, *not* spiff-

spiky, *not* -ey

spill, past and partic. **spilled** *or* **spilt**

spillikin, splinter, *not* spel(l)ican

spillway, passage for surplus water from dam (one word)

spina bifida (med.), congenital defect of spine

spinach, *not* -age

spin/-dry, -drier (hyphens)

spine (bind.), that part of the case protecting the back of a book, and bearing — **lettering**

spinney, a thicket, *not* -ny

spinning/-jenny, -wheel (hyphens)

spin-off (noun, hyphen)

Spinoza/ (Baruch), 1632–77, Dutch philosopher; **-ism**

spiraea (bot.), an ornamental shrub

Spires, W. Germany, *use* **Speyer**

spirit, abbr. **sp.**

spirito (It. mus.), life, spirit

spiritual, of the spirit

spiritualize, *not* -ise

spirituel, marked by refinement and quickness of mind; *fem.* the same in Eng. (not ital.)

spirt, *use* **spurt**

Spithead, strait between IW and Portsmouth

Spitsbergen, Arctic Ocean, *not* Spitz-

spiv/ (slang), flashy petty black--marketeer; **-ish, -very**

Split, Yugoslavia, *formerly* **Spalato**

split fractions, *see* **fractions**

split infinitive, the separation of 'to' from its verb by other words, as 'he used to often say'. Objected to by many, but the cure is often worse than the disease. See *Modern English Usage*, pp. 579–82

Spode/ (**Josiah**), 1754–1827, maker of — **china** at Stoke

Spohr (**Louis**), 1784–1859, German composer

spoil, past and partic. **spoiled** *or* **spoilt**

spolia (Lat.), spoils; — *opima*, the richest spoils; also trophy won by generals of opposing armies in single combat

spoliation, *not* spoil-

spond/**ee,** foot of two long syllables; adj. **-aic**

sponge/**-bag, -cake** (hyphens)

spongy, like sponge, *not* -gey

spontane/**ity, -ous**

spoon-feed (hyphen)

spoonful/, *pl.* **-s**

sporran, a kilt pouch

spos/*a* (It.), a bride, *pl.* **-e**; **-o**, bridegroom, *pl.* **-i**

spotlight (one word)

spp., species (*pl.*)

SPQR, *Senatus Populusque Romanus* (the Senate and Roman people); small profits and quick returns

SPR, Society for Psychical Research

Spr., Sapper

sprightly, lively, *not* spritely

spring (**season of**) (not cap.)

springbok (Afrik.), an antelope, *not* -buck

spring/**tide, -time,** season of spring (one word)

spring-tide, tide of greatest range (hyphen)

sprinkled edges (bind.), cut edges of books sprinkled with coloured ink

spruit (Afrik.), watercourse

spry/, active; **-er, -est, -ly, -ness**

s.p.s., sine prole superstite (without surviving issue)

spue, *use* spew

spurry, slender plant, *not* -ey

spurt, *not* spirt

sputnik, a Russian earth satellite

sput/**um,** expectorated matter; *pl.* **-a**

sq./, square; — **ft.,** — feet; — **in.,** — inches; — **m.,** — metres, — miles; — **yd.,** — yards (each *sing.* and *pl.*). These are not used in scientific and technical work

Sqn. Ldr. *or* **Sqn/Ldr,** Squadron Leader

squacco/, small crested heron, *pl.* **-s**

squalls (meteor.), abbr. **q.**

square (bind.), that part of the case which overlaps the edges of a book

square back (bind.), a book whose back has not been rounded (*see* **rounding**)

square brackets, *see* **brackets**

square root (two words)

squeegee, rubber-edged implement for sweeping wet surfaces, *not* squil-

squirearchy, government (influenced) by landed proprietors, *not* -rarchy

SR, Southern Railway (prior to nationalization); — Region

Sr, strontium (no point)

Sr., Sister, (Sp.) *Señor*

sr., senior, *use* sen.

sr, steradian

Sra. (Sp.), *Señora*

SRC, Science Research Council, Students' Representative Council

Sri Lanka, 'Resplendent Island', name of Republic of Ceylon since 1972

Srinagar, Kashmir, *not* Ser-

SRN, State Registered Nurse

SRO, statutory rules and orders
Srta. (Sp.), *Señorita*
SS, Saints, Secretary of State, steamship, (Fr.) *Sa Sainteté* (His Holiness), (Ger.) *Schutzstaffel* (Nazi élite corps)
SS (collar of), former badge of House of Lancaster, *not* esses
ss., subsection, (med.) half
s.s., screw steamer, (mus.) *senza sordini* (without mutes)
SSAFA, Soldiers', Sailors', and Airmen's Families Association
SSC, Solicitor to the Supreme Court
SSE, south-south-east
S.Sgt., S/Sgt., Staff Sergeant
SSR, Soviet Socialist Republic
SSRC, Social Science Research Council
SST, supersonic transport
SSW, south-south-west
St, stokes
St., Saint, in alphabetic arrangement to be placed under Saint, *not* St-; Strait, -s; Street; in Slavonic languages, *Stari*, old
st., stanza, stem, stone, strophe, (cricket) stumped
s.t., short ton (2,000 lb.)
Sta., Station
Sta (It., Sp., Port.), Santa (female saint) (no point)
Staal (Georges Frédéric Charles, baron de), 1824–1907, Russian diplomat; — **(Marguerite Jeanne Cordier De Launay, baronne de),** 1684–1750, French writer. *See also* **Staël, Stahl**
'Stabat Mater', 'the Mother was standing' (Catholic hymn)
stabbing (bind.), wire-stitching near the back edge of a closed section or pamphlet; the piercing of a book section prior to sewing or stitching
stabilize, *not* -ise
staccato/ (It. mus.), in abrupt, detached manner; as noun, *pl.* -s
stadi/um, in anc. Greece a measure of length, a racecourse, *pl*

-a; in mod. use a sports ground, *pl.* **-ums**
Stadtholder, Dutch governor, *not* Stadh-
Staël (Madame de), (in full, **Staël-Holstein, Anne Louise Germaine, baronne de,** *née* **Necker**), 1766–1817, French writer. *See also* **Staal, Stahl**
staff/, a pole; (mus.) set of lines on which music is written, *pl.* **-s** *or* **staves**; a body of persons in authority, *pl.* **-s.** *See also* **stave**
Staffs., Staffordshire
Staff Sergeant, abbr. **S.Sgt.** *or* **S/Sgt.**
stage-coach (hyphen)
stagecraft (one word)
stage/ door, — fright (two words)
stage-manage/, -r (hyphens)
staghound (one word)
Stagirite (the), Aristotle, *not* Stagy-
stag-party (hyphen)
stagy, *not* -ey
Stahl (Friedrich Julius), 1802–61, German lawyer and politician; — **(Georg Ernst),** 1660–1734, German chemist. *See also* **Staal, Staël**
staid, solemn. *See also* **stayed**
Stakhanovite, prodigious (esp. Russian) worker (cap.)
stalactite, deposit hanging from roof of cave
Stalag, German POW camp
stalagmite, deposit rising from floor of cave
stalemate (one word)
Stalingrad, Russian city, *previously* **Tsaritsyn,** *now* **Volgograd**
stamen/ (bot.), *pl.* **-s**
stamina, power of endurance
Stamp Office (two words)
stanch (verb), to check a flow. *See also* **staunch**
standardize, *not* -ise
Standards (British), *see* BSI
standing type (typ.), type not yet distributed after use
stand/-offish, -pipe (hyphens)

stand/point, -still (noun) (one word)

stannary, a tin mine

stannum, tin, symbol **Sn**

Stanstead, Suffolk

Stansted, Essex, Kent

stanz/a, a group of rhymed lines; *pl.* **-as**; abbr. **st.**

stapes (anat.), a bone in the ear; *pl.* same

star/, -red, -ring, -ry

Starcross, Devon (one word)

star-dust (hyphen)

star/fish, -light (one word)

Stars and Stripes, US flag

'Star-Spangled Banner (The)', US National Anthem

starting/-handle, -pistol, -point, -post (hyphens)

starting price (two words)

stat., statics, statuary, statute

State (cap. of political community and in place-names)

Staten Island, New York, *not* Staa-; abbr. **SI**

stater, an anc. Greek coin

stateroom (one word)

statesman/like, -ship (one word)

statics, the science of forces in equilibrium (*sing.*)

Station, abbr. **Sta.** or **Stn.**

stationary, not moving

Stationers' Hall, London (apos. after s)

stationery, paper etc.

statistics, the subject, *sing.*; numerical facts systematized, *pl.*

stator (elec.), a stationary part within which something revolves

statuary, abbr. **stat.**

status/, rank, *pl.* **-es** (not ital.)

status quo/ (Lat.), the same state as now; — — **ante,** ditto as before

statute, a written law; abbr. **stat.**

statute-book (hyphen)

statutory, *not* -ary

staunch, firm, loyal. *See also* **stanch**

stave/, staff (mus.), a snatch of

song, a piece of wood in the side of a barrel; *pl.* **-s.** *See also* **staff**

stay-at-home (adj. and noun, hyphens)

stayed, stopped, remained. *See also* **staid**

STB, *Sacrae Theologiae Baccalaureus* (Bachelor of Sacred Theology)

STC, Senior (Officers') Training Corps, Short-Title Catalogue

stchi, *use* **shchi**

STD, *Sacrae Theologiae Doctor* (Doctor of Sacred Theology), Subscriber Trunk Dialling

Ste (Fr. f.), *sainte* (female saint) (no point)

S[te] (Fr.), *Société* (Society)

steadfast, *not* sted-

steamboat (one word)

steam/-engine, -hammer (hyphens)

steam radio (colloq.) (two words)

steamroller (one word)

steamship (one word), abbr. **SS**

steam train (two words)

Steel (Flora Annie), 1847–1929, Scottish novelist

Steele (Sir Richard), 1672–1729, English essayist and dramatist

Steell (Gourlay), 1819–94, Scottish painter; — **(Sir John),** 1804–91, his brother, Scottish sculptor

steenbok (Afrik.), a small antelope, *not* steinbuck. *See also* **steinbock**

steenkirk (hist.), a neck-cloth, *not* steinkirk

steeplechase/, cross-country horse-race; **-r,** horse *or* rider

steeplejack (one word)

Steevens (George), 1736–1800, English Shakespearian commentator; — **(George Warrington),** 1869–1900, English journalist. *See also* **Stephen, Stephens, Stevens**

Stefansson (Vilhjálmur), 1879–1962, Canadian Arctic explorer

Steinbeck (John Ernst), 1902–68, US novelist
Steinberg, a hock (wine)
steinbock, Alpine ibex
steinbuck, *use* **steenbok**
steinkirk, *use* **steen-**
stem, abbr. **s.** *or* **st.**
stemm/a, a pedigree; *pl.* **-ata**
stencil/, **-led**, **-ler**, **-ling**
Stendhal, pseud. of **Marie Henri Beyle**, 1783–1842, French novelist
step/brother, **-child**, **-daughter**, **-father**, **-mother**, **-parent**, **-sister**, **-son** (one word)
Stephen (Sir Leslie), 1832–1904, English biographer and critic. *See also* **Steevens, Stephens, Stevens**
stephengraph, *use* **stev-**
Stephens (Alexander Hamilton), 1812–83, US statesman; — **(James)**, 1882–1950, Irish poet and novelist. *See also* **Steevens, Stephen, Stevens**
Stephenson (George), 1781–1848, English locomotive engineer; — **(Robert)**, 1803–59, his son, English engineer (tubular bridges). *See also* **Stev-**
steppe, treeless plain, esp. of Russia
stepping-stone (hyphen)
steradian, unit of solid angle, abbr. **sr**
stere, unit of volume, 1 cubic metre
stereophonic, of reproduction of sound by two channels, colloq. **stereo/**, *pl.* **-s**
stereotype (typ.), a plastic, lead alloy, or rubber duplicate plate cast from a matrix moulded from original relief printing material; colloq. **stereo**
sterilize, *not* **-ise**
Sterling (John), 1806–44, Scottish writer. *See also* **Stir-**
sterling, (of money), of standard value; abbr. **stg.**
Sterne (Laurence, *not* Law-**)**, 1713–68, English novelist
stet/ (typ.), a Latin word mean-

ing 'let it stand', written in the proof margin to cancel an alteration, dots being placed under the deleted portion of text; as verb, **-ted**, **-ting**
stetson, wide-brimmed felt hat
Stettin, Ger. for **Szczecin**
Steuart (Sir James Denham), 1712–80, Scottish economist; — **(John Alexander)**, 1861–1932, Scottish journalist and novelist. *See also* **Stewart, Stuart**
stevengraph, colourful picture made of silk, *not* steph-
Stevens (Alfred), 1818–75, English sculptor; — **(Thaddeus)**, 1792–1868, US abolitionist; — **(Wallace)**, 1879–1955, US poet. *See also* **Steevens, Stephen, Stephens**
Stevenson (Adlai Ewing), 1835–1914, US Vice-President; — **(Adlai Ewing)**, 1900–65, his grandson, US politician; — **(Robert)**, 1772–1850, English lighthouse engineer; — **(Robert Louis Balfour)**, 1850–94, his grandson, Scottish novelist, essayist, and poet, abbr. **R.L.S.** *See also* **Steph-**
Stewart, family name of Earl of Galloway; — **(Dugald)**, 1753–1828, Scottish metaphysician; — **(James)**, 1831–1905, Scottish African missionary. *See also* **Steuart, Stuart**
Stewartry, district of Dumfries & Galloway
St.Ex., Stock Exchange
stg., sterling
stibium, antimony, symbol **Sb**
stichometry, division into or measurement by lines of verse, *not* stycho-
stichomythia, dialogue in alternate lines of verse, *not* stycho-
Stieglitz (Alfred), 1864–1946, US photographer
stigma/, a brand; *pl.* **-s**, *but* **-ta** with ref. to Christ's wounds
stigmatize, *not* **-ise**

stilb, unit of luminance, abbr. **sb**; in SI units 1 cd/cm²

stile, over a fence, vertical piece in framework. *See also* **style**

stiletto/, a dagger; *pl.* **-s**

still birth (two words)

stillborn (one word)

still life/, *pl.* **-s** (hyphen when attrib.)

stilly, adj. and adv. from still

Stilton (cheese) (cap.)

stilus, *use* **stylus**

stimie, *use* **stymie**

stimul/ate, -able

stimul/us, *pl.* **-i**

stip., stipend, **-iary**

Stirling, town and district of Central

Stirling (J. Hutchison, *not* -inson), 1820–1909, Scottish metaphysician; — **(Sir William Alexander, Earl of)**, ?1567–1640, Scottish poet, usu. called William Alexander. *See also* **Ster-**

stirp/**s** (Lat.), lineage; *pl.* **-es**

stitching (bind.), the sewing together of all the sections of a book in a single operation. *See also* **sewing**

Stn., Station

stoa/, (Gr. archit.) long building with colonnade; *pl.* **-s**; (cap.) philos. school of Zeno and the Stoics (*see* **Stoic**)

Stock Exchange (caps.); abbr. **S/E, St. Ex.,** *or* **Stock Ex.**

Stockhausen (Karlheinz), b. 1928, German composer

stockholder (one word)

stockinet, an elastic fabric, *not* -ette, -inget

stockpile (one word)

stoep (Afrik.), terrace attached to a house, *not* stoup

Stoic/, (member) of Gr. philos. school founded by Zeno in 4th c. BC, extolling virtue and suppressing passion, **-ism**; (gen., not cap.) person having great self-control in adversity, adj. **-al**

stoichio/logy, the doctrine of elements, *not* stoechio-, stoich-

eio-; **-metric** (chem.), having elements in fixed proportions; **-metry** (chem.), the proportion in which elements occur in a compound

stoke/hold, -hole (one word)

Stoke-on-Trent, Staffs. (hyphens)

stokes, unit of kinematic viscosity, abbr. **St**; in SI units 10⁻⁴ m²/s

STOL, short take-off and landing

ston/e, abbr. **st.**; **-y**

stone (typ.), table on which pages of type are imposed

Stone House, Cumbria

Stonehouse, Ches., Devon, Glos., Northumb., Strathclyde

stonemason (one word)

stonewall (verb), to obstruct (one word)

Stonyhurst College, Lancs.

stoop, *see* **stoep, stoup**

stop/cock, -gap (one word)

stop-go, alternating progress and lack of it (hyphen)

stop/-off, -over (nouns, hyphens)

stop-press (hyphen)

stop valve (two words)

stop-watch (hyphen)

storable, *not* -cable

storage heater (two words)

storehouse (one word)

storey/, a horizontal division of a building; *pl.* **-s**; *not* story, -ies; **-ed**

storied, celebrated in story, *not* -yed

storiolog/y, scientific study of folklore, *not* story-; **-ist**

storm petrel (two words), *not* stormy —

storm/-troops, -trooper (hyphens)

Storting, legislative assembly of Norway

story (of a building), *use* **storey**

story/-book, -line, -teller (hyphens)

stoup, flagon, holy-water basin, *not* stoup

Stow (**John**), ?1525–1605, English historian

stowaway (one word)

Stowe, village and school in Bucks.; — (**Harriet Elizabeth,** *née* **Beecher**), 1811–96, US writer

Stow-on-the-Wold, Glos. (hyphens)

STP, *Sacrae Theologiae Professor* (Professor of Sacred Theology); standard temperature and pressure

Str., Strait, -s

str., stroke (oar)

Strachey, family name of Baron O'Hagan; — (**Giles Lytton**), 1880–1932, English biographer

Strad, colloq. for Stradivarius (violin) (cap., no point)

Stradbroke (**Earl of**)

Stradivarius, an instrument of the violin family made by **Antonio Stradivari** (in Latin **Antonius Stradivarius**), of Cremona, *c.*1644–1737; colloq. **Strad**

Strafford (**Earl of**)

straightforward (one word)

strait/-,-s, cap. when with name; abbr. **St.** *or* **Str.**

strait/-jacket, -laced, *not* straight- (hyphens)

Straits Settlements, now a component of Malaysia

Stranraer, Dumfries & Galloway

strappado/, form of torture, *pl.* **-s**; as verb, **-ed, -ing**

Strasburg, English form of Fr. **Strasbourg,** Ger. **Strassburg**

Stratford-upon-Avon, War., *not* -on- (hyphens)

Strath, written separately when referring to the strath itself, as Strath Spey

Strathclyde, ancient kingdom and modern region in Scotland; university, 1964; — (**Baron**)

strathspey, Scottish dance

stratosphere, the upper atmosphere, where temperature is constant

strat/um, a layer; *pl.* **-a**

strat/us, a low layer of cloud, *pl.* **-i,** abbr. **s.**

Straus (**Oskar**), 1870–1954, Austrian-born composer

Strauss (**Johann I**), 1804–49, Austrian composer of Viennese waltzes; — (**Johann II**), 1825–99, his son, the best known of the family; — (**Joseph**), 1827–70, and — (**Eduard**), 1835–1916, also sons of Johann I; — (**Johann III**), 1866–1939, son of Johann II; — (**Richard Georg**), 1864–1949, German composer of operas, songs, and orchestral works

Stravinsky (**Igor Fyodorovich**), 1882–1971, Russian-born composer

Streatfeild, family name, *not* -field

street (typ.), name of, to have initial caps., as Regent Street; spell out when a number, as Fifth Avenue; number of house in, not to be followed by any point, as 6 Fleet Street; abbr. **St.**

stretto/ (mus.), (passage played) in quicker time, *pl.* **-s**

strew, to scatter, *not* strow; partic. **strewed** *or* **strewn**

stri/a (anat., geol.), a stripe; *pl.* **-ae**

strike/-on (comp.), composition on typewriters using carbon ribbons (hyphen); *see also* **golf-ball;** — **-through** (typ.), the penetration of ink into paper

Strine, comic transliteration of Austral. speech (cap.)

stringendo (mus.), pressing, accelerating the speed

stripping-in (typ.), in filmsetting, the correction of small composition errors by the excision and careful replacement of the film containing them

strip-tease (hyphen)

stripy, having stripes, *not* -ey

strive, past **strove,** partic. **striven**

Stroganoff (cook.), (designating) dish of meat cooked in sour cream (cap.)

strontium, symbol **Sr**

Strood, Kent

Stroud, Glos., Hants

strow, *use* **strew**

'struth, colloq. for God's truth (apos.)

Struwwelpeter, German child's book

strychnine, *not* -in

Sts., Saints

Stuart (House of); — **(Leslie),** 1866–1928, pseud. of **Thomas Barrett,** English composer. *See also* **Steuart, Stewart**

stucco/, (to apply) plaster coating to wall surfaces; *pl.* **-es;** as verb, **-es, -ed**

Stück (Ger. n.), a piece

student, name for fellow of Christ Church, Oxford

studio/, *pl.* **-s**

stumbling-block (hyphen)

stupefy, *not* -ify

Sturluson, *see* Snorri

Sturm und Drang (Ger.), storm and stress (no hyphens)

Sturm-und-Drang-Periode, German romanticism of the late eighteenth century (three hyphens)

Stuttgart, W. Germany, *not* Stü-, -ard

Stuyvesant (Peter), 1592–1672, Dutch colonial governor

sty, inflamed swelling on eyelid; enclosure for pigs; *not* stye; *pl.* **sties**

stycho/metry, -mythia, *use* **sticho-**

Stygian, (as) of the River Styx

style, custom, manner, etc. *See also* **house style, stile**

stylize, *not* -ise

stylus/, needle-like device used in playing gramophone records, *pl.* **-es,** *not* sti-

stymie, to thwart, obstruct, an obstruction; *not* sti-, -my

Styx (Gr. myth.), a river in Hades

s.u. (Ger.), *siehe unten* (=see below)

Suabia, *use* **Swa-**

suable, that can be sued, *not* -eable

Suakin, Sudan, *not* -im

suaraj, *use* **swaraj**

suaviter in modo (Lat.), gently in manner. *See also* ***fortiter in re***

sub/, colloq. for subaltern, sub-editor, submarine, subscription, substitute; as verb, to act as substitute, to sub-edit, **-bed, -bing**

sub (Lat.), under

subahdar (Ind.), a native captain

subaltern, officer below rank of captain; colloq. **sub**

subaudi (Lat.), understand, supply

sub/-basement, -branch, -breed (hyphens)

sub/class, -committee, -conscious, -contract, -deacon, -dean, -divide, -division (one word)

sub-edit/, -or, -orial (hyphens)

sub finem (Lat.), towards the end, abbr. *s.f.*

subfusc/, -ous, dark, esp. of formal clothing at some universities

sub/genus, -heading (one word)

subj., subject, -ive, -ively, subjunctive

sub judice (Lat.), under consideration

subjunctive, abbr. **subj.**

sub/kingdom, -lease, -let (one word)

sub-lieutenant, abbr. **Sub-Lt.** (cap. as title)

Sublime Porte (hist.), Turkish Court and Government

sub-machine-gun (two hyphens)

submicroscopic (one word)

sub/ modo (Lat.), in a qualified sense; — *nomine,* under a specified name, abbr. **s.n.**

subnormal (one word)

sub-plot (hyphen)

subpoena/, (to serve) a writ commanding attendance; *pl.* **-s**; as verb, **-ed,** *not* -pena (one word, not ital.)

sub rosa (Lat.), under the rose, privately

subscript, (symbol) written below. *See also* **iota**

subscription, colloq. **sub**

subsection (one word); abbr. **subsec.**

subsequence, noun from **subsequent,** that which follows

sub-sequence, sequence that is part of a larger one

subsidize, to pay a subsidy to, *not* -ise

subsidy, a grant of public money

sub sigillo (Lat.), in the strictest confidence

sub silentio (Lat.), in silence

sub/sonic, -species, abbr. **subsp.** (*sing.*), **subspp.** (*pl.*), **-standard** (one word)

substance (amount of), symbol *n*

substantive, abbr. **s., sb.,** *or* **subst.**

substitute, colloq. **sub**

substrat/um, *pl.* **-a**

sub/structure, -terranean (one word)

subtil/, -e, *use* **subtle** in all senses

subtilize, to rarefy, *not* -ise

subtitle (one word)

subtl/e, fine, rarefied, elusive, cunning; **-er, -est, -ety, -y,** *not* subtil-

subtopia, suburban development regarded unfavourably

sub/total, -tropical (one word)

Suburbia, suburbs and their inhabitants (cap.)

sub/ voce or — verbo (Lat.), under a specified word; abbr. **s.v.**

subway (one word)

succeeded, abbr. **s.**

succès/ de scandale (Fr. m.), success due to being scandalous; — *d'estime,* success with more honour than profit; — *fou,* extravagant success

Succoth, Jewish Feast of Tabernacles

suchlike (one word)

sucking-pig (hyphen)

sud (Fr. m.), south; abbr. **s.**

Süd (Ger. m.), south; abbr. **S.** (cap.)

Sudan/, indep. 1956; adj. **-ese;** *not* Sou-

Sudra, Hindu labourer caste

Suë (Eugène), 1804–57, French playwright

su/e, -ed, -ing

Suède, Fr. for Sweden, adj. *suédois*

suede, dull-dressed kid, as for gloves (no accent)

Suetonius (Gaius Tranquillus), *c.*70–*c.*160, Roman biographer and antiquarian; abbr. **Suet.**

suff., suffix

suffic/it (Lat.), it is sufficient; *pl.* **-iunt**

Sufi, a Muslim mystic, *not* Sofi, Soofee. *See also* **Sophy**

sugar/-bowl, -cane (hyphens)

suggestible, open to suggestion, *not* -able

suggestio falsi (Lat.), an indirect lie

sui/ generis (Lat.), (the only one or ones) of his, her, its, *or* their own kind; — *juris,* of full age and capacity

suisse/ (Fr. adj.), of Switzerland, (m., cap.) Swiss man, (not cap.) beadle of a church; *fem.* **-sse;** **la Suisse,** Fr. for Switzerland

suite, a set of rooms, attendants, *or* musical pieces (not ital.)

suivez (mus.), follow the soloist

suk(h), *use* **souk**

Sulawesi, Indonesia, formerly **Celebes**

Suleiman the Magnificent, ?1495–1566, Sultan of Turkey

Sully Prudhomme (René François Armand), 1839–1907, French poet and critic

sulpha, class of drugs

sulph/ur, symbol **S**; **-ate, -ite, -uretted, -urize,** not -ise

sultan/, a Muslim ruler, fem. **-a**; **Sultan (the),** of Turkey, till 1922; abbr. **Sult.**

Sultanpur, India

sum (math.), symbol Σ

sumac (bot.), an ornamental tree, not sh-, -ach, -ack

Sumburgh Airport, Shetland

Sumer/, Babylonia; adj. **-ian**

summarize, not -ise

summer (not cap.)

summersault, use somer-

summer-time, the summer season (hyphen); **British Summer Time,** one hour in advance of GMT, in summer only, 1922–67, and from 1972 (three words). See also **BST**

summonsed (law), issued with a summons, not -oned

summum bonum (Lat.), the supreme good

Sumter (Fort), SC, US

sun, cap. only in astronomical contexts and in a list of planets; abbr. **S.**

Sun., Sunday

sun/bathe, -beam (one word)

sun-bonnet (hyphen)

sunburn/, -t or **-ed** (one word)

sundae, ice-cream with crushed fruit and nuts

Sunday, abbr. **Sun.**

sun/dial, -down (one word)

sun-dress (hyphen)

sunflower (one word)

sun/-glasses, -god, -hat (hyphens)

sunlight (one word)

sunn, E. Indian fibre

Sunna, traditional Muslim law, not -ah

Sunni, an orthodox Muslim, also collect. pl., not -ee; also **Sunnite**

sun/rise, -set, -shade, -shine, -spot, -stroke (one word)

sun/-tan, -trap (hyphens)

suo/jure (Lat.), in one's own right; — *loco,* in its own place

sup., superior, supine

sup. (Lat.), *supra* (above)

superannu/ate, -able (one word)

super-calendered paper, highly polished but not coated; abbr. **SC**

supercargo/, person in ship managing sales, pl. **-es** (one word)

supercede, use -sede

super-ego/ (psych.), part of mind that exerts conscience, pl. **-s** (hyphen)

superexcellen/ce, -t (one word)

superficies, a surface, pl. same

superhuman (one word)

superintendent, abbr. **supt.**

superior, abbr. **sup.**

superiors (typ.), small characters set above ordinary characters, e.g. ^{1}, ^{2}, ^{a}, ^{b}

superl., superlative

super/man, -market, -natural, -power, -script (one word)

supersede, to set aside, take the place of, not -cede

super/sonic, -structure, -tanker (one word)

supervise, not -ize

supervisor, not -er

supine, abbr. **sup.**

supple/, -ly, not supply

supplement, abbr. **suppl.**

suppositious, hypothetical, assumed

supposititious, spurious

suppository, a medicated plug, not -ary

suppressio veri (Lat.), suppression of the truth

suppressor, not -er

supr., supreme

399

supra (Lat.), above, formerly; abbr. *sup.*

suprême (Fr. f.), a method of cooking, with a rich cream sauce

Supreme Court (US)

supremo/, leader or ruler, *pl.* **-s**

supt., superintendent

suq, *use* **souk**

sur (Fr. prep.), upon; abbr. **s/** in addresses

sura, a chapter of the Koran (cap. for particular chapters)

Surabaya, Indonesia

surah, a thin silk fabric

Surat, Bombay, India

surcingle, a belt

Sûreté, Paris, the French CID

surfboard (one word)

surf-riding (hyphen)

surg/eon, -ery, -ical, abbr. **surg.**

Surgeons (Royal College of), abbr. **RCS**

surmise, conjecture, *not* -ize

surplus, abbr. **S.**

surprise, *not* -ize

sursum corda (Lat.), (lift) up your hearts

Surtees (Robert Smith), 1805–64, English fox-hunting novelist

surtout, an overcoat

surv., survey/ing, -or, surviving

Surv.-Gen., Surveyor-General

survivor, *not* -er

sus/, slang for suspect, suspicion; — **out**, to reconnoitre, **-sed, -sing**

Susanna (Apocr.), abbr. **Sus.**

susceptible, *not* -able

suspender, *not* -or

sus. per coll. (Lat.), *suspen/sio* (-*sus*, -*datur*) *per collum*, hanging (hanged, let him be hanged) by the neck

Susquehanna River, New York State, US, *not* -ana

Sutlej, Punjab river

suttee, (custom of) Hindu widow immolating herself on husband's pyre, *not* sati

suum cuique (Lat.), let each have his own

Suwannee River, Ga. & Fla., US, *not* Swa-

SV, *Sancta Virgo* (Holy Virgin), *Sanctitas Vestra* (Your Holiness)

s.v., *sub voce*, or *verbo* (under a word or heading, as in a dictionary); sailing vessel

SVA, Incorporated Society of Valuers and Auctioneers

svelte (Fr.), elegant

Svendsen (Johan Severin), 1840–1911, Norwegian composer

Sverige, Swedish for **Sweden**

s.v.p. (Fr.), *s'il vous plaît* (if you please)

SW, South Wales, south--west(ern)

Sw., Swed/en, -ish

Swabia, W. Germany, *not* Su-

Swahili, a people and language of Zanzibar and the coast opposite

Swammerdam (Jan), 1637–80, Dutch naturalist

Swanee, *use* **Suwannee**

swansdown (one word)

swan-song (hyphen)

swap, to exchange, *not* -op

SWAPO, South West Africa People's Organization

swaraj, self-govt. for India, *not* su-

swash (typ.), (of) letters with elaborate tails or flourishes

swat, to hit sharply. *See also* **swot**

swath, a line of cut grass

swathe, to bind

Swazi/, -land, S. Africa, indep. 1968

swede, a kind of turnip (not cap.)

Sweden, abbr. **Sw.**, in Swed. **Sverige**, in Fr. **Suède**. *See also* **Assemblies**

Swedenborg (Emanuel), 1688–1772, Swedish philosopher

Swedish, abbr. **Sw.**; (typ.) alphabet contains Ger. ä, ö, and 'Swedish a' (Å, å) pron. somewhat as *aw*. In alph. arrangement å, ä, ö, are put after *z*. The acute accent may be used to mark an accented syllable

sweepback (of aircraft's wings) noun and adj. (one word)

Sweet (Henry), 1845–1912, English philologist. *See also* **Swete**

sweet-and-sour (cook. adj., hyphen)

sweet/bread, -heart, -meal, -meat (one word)

sweet-william (bot.) (hyphen, not caps.)

swelled rule (typ.), a rule wider in centre than at ends. *See also* **rule**

Swete (Henry Barclay), 1835–1917, English biblical scholar. *See also* **Sweet**

Sweyn, name of three Danish kings

SWG, standard wire gauge

swimming/-bath, -pool (hyphens)

Swinburne (Algernon Charles), 1837–1909, English poet

Swinden, N. Yorks.

Swindon, Glos., Staffs., Wilts.

swing-door (hyphen)

swingeing, hard (blow)

swing-wing (of aircraft) (hyphen)

Swinton, Gr. Manch., N. Yorks., S. Yorks., Borders

switchboard (one word)

Swithun/ (St.), Bp. of Winchester 852–62, *not* -in; **-'s Day,** 15 July

Switzerland, abbr. **Switz.**; in Fr. **la Suisse,** in Ger. **die Schweiz,** in It. **Svizzera.** *See also* **Assemblies**

swivel/, -led, -ling

swizz/ (slang), a swindle, a disappointment, *pl.* **-es,** *not* swiz

swop, to exchange, *use* **swap**

sword/-dance, -stick (hyphens)

swordfish (one word)

swot, to study hard. *See also* **swat**

SWP, Socialist Workers' Party

swung dash (typ.), ∼

SY, steam yacht

Sybil, Christian name. *See also* **sibyl**

sycamine (NT, AV), the black mulberry-tree

sycamore (bot.), a kind of fig-tree, *Ficus sycomorus* (**sycomore** in Bible): (Eur. and Asia) an ornamental shade-tree, the sycamore maple, *Acer pseudoplatanus*; (US) a plane tree, *Platanus*

syce (Anglo-Ind.), a groom. *See also* **sice**

sycomore, see **sycamore**

Sydney, NSW, Australia. *See also* **Sidney**

Sydney Heads, two cliffs

Sykes, see **Si-**

syllab/ication, *not* -ification; **-ize,** *not* -ise

syllabub, *use* **si-**

syllabus/, *pl.* **-es**

sylleps/is (gram.), application of a word in differing senses to two others (he caught his train and a cold), *pl.* **-es**

syllogism, a logical argument of two premisses and a conclusion

syllogize, to argue by syllogism, *not* -ise

sylvan, *use* **si-**

Sylvester (James Joseph), 1814–97, English mathematician; — **(Josuah)**, 1563–1618, English poet

sylvicultur/e, -ist, *use* **si-**

symbolize, *not* -ise

symmetr/ic, -ical, -ize, *not* -ise

sympathique (Fr.), congenial, having the right artistic feeling for

sympathize, *not* -ise

symposi/um, a drinking-party (obs. in this sense except, with cap., as title of one of Plato's Socratic dialogues), a conference, a collection of views on a topic; *pl.* **-a**

syn., synonym, -ous

synaeres/is (gram.), contraction of two vowels, *pl.* **-es,** *not* -neresis

synagog/ue, an assembly of Jews for worship, their place of worship; adjs. **-al, -ic**

synchromesh, system of gear-wheel design

synchronize, to (cause to) coincide in time, *not* -ise; colloq. **sync,** *not* synch

syncope, (gram.) a cutting short, (med.) a fainting

syndrome, a collection of concurrent symptoms

synonym/, a word with the same meaning as another; **-ous,** abbr. **syn.**; **-ize,** *not* -ise; **-y,** *not* -e, -ey

synops/is, a summary; *pl.* **-es**

Synoptic Gospels, Matthew, Mark, and Luke, which relate the events from the same point of view

synthes/ist, -ize, *not* synthet-, -ise

syphon, *use* **si-**

Syr., Syria, -c, -n

syr., syrup

syren, *use* **si-**

Syriac, abbr. **Syr.**; has 22 letters, besides vowel points, reads from right to left, and is set as Hebrew. There are three forms of type, Estrangelo, Jacobite, Nestorian

Syringa, (bot.) the lilac genus

syringa, the mock orange

syringe/, instrument for squirting or injecting; **-ing;** *not* si-

syrin/x, pan-pipe, (anat.) a narrow tube from throat to ear--drum, the vocal organ of birds, (archaeol.) a narrow gallery in rock; *pl.* **-xes** *or* **-ges**

syrup, *not* sir-; abbr. **syr.**

syst., system

system/atic, methodical; **-atize,** *not* -ise; **-ic** (physiol.), of the bodily system as a whole

syzygy (astr.), the moon being in conjunction or opposition

Szczecin, Poland; in Ger. **Stettin**

Szigeti (**Joseph**), 1892–1973, Hungarian violinist

Szymanowski (**Karol**), 1883–1937, Polish composer

T

T, tesla, (as prefix) tera-, tritium; the nineteenth in a series

T., Tenor, Territory, Testament, (It. mus.) *tace* (be silent)

t., ton, -s, (*but* no point in scientific and technical work), tonne, -s, town, -ship, tun, -s, (Fr.) *tome* (volume), *tonneau* (ton), (Lat.) *tempore* (in the time of), (mus.) tempo (time), *tenor/e, -i* (tenor, -s), (meteor.) thunder

't (Du.), *het* (the, n.) as van 't Hoff

T, temperature

t, time

TA, Territorial Army

Ta, tantalum (no point)

Tabago, *use* **To-**

table/ alphabétique (Fr. typ. f.), index; — *des matières,* table of contents

tableau/, picturesque presentation, dramatic situation, *pl.* **-x;** — **vivant,** still silent group representing scene, *pl.* **-x vivants** (not ital.)

table-cloth (hyphen)

table d'hôte, fixed meal for guests at hotel, *pl.* **tables**

tableland (one word)

tablespoonful/, *pl.* **-s,** abbr. **tbsp.**

table tennis (two words)

table-top (hyphen)

tablier (Fr. m. hist.), apron-like part of woman's dress

taboo/, (something) forbidden, to prohibit, a prohibition, *not* -u; **-ed, -s**

tabor, small drum, *not* -our

tabouret, drum-shaped stool, *not* -oret

tabul/a (Lat.), a document, *pl.* **-ae**; *tabul/a ras/a,* a blank surface, *pl.* **-ae -ae**

tabulat/e, -or

Tac., Tacitus

tac-au-tac (fencing), parry and riposte (hyphens, not ital.); but *du tac au tac* (Fr. fig.), from

defence to attack (no hyphens, ital.)

tace (mus.), be silent; abbr. **T.**

tacet (mus.), is silent

tachism, action painting, in Fr. m. *tachisme*

tacho/, colloq. for tachometer, *pl.* **-s**

tachograph, device for recording speed and travel-time of vehicle

tachometer, instrument for measuring speed of rotation

tachymeter, instrument in surveying

Tacitus (Gaius Cornelius), *c.*55–120, Roman historian; abbr. **Tac.**

Tadjikistan, a Soviet Socialist Republic, *not* Tadzh-, Taj-. *See also* **USSR**

taedium vitae (Lat.), weariness of life

Tae-ping, *use* **Taiping** (one word)

taffeta, a fabric, *not* the many variants

tagetes (bot.), a type of marigold, *not* -ete; *pl. same*

tagliatelle (It.), pasta in ribbons, *not* -elli

Tagore (Rabindranath), 1861–1941, Indian poet

Tahiti, island, Pacific Ocean

tahr, *use* **thar**

tail (typ.), blank space at the bottom of a page. *See also* **margins**

tail-back, queue of vehicles in traffic jam (hyphen)

tail/-gate, -lamp, -light (hyphens)

taille/ (Fr. f.), engraving, also size, etc.; — *douce,* copperplate engraving

tailpiece (typ.), the design at the end of a section, chapter, or book

tail/pipe, -plane (one word)

tailstock, adjustable part of lathe holding spindle (one word)

Tain, Highland

Taine (Hippolyte Adolphe), 1828–93, French historian and lit. critic

Taipei, capital of Taiwan, q.v.

Taiping rebellion, 1850–64, *not* Tae-

Tait (Archibald Campbell), 1811–82, Abp. of Canterbury; — **(Peter Guthrie),** 1831–1901, Scottish mathematician and physicist. *See also* **Tate**

Taiwan, *formerly* **Formosa**

Tajikistan, *use* **Tadj-**

Taj Mahal, mausoleum at Agra, India, *not* — Me-

take (typ.), single batch of copy or of proofs

take-away (noun and adj., hyphen)

take-home pay (one hyphen)

take/-off, -over (hyphens)

Tal (Ger. n.), valley, not now *Th-*

Talbot of Malahide (Baron)

tales (law), a suit for summoning jurors to supply a deficiency; **tales/man,** one so summoned, *pl.* **-men**

Talfourd (Sir Thomas Noon), 1795–1854, English playwright, biographer, and author of the Copyright Act of 1842

Taliesin, 6th c., Welsh bard

talisman/, a charm, amulet; *pl.* **-s**

talis qualis (Lat.), such as it is

Talleyrand-Périgord (Charles Maurice de), 1754–1838, French politician

Tallinn, cap. of Estonian SSR, *formerly* **Reval**

Tallis (Thomas), ?1505–85, English composer, *not* Talys, Tallys; **Tallis's canon,** a hymn-tune

tally-ho/, *pl.* **-s** (hyphen)

Talmud, Hebrew laws and legends

Tal-y-llyn, Gwynedd, Powys (two hyphens)

TAM, Television Audience Measurement

Tam., Tamil

tam, colloq. for tam-o'-shanter (no point)

tambourin, a long narrow drum of Provence, (music for) a dance accompanied by it

tambourine, a small drum, with jingling metal discs, beaten with the hand

tameable, *not* -mable

Tamerlane, 1336–1405, Mongol conqueror (*but* **Tamburlaine** in Marlowe)

Tameside, district, Greater Manchester

Tamil, S. Indian language, *not* -ul; abbr. **Tam.**

Tammany Hall, Democratic Party HQ, New York, *not* Tama-

'Tam o' Shanter', poem by Burns (caps., small *o*, apos., no hyphen)

tam-o'-shanter, a woollen cap (hyphens, apos.); colloq. **tam**

tampion, plug for muzzle of gun or top of organ-pipe, *not* to-

tampon (med.), absorbent plug

tan (math.), tangent (no point)

Tananarive, Malagasy Republic, *not* Antananarivo

T. & A.V.R., Territorial and Army Volunteer Reserve

T'ang, Chin. dynasty

tangerine orange, *not* tangier-

tangible, *not* -eable

Tangier, Morocco, *not* -iers

tango/, *pl.* **-s**

tanh (math.), hyperbolic tangent (no point)

Tanjor/, -e, India, *use* **Thanjavur**

Tannhäuser, opera by Wagner, 1845

tanrec, *use* **tenrec**

tantalize, *not* -ise

tantalum, symbol **Ta**

tant/ mieux (Fr.), so much the better; — *pis,* so much the worse

Tanzania (Tanganyika and Zanzibar), formed 1964

Taoiseach, Prime Minister, Republic of Ireland

Taoism, doctrine of Lao-tzu, *not* Tâ-, Taö- (one word)

tap-dance (hyphen)

tape record/er, -ing (two words)

tapis (*sur le or* on the) (Fr.), under consideration, *or* discussion

tapisserie (Fr. f.), tapestry

taproom (one word)

tap-water (hyphen)

taradiddle, petty lie, pretentious nonsense, *not* tarra-

tarantella, an Italian dance, the music for it

tarantism, dancing mania

tarantula, a spider

Tarbert, Strathclyde, Western Isles, Co. Kerry; (river); Highland

Tarbet, Loch Lomond, Strathclyde, and Loch Nevis, Highland

tariff, duty on particular goods, *not* -if

tarlatan, a muslin, *not* -etan

Tarmac, propr. term for tar macadam; part of airfield made of this (cap.); **tarmac/,** to cover with Tarmac (not cap.); **-ked**

tarot, card-game or trump in this, *not* -oc

tarpaulin, waterproof cloth, orig. of tarred canvas, *not* -ing

Tarpeian Rock, anc. Rome

tarradiddle, *use* tara-

Tarragona, Spain

tarry, covered with or like tar, *not* tary

Tartar/, an inhabitant of Tartary, an intractable person, *not* Tatar; adj. **-ian**

tartar/, a deposit on the teeth; adj. **-ic**

tartar sauce, *not* tartare

Tartar/us (Gr. myth.), the place of punishment in Hades; adj. **-ean**

Tartary, *properly, but less usually,* **Tatary,** a region of W. Asia and E. Europe

Tartuffe (*Le*), play by Molière, 1669

Tartuff/e, religious hypocrite, adj. **-ian** (Fr. m. *tartufe*)

Tas., Tasmania

task/master, -mistress (one word)

Tasmania, abbr. **Tas.**

Tass, the Soviet news agency

taste-bud (hyphen)

ta-ta, colloq. for good-bye (hyphen)

Tatar, *use* Tartar

Tate/ (Nahum), 1652–1715, Irish poet and playwright; — (**Sir William Henry**), 1842–1921, English industrialist; — **Gallery,** London. *See also* **Tait**

tatterdemalion, a ragged fellow, *not* -ian

Tattersalls, London (-s' only in the possessive case)

tattoo/, design on the skin, *not* the many variants; **-ed, -er, -ing, -s**

tattoo/, drumbeat, call to quarters, mil. parade by night; *pl.* **-s**

Tauchnitz (Karl Christoph), 1761–1836, founded at Leipzig, 1796, publishing firm famous for editions of Latin and Greek authors; — (**Christian Bernhard, Baron von**), 1816–95, founded at Leipzig, 1837, the Librairie Bernhard Tauchnitz, famous for reprints of British and US authors, banned in Britain and US because of copyright infringements

tau cross, the T

tausend (Ger.), thousand

taut (naut.), tight, in good condition, *not* -ght

tautologize, to repeat oneself in different words, *not* -ise

tavern (**name of**), *see* **hotel**

tawny, tan-colour, *not* -ey

tax-free (hyphen)

tax haven (two words)
taxi/, -ing
tax/on, a taxonomic group, *pl.*
  -a
taxpayer (one word)
tax return (two words)
Taylor Institution, Oxford, *not*
  Taylorian, *not* Institute
Taylour, family name of Mar-
  quess of Headfort
Tayside, region of Scotland
  (one word)
tazz/a (It.), bowl or cup, *pl.* **-e**
TB, torpedo-boat, tuberculosis
  (tubercle bacillus)
Tb, terbium (no point)
Tbilisi, cap. of Georgia, USSR,
  not Tiflis
T-bone (hyphen)
tbsp., tablespoonful
Tc, technetium (no point)
TCD, Trinity College, Dublin
Tchad (Lake etc.), *use* **Chad**
Tchaikovsky (Peter Ilich),
  1840–93, Russian composer
Tchebyshev, Tchekoff, etc.,
  use **Che-**
TD, Territorial Decoration, (Ir.)
  Teachta Dála (Member of the
  Dáil)
Te, tellurium (no point)
te, tonne, **-s**
te (mus.), *not* si, ti
tea/-bag, -chest (hyphens)
tea/-cake, -cup (one word)
tea-leaf (hyphen)
teapot (one word)
tease, *not* -ze
teasel (bot.), *not* -sle, -zel, -zle
tea-shop (hyphen)
teaspoonful/, *pl.* **-s** (one word),
  abbr. **tsp.**
tech., technical, **-ly**
technetium, symbol **Tc**
technol., technological, **-ly**
techy, peevish, *use* **tetchy**
teddy bear (two words, not
  cap.)
Teddy boy (two words, one
  cap.)
tedesc/o (It. adj.), *fem.* **-a,** German (not cap.)
teed (golf)

teehee, *use* **tehee**
teenage/, -d, -r (one word)
teepee, *use* **tepee**
Teesside, Cleveland
teetotal/, abstaining from
  intoxicants, abbr. **TT; -ism,**
  -ler, -ly
teetotum/, a four-sided top spun
  with the fingers; *pl.* **-s**
Tegnér (Esaias), 1782–1846,
  Swedish poet
tehee, scornful laugh, to titter,
  not teehee
Tehran, cap. of Iran, *not* -heran
Teignmouth, Devon. *See also*
  Tyne-
Teil/ (Ger. m. *or* n.), a part; *pl.*
  -e, not now *Th-* (cap.)
Teilhard de Chardin
  (Pierre), 1881–1955, French
  writer
Tel., Telegraph, Telephone
tel, *use* **tell**
telamon/ (archit.), male figure
  used as support, *pl.* **-es**
Tel Aviv, Israel
tele/communication(s),
  -printer, -type (one word)
telephone/ book, — number
  (two words)
televis/e, *not* -ize; **-ion**
telex, (to send by) system of
  telegraphy using teleprinters
Telford, Shropshire, 'new town',
  1963
Telford (Thomas), 1757–
  1834, Scottish engineer
tell (archaeol.), artificial mound
  in Middle East, *not* tel
Tell-el-Amarna, Egypt
tell-tale (hyphen)
tellurian, (inhabitant) of the
  earth
tellurion, orrery, *not* -ium
tellurium, symbol **Te**
Telstar, communications satellite, launched 1962
Telugu, Dravidian language or
  people, *not* -oogoo
Téméraire (The Fighting),
  picture by Turner, 1839
temp., temperature, temporary
temp., *tempore* (in the time of)

temperature, symbol **T** (no point). In scientific work the unit of temperature is the kelvin (K), although ° C is still used; **—, degrees of** (typ.), to be in arabic numerals, as 10°C, 50°F; abbr. **temp.**

Templar, member of a religious order, the **Knights Templars;** student or lawyer living in the Temple, London

template, a pattern or gauge, *not* -plet

Temple Bar, London

temp/o (It. mus.), time; *pl.* **-os,** abbr. **t.**

tempora mutantur (Lat.), times are changing

temporary, abbr. **temp.**; adv. **temporarily**

tempore (Lat.), in the time of; abbr. **t.** *or* **temp.**

temporize, *not* -ise

ten., tenuto

Tenasserim, Burma, *not* Tenn-

Ten Commandments (the) (caps.)

tendentious, calculated to promote a particular point of view, *not* -cious

tenderfoot, novice (one word)

tenderize, *not* -ise

Tenerife (peak and island of), Canary Is., *not* -iffe

Teniers (David), 1582–1649, and 1610–90, Dutch painters, father and son

Tenison (Thomas), 1636–1715, Abp. of Canterbury. *See also* **Tennyson**

Tennasserim, *use* **Tena-**

Tennessee, off. abbr. **Tenn.** or (postal) **TN**

Tenniel (Sir John), 1820–1914, English cartoonist and caricaturist

tennis, it is no longer necessary to use 'lawn tennis', *but see* **ILTF; real tennis** is the game played on an indoor court

tennis/-ball, -court, -racket (hyphens)

Tennyson (Alfred, Lord),

1809–92, Poet Laureate 1850–92. *See also* **Tenison**

tenor, settled course, *not* -our; (mus.) male voice, abbr. **T.**

tenor/e (It. mus. m.), tenor voice; *pl.* **-i,** abbr. **t.**

tenrec, hedgehog-like mammal, *not* tanrec

tenson, contest between troubadours, *not* tenz-

tenuto (mus.), held on, sustained, abbr. **ten.**

tepee, Amer. Indian tent, *not* tee-

Ter., Terence, Terrace

ter (Lat.), thrice

tera-, prefix meaning 10^{12}; abbr. **T**

terat., teratology, study of malformations

terbium, symbol **Tb**

terce (eccl.), the office said at the third daytime hour. *See also* **tierce**

tercel, a male hawk, *not* tier-

tercet (prosody), a triplet, *not* tiercet

Terence (in Lat., **Publius Terentius Afer**), 190–159 BC, Roman comic playwright; abbr. **Ter.**

Teresa (St.), *not* Th-

tergiversat/e, to change one's principles; **-ion, -or**

termagant, a brawling woman (cap. in hist. use as imaginary deity)

termination, abbr. **term.**

terminator, *not* -er

terminology, abbr. **term.**

termin/us, *pl.* **-i**

terminus/ ad quem (Lat.), the finish; **—** *a quo,* the starting--point

Terpsichore (Gr.), muse of dancing

Terr., Territory

terrace, cap. when with name; abbr. **Ter.**; in Fr. f. *terrasse*

terracotta, (an object made of) unglazed kiln-burnt clay and sand; its colour (one word)

Terra del Fuego, *use* **Tierra**

terrae/ filius (Lat.), son of the soil; *pl.* — *filii*

terra firma, dry land (two words, not ital.)

terr/a incognit/a (Lat.), unexplored region; *pl. -ae -ae*; *-a sigillata,* astringent clay, or Samian ware (ital.)

terrazzo/, floor of stone chips set in concrete and smoothed, *pl. -s*

terret, ring for driving-rein, *not* -it

terre-verte, soft green earth used as pigment (hyphen, not ital.)

Territory (cap. in geog. names), abbr. **T.** *or* **Terr.**

terrorize, *not* -ise

tertio (Lat.), in the third place; abbr. 3°

tertium quid (Lat.), a third something, an intermediate course

Terylene, propr. term for synthetic polyester used in textiles (cap.)

terz/a rim/a (It.), a particular rhyming scheme; *pl. -e -e*

terzetto/ (mus.), piece for three voices or instruments, *pl. -s*

TES, *Times Educational Supplement*

tesla, unit of magnetic induction, from **Tesla** (**Nikola**), 1856–1943, Yugoslav-American physicist

tessellate(d), pave(d) with tiles, *not* -ela-

tesser/a, small square tile; *pl. -ae* (not ital.)

tessitura (mus.), the ordinary range of a voice

Testament, abbr. **T.** *or* **Test.**

testamur (Lat.), examination certificate

test/ drive (hyphen as verb), — **flight** (two words)

test/is (anat.), *pl. -es*

test/ match, — **paper,** — **pilot** (two words)

test-tube/ (hyphen), —-— **baby** (one hyphen)

tetchy, peevish, *not* techy

tête-à-tête (adv., adj., *and* noun, hyphens, accents, not ital.)

tetrameter (prosody), verse of four measures

Teufelsdröckh (**Herr**), in Carlyle's *Sartor Resartus*

Teut., Teuton, -ic

Tex., Texas (off. abbr.), Texan

textbook (one word)

textus receptus (Lat.), the received text; abbr. **text. rec.**

Teyte (**Maggie**), 1888–1976, English soprano

t.g., type genus

TGWU, Transport and General Workers' Union

Th, thorium (no point)

Th., Thomas, Thursday

th, in Welsh, a separate letter, not to be divided

thagi, *use* **thuggee**

Thailand, *formerly* **Siam,** adj. **Thai**

Thaïs, 4th c. BC, Greek courtesan

thaler, former German coin

Thalia (Gr.), muse of comedy

thalidomide, a sedative drug, withdrawn 1961

thallium, symbol **Tl**

thalweg, lowest line along river, in Ger. now *Talweg*

Thanjavur, India, *not* Tanjor, -e

Thanksgiving Day (US), fourth Thursday in November

thar, Nepalese or Himalayan goat, *not* tahr

Tharrawaddy, Burma, *not* Tharawadi

the, if part of the title of a book etc., should be italic with cap. initial. *See also* **periodicals** and *Hart's Rules*, pp. 23–4

thé (Fr. m.), tea; — *dansant*, afternoon tea with dancing

Theaetetus, dialogue by Plato, named after disciple of Socrates

theat., theatrical

theatre, *not* -er, Fr. *théâtre* (m.); *Théâtre français,* Paris (one cap.)

thec/a (anat., bot.), case, sheath, sac, *pl. -ae*

theirs (no apos.)

them/a (Gr.), a theme; *pl.* **-ata**

Theo., Theodore

theocracy, a priest- or god-governed state (cap. with hist. ref. to Jews), *not* -sy

theocrasy, mingling of several divine attributes in one god, *not* -cy

Theocritus, 3rd c. BC, Greek poet; abbr. **Theoc.**

theol., theolog/y, -ian, -ical

theologize, *not* -isc

Theophrastus, *c.*370–*c.*288 BC, Greek philosopher and botanist, abbr. **Theoph.**

theor., theorem

theoret., theoretic, -al, -ally

theorize, *not* -ise

theosoph/y, a philosophy professing knowledge of God by inspiration; **-ical, -ist, -ize**

Theoto/copuli, -kopoulos, *see* **Greco**

therapeutic/, healing; **-s,** study of healing agents; abbr. **therap.**

therefore (math.), sign ∴.

Theresa (St.), *use* Ter-

Thérèse (Fr.)

thermodynamics (one word)

thermomet/er, -ric, abbr. **thermom.**

thermonuclear (one word)

Thermopylae (**Pass of**), Greece; battle, 480 BC

THES, *Times Higher Education Supplement*

thesaur/us, *pl.* **-i**

thes/is, *pl.* **-es**

Thess., Thessaly; — (**1, 2**), Thessalonians (NT)

they'll, to be printed close up

thias/os (anc. Gr.), a gathering to worship a deity, *pl.* **-oi,** *not* **-us**

thicken/, -ed, -er, *not* -or, **-ing**

thickset (one word)

thimblerig/, a trick, to play this; **-ger, -ging** (one word)

thin/, -ner, -nish

thingamy, *not* -ummy

think-tank, advisory organization (hyphen)

third (adj.), abbr. **3rd**

Third World, non-aligned countries of Asia, Africa, and Latin America (caps.)

Thirty-nine Articles (**the**) (hyphen, two caps.)

thirty-twomo (typ.), a book based on 32 leaves, 64 pages, to the sheet; abbr. **32mo**

Thirty Years War, 1618–48 (caps., no apos.)

tho', though

thole-pin (naut.), one of two which keep an oar in position, *not* -owl, -owel (hyphen)

thol/os (Gr. archit.), a dome--shaped building, esp. a tomb, *pl.* **-oi**

Thomas, abbr. **Th.** *or* **Thos.**

Thom/ism, doctrine of St. Thomas Aquinas; **-ist**

Thompson (**Sir Benjamin, Count von Rumford**), 1753–1814, American-born founder of Royal Institution, London; — (**Sir D'Arcy Wentworth**), 1860–1948, Scottish biologist; — (**Sir Edward Maunde**), 1840–1929, English librarian and palaeographer; — (**Francis**), 1859–1907, English poet; — (**Silvanus,** *not* Sy-, **Phillips**), 1851–1916, English physicist

Thomsen's disease, muscular spasm

Thomson (**Prof. Arthur**), 1858–1935, Scottish anatomist; — (**Sir Charles Wyville**), 1830–82, Scottish zoologist; — (**James**), 1700–48, Scottish poet, 'The Seasons'; — (**James**), 1834–82, Scottish poet, 'B.V.', 'City of Dreadful Night'; — (**Prof. Sir John Arthur**), 1861–1933, Scottish zoologist and writer; — (**Joseph**), 1858–95, Scottish African traveller; — (**Prof. Sir Joseph John**), 1856–1940, English physicist; — (**Sir William, Baron Kelvin**), 1824–1907, British mathematician and physicist (hence **kelvin,**

Thomson (*cont.*)
q.v.); — **of Fleet** *and* **of Monifieth** (**Barons**)

thor/ax (anat., zool.), the part of the body between neck and abdomen or tail, *pl.* **-aces**; adj. **-acic**

Thoreau (**Henry David**), 1817–62, US author and philosopher

thorium, symbol **Th**

thorn, the Old-English Þ, þ, and Icelandic Þ, þ (distingush from **eth** and **wyn**, qq.v.); also used in phonetic script

thorough, *not* thoro'

thorough-bass (mus.), harmony above bass indicated by system of numerals (hyphen)

thorough/bred, -going (one word)

Thos., Thomas

thou, colloq. for thousand, -th (no point or apos.)

though, abbr. **tho'**

thousand-and-first etc. (hyphens)

thow(e)l-pin, *use* **thole-**

thr., through

thrall, slave, bondage; **thral-dom**, bondage

thrash, to beat soundly, to strike the waves, make way in water; — **out**, to discuss exhaustively. *See also* **thresh**

threadbare (one word)

Threadneedle Street, London (two words)

threadworm (one word)

threefold (one word)

three-point turn (one hyphen)

three-quarter(s) (hyphen)

three Rs (**the**), reading, writing, arithmetic (no point, no apos.)

three/score, -some (one word)

threescore and ten, seventy (three words)

thresh, to beat out corn. *See also* **thrash**

threshing/-floor, -machine (hyphens)

threshold, *not* -hhold

thrips, plant pest, is *sing.*

thrive, past **throve**, partic. **thrived**

thro', *use* **through**

Throckmorton (**Sir Nicholas**), 1515–71, English diplomat

throes, violent pangs

Throgmorton Avenue, *also* **Street**, London, EC

Throndhjem, *use* **Trondheim**

through, *not* thro'; **Monday through Friday** (US), from Monday to Friday inclusive, abbr. **thr.**

throw out (bind.), to mount a diagram, map, etc., upon a page-wide guard so that it may remain in view while other pages are read

Thucydides, *c.*460–*c.*400 BC, Greek historian; abbr. **Thuc.**

thug/, a brutal ruffian (cap. with hist. ref. to assassins in India); **-gee,** (hist.) the practice of the Thugs, (*not* thagi), **-gery**

thuja, accepted English form of bot. *Thuya*, the arbor-vitae genus

Thule, Greenland; — (**Southern**), S. Atlantic

thulium, symbol **Tm**

thumb/-index, -print (hyphens)

thumbscrew (one word)

thunder (meteor.), abbr. **t.**

thunderstorm (one word)

Thursday, abbr. **Th., Thur.,** *or* **Thurs.**

thuya, *see* **thuja**

THWM, Trinity High-Water Mark

thym/e, the herb; adj. **-y**, *not* -ey

Thynne, family name of Marquis of Bath

Ti, titanium (no point)

ti (mus.), *use* **te**

Tian-Shan, mountain range, USSR and China

tiara/, turban, diadem, ornamental coronet; **-ed**, *not* -'d

Tiberias, anc. Palestine, *now* **Teverya,** Israel

Tiberius (in full, **Tiberius Claudius Nero Caesar**), 42 BC–AD 37, second emperor of Rome, 14–37

tibi/a (anat.), inner bone from knee to ankle, *pl.* **-ae**

Tibullus (Albius), *c.*50–19 BC, Roman poet; abbr. **Tib.**

tic douloureux, facial neuralgia, *not* dol- (two words)

ticket/, -ed, -ing

ticket-office (hyphen)

tick-tack, regular beat, racecourse semaphore, *not* tic-tac

Ticonderoga, New York

t.i.d. (med.), *ter in die* (three times a day)

tidbit, *use* tit-

tiddly (slang), tiny, slightly drunk, *not* -ey

tiddly-wink/, counter; **-s,** the game, *not* tiddle(d)y-

tide-mark (hyphen)

tie, tying, *not* tieing

tie/-break, -pin (hyphens)

tierce, (arch.) wine-measure; (mus.) interval of two octaves and a major third; sequence of three cards. See also **terce**

tierc/el, -et, *use* terc-

Tierra del Fuego, S. America, *not* Terra —

tiers état (Fr. m.), third estate, the common people (not caps.)

Tietjens (Therese Cathline Johanna), 1830–77, German-born Hungarian soprano, *not* — Titiens

Tiffany (Charles Lewis), 1812–92, US jeweller

tiffany, a gauze muslin

tiffin (Anglo-Ind.), light lunch, *not* -ing

Tiflis, *use* Tbilisi

tigerish, *not* tigr-

Tighnabruaich, Strathclyde

tight back (bind.), the cover glued to the back, so that it does not become hollow when open

tightrope (one word)

tigrish, *use* tiger-

TIH, Their Imperial Highnesses

tike, *use* ty-

tilde, the mark as over the Sp. *n*, *ñ*; in Port. *til*

Tilsit (Treaty of), 1807, *not* -tt

Tim. (1, 2), First, Second Epistle to Timothy (NT)

timbale (Fr. cook. f.), dish of mince in pastry

timbre, characteristic quality of sounds of a voice or instrument (not ital.)

Timbuktu, Mali, *not* -buctoo; *but* **'Timbuctoo',** Tennyson's prize-poem, 1829, and usu. in figurative sense

time, symbol *t*

timeable, *not* -mable

time-and-motion (adj., hyphens)

time/ bomb, — exposure (two words)

timekeeper (one word)

time/-lag, -limit (hyphens)

time of day (typ.), to be in numerals, with full point where time includes minutes as well as hours; 9.30 a.m. (or 09.30), 10 a.m. (or 10.00), 4.30 p.m. (or 16.30); but such phrases as half-past two, a quarter to four, should be spelt out

timepiece (one word)

Times (The), established 1788 (caps.). *The* (cap. *T,* ital.) should always be printed as part of title

time/-scale, -sheet, -signal, -switch (hyphens)

timetable (one word)

timpan/o (mus.), the orchestral kettledrum, *not* ty-; *pl.* **-i.** See also **tympanum**

tin, symbol **Sn** (*stannum*)

tinct., tincture

Tindal (Matthew), 1656–1733, English theologian, *not* -all. See also **Tyn-**

tin foil (two words)

tingeing, *not* -ging

tin Lizzie, an old motor car (one cap.)

tin-pan alley, the world of popular music (one hyphen, not caps.)

tin plate (two words)

tinpot, (derog.) inferior

tinsel/, -led, -ling

Tintagel, Cornwall, *not* -il

tintinnabulation, ringing of bells

tip in (bind.), to insert a plate etc. by pasting its inner margin to the next page

tip-off (noun, hyphen)

Tipperary, town and county, Ireland

tippet, a cape, *not* tipet

tipstaff/, a bailiff; *pl.* **-s**

tip/toe, -toeing, -top (one word)

TIR, Transport International Routier

tirailleur (Fr. m.), a sharp-shooter (not ital.)

Tiranë, Albania, *not* -na

tire, of a wheel, *use* **tyre**

tiré à part (Fr. typ. m.), an offprint

tiro/, a novice, *pl.* **-s**, *not* tyro

Tirol(o), *see* **Tyrol**

'tis, for 'it is' (apos., close up)

tisane, *use* **ptisan**

Tisiphone (myth.), one of the Furies

Tit., Titus (NT)

tit., turn over

titanic, of titanium; colossal (not cap., except with ref. to Titans)

Titanic (**the**), liner sunk by iceberg, 1912

titanium, symbol **Ti**

titbit, *not* tid- (one word)

Titel/ (Ger. typ. m.), the title; *-blatt* (n.), title-page; *-zeile* (f.), headline (caps.)

tit for tat (three words)

titi, S. Amer. monkey (one word)

ti-ti, NZ mutton-bird (hyphen)

Titian (**Tiziano Vecellio**), 1477–1576, Venetian painter

Titiens, *use* **Tietjens**

titillate, to excite pleasurably

titivate, to smarten up, *not* titt-

title-deed (hyphen)

title-page (typ.), *see* **preliminary matter**

title/-piece, -role (hyphens)

titles (**cited**), of articles in periodicals, chapters in books, shorter poems, and songs, to be roman quoted, not italic; of series of books etc., roman (no quotation marks); of books, periodicals, newspapers, plays, long (book-length) poems, paintings, and sculptures (but identificatory descriptions to be roman), to be italic

title-sheet (typ.), that containing the preliminary matter

titles of honour, as LL D, FRS, are usually in caps. Frequently even s. caps give a better general effect. *See also* **capitalization, compound ranks**

Tito/, pseud. of Josip Broz, 1892–1980, president of Yugoslavia; **-ism**

titre (chem.), strength of solution determined by titration, *not* -ter

titre (Fr. typ. m.), title

tittivate, *use* **titi-**

tittup/, to behave or move in a lively way, **-ed, -ing, -py**

Titus/ (**Epistle to**) (NT), abbr. **Tit.;** — (Roman praenomen), abbr. **T.**

T/-joint, -junction (hyphens)

TKO, technical knock-out

Tl, thallium (no point)

TLS, *Times Literary Supplement*

TLWM, Trinity Low-Water Mark

TM, transcendental meditation

Tm, thulium (no point)

tmes/is (gram.), division of compound word by intervening word(s), *pl.* **-es**

TN, Tennessee (off. postal abbr.)

TNT, trinitrotoluene

TO, Telegraph Office, turn over

toad-in-the-hole (hyphens)

toad/stone, -stool (one word)

to and fro (adv., three words)

toast/-master, -rack (hyphens)

Tobago, W. Indies, *not* Ta-. *See also* **Trinidad**

Tobermore, Co. Londonderry

Tobermory, Isle of Mull, Strathclyde

Tobit (Apocr.), not to be abbreviated

toboggan/, *not* -ogan; **-er, -ing**

toby/ **collar,** — **jug** (two words, not cap.)

Toc H, Talbot House (no point)

Tocqueville (Alexis Charles Henri Maurice Clérel, comte de), 1805–59, French statesman and polit. writer

tocsin, alarm signal

today (one word)

to-do, commotion (hyphen)

toe/ (verb), **-d, -ing**

toe/**-cap, -nail** (hyphens)

toffee, *not* -y

toga/, anc. Roman mantle; *pl.* **-s**

Togo/, Cent. W. Africa, indep. 1960, adj. **-lese**

toile (Fr. f.), linen-cloth

toilet, *not* -ette

toilet/**-paper, -roll** (hyphens)

toilet/ **powder,** — **soap** (two words)

toilette (Fr. f.), toilet

toing and froing (three words)

Toison d'or (Fr. f.), the golden fleece

Tokay, a Hungarian wine; in Hung. **Tokaj**

Tokyo, cap. of Japan, *not* -io; *formerly* Yeddo

Toler, in family name of Earl of Norbury

Tolkien (John Ronald Reuel), 1892–1973, English writer, *not* -ein

Tolstoy (Count Leo), 1828–1910, Russian novelist, *not* -oi

tomalley, so-called liver of lobster, *not* -ly

tomato/, *pl.* **-es**

tomboy (one word)

tom-cat (hyphen)

tome (Fr. m.), a volume; abbr. **t.**

tommy, British private soldier (not cap.)

tommy/**-gun, -rot** (hyphens, not cap.)

tomorrow (one word)

Tompion (Thomas), ?1639–1713, English clock-maker

tompion, *use* **tampion**

tomtit, a small bird (one word)

tom-tom, Indian etc. drum (hyphen)

ton/ (weight), **-s,** abbr. **t.**

ton (Fr. m.), style

Tonbridge, *but* **Tunbridge Wells,** Kent

tond/**o** (It.), painting or relief of circular form, *pl.* **-i** (not ital.)

tone-poem (mus.) (hyphen)

Tongking, Vietnam, *not* Tun-, Tonkin, Tonquin

tongu/**e** (verb), **-ed, -ing**

tonic sol-fa (one hyphen)

tonight (one word)

tonn., tonnage

tonne, metric ton (1,000 kilograms), abbr. **t.** *or* **te**

tonneau, rear of motor car

tonneau (Fr. m.), ton, tun, or cask; abbr. **t.**

tonsil/, **-lar, -lectomy, -litis**

tooling (bind.), impressing a design or lettering on the (usu. leather) binding by hand

tooth/**ache, -brush, -comb, -paste, -pick** (one word)

topcoat (one word)

topgallant, mast or sail (one word)

top-heavy (hyphen)

topi (Anglo-Ind.), a hat, *not* topee

topmast (one word)

top-notch, first-rate (hyphen)

topog., topograph/y, -ical

top/**os,** stock theme, *pl.* **-oi**

topsail (one word)

tops and tails (typ.), prelims and index etc.

topside (naut.), ship's side between water-line and deck; also of beef (one word)

topsoil (one word)

topsy-turvy (hyphen)

Tor Bay, Devon

Torbay, town, Devon

torc, *use* **torque**

torchlight (one word)

Torino, It. for **Turin**

tormentor, *not* -er

tornado/, *pl.* **-es**
torniquet, *use* **tour-**
torpedo/, *pl.* **-es**
Torphichen (Baron)
torque, a metal ornament, (mech.) turning effect, *not* torc
Torquemada (Tomás de), 1420–98, Spanish Inquisitor
torr (phys.), unit of pressure, with numbers **Torr** (no point)
Torres Vedras, fortified town near Lisbon, Portugal; battle, 1810
torso/, (representation of) human trunk, something unfinished or mutilated; *pl.* **-s**
tortoiseshell (one word)
tortue/ (Fr. f.), turtle; — **claire,** clear turtle soup
Toscanini (Arturo), 1867–1957, Italian conductor
totalizator, betting device, *but* **Horserace Totalisator Board;** colloq. **tote**
totalize, to collect into a total, *not* -ise
t'other, the other, *not* tother
totidem verbis (Lat.), in so many words
toties quoties (Lat.), the one as often as the other
toto caelo (Lat.), diametrically opposed, not — *coe-*
touché (Fr.), acknowledging a hit
touch/-line, -type (hyphens)
Toulouse-Lautrec (Henri Marie Raymonde de), 1864–1901, French painter
toupee, a wig, *not* -ée, -et
tour/ (Fr. m.), a tour; (f.) tower; **tour à tour,** alternately, in turn; — **de force,** a feat of strength or skill; — **de main,** sleight of hand
touraco/, large Afr. bird, *pl.* **-s,** *not* tu-, -cou, -ko
touring-car (hyphen)
tourmaline, a mineral, sometimes cut as a gem, *not* -in
Tournai, Belgium, in Fl. **Doornik**
Tournay, France

tournedos, small fillet of beef with suet, *pl. same*
tourney, (take part in) a tournament
tourniquet, bandage etc., for stopping flow of blood, *not* torn-
Toussaint L'Ouverture (François), 1743–1803, Haitian Negro general and liberator; in Fr. **Louverture** (no apos.)
tout/ **à coup** (Fr.), suddenly; — **à fait,** entirely; — **court,** abruptly, or simply; — **de même,** all the same; — **de suite,** immediately; — **d'un coup,** all at once; — **ensemble (le),** the general effect; — **le monde,** all the world, or everybody (no hyphens)
tovarish/, comrade, *pl.* **-i;** Anglicized from Russ. *tovarishch*
towel/, **-ling**
town, abbr. **t.**
town/ **councillor,** *not* -ilor; abbr. **TC;** — **hall,** — **house** (two words)
Townshend (Marquess)
township, abbr. **t.**
toxaemia, blood-poisoning, *not* -xemia
toxicol., toxicolog/y, -ical
toxin, a poison, *not* -ine
toxophilite, (student, lover) of archery
Tpr., Trooper
Tr., trustee
tr., translat/ed, -ion, -or
Trabzon, Turkey, *formerly* Trebizond
tracasserie, fuss, annoyance (not cap.)
traceable, *not* -cable
trache/a, the windpipe; *pl.* **-ae**
track suit (two words)
Tractarian/; **-ism,** the Oxford or High-Church Movement of mid-nineteenth century (cap.)
tractor, *not* -er
trad, colloq. for traditional (jazz); also as abbr. (point)
tradable, *not* -eable
trade/ **mark,** — **name** (two

414

words); *see* **proprietary terms**

trade/ union (*not* trades union), *pl.* — **unions** (two words), abbr. **TU**, *but* **Trades Union Congress**, abbr. **TUC**; **-union-ism**, **-unionist** (hyphens)

trade wind (two words)

traduction (Fr. f.), translation

traffic/, **-ked**, **-ker**, **-king**; — **jam** (two words)

traffic-light (hyphen)

trag., tragedy, tragic

tragedi/an, a writer of tragedies, a tragic actor; **-enne**, a tragic actress

tragicomedy, a drama of mixed tragic and comic elements (one word)

trahison des clercs (Fr. f.), betrayal of standards by the intellectuals

traipse, to trudge, *not now* trapes

trait, a characteristic

trait d'union (Fr. typ. m.), the hyphen

trammel/, to entangle; **-led**, **-ling**

trampoline, sprung canvas sheet used by acrobats, gymnasts, etc., *not* -in

tranche, portion of income etc. (not ital.)

tranny, colloq. for transistor (radio), *not* -ie

tranquil/, **-lity**, **-lize**, *not* -ise, **-ly**

trans., transactions, transitive, translat/ed, **-ion**, **-or**

transact/, **-or**

transalpine (one word, not cap.)

transatlantic (one word, not cap.)

transcontinental (one word)

transexual, *use* transs-

transf., transferred

transfer/, **-able**, **-ee**, **-ence**, **-red**, **-rer**, **-ring**

transgress/, to pass beyond the limit of; **-ible**, **-or**

tranship/, **-ment** (one word), *not* transs-

transition, literary monthly, 1927–38 (not cap.)

transitive (gram.), (of verb) taking a direct object; abbr. **trans.** *or* **transit.**

Transkei, S. Afr., indep. 1976

translat/ed, **-ion**, **-or**, abbr. **tr.** *or* **trans.**; **-able**

transmissible, *not* -mittable

transonic, relating to speeds close to that of sound, *not* transs-

transpontine, on the south side of the Thames. *See also* **cispontine**

transpose (typ.), to interchange letters, words, lines of type, etc.; *see* **proof-correction marks**

transsexual, *not* transexual

transsonic, *use* transonic

Transvaal, prov. of S. Africa; abbr. **Tvl.**

Transylvania, Romania

trapes, to trudge, *use* **traipse**

trattoria (It.), Italian-style restaurant

trauma/, wound or shock, *pl.* **-s**

travail/, **-ed**, **-ing**

travel/, **-led**, **-ler**, **-ling**

travel agent (two words)

Travellers' Club, London (apos. after *s*)

travois, N. Amer. Ind. vehicle, *pl. same*

Trawsfynydd, Gwynedd

TRC, Thames Rowing Club

tread, past **trod**, partic. **trodden**

treadmill (one word)

Treas., treasurer, treasury

trecent/o, 1300–99, and It. art and literature of that century; **-ist** (not ital.)

tre corde (mus.), direction in piano-music to release soft pedal. *See also* **una corda**

Treitschke (**Heinrich von**), 1834–96, German historian

trek/ (Afrik.), journey (verb and noun); *not* -ck; **-ked**, **-ker**, **-king**

trellis/, *not* -ice; **-work** (hyphen)

trematode (noun *or* adj.), (of) flatworm, *not* -oid

tremolo/ (mus.), tremulous effect or device producing this, *pl.* **-s**

tremor, *not* -our

Trengganu, *see* **Malaya (Federation of)**

trente-et-quarante (Fr. m.), a game of chance (hyphens)

trepan/, (to use) a surgeon's saw, a borer; to trap; **-ation, -ned, -ning**

Tresco, Scilly

Trescowe, Cornwall

Tresilian Bay, S. Glamorgan

Tresillian, Cornwall

trestle, a table support, *not* tressel

Trèves, *use* **Trier**

trevet, *use* **trivet**

TRH, Their Royal Highnesses

tribrach (prosody), foot of three short syllables (◡◡◡)

trichology, study of the hair

tricolour, (flag) having three colours, esp. the French flag; in Fr. m. *drapeau tricolore*

tricorn, (imaginary animal) with three horns, *not* -ne

trienni/al, lasting, or occurring every, three years; **-um**, period of three years, *pl.* **-ums**

Trier, W. Germany, *not* Trèves

trig., trigonometry

trigesimo-secundo, *use* **thirty-twomo**, q.v.

trigon., trigonometr/y, -ical

trillion, in Britain, France (since 1948), and Germany, a million million millions; in US, a million millions. *See also* **billion**

trimeter (prosody), verse of three measures

Trin., Trinity

Trinidad and Tobago, W. Indies, indep. 1962

Trinity Sunday, the one after Whit Sunday

triphthong, three vowels in a single syllable, as eau in beau

tripos, honours examination for BA at Cambridge

triptych, set of three painted or carved panels, hinged together

triptyque, a customs permit for a motor car

triquetr/a, ornament of three interlaced arcs, *pl.* **-ae**; adj. **-al**

trireme, anc. Gr. ship with three banks of oars

Tristan da Cunha, S. Atlantic, *not* — d'Acunha

Tristan und Isolde, opera by Wagner, 1865

Tristram, knight of the Round Table (usual Eng. form of Tristan)

tritium, symbol **T**

triturat/e, to grind finely; **-or**

triumvir/, one of a committee of three; Eng. *pl.* **-s**; Lat. *pl.* **-i**; collective noun **-ate**

trivet, iron bracket hooked to a grate, *not* tre-

trivia, trifles, is *plural*

trocar (med.), instrument for withdrawing fluid from the body, *not* troch-

troche (med.), a medicated lozenge

troch/ee, a foot of two syllables (–◡); adj. **-aic**

troika, Russian vehicle drawn by three horses abreast; (fig.) a committee of three

trolley, *not* trolly

trollop, a slatternly woman

Trollope (Anthony), 1815–82, and — **(Frances)**, 1780–1863, his mother, English novelists

trompe-l'œil (Fr. m.), an illusion, esp. in still-life painting or plaster ornament

Trondheim, Norway, *formerly* **Trondhjem**, *not* Th-, Dront-

troop/, assembly of soldiers etc.; in *pl.* of armed forces generally; **-er**, abbr. **Tpr.** *See also* **troupe**

tropaeolum/, a trailing plant; *pl.* **-s**. As bot. genus, *Tropaeolum*

tropical (not cap.)

troposphere, the lower atmosphere, where temperature falls

with rise in height. *See also*
stratosphere
troppo (mus.), too much
Trotsky/ (**Leon**), pseud. of Lev
Davidovich Bronstein, 1879–
1940, Russian revolutionary, *not*
-tz-, -ki; **-ism, -ist, -ite**
trottoir (Fr. m.), footway, pavement
troup/e, company of performers; **-er.** *See also* **troop**
trousseau/, bride's outfit; *pl.* **-s**
(not ital.)
trouvaille (Fr. f.), a lucky find
trouvère, medieval poet in N.
France
troy weight (not cap.): 1 troy
pound (approx. 373 g.) = 12
troy ounces; 1 troy ounce = 20
pennyweights
TRRL, Transport and Road
Research Laboratory
Trs., Trustees
Trucial States, Arabia, seven
independent sheikhdoms, *now*
United Arab Emirates
Trudeau (**Pierre Elliott**), b.
1919, Canadian statesman
trudgen, stroke in swimming
Truman (**Harry S**), 1884–
1972, US politician, President
1945–53 (no point after the S)
trumpet/, -ed, -ing
Truron:, sig. of Bp. of Truro
(colon)
Trustee, abbr. **Tr.,** *pl.* **Trs.**
TS (paper), tub-sized, typescript
tsar, emperor of Russia; **tsare-**
vich, his eldest son, *not* cesarevitch; **tsarevna,** his dau.;
tsarina, his wife (in Russ.
tsaritsa) (caps. as titles); *not*
cs-, cz-, tz-
Tsaritsyn, USSR, *now* **Volgo-**
grad
Tsarskoe Selo, near St. Petersburg (*now* Leningrad), imperial
residence; renamed Detskoe
Selo and, later, Pushkin; *not*
Tz-, Z-
TSB, Trustee Savings Bank
Tschaikowsky (**P. I.**), *use*
Tchaikovsky

tsetse, African fly, *not* tzetze
TSH, Their Serene Highnesses
T/-shirt, -square (hyphens)
tsp., teaspoonful
tsunami (Jap.), tidal wave (not
ital.)
t.s.v.p., tournez s'il vous plaît (Fr.),
please turn over, PTO
Tswana, (language of) Bechuana people
TT, teetotal, Tourist Trophy,
tuberculin tested
TTS (typ., comp.), teletypesetting
TU, trade union, -s
tub/a, bass saxhorn; *pl.* **-as**
Tübingen, town and university,
W. Germany
TUC, Trades Union Congress
tuck/-box, -shop (hyphens)
Tuesday, abbr. **Tu., Tue.,** *or*
Tues.
Tuileries, Paris (one *l*)
tulle, fine silk fabric
tumbrel, a cart, *not* -il
tumour, a morbid growth, *not*
-or
tumul/us, anc. burial mound;
pl. **-i** (not ital.)
tun/ (cask), **-s,** abbr. **t.**
Tunbridge Wells, *but* **Ton-**
bridge, Kent
tuneable, *not* tunable
Tungking, *use* **Tongking**
tungsten, wolfram, symbol **W**
Tunisia, N. Africa, indep. 1956,
cap. **Tunis**
tunnel/, -led, -ling
tu quoque! (Lat.), so are you!
tura/co, -cou, -ko, *use* **touraco**
turbo/-jet, -prop (hyphens)
Turco/ (hist.), Algerian soldier
in French army; *pl.* **-s;** *not* -ko
Turco-, comb. form of Turkish,
not Turko-
Turcoman, *use* **Turkoman**
Turcophile, one friendly to the
Turks
Turgenev (**Ivan Sergeevich**),
1818–83, Russian novelist, *not*
the many variants
Turin, Italy, in It. **Torino**
Turk., Turk/ey, -ish

turkey red (not cap.)
Turki, (of) group of Ural-Altaic languages
Turkish, language of Turkey; *see* **accents**
Turkistan, a region of cent. Asia, *not* Turke-
Turkmenistan, a Soviet Socialist Republic. *See also* **USSR**
Turko-, *use* **Turco-**
Turkoman/, an inhabitant or language of Turkmenistan, *not* Turco-; *pl.* **-s**
Turku, Finland, in Swed. **Åbo**
turn/, -ed sort (typ.), a sort correctly or mistakenly printed upside-down or on its side; **-ed commas,** name for **quotation marks,** q.v.
Turner (**Joseph Mallord William,** *not* Mall/ad, -ard), 1775–1851, English painter
turning/-circle, -point (hyphens)
Turnour, family name of Earl Winterton
turn-out (noun, hyphen)
turn over, abbr. **TO**
turnover (noun and adj., one word)
turn-round (noun, hyphen)
turntable (one word)
Tuskar Rock, lighthouse, Co. Wexford
Tuskegee Institute, Ala., US
tusser, Ind. or Chin. silk(worm), *not* -ah, -ore
tussock, clump of grass, *not* -ac(k)
Tutankhamun, Egyptian pharaoh, *not* -amen
tutti/ (mus.), (passage performed with) all instruments or voices together, *pl.* **-s**
tutti-frutti, confection (hyphen)
tutu, ballet-dancer's skirt, NZ shrub (one word)
tuum (Lat.), thine. *See also* *meum*
Tuvalu/, W. Pacific, *formerly* Ellice Is., indep. 1976; adj. **-an**

tu-whit, tu-whoo, (to make) an owl's cry
tuyère, furnace-nozzle
TV, television
TVA (US), Tennessee Valley Authority
Tvl., Transvaal
TWA, Trans World Airlines
Twain (**Mark**), *see* **Clemens**
'twas, for 'it was' (apos., close up)
Tweeddale, district of Borders; — (**Marquis of**)
Twelfth/ Day, 6 Jan.; — **Night,** 5 Jan.
twelvemo, *see* **duodecimo**
twenty-fourmo (typ.), a book based on 24 leaves, 48 pages, to the sheet; abbr. **24mo**
twentymo (typ.), a book based on 20 leaves, 40 pages, to the sheet; abbr. **20mo**
'twere, 'twill, for 'it were', 'it will' (apos., close up)
twerp, stupid person, *not* -irp
twilit, illuminated (as if) by twilight
Twisleton-Wykeham-Fiennes, family name of Baron Saye and Sele
'twixt, for 'betwixt' (apos.)
twofold (one word)
two/-foot, -inch, -mile, -pound, -stroke, -ton (adjs., hyphens)
twosome (one word)
'twould, for 'it would' (apos., close up)
TX, Texas (off. postal abbr.)
TYC, Thames Yacht Club
tyke, an objectionable fellow; a Yorkshireman, *not* ti-
Tyler (**John**), 1790–1862, US President 1841–5; — (**Wat**), d. 1381, English rebel
Tylers and Bricklayers, livery company
Tylor (**Sir Edward Burnett**), 1832–1917, English anthropologist
tympan/um, the ear-drum; *pl.* **-a.** *See also* **timpano**
Tyndale (**William**), 1484–

1536, English priest, translated
Bible, burned at stake

Tyndall (John), 1820–93, English physicist

Tyne and Wear, metropolitan county

Tynemouth, Tyne & Wear. *See also* **Teign-**

Tynwald, Manx assembly

typ., typograph/er, -ic, -ical, -ically

type, a piece of metal having on one end a letter or character in relief, used in letterpress printing; also called **sort**. *See also* **height to paper**

type/face, -founder, -script, -setting (one word)

type metal, an alloy of lead, antimony, and tin

typescript, abbr. **TS**

typewriter, the machine

typist, the user of a typewriter

typograph/ic (typ.), of a p
ing process based on raised
letters etc. which alone com
into contact with ink and pap
see also **letterpress; -y,** the
design of printed matter

tyrannize, *not* -ise

tyre (of a wheel), *not* tire

tyro, *use* tiro

Tyrol/, region of Austria and
Italy, in Ger. **Tirol**, in It.
Tirolo; adj. -ese

tzar etc., *use* ts-

Tzarskoye Selo, *use*
Tsarskoe —

tzetze, *use* tsetse

U

U, upper-class (*see also* **non-U**), uranium, the twentieth in a series

U, Burmese title (not an abbr.)

U., Unionist, film-censorship classification

u, unified atomic mass unit

u., (Ger.) *und* (and), *unter* (among), (meteor.) ugly, threatening weather

Ü, ü, in German etc., may *not* be replaced by *Ue, ue* (except in some proper names), or *U, u*

u.a. (Ger.), *unter anderem* (among other things)

UAE, United Arab Emirates

UAR, United Arab Republic, 1958–71

Übersetzung (Ger. f.), translation

ubique (Lat.), everywhere

ubi supra (Lat.), in the place above (mentioned); abbr. *u.s.*

U-boat, German submarine (hyphen)

UC, University College

u.c. (typ.), upper case (*but* in proof correction (q.v.) and copy preparation specify 'cap' or 'uc'), (It. mus.) una corda (q.v.)

UCATT, Union of Construction, Allied Trades, and Technicians

UCC, Universal Copyright Convention

UCCA, Universities' Central Council on Admissions

UCD, University College, Dublin

UCH, University College Hospital, London

UCL, University College London

UCLA, University of California at Los Angeles

UCS, University College School

UCW, University College of Wales

Udaipur, India, *not* Odeypore, Oodey-, Ude-

udal/, freehold right based on continuous occupation, *not* odal; **-ler, -man** (one word)

Udall (**Nicholas**), 1505–56, English playwright

UDC, Universal Decimal Classification, Urban District Council

UDI, unilateral declaration of independence

UDR, Ulster Defence Regiment

UEA, University of East Anglia

UEFA, Union of European Football Associations

UFAW, Universities Federation for Animal Welfare

UFC, United Free Church (of Scotland)

Uffizi Gallery *and* **Palace,** Florence, *not* -izzi

UFO, unidentified flying object; also **ufo/,** *pl.* **-s**

Ugand/a, Africa, indep. 1962, republic 1963; adj. **-an**

UGC, University Grants Committee

Ugley, village in Essex

ugli/, a citrus fruit, *pl.* **-s**

UHF, ultra-high frequency

uhlan, a Prussian lancer

Uhland (**Johann Ludwig**), 1787–1861, German poet

UHT, ultra heat tested (long-keeping milk)

u.i., ut infra (as below)

uitlander (Afrik.), foreigner

UJD, *Utriusque Juris Doctor* (Doctor of Laws)

UK, United Kingdom

UKAEA, United Kingdom Atomic Energy Authority

ukase, in Russ. *ukaz,* an edict, *not* oukaz

ukiyo-e, school of Japanese art

Ukraine, *properly* **Ukrainian Soviet Socialist Republic,** one of the republics forming the Union of Soviet Socialist Republics (*see* **USSR**); *not* Little Russia; adj. **Ukrainian**

ukulele, small guitar, *not* uke-
Ullswater, Cumbria, *not* Ulles-
uln/a, lower-arm bone, *pl.* **-ae**
(not ital.)
Ulster/, Ireland; **-man,**
-woman (one word)
ulster, a coat (not cap.)
ult., *see* **ultimo**
ultima ratio/ (Lat.), the last
resource; — — *regum,* resort
to arms
ultima Thule, distant unknown
region (one cap., ital.)
ultimatum/, final proposal;
pl. **-s**
ultimo, in, *or* of, the last month;
better not abbreviated
ultimum vale (Lat.), the last
farewell
ultimus haeres (Lat., the final
heir), the crown or the state
ultra/ (Lat.), beyond, extreme;
— *vires,* beyond legal power
ultra-high (of frequency) (hyphen)
ultra/ist, -ism
ultra/sonic, -violet (one word)
Ulyanov, *see* **Lenin**
Umbala, -balla, *use* **Ambala**
umbel/ (bot.), flower-cluster;
-lar, -late
umble pie, *use* **humble** —
umbo/, boss of shield, *pl.* **-s**
umbr/a (astr.), a shadow; *pl.*
-ae
UMIST, University of Manchester Institute of Science and
Technology
umlaut, *see* **Ä, ä; Ö, ö; Ü, ü;**
and **accents**
Umritsur, *use* **Amritsar**
UN, United Nations, *not* UNO,
Uno
un- (words in), for further spellings *see* the corresponding positive forms
'un, colloq. for one (as in good
'un) (apos.)
UNA, United Nations Association
una corda (mus.), direction in
piano-music to use soft pedal. *See
also* **tre corde**

unadvisable, (of person) not
open to advice. *See also* **inadvisable**
un-American (hyphen, cap.)
unanim/ous, -ity (a, *not* an)
unappealing (one *l*)
unauthorized, *not* -ised
una voce (Lat.), unanimously
unbaptized, *not* -ised
unberufen (Ger.), *absit omen*
unbiased, *not* -ssed
unbribable, *not* -eable
unbusinesslike (one word)
uncared-for (hyphen)
unchangeable, *not* -gable
unchristian (one word)
uncircumcised, *not* -ized
unclench, cf. **clench**
unclinch, cf. **clinch**
uncome-at-able, inaccessible
(two hyphens)
uncooperative (one word)
uncoordinated (one word)
UNCSTD, United Nations Conference on Science and Technology for Development
UNCTAD, United Nations Conference on Trade and Development
unctuous, greasy, *not* -ious
undeniable, *not* -ny-
under-, prefix joined to nouns,
adjectives, adverbs, and verbs,
normally forms one word (*see*
the cases given below), *but note*
under-part, under-secretary,
under-sexed, under-shrub,
under-side, under-surface
**under/achieve, -act, -car-
riage, -clothes, -coat, -cover**
(adj.), **-current, -cut,
-developed, -dog, -em-
phasis, -emphasize,** *not* -ise,
**-employed, -estimate, -ex-
pose, -ground, -growth,
-hand** (one word)
underlay (typ.), to make formes,
blocks, etc., type-high by placing card or paper underneath
underlie, *but* **underlying**
underline (typ.), caption to illustration, diagram, etc. *See also*
proof-correction marks

under/manned, -mentioned (one word)

under-part, subordinate part or role (hyphen)

under/privileged, -rate (one word)

underproof, containing less alcohol than proof spirit does, abbr. **u.p.** (one word)

under-runners (typ.), excess of marginal notes which are continued below body of text, generally at foot of page. *See also* **shoulder-notes**

undersea (one word)

under-secretary (hyphen, caps. as title)

undersell (one word)

under/-sexed, -shrub, -side (hyphens)

under/signed, -sized, -skirt, -staffed (one word)

under-surface (hyphen)

under/tone, -value, -water (adj.) (one word)

under way, in motion (two words)

underweight (one word)

underwrite/, -r (one word)

undies, colloq. for (women's) underclothing

UNDRO, United Nations Disaster Relief Organization

unenclosed, *not* unin-

un-English (hyphen, cap.)

Unesco, United Nations Educational, Scientific, and Cultural Organization

unexception/able, that cannot be faulted; **-al,** not unusual

Ungarn, Ger. for **Hungary**

unget-at-able, inaccessible (two hyphens)

unguent, ointment

ungul/a, hoof, talon, *pl.* **-ae**

unheard-of (hyphen)

Uniat, (member of) Eastern Church in communion with Rome, *not* -ate

Unicef, United Nations (International) Children's (Emergency) Fund

unidea'd, having no ideas, *not* -aed

Unido, United Nations Industrial Development Organization

Unionist, abbr. **U.**

unionize, bring into trade-union organization (one word, *not* -ise)

un-ionized, not ionized (hyphen, *not* -ised)

Union Jack (two words)

unisex, tendency of sexes to be indistinguishable in dress etc.; (adj.) of clothes wearable by either sex (one word)

Unit., Unitarian, -ism

United Arab/ Emirates, group of seven Emirates on Persian Gulf, abbr. **UAE;** — — **Republic,** 1958–71; abbr. **UAR**

United/ Free Church of Scotland (caps.), abbr. **UFC;** — **Kingdom** (caps.), abbr. **UK;** — — **Presbyterian/,** abbr. **UP;** — — **Church** (caps.), abbr. **UPC;** — **Service Club** (*not* Services). *See also* **US, USAF, USI, USIS, USN, USPG, USS**

units, *see* **BSI, SI**

Univ., University; (in Oxford) University College

Univers (typ.), a sanserif (q.v.) fount; *not* a generic term for the sanserif face

universal (**a,** *not* an); abbr. **univ.**

Universal Copyright Convention, adopted 1952, effective in sixteen countries (incl. US) from 1955, in UK 1957, and in USSR 1973. Basically each member-nation (now more than sixty) extends benefit of its own copyright laws to works by citizens of other member-nations, regardless of place of original publication. *See also* **Berne Convention, copyright notice;** abbr. **UCC**

universalize, *not* -ise

Universal Time, abbr. **UT**

University College London (no comma), abbr. **UCL**

unjustified setting (typ.), composition with a single invariable word-space, giving an uneven right-hand margin (as in this book)

Unknown Soldier, unidentified soldier representing casualties in war (caps.)

unladylike (one word)

unlicensed, *not* -ced

unm., unmarried

unmanageable, *not* -gable

unmistakable, *not* -eable

unmovable, *not* -eable

unnameable, *not* -mable

UNO, *or* **Uno,** United Nations Organization, *use* **UN**

uno animo (Lat.), unanimously

unparalleled, *not* -elled

unperson, person whose identity is denied or ignored (one word)

unpractical, not suitable for actual conditions, *not* im-. *See also* **impracticable**

unputdownable (of book) (one word)

unridable, *not* -eable

unrivalled, *not* -aled

Unrra, United Nations Relief and Rehabilitation Administration

Unrwa, United Nations Relief Works Agency

unsaleable, *not* -lable

unscalable, *not* -eable

unselfconscious (one word)

unserviceable, *not* -cable

unsewn binding (bind.), method in which the gatherings of each volume are clamped together, their back folds sheared off, and the resultant leaf-edges glued to a flexible backing

unshakeable, *not* -kable

unskilful, *not* -skill-

untraceable, *not* -cable

un/trammelled, -travelled (one word, *not* -eled)

unwieldy, cumbersome, *not* -ldly

u.ö. (Ger.) *und öfters* (and often)

UP, United Presbyterian; United Press; Uttar Pradesh (*formerly* United Provinces), India

u.p., underproof; (colloq.) (it is) all up (with someone)

up., upper

up-and/-coming (colloq.), likely to succeed; — - —**over** (adj.), of door (hyphens)

Upanishad, a Sanskrit philosophical treatise

upbeat (one word)

up-country (hyphen)

up-end, to set on end (hyphen)

up/field, -hill (one word)

up/ish, -ity, *use* upp-

up-market (adj., hyphen)

UPOW, Union of Post Office Workers

upper case (typ.), in handsetting, the case containing capitals, small capitals, numerals, and signs; hence a generic term for the capital letters of a fount. *See also* **manuscript, capitalization**

Upper Volta, W. Africa, indep. 1960

upp/ish, self-assertive; **-ity,** arrogant, snobbish, *not* upi-

Uppsala, Sweden, *not* Ups-

upright (one word)

upside-down (one hyphen)

up/stage, -stairs, -stream (one word)

upsy-daisy, encouragement spoken to child, *not* ups-a- (hyphen)

uptight (colloq.), nervously tense (one word)

up to date (hyphens when attrib.)

UPU, Universal Postal Union

upwind (one word)

uraemia (path.), *not* ure-

Urania (Gr.), muse of astronomy

uranium, symbol **U**

Uranus, Greek mythology and astronomy

urari, *use* **curare**

urbanize, *not* -ise

urbi et orbi (Lat.), to the city (Rome) and the world

Urdu, an off. language of Pakistan

urethr/a (anat.), *pl.* **-ae**

URI, upper respiratory infection

urim and thummim, Exod. 28:30, are *plurals*

Uruguay/, abbr. **Uru.**; adj. **-an**

US, United Service, — States

u.s., *ubi supra* (in the place above [mentioned]), *ut supra* (as above)

u/s, unserviceable

USA, United States of America

usable, *not* -eable

USAF, United States Air Force

USCL, United Society for Christian Literature

USDAW, Union of Shop, Distributive, and Allied Workers

usf. (Ger.), *und so fort* (and so on)

Usher Hall, Edinburgh

USI, United Service Institution (*not* Services)

USIS, United States Information Service

USN, United States Navy

USPG, United Society for the Propagation of the Gospel

usquebaugh (Gaelic), whisky

USS, United States Ship; Universities Superannuation Scheme

Ussher (**James**), 1581–1656, Irish divine, Abp. of Armagh

US spellings. Noteworthy differences from British usage such as aluminum, maneuver, and pajamas are given in their alphabetical places. The following general categories of US usage should also be noted:

1. *e* for *ae* and *oe*, initially as in esthete, eon, and estrogen, and medially as in fetus and toxemia.

2. *-ense* for *-ence* in the words defense, license (noun and verb), offense, and pretense.

3. *-er* for *-re*, as in center, fiber, and theater (but others such as acre, lucre, and ogre are the same as in British usage).

4. *o* for *ou*, as in mold; in

particular *-or* for *-our*, as in color, harbor, and tumor.

5. *z* occasionally for soft *s*, as in cozy and (despite etymology) analyze.

6. Final *-e* is often omitted before a suffix beginning with a vowel where in British usage it is retained, as in milage and salable. But after soft *c* and *g* it is always retained, as in British usage.

7. Final *-l* in words stressed on any syllable but the last is not usually doubled as in British usage when followed by a suffix beginning with a vowel, as in counselor, teetotaler, rivaled, and traveling; note also clarinetist.

8. Final *-ogue* is sometimes shortened to *-og*, as in analog, catalog, epilog, and pedagog

USSR, Union of Soviet Socialist Republics, comprising the Russian Soviet Federative Socialist Republic (RSFSR) and the Armenian, Azerbaijan, Estonian, Georgian, Kazakh, Kirghiz, Latvian, Lithuanian, Moldavian, Tadjik, Turkmen, Ukrainian, Uzbek, and White Russian (Belorussian) Soviet Socialist Republics.

usu., usual, -ly

usucaption (law), acquisition of title by continued occupation of property, *not* -capion

usurper, *not* -or

usw. (Ger.), *und so weiter* (and so on)

UT, Universal Time, Utah (q.v.)

Utah, off. postal abbr. **UT**

Utakamand, *use* **Ootacamund**

Utd., United

ut dictum (Lat.), as directed; abbr. **ut dict.**

U Thant, 1909–74, Burmese statesman, Sec.-Gen. of UN 1962–71

utilize, *not* -ise

ut infra (Lat.), as below; abbr.
  u.i.
uti possidetis (Lat.), as you now
  possess (opposed to *status quo
  ante*)
Utopia/, -n (caps.)
UTS, ultimate tensile strength
ut supra (Lat.), as above; abbr.
  u.s.
Uttar Pradesh, India, *formerly* United Provinces, abbr.
  UP
U/-tube, -turn (hyphens)

ut videtur (Lat.), as it seems
UU, Ulster Unionist
UV, ultraviolet
U/W, underwriter
UWIST, University of Wales
  Institute of Science and Tech-
  nology
UWT, Union of Women
  Teachers
uxor (Lat.), wife; abbr. *ux.*
Uzbekistan, a Soviet Socialist
  Republic, *not* Uzbeg-. *See also*
  USSR

V

V, five, vanadium, (elec.) volt(s), (Ger.) *Vergeltungswaffe* (reprisal weapon: V1, flying bomb; V2, rocket); not used in the numeration of series

V., Vice, Volunteers

℣, sign for versicle

v., verse, verso, versus, very, (meteor.) unusual visibility, (Ger.) *von* (of)

v (mech.), velocity

v. (Lat.), *vice* (in place of), *vide* (see), (It. mus.) *violino* (violin), *voce* (voice)

VA, Vicar Apostolic, vice--admiral, (Order of) Victoria and Albert (for ladies), Virginia (off. postal abbr.)

Va., Virginia (off. abbr.); (mus.) viola

v.a., verb active; (Lat.) *vixit ... annos* (lived [so many] years)

vaccinat/e, to inoculate with vaccine; **-ion, -or**

vacillat/e, to move from side to side, to waver; **-ion, -or**

vacu/um, *pl.* **-ums,** or **-a** in scientific and technical use

vacuum/ cleaner, — flask, — pump (two words)

VAD, Voluntary Aid Detachment (for nursing)

vade-mecum/, a handbook or other article carried on the person; *pl.* **-s** (hyphen)

vae victis! (Lat.), woe to the vanquished!

vaille que vaille (Fr.), whatever it may be worth, at all events

vainglor/y, -ious (one word)

Vaisya, Hindu farmer and merchant caste

valance, short curtain or drapery. *See also* **valence**

vale! (Lat.), farewell!, *pl.* *valete!*; **vale** (noun), a farewell (not ital.)

valence (chem.), combining

power of an element. *See also* **valance, valency**

Valencia, Ireland, Spain

Valenciennes, rich kind of lace (cap.)

valency (chem.), unit of combining power of an element. *See also* **valence**

Valentia (Viscount), *not* -cia

valentine, a sweetheart; a card or message sent to or received by one (not cap.)

Valentine's Day (St.), 14 Feb. (apos.)

valet, a gentleman's personal attendant (not ital.)

valeta, *use* **vel-**

Valetta, Malta, *use* **Vall-**

Valhalla (Norse myth.), palace in which souls of dead heroes feasted, *not* W-. *See also* **Hel**

Valkyrie (Norse myth.), each of Odin's twelve handmaidens, *pl.* same, *not* W-

Valladolid, Spain

Valletta, Malta, *not* Vale-, -eta

Vallombrosa, N. Italy, *not* Vallam-

valorize, artificially fix price, *not* -ise

valour, *but* **valorous**

Valparaiso, Chile

valse/ (Fr. f.), waltz; **— à deux temps, à trois temps** (no hyphens), variations of the waltz

value added tax (three words), abbr. **VAT**

van, van den, van der, as prefix to a proper name, usu. *not* initial cap., except at beginning of sentence. Signatures to be copied

vanadium, symbol **V**

Van Allen belt *or* **layer,** of radiation surrounding the earth

Vanbrugh (Sir John), 1664–1726, English architect and playwright

Van Buren (Martin), 1782–1862, US President 1837–41

426

V. & A., Victoria and Albert Museum

Vandal, (member) of Germanic people of 4th–5th cc. (cap.); **vandal,** wilful destroyer of art or property (not cap.)

vandalize, *not* -ise

van de Graaff (Robert Jemison), 1901–67, US physicist

Van Diemen's Land, *not* — Dieman's — (apos.); now **Tasmania**

Van Dijck (Christoffel), 1601–?69, typefounder; his work (now largely lost) served as model for Caslon (q.v.) and for the Monotype van Dijck (1935)

Van Dyck (Sir Anthony), 1599–1641, Flemish painter (two words); Anglicized form **Vandyke/** (one word), used to denote a work by him, and in — **beard,** — **brown;** **vandyke,** each of series of points bordering lace etc. (not cap.)

Vane-Tempest-Stewart, family name of Marquess of Londonderry (hyphens)

Van Eyck (Hubert), 1366–1426, and — **(Jan),** 1385–1440, his brother, Flemish painters

Van Gogh (Vincent), 1853–90, Dutch painter

Vanhomrigh (Esther), 1692–1723, Swift's 'Vanessa'

Van Nostrand Reinhold Co., publishers

Van 't Hoff (Jacobus Hendricus), 1852–1911, Dutch physicist and chemist

Vanuatu, *formerly* **New Hebrides,** indep. 1980

vaporiz/e, *not* -ise; **-er,** *not* -or

vapour, *but* **vapor/ific, -iform, -imeter, -ous**

Var, river and dép. France

var., (biol.) variety

Varanasi, *see* **Benares**

vari/a lectio/ (Lat.), a variant reading, abbr. **v.l.**; *pl.* **-ae -nes,** abbr. **vv.ll.**

variegated, diversified (usu. in colour)

variety (biol.), abbr. **var.** *See also* **botany**

variorum edition, one with notes by various commentators

variorum notae (Lat.), notes by commentators

vas/ (anat.), a duct; *pl.* **-a**

vascul/um, a botanist's specimen-case; *pl.* **-a**

vas deferens, spermatic duct, *pl.* **vasa deferentia**

vasectomy, excision of vas deferens

Vaseline, propr. term for type of petroleum jelly (cap.)

vassal, feudal retainer

Vassar College, New York

VAT, value added tax

Vat., Vatican

Vaudois, *use* **Waldenses**

Vaughan Williams (Ralph), 1872–1958, English composer

Vauvenargues (Luc de Clapiers, marquis de), 1715–47, French writer

v. aux., verb auxiliary

Vaux of Harrowden (Baron)

vb., verb

VC, vice-chairman, — -chancellor, — -consul, Victoria Cross

VCH, Victoria County History

v. Chr. (Ger.), *vor Christo* or *vor Christi Geburt,* BC

VD, venereal disease, Volunteer (Officers') Decoration

v.d., various dates

v. dep., verb deponent

VDH, valvular disease of the heart

VDU (comp.), visual display unit

VE, victory in Europe (VE day 8 May 1945)

v^e (Fr.), *veuve* (widow)

veau (Fr. m.), calf, (cook.) veal, (bind.) calf, calfskin

VEB, *Volkseigener Betrieb,* state--owned company (GDR)

Veda (*sing.* and *pl.*), anc. Hindu scriptures (cap.)

Vedda, aboriginal of Sri Lanka

vedette, mounted sentinel, patrol boat, *not* vi-

veg, colloq. for vegetable(s) (no point)

Vega (Garcilaso de la), *see* **Garcilaso**

Vega Carpio (Lope Felix de), known as **Lope de Vega,** 1562–1635, Spanish playwright and poet

vehmgericht, German medieval system of tribunals (not ital.); adj. **vehmic**

veille (Fr. f.), the day before, eve. *See also vieille,* **vielle**

Velázquez (Diego Rodríguez de Silva y), 1599–1660, Spanish painter, *not* Velas-

veld (Afrik.), open country, *not* -dt

veleta, a dance in triple time; *not* val-

Velikovsky (Immanuel), 1895–1979, Russian-born writer on astronomy

vellum/, fine parchment; — -**paper,** that imitating vellum

veloce (mus.), quickly

velocity (mech.), symbol *v*

velour, plushlike fabric, *not* -ours

velouté (cook.), sauce of white stock mixed with cooked butter and flour (not ital.)

vel/um, membrane, *pl.* -**a**

velveteen, velvet-like cotton fabric

velvety, *not* -tty

Ven., Venerable (used for archdeacons only)

ven/a cav/a, each of veins carrying blood to heart, *pl.* -**ae** -**ae**

venal/, (of person) bribable; (of conduct) sordid; -**ity,** -**ly.** *See also* **venial**

venation, arrangement of veins in leaf

Vendée (La), dép. France; adj. **Vendean,** of the royalist party, 1793–5

vend/ee, person to whom one sells; -**er,** *use* -**or**

vendetta/, a blood feud, *pl.* -**s** (not ital.)

vendeuse (Fr. f.), saleswoman

vendible, that may be sold, *not* -able

Vendôme, dép. Loir-et-Cher, France; **Colonne** —, and **Place** —, Paris

vendor, seller, *not* -er

venepuncture (med.), puncture of a vein, *not* veni-

venerat/e, -or

venere/al, -ology

venery (arch.), hunting; sexual indulgence

venesection, phlebotomy, *not* veni-

Venet., Venetian

venetian blind (not cap.)

Venezuela, S. American rep.; abbr. **Venez.**

venial, (of sin) pardonable. *See also* **venal**

Venice, Italy, in Fr. **Venise,** Ger. **Venedig,** It. **Venezia**

Venn diagram, set of circles representing logical categories (one cap.)

venose, having prominent veins

ven/ous, of the veins, -**osity**

ventilat/e, -or

ventre à terre (Fr.), at full speed

ventriloquize, *not* -ise

Venus's/ comb, a white-flowered weed; — **fly-trap,** a herb; — **girdle,** ribbon-like jellyfish; — **slipper,** lady's slipper (one cap., -us's)

veranda, *not* -ah

verb, abbr. **vb.**

verb. (Ger.), *verbessert* (improved, revised)

verbalize, to put into words, *not* -ise

verbatim (Lat.), word for word (not ital.)

verboten (Ger.), forbidden

verbum satis sapienti (Lat.), a word to the wise suffices; abbr. *verb. sap.*

verd-antique, a stone, *not* verde- (hyphen)

verderer, forester, *not* -or

428

verdigris, green rust on copper, *not* verde-

Verein (Ger. m.), Association (cap.)

Vereinigte Staaten, Ger. for **United States** (of America)

Vereshchagin (Vasili Vasilievich), 1842–1904, Russian painter

Verey light, *use* **Very**

Verfasser (Ger. m.), author, abbr. **Verf.**

verger, an attendant in a church. *See also* **virger**

Vergil (Polydore), 1470–1555, Italian humanist

Vergilius (Publius — Maro), 70–19 BC, Roman poet, usu. Anglicized as **Virgil**

verglas, thin layer of ice or sleet (not ital.)

vergleiche (Ger.), compare; abbr. *vgl.*

verismo/, realism in art, *pl.* **-s** (not ital.)

Verlag (Ger. m.), publishing house

vermilion, *not* -llion

Vermont, off. abbr. **Vt.** or (postal) **VT**

vermouth, an appetizer

Verner's law (phonet.), relating to voicing of fricatives in Germanic languages (one cap.)

Veronese (Paolo), 1528–88, Italian painter, real name **Cagliari**

Verrocchio (Andrea del), 1435–88, Italian painter and sculptor, *not* the many variants

verruc/a, wart, *pl.* **-ae**

Versailles, near Paris

vers de société (Fr. m.), society verses

verse, abbr. **v.** or **ver.,** *pl.* **vv.**

versicle, short verse in liturgy said or sung by minister, followed by people's (or choir's) response; (typ.) the sign ℣ used in liturgical works. *See also* ℟

vers libre (Fr. m.), free verse

verso/ (typ.), the left-hand page, usu. having an even page num-

ber, 2, 4, 6, etc.; abbr. **ᵛ, v.** (not ital.); *pl.* **-s**

versus (Lat.), against (not ital.); abbr. **v.** *or* **vs.**

vert (her.), green

vert., vertical

vertebr/a, a segment of the backbone; *pl.* **-ae** (not ital.)

vertebrate, animal with a spinal column, of phylum Vertebrata

vert/ex, highest point, angular point of triangle etc.; *pl.* **-ices**

vertical, abbr. **vert.**

vertig/o, giddiness, *pl.* **-os**; adj. **-inous**

vertu, *use* **virtu**

Vertue (George), 1684–1756, English engraver

verve, spirit (not ital.)

vervet, small Afr. monkey

Verwoerd (Hendrik Frensch), 1901–66, Prime Minister of Republic of South Africa, 1958–66

Very light, a flare fired from a pistol, *not* Verey

Very Revd, Very Reverend (for deans, provosts, and former moderators)

vespiary, a nest of wasps

Vespucci (Amerigo), 1454–1512, Italian navigator

vestigia (Lat. *pl.*), traces

vet/, colloq. for veterinary surgeon (no point); to examine carefully; **-ted, -ting**

veteran, car made before 1916. *See also* **vintage**

veto/, ban; *pl.* **-es**

veuf (Fr. m.), widower

veuve (Fr. f.), widow; abbr. **vᵉ** *or* **Vve**

Vevey, Switzerland, *not* -ay

vexata quaestio (Lat.), a disputed question

vexillology, study of flags (three *l*s)

vexill/um, anc. Rom. military standard, *pl.* **-a** (not ital.)

v.f., very fair

VG, Vicar-General

v.g., very good

vgl. (Ger. for cf.), *vergleiche*, compare

VHF, very high frequency

v.i., verb intransitive

v.i. (Lat.), *vide infra* (see below)

via, by way of, *not* -à (not ital.)

via media (Lat.), a middle course (ital.)

vibrato/ (mus.), rapid variation of pitch, *pl.* **-s**

vibrator, *not* -er

Vic., vicar, -age, Victoria

vicar/, in Ch. of England, an incumbent of a parish who is not a **rector,** q.v.; adj. **-ial**

Vicar/-Apostolic, abbr. **VA;** — **-General, VG**

vicarious, delegated, exercised or suffered by one person for another

Vicar's College, a cathedral residence

vice, a tool, US **vise**

vice, in place of (ital.)

vice/, abbr. **V.;** — **-admiral,** abbr. **VA;** — **-chairman, VC;** — **-chamberlain;** — **-chancellor,** — **-consul, VC;** — **-president, VP** (caps. as titles)

vicegerent, a deputy (one word)

viceregent, cap. as title (one word)

vicereine, woman viceroy or viceroy's wife (cap. as title)

viceroy (cap. as title); adj. **viceregal**

vice versa (Lat.), the order being reversed (no hyphen or accent, not ital.), abbr. **v.v.**

vichyssoise, chilled soup of leeks and potatoes

vicomt/e (Fr.), Viscount; *fem.* **-esse** (not cap.)

victimize, *not* -ise

Victoria/, abbr. **Vic.;** — **and Albert** (Order of), for ladies, abbr. **VA;** — **Cross,** abbr. **VC;** — **Nyanza,** *use* **Lake Victoria**

victoria, a carriage, a S. Amer. water-lily, a pigeon (not cap.)

victor ludorum, most successful competitor (not ital.)

victual/, -led, -ler, -ling

vicuña, S. Amer. mammal, cloth made from its wool or an imitation of it

vide/ (Lat.), see, abbr. *v.;* — *ante,* see above; — *infra,* see below, abbr. *v.i.; videlicet,* namely (one word), abbr. **viz.;** *vide/ post,* see below; — *supra,* see above, abbr. *v.s.*

vide (Fr. mus.), open (of strings)

video/, recording or broadcasting of photographic images, *pl.* **-s; -cassette, -recorder** (hyphens)

videophone, telephone transmitting picture of speaker (one word)

videotape, (to use) tape for recording television broadcasts (one word)

vidette, *use* **ve-**

videtur (Lat.), it seems

vide ut supra (Lat.), see as above

vidimus (Lat. 'we have seen'), certified copy of accounts etc. (not ital.)

vie (rival), **vying**

vieille (Fr. f.), an old woman. *See also* **veille**

vielle, a hurdy-gurdy. *See also* **veille**

Vienn/a, in Ger. **Wien,** in Fr. **-e;** adj. **-ese**

Vienne, town in dép. Isère, France; river, tributary of R. Loire. *See also* **Haute-Vienne**

viennoise (à la) (Fr.), in Viennese style (not cap.)

vient de paraître (Fr.), just published

Vientiane, Laos

Vierkleur (Afrik.), flag of Transvaal Republic

vi et armis (Lat.), by force and arms

Vietcong, former guerrilla force in S. Vietnam

Vietnam/, SE Asia; adj. **-ese**

vieux jeu (Fr. m.), an outworn subject

viewfinder (one word)

view halloo (hunt.), *not* the many variants

viewpoint (one word)

vif/y (Fr. mus.), lively, briskly

vigesimo, *use* **twentymo**

vigesimo-quarto, *use* **twenty--fourmo**

vigilante/, member of vigilance body, *pl.* **-s** (not ital.)

vignettes (typ.), illustrations with undefined edges

vigour, *but* **vigorous**

Viking, Scandinavian trader and pirate of eighth to tenth cc.

vilayet, Turkish province; abbr. **vila.**

vilif/y, to disparage; **-ied, -ier, -ying**

village, abbr. **vil.**

villain/, an evil-doer, (colloq.) a rascal; **-ous, -y.** *See also* **villein**

Villa-Lobos (Heitor), 1887–1959, Brazilian composer

villeggiatura (It. f.), country holiday

villégiature (Fr. f.), country holiday

villein/, a serf; **-age.** *See also* **villain**

Villiers, family name of Earls of Clarendon and Jersey; — **de l'Isle Adam (Jean Marie Mathias Philippe Auguste, comte de),** 1838–89, French writer

Vilnius, cap. of Lithuania, USSR

vinaigrette, smelling-bottle, *not* vineg(a)r-

Vinci (Leonardo), 1690–1730, Italian composer; — **(Leonardo da),** *see* **Leonardo**

vin du pays (Fr. m.), wine of the neighbourhood

vineg(a)rette, *use* **vinaigrette**

vingt-et-un (Fr. m.), a card--game

vintage, car made between 1917 and 1930. *See also* **veteran**

vinyl, chem. radical, plastic containing it

viol/ate, -able, -ator, *not* -er

violino (It.), violin; abbr. *v.*

Viollet-le-duc (Eugène Emmanuel), 1814–79, French architect

violoncell/o, *not* violin-; *pl.* **-os;** usu. shortened to **cello; -ist**

VIP, very important person

virago/, termagant; *pl.* **-s**

Virchow (Rudolf), 1821–1902, German pathologist

virelay, short (esp. old French) lyric poem

virger, spelling of *verger* at certain cathedrals, such as St. Paul's and Winchester

Virgil, *see* **Vergilius**

Virginia/, off. abbr. **Va.** or (postal) **VA;** — **creeper** (bot.), *not* -ian

virginibus puerisque (Lat.), for girls and boys

virg/o intact/a, virgin with hymen intact, *pl.* **-ines -ae** (not ital.)

virgule (typ.), solidus

virgule (Fr. typ. f.), comma

virtu/, love of fine arts, virtuosity; — **(articles of),** artistic articles of interesting workmanship, antiquity, or rarity, *not* ve-, -ue (not ital.)

virtuos/o, one skilled in an art, *pl.* **-i**

virus/, a submicroscopic infective agent; *pl.* **-es**

vis/ (Lat. f.), force, *pl.* **vires;** — *a tergo,* force from behind

visa/, permit or endorsement on a passport (not ital.); *pl.* **-s;** as verb, **-ed,** *not* -'d

vis-à-vis (Fr.), face to face (hyphens)

visc/era, *pl.*, interior organs, esp. in the abdomen; *sing.* **-us**

viscount/, -ess (cap. as title); abbr. **Visc.**

viscount/cy, -ship, the rank or jurisdiction of a viscount; **-y** only in hist. use

viscous, sticky

vise, a tool, US for **vice**

Vishnu, Hindu god

visier, *use* **vizier**

vis/ inertiae (Lat.), force of
inanimate matter; — *major*,
superior force; — *medicatrix
naturae*, nature's power of
healing

visit/, -ed, -ing, -or, *not* -er

visitors' book, *not* -or's

visor, part of helmet, peak of
cap, sun-shield in car etc., *not*
viz-

vista/, a view; *pl.* -s

visualize, *not* -ise

vis viva (Lat.), living force

vitalize, *not* -ise

vitell/us, yolk of egg, *pl.* -i

vitiat/e, to spoil; -or

viticulture, culture of vines

Vitoria, Spain; battle, 1813

vitriol/ (oil of), sulphuric acid;
— (blue), copper sulphate; —
(green), ferrous sulphate; —
(white), zinc sulphate

vituperat/e, to revile; -or

viva/, short for viva voce (q.v.),
pl. -s; as verb, -ed, *not* -'d

viva! (It.), long live!

vivace (mus.), lively, quickly

vivandi/er (Fr.), *fem.* -*ère*, army
sutler

vivant rex et regina! (Lat.),
long live the King and Queen!

vivari/um, enclosure for living
things; *pl.* -a

vivat/ regina! (Lat.), long live
the Queen! — *rex!* ditto the
King!

viva/ voce, orally, *not* vivâ; also
noun, an oral examination, *pl.*
— voces, and (with hyphen)
verb (not ital.). *See also* viva

vive/! (Fr.), long live! — *la
bagatelle!* ditto trifles! — *la
différence!* ditto the difference! — *la République!* ditto
the Republic!

vivisect/, -ion, -or

vixit . . . annos (Lat.), lived (so
many) years; abbr. v.a.

viz., *videlicet* (namely) (not ital.;
comma before, but *prefer*
namely)

vizier, a Muslim official, *not* -ir,
-sier

vizor, *use* visor

VJ, victory over Japan (VJ day
15 Aug. 1945; in US 2 Sept.
1945)

v.l., *varia lectio* (a variant reading)

Vlach, Bulgarian form of Wallachian

Vlaminck (Maurice de),
1876–1958, French painter

Vlissingen, Du. for Flushing

v.M. (Ger.), *vorigen Monats* (last
month)

VMH, Victoria Medal of
Honour

v.n., verb neuter

VO, Veterinary Officer, (Royal)
Victorian Order

voc., vocative

vocab., vocabulary

vocalize, *not* -ise

vocative, abbr. voc.

voce (It. mus.), voice; abbr. *v.*

vogue la galère! (Fr.), happen
what may!

Vogüé (Eugène Melchior, vicomte de), 1848–1910, French
essayist

voice/-box; -over, narration
without picture of speaker;
-print (hyphens)

voilà/ (Fr.), see there! — *tout*,
that is all

voile, semitransparent dress-
-material (not ital.)

vol., volume

Volapük, an artificial international language, invented by
J. M. Schleyer, 1879

volatilize, *not* -ise

vol-au-vent (Fr. cook. m.), filled
puff-pastry case (not ital.)

volcano/, *pl.* -es

Volgograd, USSR, *formerly*
Tsaritsyn, *then* Stalingrad

Völkerwanderung (Ger. f.),
4th–6th-c. migration of Germanic peoples

Volksausgabe (Ger. f.), popular
edition

Volkslied/ (Ger. n.), a folk-song;
pl. -er (cap.)

Volkswagen, German make of
car, abbr. VW

vols., volumes

volt/ (elec.), abbr. **V,** SI unit of potential difference; — **-ampere** (hyphen)

Voltairean, *not* -ian

voltameter, for measuring electric current by electrolysis

volte-face (Fr.), a turning about (not ital.)

volti subito (It. mus.), turn over quickly; abbr. *v.s.*

voltmeter, for measuring electric potential

volume, abbr. **vol.,** *pl.* **vols.**

voluntarism, doctrine of financial independence of Church; (philos.) belief in dominance of will; belief in importance of voluntary action; *not* -aryism

Volunteers, abbr. **V.**

von, as prefix to a proper name, usu. *not* initial cap., except at beginning of sentence. Signatures to be copied

voodoo, witchcraft, *not* vudu

voortrekker (Afrik.), pioneer

vor/ Christi Geburt, Christo (Ger.), BC; abbr. *v. Chr.*

vort/ex, whirlpool, whirling mass, engrossing system; *pl.* **-exes** (*but* **-ices** in scientific and technical use)

votable, *not* -eable

vouch/er, a coupon; one who vouches, in law **-or**

voussoir (archit.), tapered stone in arch (not ital.)

vowelize, *not* -ise

vox/ (Lat. f.), voice, *pl.* *voces*; — *et praeterea nihil,* voice and nothing else; — *humana,* an organ-stop; — *populi,* public sentiment

voyeur, one who pries on sexual activity (not ital.)

voyez! (Fr.), see! look!

VP, vice-president

VR, variant reading, *Victoria Regina* (Queen Victoria), Volunteer Reserve

vraisemblance, appearance of truth (not ital.)

VRD, Royal Naval Volunteer Reserve Officers' Decoration

v. refl., verb reflexive

VRI, *Victoria Regina et Imperatrix* (Victoria Queen and Empress)

VS, Veterinary Surgeon

vs., versus

v.s. (Lat.), *vide supra* (see above); (It. mus.) *volti subito* (turn over quickly)

VSO, Voluntary Service Overseas

VSOP, very special old pale

VT, Vermont (off. postal abbr.)

Vt., Vermont (off. abbr.)

v.t., verb transitive

VTOL, vertical take-off and landing

vudu, *use* **voodoo**

Vuillard (**Jean Édouard**), 1868–1940, French painter

Vuillaume (**Jean Baptiste**), 1798–1875, most important of French family of makers of bowed instruments

vulcanize, to treat rubber with sulphur at high temperature, *not* -ise

Vulg., the Vulgate

vulg., vulgar, -ly

vulgarize, *not* -ise

Vulgate, the Latin Bible of the RCC; abbr. **Vulg.; vulgate,** accepted text of an author (not cap.)

vulgo (Lat.), commonly

vv., verses, (mus.) first and second violins

v.v., vice versa

Vve (Fr.), *Veuve* (widow)

vv. ll., *variae lectiones* (variant readings)

VW, Very Worshipful, Volkswagen

v.y. (bibliog.), various years

Vyborg, USSR

vying, *see* **vie**

Vyrnwy, lake and river, Powys

W

W, watt, -s; wolfram (tungsten); not used in the numeration of series

W., Wales, warden, Wednesday, Welsh, west, -ern

w., week, -s, (cricket) wide, wife, (meteor.) wet dew

WA, Washington State (off. postal abbr.), Western Australia

WAAF, Women's Auxiliary Air Force, *earlier and later* **WRAF**

wabbl/e, -y, -ing, *use* **wo-**

waddy, Australian Aboriginal's war-club

Wade–Giles, *see* **Chinese**

wadi/ (Arab.), the dry bed of a torrent, *not* -y; *pl.* **-s**

w.a.f., with all faults

W. Afr., West Africa

wag/, a joker, **-gery, -gish**

wage-earner (hyphen)

waggly, waving, unsteadily, *not* -ey

wagon/, -er, -ette, *not* wagg-

wagon/ (Fr. m.), a railway carriage; **wagon-lit,** sleeping-car, *pl.* **wagons-lits** (hyphen, not ital. in Eng. usage)

wagtail, a bird (one word)

Wahabi, a sect formed by Abd-el-Wahhab (1691–1787) to restore primitive Islam, *not* Wahh-, -bees

Wahrheit (Dichtung und) (Ger.) (Fiction and Truth), by Goethe (1811–33)

Waiapu (Bp. of), New Zealand

Wai-hai-wei, *use* **Weihaiwei** (one word)

Waikiki Beach, Hawaii

Wainfleet, Lincs.

wainscot/, panelled woodwork on an interior wall; **-ed, -ing**

waist/band, -belt (one word)

waiting/-list, -room (hyphens)

wake, past **woke,** partic. **woken**

Wakefield:, sig. of Bp. of Wakefield (colon)

Wakley (Thomas), 1795–1862, English doctor, founded *The Lancet* in 1823. *See also* **Walkley**

Wal., Walloon

Walachian, one of a non-Slav people of SE Europe, *not* Wall-

Waldenses, religious sect, *not* Vaudois

Waldteufel (Émile), 1837–1912, French composer

wale, a flesh mark, *use* **weal**

wale-knot, *not* wall-

Waler, a NSW horse

Wales, abbr. **W.**

Walhalla, *use* **V-**

walkabout, Austral. Aboriginal's period of wandering in the bush; informal stroll among crowd by royal person etc. (one word)

walkie-talkie, portable transmitter and receiver, *not* -y -y

walking-stick (hyphen)

Walkley (Arthur Bingham), 1855–1926, English dramatic critic. *See also* **Wakley**

walk-over, an easy victory (hyphen); abbr. **w.o.**

Walküre (Die), second part of Wagner's *Ring des Nibelungen,* 1870

Walkyrie, *use* **V-**

walla, *use* **wallah**

wallaby, a small kangaroo, *not* the many variants

Wallace (Alfred Russel, *not* -ell), 1823–1913, English naturalist; — **(Sir Donald Mackenzie),** 1841–1919, English writer; — **(George Corley),** b. 1919, US politician; — **(Henry Agard),** 1888–1965, US politician; — **(Lewis, 'Lew'),** 1827–1905, US general and author; — **(Sir Richard),** 1818–90, English art collector and philanthropist (The Wallace Collection, London); —

(**Prof. Robert**), 1853–1939, Scottish agricultural writer; — (**Sir William**), 1272–1305, Scottish hero; — (**William Vincent**), 1812–65, Irish composer. *See also* **Wallas, Wallis**

Wallachian, *use* **Wala- wallah** (Anglo-Ind., *now* slang), a man (usu. in some specified connection)

Wallas (**Graham**), 1858–1932, English socialist writer. *See also* **Wallace, Wallis**

wall-eye(d) (hyphen)

wallflower (one word)

Wallis (**Sir Barnes**), 1887– 1979, British inventor; — (**George Harry**), 1847–1936, English art writer; — (**John**), 1616–1703, English mathematician, a founder of the Royal Society. *See also* **Wallace, Wallas**

wall-knot, *use* **wale-knot**

Walloon, (a speaker of) a French dialect of S. Belgium and parts of N. France; abbr. **Wal.**

Wallop, family name of Earl of Portsmouth

wall-painting (hyphen)

wallpaper (one word)

Wall Street, New York

Walpurgis night, the one preceding 1 May

Walton (**Izaak,** *not* Isaac), 1593–1683, English author of *Compleat Angler*

waltz, a dance (not ital.); in Fr. f. *valse*; in Ger. m. *Walzer* (ital.)

Walvis Bay, Namibia

W. & M., William and Mary (King and Queen)

wanderlust (one word)

wapiti, American elk, *not* wapp-

War., Warwickshire

war/ baby, — bride, — crime (two words)

Warboys, Cambs.

war/-cloud, -cry (hyphens)

Ward (**Artemas**), 1727–1800, American Revolutionary gen-

eral; — (**Artemus**), pseud. of Charles Farrar Browne, 1834–67, US humorist; — (**Mrs Humphry,** *not* -rey), 1851–1920, English novelist (Mary Augusta Arnold)

war damage (two words)

war-dance (hyphen)

warden, abbr. **W.**

war/-game, -god (hyphens)

warhead (one word)

war-horse (hyphen)

Warlock (**Peter**), pseud. of Philip Heseltine, 1894–1930, English composer

war-lord (hyphen)

warmonger (one word)

War Office, abbr. **WO,** *now see* **M.o.D.**

war/-paint, -path (hyphens)

warrant/er, one who authorizes or guarantees; **-or** (law), one who gives warranty

warrant-officer (hyphen)

war/ship, -time (one word)

Warwickshire, abbr. **War.**

Wash., Washington State (off. abbr.)

wash-basin (hyphen)

washboard (one word)

wash-drawing, one made with a brush and black or grey water- -colour

washhouse (one word)

Washington, State of US, off. abbr. **Wash.** or (postal) **WA**

Washington, DC, US capital

wash-out (noun, hyphen)

wasn't, to be printed close up

WASP, white Anglo-Saxon Protestant

Wassermann, blood test for syphilis, from **August von Wassermann,** 1866–1925, German bacteriologist

wastable, *not* -eable

waste-paper basket (one hyphen)

watch/-case, -chain, -dog, -fire (hyphens)

watch/maker, -man (one word)

watch-tower (hyphen)

watchword (one word)

water/-bird, -bottle, -closet, abbr. **WC, -colour** (hyphens)

water/course, -fall, -fowl, -front (one word)

water-hole (hyphen)

watering-place (hyphen)

water/-jump, -level, -lily, -line, -main (hyphens)

water/logged, -man (one word)

watermark (typ.), a design in the paper itself (one word)

water/-melon, -mill (hyphens)

water polo (two words)

waterproof (one word)

water-rate (hyphen)

water/shed, -side, -spout, -tight (one word)

water torture (two words)

water-tower (hyphen)

water/way, -works (one word)

Watling Street, a Roman road in England

watt (phys.), unit of power; abbr. **W**

Watteau (Jean Antoine), 1684–1721, French painter, whence — **hat** etc.

Watts-Dunton (Walter Theodore), 1832–1914, English man of letters

Waugh (Alec), b. 1898, — **(Auberon),** b. 1940, his nephew, and — **(Evelyn Arthur St. John),** 1903–66, his father, English writers

waul, a cat-cry, *not* -wl

wavey, *use* **wavy**

wave/band, -guide (one word)

wavelength (one word), symbol λ (lambda)

wave number (two words)

wavy, *not* wavey

wax/cloth, -work (one word)

Waynflete (William of), 1395–1486, Bp. of Winchester, Lord Chancellor

Waziristan, formerly NW Frontier Province of India, *not* Wazar-

Wb, weber

WBA, West Bromwich Albion

WC, water-closet, West-Central postal district of London, without charge

WCC, World Council of Churches

W/Cdr., Wing-Commander

WD, War Department

w/e, week ending

WEA, Workers' Educational Association

weak, abbr. **wk.**

weal, a flesh mark, *not* wale

wear, a dam, *use* **weir**

weather/cock, -man (one word)

weazen, *use* **wizened**

Webb (Beatrice), 1858–1943, wife of — **(Sidney James, Baron Passfield),** 1859–1947, English economists and sociologists; — **(Mary),** 1881–1927, English novelist

weber (phys.), unit of magnetic flux, abbr. **Wb**

web/-fed (typ.), presses which receive paper from a reel and not as separate sheets; — -**letterpress,** — -**offset**

Webster (Daniel), 1782–1852, US statesman and orator; — **(Noah),** 1758–1843, US lexicographer

wedding breakfast (two words)

wedding/-cake, -day (hyphens)

wedding march (two words)

wedding-ring (hyphen)

Wedgwood ware, *not* Wedge-, a superior kind of pottery, invented by **Josiah Wedgwood,** 1730–95

Wednesday, abbr. **W.** *or* **Wed.**

week/, -s, abbr. **w.** *or* **wk.**

week/day, -end (one word)

Weelkes (Thomas), *c.*1575–1623, English composer

weepie (colloq.), a sentimental film, play, etc.

weepy (colloq.), tearful

weever, a fish

weevil, a beetle

w.e.f., with effect from

Wehrmacht (Ger. hist. f.), German armed forces (cap., ital.)

Weidenfeld (George) & Nicolson, Ltd., publishers

weighbridge (one word)

weight, abbr. **wt.**

weights, use numerals; abbreviations as cwt., g., lb., oz., *not* to have *s* added for the plural, but add *s* in qrs. See *Hart's Rules,* p. 3

Weihaiwei, China, *not* Wai-hai-wei

Weil (Adolf), 1848–1916, German physician; — **(Simone),** 1909–43, French philosopher

Weill (Kurt), 1900–50, German-born US composer

Weimaraner, breed of dog

Weingartner (Paul Felix), 1863–1942, German conductor and composer

weir, a dam across a river, *not* wear

Weismann/ (August), 1834–1914, German zoologist; **-ism,** a theory of heredity

Weissnichtwo (Ger. for Know-not-where), in Carlyle's *Sartor Resartus*

Weizmann/ (Chaim), 1874–1952, Polish-born chemist and Zionist leader; — **Institute,** Israel

Weizsäcker (Carl Friedrich, Freiherr von), b. 1912, German philosopher and physicist; — **(Julius),** 1828–89, German historian; — **(Karl),** 1822–99, German theologian

welch, *use* welsh

Welch Fusiliers (Royal) (*but* **Welsh Guards**)

Welfare State (two words, caps.)

welk, *use* whelk

we'll, to be printed close up

well-, prefix joined to participles in -ed or -ing takes hyphens when the compound is used attrib., and to preserve the unity of the sense when it is used predic., e.g. a well-known book; the book is well known; the action was not well-advised; he has not been well advised

well/-being, -doer (hyphens)

Welles (Orson), b. 1915, US actor and film-director

well-known, *see* well-

wellnigh (one word)

well-to-do (hyphens)

well-wisher (hyphen)

Welsh, abbr. **W.;** alphabet has 26 letters; *ch, dd, ff, ng* (following *g*), *ll, th,* being each counted as one. No *j, k, q, v, x, z. See also* **accents**

welsh/, to default in payment, *not* -lch- **-er**

Welsh rabbit, melted cheese on toast, *not* — rarebit

Welt/anschauung (Ger.), world-philosophy; *-politik,* participation in international politics; *-schmerz,* world-sorrow

Wemyss/ Bay, Strathclyde; — **Castle,** Fife; — **(Earl of)**

wen, *see* wyn

we're, weren't, to be printed close up

werewol/f (myth.), a human capable of turning at times into a wolf; *pl.* -ves, *not* werw-

west/, -ern, abbr. **W.** *See also* **capitalization, compass**

West Africa, abbr. **W. Afr.**

westbound (one word)

West Bridgford, Notts., *not* Bridge-

West End, London (caps.)

westeria, *use* wistaria

Westermarck (Edward Alexander), 1862–1939, Finnish anthropologist

Western Australia, abbr. **WA**

Westhoughton, Greater Manchester (one word)

Westmeath, Co., Ire.; — **(Earl of)** (one word)

Westmorland, former county; — **(Earl of),** *not* -eland

Westonzoyland, Som. (one word)

West Virginia, off. abbr. **W. Va.** or (postal) **WV**

wetlands (one word)

w.f. (typ.), wrong fount, q.v. (without points as a proof-correction mark). *See also* **proof-correction marks**

WFEO, World Federation of Engineering Organizations

WFTU, World Federation of Trade Unions

W.G., W. G. Grace, 1848–1915, English cricketer

Wg.-Comdr., Wg/Cdr., Wing-Commander

Wh, watt-hour

whalebone (one word)

whallabee, *use* **wallaby**

whar/f, landing-stage; abbr. **whf.**; *pl.* **-fs**

Wharfedale, Yorks.

Wharncliffe (Earl of)

what-d'you-call-it? (colloq.) (hyphens, apos.)

Whately (Richard), 1787–1863, Abp. of Dublin, *not* -ey

whatever (one word)

Whatman paper, a first-quality English handmade drawing-paper (cap.)

whatnot, indefinite thing; a piece of furniture with shelves (one word)

what not, many other similar things, e.g. pens, pencils, and what not (two words)

what's-its (*or* **his**)**-name** (hyphens)

whatsoever (one word)

wheatear, a bird (one word)

Wheatstone bridge (elec.)

wheel/barrow, -base, -chair (one word)

whelk, pimple, mollusc, *not* we-

when/ever, -soever (one word)

whereas (law), a word which introduces the recital of a fact (one word)

where/ver, -soever (one word)

whether or/ not, *not* — — no

whf., wharf

which (gram.), now refers (except as a liturgical archaism) exclusively to things; **who,** to persons

whidah, Afr. weaver-bird, *not* why-

whiffletree, *use* **whipple-**

Whig/, -gish, -gism (cap.)

whimbrel, a bird, *not* wim-

whimsy, caprice, *not* -ey

whinny, (to give) a gentle neigh, *not* -ey

whip/cord, -lash (one word)

whipper/-in (hunting), *pl.* **-s-in**

whipper-snapper (hyphen)

whippletree, the crossbar to which the traces of a harness are attached, *not* whiffle- (one word)

whippoorwill, Amer. nightjar (one word)

whirl/pool, -wind (one word)

whirr, (make) continuous buzzing sound, *not* whir

whisky (Scotch); **whiskey** (Irish and US)

Whistler (James Abbott McNeill, *not* -eil), 1834–1903, US painter and etcher

Whitaker & Sons, publishers of *Almanack* (not -*ac*) etc.

White, a white person (cap.)

white (typ.), any space of paper not printed upon; **line of —,** a line not printed upon

whitebait (one word)

Whitechapel, London (one word)

Whitefield (George), 1714–70, English preacher, *not* Whitf-; *but pron.* wit-

Whitehall, London (one word)

Whitehorse, Yukon (one word)

Whiteing (Richard), 1840–1928, English journalist and novelist

white-out, polar blizzard (hyphen)

White Russian Soviet Socialist Republic, a Soviet Socialist Republic (*but use* **Belorussia**). *See also* **USSR**

White's Club, London

whitewash (one word)

Whitey, derog. for white person (cap.). *See also* **whity**

whitish, *not* -eish

Whitman (**Walt**), 1819–92, US poet

Whit/ Monday, Sunday, seventh after Easter

Whittier (**John Greenleaf**), 1807–92, US poet

Whittlesey, Cambs., *not* -sea

whity, whitelike, *not* -ey. *See also* **Whitey**

whiz/, *not* whizz; **-zed, -zing**

whiz-kid (colloq.) (hyphen)

who, *see* **which**

WHO, World Health Organization

whoa, command to stop

who'd, to be printed close up

whodunit, a novel or play of crime detection, *not* -nnit (one word)

whole-bound (bind.), full--bound (q.v.)

who'll, to be printed close up

whooping cough, *not* hooping

whore-house, a brothel (hyphen)

who's, who is, to be printed close up (apos.)

whose, of whom *or* of which

Who's Who, Who Was Who, reference books

whydah, *use* **whidah**

Whymper (**Edward**), 1840–1911, British climber and writer

Whyte-Melville (**George John**), 1821–78, Scottish novelist (hyphen)

WI, West Ind/ies, -ian; Windward Islands; Wisconsin (off. postal abbr.) ; Women's Institute

wich/-elm, -hazel, *use* **wy-**

Wicliffe, *use* **Wyclif**

widdershins, *use* **withershins**

wide (cricket), abbr. **w.**

wide awake, fully awake, (colloq.) alert, knowing (hyphen when attrib.)

wideawake, a kind of hat (one word)

widespread (one word)

widgeon, a bird, *not* wig-

widow (typ.), a break-line at top of page or column (to be avoided)

Wieland (**Christoph Martin**), 1733–1813, German poet and novelist

Wien, Ger. for **Vienna**

Wiener (**Norbert**), 1894–1964, US mathematician and writer on cybernetics

Wiener-Neustadt, Austria (hyphen)

Wiener schnitzel, veal cutlet dressed with breadcrumbs (two words, one cap.)

Wieniawski (**Henri**), 1835–80, Polish violinist and composer; — (**Joseph**), 1837–1912, his brother, Polish pianist and composer

Wiesbaden, W. Germany

wife, abbr. **w.**

wigeon, *use* **widg-**

Wiggin (**Kate Douglas**), 1856–1923, US educator and novelist

Wight (**Isle of**), abbr. **IOW, I.o.W.,** *or* **IW**

Wigorn:, former sig. of Bp. of Worcester (colon)

Wigton, Cumbria

Wigtown, district and town, Dumfries & Galloway

Wilamowitz-Moellendorf (**Ulrich von**), 1848–1931, German classical scholar

wildcat, a hot-tempered person; reckless or sudden (one word)

Wilde (**Henry**), 1833–1919, English physicist; — (**Oscar Fingall O'Flahertie Wills**), 1856–1900, Irish playwright and poet

wildebeest, a gnu; in Afrik. *wildebees/*, *pl.* **-te**

wild/fire, -fowl (one word)

wild-goose chase (one hyphen)

wildlife (one word)

wilful/, -ly, -ness, *not* will-

Wilhelmj (**August Emil Daniel Ferdinand**), 1845–1908, German violinist

Wilhelmshaven, former German naval station (one word)

Wilhelmstrasse, Berlin, the former German Downing Street

Williams & Glyn's Bank, Ltd.

Willkie (Wendell Lewis), 1892–1944, US politician

will-o'-the-wisp, the *ignis fatuus*, elusive person, delusive hope (apos., hyphens)

will-power (hyphen)

willy-nilly (hyphen)

Wiltshire, abbr. **Wilts.**

Wimborne (Viscount)

Wimborne Minster, Dorset

wimbrel, *use* wh-

Wimpy, propr. term for a kind of hamburger (cap.)

Wimsey (Lord Peter), in novels by D. L. Sayers

wincey, a cloth, *not* -sey

Winchelsea, E. Sussex

Winchilsea (Earl of) ; — (Anne Finch, Countess of), 1661–1720, English poet

Winckelmann (Johann Joachim), 1717–68, German art critic

wind (naut.), Beaufort scale: 1, light air; 2, light breeze; 3, gentle —; 4, moderate —; 5, fresh —; 6, strong —; 7, near gale; 8, gale; 9, strong —; 10, storm; 11, violent —; 12, hurricane

wind/bag, -fall (one word)

Windhoek, SW Africa

Wind. I., Windward Islands

window/-box, -ledge, -seat, -tax (hyphens)

wind/pipe, -screen (one word)

wine/bottle, -glass, -press, -skin (one word)

winey, *use* winy

Winged Victory (anc. Gr.), (statue of) (Athena) Nike (caps.)

wino/ (slang), an alcoholic, *pl.* -s

winsey, *use* -cey

wint/er, -ry (not cap.)

winter-green, an aromatic plant (hyphen)

winterize, to adapt for use in cold weather, *not* -ise

Winton:, sig. of Bp. of Winchester (colon)

winy, winelike, *not* -ey

wipe-out (noun, hyphen)

WIPO, World Intellectual Property Organization

Wis., Wisconsin (off. abbr.)

Wisbech, Cambs., *not* -each

Wisconsin, off. abbr. **Wis.** or (postal) **WI**

Wisdom of Solomon (Apocr.), abbr. **Wisd.**

wiseacre (one word)

wishbone (one word)

wishy-washy (hyphen)

Wislicenus (Johannes), 1835–1902, German chemist

wistaria (bot.), *not* wisteria

witch/-elm, -hazel, *use* wych-

witchetty, (Austral.) larva of beetle or moth, *not* -ety

witenagemot, Anglo-Saxon Parliament

withal, *not* -all

with/e, *pl.* **-es,** *or* **with/y,** *pl.* **-ies,** a flexible twig, often of willow, *not* wy-

withershins (Sc.), anticlockwise; opp. **deasil;** *not* widder-

withhold, etc. (one word, two *h*s)

without, abbr. **w/o**

witness-box (hyphen)

Wittenberg, E. Germany

Witwatersrand University, Johannesburg, S. Africa

wivern, *use* wy-

wizened, shrivelled, *not* weaz-, -en

wk., weak, week, -s, work

W. long., west longitude

Wm., William

WMO, World Meteorological Organization

WNW, west-north-west

WO, War Office (*now see* **M.o.D.**), Warrant Officer, Wireless Operator

w.o., walk-over

w/o, without

wobbl/e, -y, *not* wa-

Wodehouse, family name of Earl of Kimberley ; — **(Sir**

Pelham Grenville), 1881–
1975, English humorous novelist. *See also* **Woodhouse**
woebegone, dismal-looking
(one word)
Wolcot (Dr John), 1738–1819,
'Peter Pindar', English writer
Wolcott (Oliver), 1726–97,
and his son, 1760–1833, American statesmen
Wolf (Friedrich August),
1759–1824, German classical
scholar; — **(Hugo),** 1860–1903,
Austrian composer
Wolfe (Charles), 1791–1823,
Irish poet; — **(Humbert),**
1885–1940, English poet; —
(James), 1727–59, British gen-
eral, took Quebec; — **(Thomas
Clayton),** 1900–38, US novelist
**Wolff (Sir Henry Drum-
mond Charles),** 1830–1908,
British politician and diplomat;
— **(Christian von),** 1679–
1754, German philosopher and
mathematician; — **(Joseph),**
1795–1862, German-born
English-domiciled traveller; —
(Kaspar Friedrich), 1733–
94, German embryologist; adj.
Wolffian
Wolf-Ferrari (Ermanno),
1876–1948, Italian composer
wolfhound (one word)
**Wollstonecraft (Mary, Mrs
Godwin),** 1759–97, English
writer, mother of **Mary —
Godwin, Mrs Shelley,** 1797–
1851, English writer
wolverine, American animal,
not wool-, -ene
womanize, *not* -ise
womankind, *not* women- (one
word)
Women's/ *Journal*; — *Own*
Women's Lib (caps., no point)
won't, to be printed close up
woo/, **-ed, -er, -s**
Wood (Anthony à), 1632–95,
English antiquary
Woodard Foundation, of a
number of English public
schools, named after **Nathaniel**

Woodard, 1811–91, English
Anglican priest, *not* Woodw-
woodbine, honeysuckle, *not*
-bind
woodchuck, N. American mar-
mot (one word)
woodcock/, bird, m. and f.; *pl.*
-s
woodcut (typ.), design cut in
the side grain of a type-high
block of wood (one word)
wood engraving (typ.), design
cut in the end grain of a type-
-high block of wood (two words)
Woodhouse, surname of Emma
in Jane Austen's *Emma*. See also
Wodehouse
wood/**land, -man** (one word)
wood-nymph (hyphen)
woodpecker (one word)
wood/**pile, -wind** (mus.),
-work, -worm (one word)
wood-shed (hyphen)
wool/, **-len, -ly**
Woolacombe, Devon
Wooler, Northumberland
Woolf (Adelaide Virginia,
née **Stephen),** 1882–1941, English novelist and essayist; —
(Leonard Sidney), 1880–
1969, her husband, English
writer
**Woollcott (Alexander Hum-
phreys),** 1887–1943, US
author and dramatic critic
Woolloomooloo, Sydney, Australia
Woolsack (House of Lords) (one
word)
wool-sorters' disease,
anthrax, *not* -er's
woolverine, *use* wolv-
Worcester:, sig. of Bp. of
Worcester (colon)
Worcestershire, former
county, abbr. **Worcs.**
Worde, *see* **Wynkyn**
Word of God (the) (caps.), but
in NT l.c. *w*
work, abbr. **wk.**
workaday (one word)
workaholic, a compulsive
worker

work and turn (typ.), printing of the two sides of a sheet of paper from one forme

work/-bench, -box (hyphens)

workday (one word)

work-force (hyphen)

work/house, -man (one word)

working class (hyphen when attrib.)

work off (typ.), actually to print the paper

work/sheet, -shop (one word)

work-to-rule (noun, hyphens)

World (the New), America; — **(the Old)**, Europe, Asia, and Africa, known to the ancients (two caps.). *See also* **Third World**

World War I, 1914–18; — — **II**, 1939–45; *also* **First World War, Second World War**

world-wide (hyphen)

wormeaten (one word)

worm's-eye view (apos., one hyphen)

wormwood, (bot.) *Artemisia*, (fig.) bitterness

Wormwood Scrubs, London

worry beads (two words)

worry-guts (colloq.) (hyphen)

worship/, -ped, -per, -ping

worthwhile (one word attrib., two words predic.)

Wotton (Sir Henry), 1568–1639, English diplomat and poet

would-be, adj. (hyphen)

wouldst, to be printed close up, no apos.

Woulfe/ (Peter), 1727–1803, English chemist; hence — **bottle,** for passing gas through liquid

wove paper, that which does not show wire marks; distinct from laid —

Wozzeck, opera by Berg, 1925, based on **Woyzeck,** play by Büchner, 1837

WP, weather permitting

w.p.b., waste-paper basket

WPC, woman police constable

w.p.m., words per minute

WR, Western Region, West Riding

WRAC, Women's Royal Army Corps

wrack, destruction; a seaweed. *See also* **rack**

WRAF, Women's Royal Air Force

Wrangel Island, Arctic Ocean

Wrangell Island, Alaska

wrangler (Cambridge University), one placed in first class of mathematical tripos

wrap-round (bind.), a folded section placed outside another section, so that sewing passes through both

wrasse, a fish, *not* -ass

wrath (noun), great anger. *See also* **wroth**

Wray, *see* **Ray**

wreath (noun)

wreathe (verb)

wrick, *use* **rick**

wristband (one word)

wrist-watch (hyphen)

writable, *not* -eable

write/-down, -off, -up (nouns, hyphens)

writer's cramp, *not* -ers' cramp

writing/-desk, -paper, -table (hyphens)

WRNS, Women's Royal Naval Service

Wrocław, Poland; in Ger. **Breslau**

wrongdo/er, -ing (one word)

wrong fount (typ.), said of letter(s) set in wrong size or cut of type; abbr. **w.f.** (without points as a proof-correction mark). *See also* **proof-correction marks**

wroth (poet. or joc.), angry. *See also* **wrath**

Wrottesley (Baron)

WRVS, Women's Royal Voluntary Service, *formerly* **WVS**

wry/bill, -mouth, -neck (one word)

WS (Sc.), Writer to the Signet (= attorney)

WSW, west-south-west

WT (*or* **W/T**), wireless telegraphy

wt., weight

wunderkind (Ger.), one who achieves success in youth (not ital.)

Württemberg, State, W. Germany (two *t*s)

Wuthering Heights, by Emily Brontë, 1846

WV, West Virginia (off. postal abbr.)

W. Va., West Virginia (off. abbr.)

WVS, Women's Voluntary Service, *now* **WRVS**

w/w, weight for weight

WY, Wyoming (off. postal abbr.)

Wyandotte, N. American Indians; **wyandotte,** a fowl, *not* -ot

wych/-elm, -hazel, *not* wich-, witch-

Wycherley (**William**), 1640–1716, English playwright

Wyclif (**John**), *c.* 1324–84, English religious reformer and translator of the Bible; *not* the many variants

Wycliffe/ College, Stonehouse; — **Hall,** Oxford, a theological college

Wykeham/ (**William of**), 1324–1404, English prelate; **-ist,** member of Winchester College (cap.)

Wymondham, Leics., Norfolk

wyn, the Old-English Þ, Þ (distinguish from **eth** and **thorn,** qq.v.); in modern editions *w* is normally substituted

Wyndham/, family name of Baron Leconfield; — (**Sir Charles**), 1837–1919, English actor; — (**George**), 1863–1913, English statesman; —'s **Theatre,** London

Wyndham Lewis (**Dominic Bevan**), 1894–1969, British writer; — (**Percy**), 1882–1957, English writer and painter

Wyndham-Quin, family name of Earl of Dunraven

Wynkyn de Worde, 1471–1534, early printer in London

Wyoming, off. abbr. **Wyo.** or (postal) **WY**

Wyredrawers (**Gold and Silver**), a livery company

wyth/e, -y, *use* wi-

Wythenshawe, Manchester

wyvern, heraldic dragon, *not* wi-

X

X, cross, ten, the twenty-first in a series, film-censorship classification

X (usu. *XP* or **Xt.,** qq.v.), the Gr. letter chi, for *Christos,* Christ

x (math.), the first unknown quantity

x. (meteor.), hoar-frost

Xanthippe, wife of Socrates; used allusively for a shrewish woman

Xavier (St. Francis), 1506–52, Jesuit missionary

X^bre (Fr.), December

XC, 90

XCIX, 99

x.cp., ex (without) coupon

x.d. *or* **ex div.,** ex (without) dividend

Xe, xenon (no point)

xebec, small Mediterranean boat, *not* z-

Xenocrates, 396–314 BC, Greek philosopher

xenon, symbol **Xe**

Xenophanes, *c.*570–*c.*500 BC, Greek philosopher and poet

Xenophon, *c.*438–*c.*354 BC, Greek historian; abbr. **Xen.**

Xeres, Spain, *use* **Jerez**

Xérez (Francisco de), b. 1504, Spanish historian of conquest of Peru

xerography, electrostatic process for printing from or copying documentary material or film. A dry resinous powder is attracted to the positively charged areas of a selenium plate, the non-image areas having been discharged upon exposure to the illuminated original. The powder is then attracted to positively-charged paper and stabilized by heat

Xerox, propr. term for a xerographic copying machine (cap.); **xerox,** to produce (copies) from such a machine (not cap.)

Xerxes, 519–465 BC, King of Persia

x-height (typ.), distance between top and bottom of those lower-case letters of a given fount that have no ascenders or descenders (e.g. n, x); the area of the shoulder occupied by these

Xhosa, Bantu language and people

x.i. *or* **ex int.,** ex (without) next interest

XL, 40

XLIX, 49

Xmas, Christmas (no point)

Xn., Christian; **Xnty.,** Christianity

x.n. *or* **ex n.,** (ex new) ex (without) the right to new shares

XP (as monogram ☧), the Greek letters *chi rho,* first two of *Christos*

X-ray (hyphen)

Xt., Christ

XX, 20

XXX, 30

xylography, the printing of wood-block books

xylonite, a celluloid

xylophon/e (mus.), a percussion instrument of wooden bars vibrating when struck; **-ist,** the player

xyst/us, anc. Gr. portico used by athletes, *pl.* **-i,** *not* -os

Y

Y, yen (q.v.), yttrium, the twenty-second in a series

Y, Netherlands, *use* **IJ**

y in Dutch, use **ij**, as Nijmegen, IJmuiden

y., year, -s, (meteor.) dry air

y (math.), the second unknown quantity

yacht/, -sman (one word)

yager, *see* **Jäger**

Yahoo, in Swift's *Gulliver's Travels*, an animal with human form but brutish instincts

Yahveh, the probable pronunciation of the Hebrew consonants YHWH which are traditionally transliterated as **Jehovah**; *not* Jahveh

Yakutsk, Siberia, *not* J-

Yale, US University, at New Haven, Conn.; propr. term for a kind of lock (cap.)

Yangtze Kiang, Chinese river (*kiang* = river), in Pinyin **Yangzijiang**

Yankee, a citizen of the New England States or of the North of US, or of the US generally, *not* -i

Yaoundé, Cameroon

yaourt, *use* **yoghurt**

yapp (bind.), a soft case with overlapping edges

yarborough, hand in whist or bridge with no card above 9 (not cap.)

yard/, -s, abbr. **yd., yds.**; number of, to be in numerals

yard-arm (naut.) (hyphen)

Yarde-Buller, family name of Baron Churston

yardstick (one word)

Yarkand, Cent. Asia, *not* -end, -und

yarl, *use* **jarl**

yarmulk/a, Jewish skull-cap, *pl.* -as

Yaroslavl, USSR, *not* Jaroslav

yashmak, Muslim woman's veil

yataghan, Muslim sword, *not* -gan

Yates (Dornford), pseud. of **Cecil William Mercer,** 1885–1960, English novelist

YB, Year Book, q.v.

Yb, ytterbium (no point)

YC, Young Conservative

yclept (arch. or joc.), called, named (one word)

yd., yds., yard, -s

y^e, in 15th- to 17th-c. works, the second letter to be superior (no point)

Yeames (William Frederick), 1835–1918, English painter

year/, -s, abbr. **y.** *or* **a** *or* **yr., yrs.**

year-book (hyphen); **Year Book** (law reports) (caps., two words), abbr. **YB**

year/-long, -round (hyphens)

years, *see* **numerals** II

Yeats (William Butler), 1865–1939, Irish poet

Yeats-Brown (Francis), 1886–1944, English author

Yeddo, Japan, *see* **Tokyo**

yellow-hammer, a bird, *not* —-ammer (hyphen)

Yellowknife, NW Territories, Canada

Yellowplush (one word) *Papers,* by Thackeray, 1841

Yellowstone/ (one word) **Park; — River**

Yemen/, S. Arabia; — **(North),** off. — **Arab Republic;** — **(South),** off. **People's Democratic Republic of Yemen;** adj. **-i**

yen, the monetary unit of Japan, *pl. same*; abbr. **Y** *or* **¥**

Yeniseisk, Siberia, on R. **Yenisei**

Yeo., Yeomanry

Yerevan, cap. of Armenia, USSR, *not* E-, -iv-

yes-man (colloq.) (hyphen)

445

yeux, see œil

Yevtushenko (Yevgeny), b. 1933, Russian writer

Yggdrasil (Scand. myth.), the tree binding heaven, earth, and hell, Anglicized from ON *yggdrasill*; *not* Ygd-

YHA, Youth Hostels Association

Yiddish/, mixed German and Hebrew dialect; **-er,** speaker of this

yield point (two words)

yippee, exclamation of delight

ylang-ylang, (perfume from) Malayan tree (hyphen)

YMCA, Young Men's Christian Association

Ymuiden, *use* IJm-

Ynca, *use* Inca

Ynys Môn, Welsh for **Anglesey**

yobbo/, colloq. for lout, hooligan, *pl.* **-s**

yodel/, song, to sing, with inarticulate partly falsetto voice; **-led, -ling,** *not* -dle, jodel

yog/a, Hindu system of philosophic meditation; **-i,** a devotee of yoga; adj. **-ic**

yogh. the Middle English ȝ, ʒ; normally use ȝ, ʒ

yoghurt, sour fermented milk, *not* yaourt, yogurt, -ourt

Yokohama, Japan (one word)

Yom Kippur (Jewish relig.), Day of Atonement

Yonge (Charles Duke), 1812–91, English historian; — **(Charlotte Mary),** 1823–1901, English novelist; *pron.* as young. *See also* **Young**

Yorke, family name of Earl of Hardwicke

Yorkshire, former county, abbr. **Yorks.**

Yorktown, Va., US (one word)

Yosemite Valley, Calif., US

Youcon, *use* Yukon

you'd, to be printed close up

Youghal, Co. Cork, Ireland

Youl (Sir James Arndell), 1809–1904, Tasmanian colonist

you'll, to be printed close up

Young (Brigham), 1801–77,

US Mormon leader. *See also* Yonge

younger, abbr. **yr.**

Young Men's Christian Association, abbr. **YMCA**; ditto **Women's, YWCA**

your, abbr. **yr.**

you're, to be printed close up

yours (no apos.); abbr. **yrs.**

you've, to be printed close up

Yo-Yo/, propr. term for kind of toy, *pl.* **-s** (hyphen, caps.)

Ypres/, Belgium, in Fl. **Ieper;** — **(Earl of)**

Yquem/, vineyard, dép. Gironde; — **(Château-d'),** a Sauterne (hyphen)

yr., year, younger, your

YRA, Yacht Racing Association, *now* **RYA,** q.v.

Yriarte (Charles), 1832–98, French writer

yrs., years, yours

Ysaye (Eugène), 1858–1929, Belgian violinist

-yse. In verbs such as analyse, catalyse, paralyse, -lys- is part of the stem (corresponding to the Greek element *-lusis*), and not a suffix like -ize. The spelling -yze is therefore incorrect, though common in US

Yseult etc., *use* **Iseult**

Yssel, *use* IJs-

YT, Yukon Territory, Canada

y^t, in 15th- to 17th-c. works, the second letter to be superior (no point)

ytterbium, symbol **Yb**

yttrium, symbol **Y**

Yucatán, Mexico (accent)

Yugoslavia, *not* J-; adj. **Yugoslav,** *not* -avian

Yuit, Eskimo of NE Siberia. *See also* **Innuit**

Yukon/ River, Alaska, US, and Canada; — **Territory,** Canada, *not* Youcon, -kon

yule/-log, -tide (hyphens)

Yvetot, dép. Seine-Maritime, France

YWCA, Young Women's Christian Association

Z

Z, the twenty-third in a series
z. (meteor.), haze
z (math.), the third unknown
quantity
zabaglione, Italian sweet with
egg-yolks, sugar, and wine (not
ital.)
Zach., Zachary
Zacynthus, anc. name of
Zante
Zaehnsdorf (Joseph), 1819–
86, Austrian-born English-
-domiciled bookbinder
zaffre, impure cobalt oxide used
as blue pigment, *not* -er
Zagreb, Yugoslavia, *not* -ab,
Agram
Zaharoff (Sir Basil), 1849–
1936, Greek-born British
financier
Zaïre, tragedy by Voltaire, 1732
Zaïre/, Cent. Africa, indep.
1960,*formerly* **Democratic Re-**
public of Congo; river,*for-*
merly **Congo**; adj. **-an**; also
(with small initial) the monet-
ary unit
Zambezi, African river, *not* -si
Zambia, *formerly* N. Rhodesia,
indep. republic 1964
Zamenhof (Ludwig Laza-
rus), 1859–1917, Pol. physi-
cian, inventor of Esperanto
zamindar, *use* **zemin-**
zanana, *use* **ze-**
Zangwill (Israel), 1864–1926,
English novelist and playwright
Zante, Gr. island, anc. name
Zacynthus
ZANU, Zimbabwe African
National Union
Zanzibar, *see* **Tanzania**
ZAPU, Zimbabwe African
People's Union
Zaragoza, Spain, *use* **Sara-**
gossa
zarape, *use* **serape**
Zarathustra, 6th c. BC, founder
of Parsee religion

zariba (Arab.), fortified enclosure, *not* the many variants
Zarskoe Selo, *use* **Tsarskoe**
Selo (q.v.)
zax, *use* **sax**
z.B. (Ger.), *zum Beispiel* (for
example)
Zealand, Denmark, *not* Zeeland;
in Dan. **Sjælland**
zealot, zealous person, fanatic
(cap. with reference to Jewish
sect)
zebec, *use* **xebec**
Zech., Zechariah
Zeeland, Holland, *not* Zea-
Zeitgeist (Ger. m.), the spirit of
the time (cap.)
Zeitschrift (Ger. f.), periodical
(cap.)
Zeltinger, a Moselle wine
zemindar, official in India under Mogul empire, Indian landowner paying tax to Britain, *not*
zamin-
Zen, meditative form of Buddhism (cap.)
zenana (Ind.), the women's
apartments, *not* za-
Zener cards, used in ESP research (one cap.)
zenith, the highest point, opp. to
nadir
Zeph., Zephaniah
zero/, *pl.* -s
Zetinje, *use* **Cetinje**
Zetland/, former off. name of
Shetland county; — (**Mar-**
quess of)
zeugma/ (gram.), figure in
which a verb or adj. is used with
two nouns of which it is strictly
applicable to only one, e.g. with
weeping eyes and hearts, *pl.* -s;
adj. **-tic**
zho, *use* **dzho**
Zhou Enlai, 1898–1976,
Chinese statesman
zibet, Asian or Indian civet, *not*
-eth

447

ziggurat, anc. Mesopotamian tower, *not* zikk-

zigzag/, -ged, -ging (one word)

Zimbabwe/, name of Rhodesia 1980–; adj. **-an**

zinc, symbol **Zn**

zinco/ (typ.), a relief block, usu. in line, made from zinc; *pl.* **-s**

zincography, the art of engraving and printing from zinc

zingara (à la) (Fr. cook.), in gypsy style

zingar/o (It.), a gypsy, *pl.* **-i**; *fem.* **-a,** *pl.* **-e**

Zinjanthropus, E. Afr. fossil hominid

Zion/, -ism, -ist, *not* Si-

zip/, -ped, -per, -ping, -py; —-bag, -fastener (hyphens)

Zip code (US), postal delivery code (one word.)

zirconium, symbol **Zr**

zloty/, monetary unit of Poland, *pl.* **-s**

Zn, zinc (no point)

zo, *use* **dzho**

zodiacal signs, *see* **astronomy**

Zoffany (**John**), 1734–1810, German-born English-domiciled painter

Zollverein (Ger. m.), customs union

zombie, revived corpse, stupid person, *not* -i, -y

zoolog/y, -ical, -ist, abbr. **zool.**; genera, species, and subspecies to be italic, other divisions roman: Carnivora (order); Felidae (family); *Felis* (genus); *Felis catus* (species). Specific epithets, even when derived from names of persons, should be l.c.: *Myotis daubentoni.* See *Hart's Rules,* pp. 6–7

zo/on, an animal; *pl.* **-a**

zōon politikon (anc. Gr.), the political animal, man

Zoroastrian, of religion founded by Zarathustra

zouave, French soldier (not ital.)

ZPG, zero population growth

Zr, zirconium (no point)

ZS, Zoological Society

zucchetto/ (RCC), a skull-cap, *pl.* **-s**

zucchini (It. pl.), courgettes

zugzwang (chess), a blockade, a position in which any move is undesirable yet some move must be made

Zuider Zee, Netherlands, *now* IJsselmeer

Zululand (one word), annexed by S. Africa, 1897

Zurich, Switzerland; in Ger. **Zürich**

zwieback (Ger.), a kind of biscuit rusk (not ital.)

Zwingli (**Ulrich**), 1484–1531, Swiss Protestant reformer